Lecture Notes in
Mathematics

Edited by A. Dold and B. Eckmann

T0214464

1105

Rational Approximation and Interpolation

Proceedings of the United Kingdom –
United States Conference held at Tampa, Florida,
December 12–16, 1983

Edited by P.R. Graves-Morris, E.B. Saff and R.S. Varga

Springer-Verlag
Berlin Heidelberg New York Tokyo 1984

Editors

Peter Russell Graves-Morris
Mathematical Institute, University of Kent
Canterbury, Kent CT2 7NF, England

Edward B. Saff
Center for Mathematical Services, University of South Florida
Tampa, Florida 33620, USA

Richard S. Varga
Institute for Computational Mathematics, Kent State University
Kent, Ohio 44242, USA

AMS Subject Classification (1980): 41-06

ISBN 3-540-13899-4 Springer-Verlag Berlin Heidelberg New York Tokyo
ISBN 0-387-13899-4 Springer-Verlag New York Heidelberg Berlin Tokyo

This work is subject to copyright. All rights are reserved, whether the whole or part of the material is concerned, specifically those of translation, reprinting, re-use of illustrations, broadcasting, reproduction by photocopying machine or similar means, and storage in data banks. Under § 54 of the German Copyright Law where copies are made for other than private use, a fee is payable to "Verwertungsgesellschaft Wort", Munich.

© by Springer-Verlag Berlin Heidelberg 1984
Printed in Germany

Printing and binding: Beltz Offsetdruck, Hemsbach/Bergstr.
2146/3140-543210

[UK/US]

Preface

This volume contains the proceedings of the Conference on Rational Approximation and Interpolation, which took place December 12-16, 1983 at the University of South Florida, Tampa, Florida. The conference was held under the auspices of the U.K. - U.S. Cooperative Science Program, an informal agreement between the U.S. National Science Foundation and the U.K. Science and Engineering Research Council to promote and support mutually beneficial scientific activities. The primary purpose of the conference was to bring together pure and applied mathematicians, physicists and engineers to exchange information and set objectives for future research efforts dealing with rational approximation and interpolation.

P.R. Graves-Morris and E.B. Saff were the primary organizers of the conference. There were 28 participants from the U.S., 14 from the U.K. and 14 others representing 11 additional countries. The number of conference members was kept limited so as to promote discussion among members with diverse backgrounds, in accordance with the aims of the U.K. - U.S. Cooperative Science Program.

The contributions to this volume include original research papers as well as a few survey articles. All of these papers were refereed and we are grateful to many advisors for their diligence. It is hoped that this volume reflects the breadth of interest in rational approximation and interpolation, and serves as a source of inspiration for further research.

We wish to thank the U.S. National Science Foundation and the U.K. Science and Engineering Research Council for sponsoring the participants from their respective countries. We are also indebted to the University of South Florida (USF) Division of Sponsored Research for the support of the other conference partcipants. The conference planning and activities were facilitated by the USF Center for Mathematical Services, the USF organizing committee consisting of Prof. M. Blake and Prof. J. Snader, and the conference co-host,

Prof. R.S. Varga. The efforts of these individuals far exceeded the norm. The secretarial help provided by Mary Baroli is also deserving of accolade as were the efforts of several staff members and students in the USF Department of Mathematics. We are further indebted to several companies in the Tampa area for having provided additional support for the conference functions.

P.R.G.-M., E.B.S., R.S.V.

LIST OF CONTRIBUTORS AND PARTICIPANTS

J. MILNE ANDERSON, Mathematics Department, University College, London WC1E6BT, United Kingdom

CHRISTOPHER T. H. BAKER, Department of Mathematics, The University, Manchester M13 9PL, United Kingdom

GEORGE A. BAKER, JR., Theoretical Division, Los Alamos National Laboratory, University of California, Los Alamos, New Mexico 87545

MICHAEL BARNSLEY, School of Mathematics, Georgia Institute of Technology, Atlanta, Georgia 30332

AI-PING BIEN, Institute of Applied Mathematics, National Cheng-Kung University, Tainan, Taiwan 700, Republic of China

H.-P. BLATT, Katholische Universität Eichstätt, Mathematish-Geographische Fakultät, Ostenstrasse 26-28, 8078 Eichstätt, West Germany

N. K. BOSE, Departments of Electrical Engineering and Mathematics, 348 Benedum Hall, University of Pittsburgh, Pittsburgh, Pennsylvania 15621

DIETRICH BRAESS, Institut für Mathematik, Ruhr-Universität, D-4630 Bochum, West Germany

CLAUDE BREZINSKI, Laboratoire d'Analyse Numérique et d'Optimisation, Universite de Lille I, 59655-Villeneuve d'Ascq Cedex, France

MARCEL G. de BRUIN, Department of Mathematics, University of Amsterdam, Roetersstraat 15, 1018 WB Amsterdam, The Netherlands

A. BULTHEEL, Department of Computer Science, K. U. Leuven, Celestijnenlaan 200A, B-3030 Leuven, Belgium

AMOS J. CARPENTER, Department of Mathematical Sciences, Butler University, 4600 Sunset Avenue, Indianapolis, Indiana 46208

FUHUA CHENG, Institute of Computer and Decision Sciences, National Tsing Hua University, Hsinchu, Taiwan 300, Republic of China

J. S. R. CHISHOLM, Mathematical Institute, University of Kent, Canterbury, Kent, CT2 7NF, England

CHARLES K. CHUI, Department of Mathematics, Texas A & M University, College Station, Texas 77843

J. G. CLUNIE, Mathematics Department, Open University, Milton Keynes MK7 6AA, United Kingdom

STEPHEN G. DEMKO, School of Mathematics, Georgia Institute of Technology, Atlanta, Georgia 30332

ALBERT EDREI, Department of Mathematics, Syracuse University, Syracuse, New York 13210

S. W. ELLACOTT, Department of Mathematics, Brighton Polytechnic, Brighton BN2 4GJ, England

MICHAEL E. FISHER, Baker Laboratory, Cornell University, Ithaca, New York 14853

W. H. FUCHS, Mathematics Department, Cornell University, Ithaca, New York 14853

J. L. GAMMEL, Department of Physics, Saint Louis University, St. Louis, Missouri 63103

I. GLADWELL, Department of Mathematics, University of Manchester, Manchester M13 9PL, United Kingdom

W. B. GRAGG, Department of Mathematics, University of Kentucky, Lexington, Kentucky 40506

P. R. GRAVES-MORRIS, Mathematical Institute, University of Kent, Canterbury, Kent CT2 7NF, England

JOHN A. GREGORY, Department of Mathematics and Statistics, Brunel University, Uxbridge UB8 3PH, England

M. GUTKNECHT, Eidgenössiche Technische Hochschule, Seminar für Angewandte Math., ETH-Zentrum HG, Zürich CH-8092, Switzerland

M. HASSON, Department of Mathematics, Carnegie-Mellon University, Pittsburgh, Pennsylvania 15213

T. HÅVIE, UNIT/NTH, Alfred Getz vei 1, N7034 Trondheim, Norway

JEANETTE VAN ISEGHEM, Laboratoire d'Analyse Numérique et d'Optimisation, Université de Lille I, 59655-Villeneuve d'Ascq Cedex, France

ARIEH ISERLES, King's College, University of Cambridge, Cambridge CB4 1LE, England

LISA JACOBSEN, Matematisk Institutt, NLHT Trondheim, Norway

WILLIAM B. JONES, Department of Mathematics, University of Colorado, Boulder, Colorado 80309

THOMAS KÖVARI, Department of Mathematics, Imperial College, 180 Queen's Gate, London SW7 2BZ, England

MICHAEL A. LACHANCE, Department of Mathematics, University of Michigan, Dearborn, Michigan 48128

G. LÓPEZ LAGOMASINO, Faculty of Physics and Mathematics, University of Havana, Havana, Cuba

D. LEVIATAN, School of Mathematics, Tel Aviv University, Ramat Aviv, 69978 Tel Aviv, Israel

G. G. LORENTZ, Department of Mathematics, University of Texas, Austin, Texas 78712

R. A. LORENTZ, GMD, Schloss Birlinghoven, 5205 St. Augustin 1, West Germany

ARNE MAGNUS, Department of Mathematics, Colorado State University, Ft. Collins, Colorado 80523

J. C. MASON, Department of Mathematics and Ballistics, Royal Military College of Science, Shrivenham, Swindon, Wiltshire, England

ATTILA MÁTÉ, Department of Mathematics, Brooklyn College of the City University of New York, Brooklyn, New York 11210

C. A. MICCHELLI, Mathematical Sciences Department, IBM Thomas J. Watson Research Center, Yorktown Heights, New York 10598

H. N. MHASKAR, Department of Mathematics, California State University, Los Angeles, California 90032

B. NELSON, Department of Mathematics, University of Manchester, Manchester M13 9PL, England

PAUL NEVAI, Department of Mathematics, The Ohio State University, Columbus, Ohio 43210

SYVERT P. NØRSETT, Mathematics Department, University of Trondheim, Trondheim, Norway. Visiting Professor, Department of Computer Science, University of Waterloo, Waterloo, Ontario N2L 3G1, Canada

J. NUTTALL, Department of Physics, University of Western Ontario, London, Ontario N6A 3K7, Canada

M. J. D. POWELL, DAMTP, Silver Street, Cambridge CB3 9EW, England

D. C. POWER, McDonnell Aircraft Company, McDonnell Douglas Corporation, St. Louis, Missouri 63166

T. RIVLIN, Mathematical Sciences Department, IBM Thomas J. Watson Research Center, Yorktown Heights, New York 10598

D. E. ROBERTS, Mathematics Department, Napier College, Colinton Rd., Edinburgh, Scotland

S. RUSCHEWEYH, Mathematisches Institut, Universität Würzburg, D-8700 Würzburg Am Hubland, West Germany

A. RUTTAN, Department of Mathematics, Kent State University, Kent, Ohio 44242

E. B. SAFF, Department of Mathematics, University of South Florida, Tampa, Florida 33620

A. SHARMA, Mathematics Department, University of Alberta, Edmonton T6G 2G1, Canada

XIE-CHANG SHEN, Department of Mathematics, Peking University, Beijing, China

A. D. SNIDER, Department of Electrical Engineering, University of South Florida, Tampa, FLorida 33620

ALLAN STEINHARDT, School of Electrical and Computer Engineering, Oklahoma State University, Stillwater, Oklahoma 74078

F. STENGER, Department of Mathematics, University of Utah, Salt Lake City, Utah 84112

D. F. STYER, Hill Center for the Mathematical Sciences, Busch Campus, Rutgers University, New Brunswick, New Jersey 08903

R. M. THOMAS, Department of Mathematics, U.M.I.S.T., Manchester M60 1QD, United Kingdom

W. J. THRON, Department of Mathematics, Campus Box 426, University of Colorado, Boulder, Colorado 80309

VILMOS TOTIK, Bolyai Institute, University of Szeged, 6720 Szeged, Hungary

LLOYD N. TREFETHEN, Courant Institute of Mathematical Sciences, New York University, New York, New York 10012

STEWART R. TRICKETT, Department of Applied Mathematics, University of Waterloo, Waterloo, Ontario N2L 3G1, Canada

G. TROJAN, Department of Physics, University of Western Ontario, London 72, Ontario N6A 3K7, Canada

JOSEPH L. ULLMAN, Department of Mathematics, University of Michigan, Ann Arbor, Michigan 48109-1003

R. S. VARGA, Department of Mathematics, Kent State University, Kent, Ohio 44242

V. V. VAVILOV, Faculty of Mechanical and Mathematics, University of Moscow, Moscow, U.S.S.R.

HANS WALLIN, Department of Mathematics, University of Umeå, S-901 87 Umeå, Sweden

G. A. WATSON, Department of Mathematical Sciences, University of Dundee, Dundee DD1 4HN, Scotland

J. WILLIAMS, Department of Mathematics, University of Manchester, Manchester M13 9PL, England

CONTENTS

BLOCK STRUCTURE

CIRCUIT THEORY

CONVERGENCE THEORY

CRITICAL PHENOMENA

LOCATION OF ZEROS AND POLES

NUMERICAL METHODS

ORTHOGONAL POLYNOMIALS

THE FABER OPERATOR

J. M. Anderson*
Mathematics Department
University College
London W.C.1.E.6.B.T.
U.K.

Abstract. The boundedness of the Faber operator T and its inverse T^{-1}, considered as mappings between various spaces of functions, is discussed. The relevance of this to problems of approximation, by polynomials or by rational functions, to functions defined on certain compact subsets of $\mathbb{C}$ is explained.

1. Introduction

Let D denote the closed unit disk $\{w: w \in \mathbb{C}, |w| \leq 1\}$ and $A(D)$ the well-known disk algebra of functions analytic in the interior D^o of D and continuous on D, with the supremum norm. The sets K we wish to consider are compact subsets of $\mathbb{C}$ whose interior K^o is a simply connected domain and whose boundary ∂K is a rectifiable Jordan curve. The corresponding algebra is denoted by $A(K)$.

Associated with K there is a sequence of polynomials $\{F_n(z)\}$ introduced first by Faber in his thesis [10] and subsequently known as Faber polynomials. These are defined as follows. Let $z = \psi(w)$ be the Riemann mapping function of $\mathbb{C}\backslash D$ onto $\mathbb{C}\backslash K$ with $\psi(\infty) = \infty$, of the form

$$z = \psi(w) = \alpha w + b_0 + \sum_{n=1}^{\infty} b_n w^{-n} .$$

The number $|\alpha|$ is called the transfinite diameter of K. It is strictly positive for the domains K we are considering. We shall

*The author thanks the Department of Mathematics of the University of California at San Diego for its kind hospitality while this report was being written.

assume, by scaling and rotation, if necessary, that $\alpha = 1$. The Faber polynomials $F_n(z)$ associated with K are defined by

$$\frac{\psi'(w)}{\psi(w)-z} = \sum_{n=0}^{\infty} F_n(z) w^{-(n+1)}, \qquad |w| > 1,$$

or, alternatively, by

$$F_n(z) = \frac{1}{2\pi i} \int_{|w|=1} \frac{w^n \psi'(w)}{\psi(w)-z}\, dw .$$

To see that these are indeed polynomials we note that

$$\frac{d^k}{dz^k} F_n(z) \Big|_{z=0} = \frac{k!}{2\pi i} \int_{|w|=1} \frac{w^n \psi'(w)}{(\psi(w))^{k+1}}\, dw .$$

For $k = n$ the right side above is $n!$, since $\alpha = 1$ and for $k > n$ it is zero. Thus $F_n(z)$ is a monic polynomial,

$$F_n(z) = z^n + \text{lower order terms} .$$

Let $\Pi(n)$ denote the set of all polynomials of degree at most n and set $\Pi = \bigcup_{n=1}^{\infty} \Pi(n)$. The Faber operator is defined, initially on Π, by

$$(1.1) \qquad (Tp)(z) = \frac{1}{2\pi i} \int_{|w|=1} \frac{p(w)\psi'(w)}{\psi(w)-z}\, dw .$$

Note that if $p(w) = w^n$ then $(Tp)(z) = F_n(z)$. Clearly, $T(\Pi(n)) = \Pi(n)$ and the mapping

$$T: \Pi(D) \longrightarrow \Pi(K)$$

is injective. Here $\Pi(D)$ and $\Pi(K)$ denote Π considered as a subspace of $A(D)$ or $A(K)$. Sets K (of the kind we are considering, of course) where the operator T is bounded are called Faber sets. For the moment we are considering only the supremum norms on $A(K)$ and $A(D)$ respectively and so T is bounded if and only if

$$\|Tp\|_\infty \le \|T\| \cdot \|p\|_\infty .$$

The bounded operator T, given by (1.1) can be extended to a bounded linear mapping of $A(D)$ into $A(K)$ given by

$$(1.2)\quad (Tf)(z) = \frac{1}{2\pi i}\int_{|w|=1} \frac{f(w)\psi'(w)}{\psi(w)-z}\,dw\ .$$

This operator, again denoted by T, is injective. This follows from the fact, established in [13], §3, that, if $f \in T(A(D))$, then we may associate with f a Faber expansion

$$f(z) = \sum_{k=0}^{\infty} a_k F_k(z)\ ,$$

with $f(z) \equiv 0$ if and only if $a_k = 0$ for all $k \geq 0$.

Care must be taken with formula (1.2) since in general $f(w)$ is defined only for $|w| \leq 1$ and $\psi'(w)$ only for $|w| \geq 1$. It is here, for example that we make use of the fact that ∂K is rectifiable so that $\psi' \in H^1$ ([8] Theorem 3.12). Since the boundedness or unboundedness of T depends only on its behavior on a dense set, namely on Π, we need consider (1.2) only for functions which are defined for $|w| > 1$ as well. This might permit the condition $\psi' \in H^1$ to be relaxed somewhat, by the use of Abel limits, say; but this seems of little interest.

When K is a Faber set the inverse T^{-1} of T, defined on the range of T, is given by

$$(T^{-1}f)(\zeta) = \frac{1}{2\pi i}\int_{|w|=1} \frac{(f\circ\psi)(w)}{w-\zeta}\,dw$$

for $\zeta \in D^o$. For $|\zeta| = 1$ this would be the Hilbert transform of the composite function $(f\circ\psi)(w)$. The mapping T^{-1} is bounded if and only if T is surjective and in that case T is an isomorphism between $A(K)$ and $A(D)$. Such a set K for which T and T^{-1} are bounded is called an inverse Faber set. Of course, T is a Banach space isomorphism and not an algebra isomorphism; products are not preserved.

2. Polynomial and Rational Approximation

We define, as usual, the best polynomial approximation $E_n(f)$ and the best rational approximation $r_n(f)$ to a function $f(z) \in A(K)$ by

$$E_n(f) = E_n(f,K) = \inf\{\|f-p_n\|_\infty,\ p_n \in \Pi(n)\}$$

$$r_n(f) = r_n(f,K) = \inf\{\|f-r_n\|_\infty,\ r_n \in R(n)\}\ .$$

Here $R(n) = R(n,K)$ denotes the set of rational functions of degree at most n with poles off K. The Faber operator (1.2) maps $R(n,D)$ onto $R(n,K)$. This follows from the elementary contour integral calculation

$$T\left(\frac{1}{w-w_k}\right)(z) = \frac{1}{2\pi i}\int_{|w|=R>1}\frac{\psi'(w)}{\psi(w)-z}\,\frac{dw}{w-w_k} = \frac{\psi'(w_k)}{z-\psi(w_k)} .$$

The following result follows immediately from the above considerations, but we state it as a theorem since it shows why inverse Faber sets are of interest.

Theorem 1. *Let K be an inverse Faber set. Then*

$$E_n(f,K) \le c\, E_n(T^{-1}f,D) \le c\, E_n(f,K) ,$$

$$r_n(f,K) \le c\, r_n(T^{-1}f,D) \le c\, r_n(f,K) ,$$

for all $f \in A(K)$. Here c is a generic constant, not necessarily the same at each occurrence.

If $g(w) \in A(D)$ has modulus of continuity $\omega(\delta)$ defined by

$$\omega(\delta,g) = \max\{|g(w_1)-g(w_2)|,\ w_1,w_2 \in D,\ |w_1-w_2| \le \delta\}$$

then, by a well-known theorem of Jackson,

$$E_n(g,D) \le c\,\omega\left(\frac{1}{n},g\right) .$$

Hence if K is an inverse Faber set, $E_n(f,K)$ is precisely of the order of $\omega\left(\frac{1}{n},T^{-1}f\right)$.

To get a reasonably good polynomial approximation to $g(w) = \sum_{k=0}^{\infty} a_k w^k$ in $A(D)$, Kövari has shown that we may take the de la Vallée-Poussin means, defined as follows. Set

$$S_n(w) = \sum_{k=0}^{n} a_k w^k , \qquad V_n(w) = \frac{1}{n}\sum_{k=n}^{2n-1} S_k(w) .$$

Note that $V_n(w)$ is a polynomial of degree at most $2n-1$. Then ([12], p.367)

$$\|g-V_n\|_\infty \le 4E_n(g,D) .$$

In most cases, for example when dealing with lip α problems, the difference between n and $(2n-1)$ causes only an additional multiplicative factor. We see that, if K is an inverse Faber set and $f \in A(K)$ with

$$(T^{-1}f)(w) = g(w) = \sum_{k=0}^{\infty} a_k w^k$$

then

$$(TV_n)(z) = \frac{1}{n} \sum_{k=n}^{2n-1} \left(\sum_{r=0}^{k} a_r F_r(z) \right)$$

affords an approximation to $f(z)$ which is reasonably good and, more importantly, easily obtainable. In particular

$$\|f-T(V_n)\|_\infty \le c\, E_n(f,K) .$$

The theory of Faber sets was, to my knowledge, first developed by Al'per [1] who showed, somewhat implicitly, that if the set K satisfies the condition

$$\text{(2.1)} \quad \sup_{|u|=1} \int_{|w|=1} \left| \frac{\psi'(w)}{\psi(w)-\psi(u)} - \frac{1}{w-u} \right| |dw| = \alpha_0 < \infty$$

then K is an inverse Faber set and the above considerations apply. Condition (2.1) is very restrictive; it does not permit ∂K to have any angles. However, (2.1) can be used to show that, if ∂K is a reasonably smooth Jordan curve,--of class $C^{1+\delta}$ for some $\delta > 0$ --then K is an inverse Faber set.

3. Bounded Boundary Rotation

Suppose now that ∂K has a right and left tangent at every point and that, at the point $z = \psi(e^{i\theta})$, the angle between the positive real axis and the right (resp. left) tangent to ∂K is denoted by $u(\theta-)$ (resp. $u(\theta+)$). If

$$\text{(3.1)} \quad V = \int_0^{2\pi} |du(\theta)| < \infty$$

then K is said to have bounded boundary rotation equal to V. Note that (3.1) implies the existence of $u(\theta+)$ and $u(\theta-)$ for all θ and that, if K is of bounded boundary rotation, then its boundary ∂K is rectifiable, though ∂K may possess angles and cusps. If K, in addition is convex, then $V = 2\pi$.

It was Kövari who first noted the relevance of bounded boundary rotation to Faber sets. He showed ([12], p.368)

Theorem 2. Let K be a closed Jordan domain of bounded boundary rotation V. Then K is a Faber set and $\|T\| \leq A \frac{V}{\pi}$, where T is defined by (1.2) and A is an absolute constant.

Faber sets have been further investigated by Andersson (see [3], [4], and for a related topic [5]). In particular, Andersson has shown that the bounded boundary rotation of ∂K is not necessary for K to be a Faber set. There are sets K, of bounded secant variation but infinite bounded boundary rotation, which are Faber sets.

As a complement of Theorem 2 Clunie and I have proved, [2].

Theorem 3. Let K be a closed Jordan domain of bounded boundary rotation. Then K is an inverse Faber set if the boundary ∂K has no external cusps.

The bound on $\|T^{-1}\|$ which we obtain depends on the total boundary rotation V and on the smallest exterior angle α, but in a complicated way depending on the geometry of K. No nice estimate like that of Theorem 2 is to be expected, though in the case of convex domains explicit bounds, in terms of α alone, could be given.

4. Besov Spaces

There is a further interesting use of the Faber operator in a situation which has not been fully exploited yet and seems to offer the possibility of further study. This arises from the characterization of Hankel operators belonging to the Schatten-von Neumann classes $\mathfrak{S}(p)$, due to Peller, Rochberg and Semmes [14], [15], [16], [17]. We discuss this briefly from the point of view of rational approximation.

The space B.M.O.A (analytic functions of bounded mean oscillation) can be defined in many ways; but for our purposes it consists of those functions $f(z)$, analytic in D, which are the projection of a function $g(e^{i\theta}) \in L^\infty(\partial D)$. We write $f = Pg$. More precisely, the function $f(z) = \sum_{n=0}^{\infty} a_n z^n$ belongs to B.M.O.A. if there is a function

$$g(e^{i\theta}) \sim \sum_{n=-\infty}^{\infty} \hat{g}(n)e^{in\theta}$$

in $L^\infty(\partial D)$ with $a_n = \hat{g}(n)$ for all $n \geq 0$. As norm we take

$$\|f\|_* = \inf \{\|g\|_\infty : f = Pg\} .$$

For further information on B.M.O.A. see [11]. In analogy with $r_n(f)$ we define

$$\rho_n(f) = \rho_n(f,D) = \inf \{\|f-r_n\|_* , \quad r_n \in R(n)\} .$$

For $p > 1$ the Besov space B^p consists of those functions, analytic in D with

$$\int_{-\pi}^{\pi} t^{-2} \int_{-\pi}^{\pi} |f(e^{i(s+t)})-f(e^{is})|^p \, ds dt < \infty .$$

Alternatively $f(z) \in B^p$ for $p > 1$ if and only if

$$(4.1) \qquad \iint_D |f'|^p (1-|z|^2)^{p-2} \, dxdy < \infty .$$

For a proof of this equivalence see [18], Chapter V, §5. Similar definitions hold for $p = 1$ involving the second difference or second derivative or for $0 < p < 1$ involving higher derivatives; see e.g. [17]. For simplicity I consider only the case $p > 1$. The following theorem is due to Peller [14].

Theorem 4. *If* $p > 1$ *and* $f(z) \in$ B.M.O.A., *then*

$$\sum_{n=0}^{\infty} (\rho_n(f))^p < \infty$$

if and only if $f(z) \in B^p$.

This theorem which concerns, in reality, the s-values of Hankel operators is seen, in this formulation, to fall in to the category of approximation theorems which say that good approximation to a function occurs if and only if the function satisfies a certain smoothness condition. The measure of "good approximation," however, is not in the supremum norm, but in the more complicated B.M.O.A. norm. The prior condition that $f(z)$ belong to B.M.O.A. is not essential; a much weaker condition would suffice. Theorem 4 can also be viewed as a sharpening of a result of Brudny [7]. However, Brudny's theorem, although less precise, does involve approximation in the supremum norm.

It would be nice to see how far all this could be carried over to a general closed domain K. A start on this project, using the Faber operator, has already been made by Peller [15]. The notion of Besov space which Peller uses is that introduced by Dynkin [9] for domains K which are Lipschitzian in the sense of [18]. We do not give the definition of these Besov spaces, denoted by $A_p^s(K)$, (but see [9] or [15] for details) since, as a practical matter, it is difficult to determine when a function belongs to $A_p^s(K)$. The concept of "smoothness on the boundary" which is most appropriate in this context is that of Brudny [6], which is expressed in terms of local approximations. Dynkin has proved ([9], Theorem 10) that for a domain K with Lipschitzian boundary the Faber operator T is an isomorphism between the Besov space B^p and his space $A_p^{1/p}(K)$ for $1 \leq p < \infty$. Using this, Peller has shown [15] how Theorem 4 can be generalized to domains with Lipschitzian boundary. To explain this we need one further piece of notation.

For a Lipschitzian K we denote by H(K) the space of functions f(z) analytic in K^o such that

$$(4.2) \quad f(z) = \frac{1}{2\pi i} \int_{\partial K} g(\zeta) \frac{d\zeta}{\zeta - z}, \quad z \in K$$

where $g(\zeta) \in L^\infty(\partial K)$. As norm we take

$$\|f\|_{**} = \inf \{\|g\|_\infty : \ (4.2) \text{ holds}\},$$

and we define $\tilde{\rho}_n(f,K)$ by

$$\tilde{\rho}_n(f) = \tilde{\rho}_n(f,K) = \inf \{\|f-r_n\|_{**};\ r_n \in R(n)\}.$$

Peller's theorem is

Theorem 5. *Suppose that* f(z) *belongs to* H(K) *for some Lipschitzian* K *and* $1 \leq p < \infty$. *Then* $f(z) \in A_p^{1/p}(K)$ *if and only if* $\sum_{n=0}^{\infty} (\tilde{\rho}_n(f))^p < \infty$.

The space H(K) obviously plays the role played by B.M.O.A. in Theorem 4. Although approximation in the norm $\| \ \|_{**}$ is difficult to calculate, the norm is natural and intrinsic to the situation. However, I feel that perhaps another generalization of B^p could be given which is easier to deal with than Dynkin's space $A_p^{1/p}(K)$; for this reason I have not given the definition of $A_p^{1/p}(K)$. The smoothness condition for the Besov space B^p, as expressed by (4.1) is a "Bergman" condition and such conditions, involving, as they do, an area integral are usually much easier to deal with for a general closed Jordan domain K than conditions involving a boundary integral.

I limit myself to discussion of one possibility. For a closed domain K let λ_K denote the hyperbolic or Poincaré metric in K^o defined by

$$\lambda_K(\theta(w))|\theta'(w)| = (1 - |w|^2)^{-1} .$$

Here $\theta(w)$ is the Riemann mapping function from D^o to K^o. We assume that ∂K is rectifiable so that $\theta'(w) \in H^1(D)$ and define $B^p(K)$ by

$$B^p(K) = \{f(z): f(z) \text{ analytic in } K^o, \iint_K |f'(z)|^p \lambda_K(z)^{2-2p}\, dxdy < \infty\}$$

for $p > 1$.

Conjecture. _Let K be closed Jordan domain with rectifiable boundary. Then the Faber operator T, given by (1.2) is an isomorphism of B^p onto $B^p(K)$ for $1 < p < \infty$._

Note that the conjecture involves no mention of cusps. However, the conjecture does involve consideration of the outer and inner mapping functions $\psi(w)$ and $\theta(w)$, which creates difficulties that I am at present unable to deal with.

References

1. Al'per, S. Ya., On the uniform approximation to functions of a complex variable on closed domains (in Russian), Izv. Akad. Nauk. S.S.S.R. Ser. Mat., 19 (1955), 423-444.

2. Anderson, J. M., and Clunie, J., Isomorphisms of the disc algebra and inverse Faber sets, to appear.

3. Andersson, J.- E., "On the degree of polynomial and rational approximation of holomorphic functions," Dissertation, Univ. of Göteborg, 1975.

4. Andersson, J.- E., On the degree of weighted polynomial approximation of holomorphic functions, Analysis Mathematica, 2 (1976), 163-171.

5. Andersson, J.- E., On the degree of polynomial approximation in $E^p(D)$, J. Approx. Theory, 19 (1977), 61-68.

6. Brudny, Ju. A., Spaces defined by means of local approximations, Trans. Moscow Math. Soc., 24 (1971).

7. Brudny, Ju. A., Rational approximation and imbedding theorems, Soviet Math. Dokl., 20 (1979), 681-684.

8. Duren, P. L., Theory of H^p-spaces, Academic Press, New York, 1970.

9. Dynkin, E. M., A constructive characterization of the Sobolev and Besov classes, Proc. Steklov Inst. Math. (1983), 39-74.

10. Faber, G., Über Polynomische Entwicklungen, Math. Ann., 57 (1903), 389-408.

11. Koosis, P., Introduction to H^p spaces, London Math. Soc. Lecture Note Series, 40, Cambridge University Press, 1980.

12. Kövari, T., On the order of polynomial approximation for closed Jordan domains, J. Approx. Theory, 5 (1972), 362-373.

13. Kövari, T., and Pommerenke, Ch., On Faber polynomials and Faber expansions, Math. Zeitschr., 99 (1967), 193-206.

14. Peller, V. V., Hankel operators of class $\mathfrak{S}_p$ and their applications (rational approximation, Gaussian processes, the problem of majorizing operators), Math. U.S.S.R. Sbornik, 41 (1982), 443-479.

15. Peller, V. V., Rational approximation and the smoothness of functions, Zap. Nauch. Sem. L.O.M.I., 126 (1983), 150-159.

16. Rochberg, R., Trace ideal criteria for Hankel operators and commutators, Indiana Univ. Math. J., 31 (1982), 913-925.

17. Semmes, S., Trace ideal criteria for Hankel operators, $0 < p < 1$, preprint.

18. Stein, E. M., Singular Integrals and Differentiability Properties of functions, Princeton Univ. Press, Princeton, N.J., 1970.

SURVEY ON RECENT ADVANCES IN INVERSE

PROBLEMS OF PADÉ APPROXIMATION THEORY

G. López Lagomasino
Fac. of Phys. and Mathematics
Univ. of Havana, Havana
Cuba

V. V. Vavilov
Fac. of Mech. and Mathematics
Univ. of Moscow
Moscow, U.S.S.R.

Abstract. Suppose a formal power series is given and that we know some facts about the (asymptotic) behavior of the poles (or part of them) for a certain subsequence of the associated Padé approximants. Inverse problems deal with finding out, with just this information, as much as possible about the analytical properties of the function corresponding to the power series. In the past two years, very important results have been obtained in this direction.

1. Introduction

Let

$$(1) \qquad f(z) = \sum_{\nu=0}^{\infty} f_\nu z^\nu$$

be a formal power series. Given two non-negative integers n , m , the classical Padé approximant $\pi_{n,m}$ of type (n,m) is defined to be the quotient p/q of any two polynomials p,q $(q \neq 0)$ such that

$$(2) \qquad \deg\ p \leq n \ , \ \deg\ q \leq m \ , \ (qf-p)(z) = Az^{n+m+1} + \dots \ .$$

In the study of the Padé approximants there are two questions which arise in a natural way. The first, or direct problem, can be formulated in the following terms. Suppose that (1) actually represents an analytic function in a neighborhood of $z = 0$ and we know certain global properties of its extension, such as its region of meromorphicity, the existence and number of poles in this region, other singularities, etc. The question is whether or not the poles of some sequences of Padé approximants, properly selected, tend to the singularities of f, and to which ones. In other words, is there any relation between the asymptotic behavior of these poles and the singularities of f ?

The second problem, or inverse, can be stated as follows. Suppose we know the behavior of the poles of a certain sequence of Padé approximants constructed from a formal power series. Is it possible to determine if this power series actually represents an analytic function in a neighborhood of $z = 0$? If so, what can be said about the meromorphic extension of this function and the location of its singularities? Do they correspond to limits of sequences of such poles? It should be emphasized that inverse type problems play a very important role in applications of Padé approximants.

Direct and inverse problems are intimately related. At present, much more is known about direct results. Nevertheless, in the past few years a group of mathematicians headed by A. A. Gončar have obtained very important answers in the inverse direction. Our aim is to survey recent results concerning the inverse problem; but, for a better understanding, it will be necessary to say some things about the leading direct results. In §2, we will consider the situation when the degree of the denominator of the sequence of Padé approximants is fixed. In §3 we discuss what problems have been solved and which remain open in a more general setting of interpolation points and fixed poles. Diagonal sequences will be discussed in §4.

2. Rows of the Padé Table

(I). A very well-known result in the Padé approximants theory is that of de Montessus de Ballore [4] which can be stated as follows:

Theorem. *Let (1) be a power series convergent in a neighborhood of $z = 0$. Let $D_m = \{z : |z| < R_m\}$ denote the greatest disk centered at the origin inside of which (1) has a meromorphic extension with no more than m poles (counting multiplicities). If D_m contains k distinct poles $z_1, z_2, \ldots, z_k$ of multiplicities $p_1, p_2, \ldots, p_k$ respectively, and*

$$(3) \qquad \sum_{i=1}^{k} p_i = m \quad ,$$

then the sequence of Padé approximants $\{\pi_{n,m}\}$, $n \in \mathbb{N}$, m _fixed, converges uniformly to_ f _on each compact set_ K _contained in_ $D'_m = D_m \setminus \{z_1, z_2, \ldots, z_k\}$ _as_ $n \to \infty$. _That is_

$$\pi_{n,m} \rightrightarrows f \ , \quad K \subset D'_m \ , \quad n \in \mathbb{N} \ .$$

In his proof, de Montessus used the fact, discovered by J. Hadamard [12], that under such hypotheses the poles of the denominators of the sequence $\{\pi_{n,m}\}$, $n \in \mathbb{N}$, m fixed, tend to the poles of f in D_m . More precisely, if for each $n \in \mathbb{N}$, the denominator $q_{n,m}$ of $\pi_{n,m}$ is taken as monic, then under the condition (3) of de Montessus' theorem it follows that

$$(4) \qquad \overline{\lim_{n \to \infty}} \|q_{n,m} - w_m\|^{1/n} = \frac{\max\{|z_i| : 1 \le i \le k\}}{R_m} \quad (< 1) \ ,$$

where

$$w_m(z) = \prod_{i=1}^{k} (z - z_i)^{p_i} \ ,$$

and $\|\cdot\|$ represents any of the equivalent norms in the $m+1$ dimensional space of coefficients corresponding to the space of polynomials of degree $\le m$. Relation (4) describes the asymptotic behavior of the poles of the sequence $\{\pi_{n,m}\}$, $n \in \mathbb{N}$, m fixed. Thus, (4) is considered to be a direct result for those rows of the Padé table which satisfy (3). Nevertheless, A. A. Gončar proved (see [8], Theorem 2 and Lemma 1) that, even in the case when $p_1 + p_2 \ldots + p_k < m$, each pole of f in D_m "attracts" at least as many poles of $\pi_{n,m}$ as its order of multiplicity, and with a geometric rate of convergence not greater than the modulus of the pole over R_m .

In the inverse direction we have the following unpublished result of A. A. Gončar (for the proof in a more general setting see R. K. Kovačeva's paper [13]):

Suppose (1) _represents a formal power series and there exists a polynomial_ $w_m(z) = \prod_{i=1}^{k} (z - z_i)^{p_i}$, $w_m(0) \ne 0$, $\sum_{i=1}^{k} p_i = m$, _such that_

$$\overline{\lim_{n\to\infty}} \|q_{n,m} - w_m\|^{1/n} = \theta < 1 \ .$$

Then f *is analytic in a neighborhood of* $z = 0$ *and* $z_1, z_2, \ldots, z_k$ *are exactly the poles of* f *in* $D_m = \{z : |z| < \{\max_{i=1}^{k} |z_i|\}/\theta\}$ *with multiplicities* $p_1, p_2, \ldots, p_k$ *respectively.*

(II). A natural question arises - what happens if you only know that $\lim q_{n,m} = w_m$ without any extra information about the speed with which this convergence takes place? An answer for a particular case is Fabry's classical theorem [5] which can be stated as follows:

Theorem. Suppose that (1) *is a formal power series for which*

$$\lim_{\nu\to\infty} \frac{f_\nu}{f_{\nu+1}} = \alpha \neq 0 \ .$$

Then f *is analytic in* $|z| < |\alpha|$ *and* α *is a singular point of* f .

This is an inverse result concerning the first row $\{\pi_{n,1}\}$, $n \in \mathbb{N}$, of the Padé table associated with (1), since $q_{n,1}(z) = (z - f_n \cdot f_{n+1}^{-1})$ whenever $f_n \cdot f_{n+1} \neq 0$.

All the preceding results led A. A. Gončar to formulate the following natural conjecture:

C-1. *Suppose* (1) *is a formal power series such that*

$$\lim_{n\to\infty} q_{n,m}(z) = w_m(z) = \prod_{k=1}^{m} (z - z_k) \ ,$$

where

$$0 < |z_1| \leq |z_2| \leq \ldots \leq |z_\mu| < |z_{\mu+1}| = \ldots = |z_m| \ ,$$

then $D_{m-1} = \{z : |z| < |z_m|\}$, f *has exactly* μ *poles in* D_{m-1} *which coincide with* $z_1, z_2, \ldots, z_\mu$ *counting multiplicities, and* $z_{\mu+1}, \ldots, z_m$ *are singular points of* f .

This conjecture was proved to be true completely in the case when $\mu = m - 1$ by V. V. Vavilov, G. López and V. A. Prohorov [14]. In that paper, in fact, it was proved that $D_{m-1} = \{z : |z| < |z_m|\}$ and

$z_1, z_2, \ldots, z_\mu$ are exactly the poles of f in D_{m-1} (even when $\mu < m-1$). The condition $\mu = m-1$ was only used to prove that z_m was a singularity of f, because it made it possible to reduce that assertion to Fabry's result. The question of whether $z_{\mu+1}, z_{\mu+2}, \ldots, z_m$ are singular points when $\mu < m-1$ remains to a certain extent unsolved (see part VI below).

(III). Previously, we saw that each pole of f in D_m demands a very good behavior (geometric rate of convergence) of a certain number of the poles of $\{\pi_{n,m}\}$, $n \in \mathbb{N}$, m fixed, (not less than the multiplicity of the pole) in a neighborhood of the considered pole of f. A new question arises. Can certain of the poles of $\{\pi_{n,m}\}$, $n \in \mathbb{N}$, m fixed, converge geometrically to some point which is not a pole of f in D_m? And more generally, do the limits of sequences of poles of $\{\pi_{n,m}\}$, $n \in \mathbb{N}$, m fixed, necessarily point out singular points of f in D_m?

To formulate these questions more exactly, we will use some terms which were introduced in [10] by A. A. Gončar. They characterize the behavior of the poles of $\{\pi_{n,m}\}$, $n \in \mathbb{N}$, m fixed.

By $P_{n,m} = \{\xi_{n,1}, \xi_{n,2}, \ldots, \xi_{n,m_n}\}$, $m_n \leq m$, shall be denoted the set of zeros of $q_{n,m}$. Let $|\cdot|_1 = \min(|\cdot|, 1)$ and $\xi \in \mathbb{C}$ be fixed. We shall set

$$\Delta(\xi) = \overline{\lim_{n \to \infty}} \prod_{j=1}^{m_n} |\xi_{n,j} - \xi|_1^{1/n} .$$

If $m_n = 0$, then the product on the right-hand side is taken to be equal to 1. We shall assume that

$$|\xi - \xi_{n,1}| \leq |\xi - \xi_{n,2}| \leq \cdots \leq |\xi - \xi_{n,m_n}| .$$

On the other hand, let

$$\delta_j(\xi) = \overline{\lim_{n \to \infty}} |\xi_{n,j} - \xi|_1^{1/n} ,$$

where $j = 1, 2, \ldots, m'$, $m' = \underline{\lim}\, m_n$. For $j = m'+1, \ldots, m$ we take $\delta_j(\xi) = 1$. If $\Delta(\xi) = 1$ (then $\delta_j(\xi) = 1$, $j = 1, 2, \ldots, m$), we put $\mu(\xi) = 0$. If $\Delta(\xi) < 1$, then for a certain μ, $1 \leq \mu \leq m$, we have $\delta_1(\xi) \leq \delta_2(\xi) \leq \cdots \leq \delta_\mu(\xi) < 1$ and

$\delta_{\mu+1}(\xi) = 1$ (or $\mu = m$) ; in this case we put $\mu(\xi) = \mu$. We shall let $P_m = \{\xi \in \mathbb{C}^* = \mathbb{C}\setminus\{0\} : \mu(\xi) \geq 1\} = \{\xi \in \mathbb{C}^* : \Delta(\xi) < 1\}$. The following characteristic will also be used: suppose $\lim \xi_{n,1} = \xi$, then there exists a λ such that $\lim \xi_{n,\lambda} = \xi$ but $\{\xi_{n,\lambda+1}\}$, $n \in \mathbb{N}$, is not convergent (or $\lambda = m$), in this case we put $\lambda(\xi) = \lambda$; if $\{\xi_{n,1}\}$, $n \in \mathbb{N}$, doesn't have a limit then we take $\lambda(\xi) = 0$. Finally, let $L_m = \{\xi \in \mathbb{C}^* : \lambda(\xi) \geq 1\}$.

(IV). In [10], A. A. Gončar proved the following:

Theorem. *Let* (1) *be a formal power series,* $m \in \mathbb{N}$ *and* $\xi \in \mathbb{C}^*$ *fixed. The following conditions are equivalent:*

(i) $\xi \in D_m$ *and* f *has a pole at* ξ .

(ii) $\Delta(\xi) < 1$ *(or what is the same* $\mu(\xi) \geq 1$) .

Moreover, if either one of these conditions is fulfilled, then

$$R_m = \frac{|\xi|}{\Delta(\xi)}$$

and $\mu(\xi) = \tau$, *where* τ *is the multiplicity of the pole* ξ *of* f .

The "direct" portions of this theorem ((i) → (ii), $\Delta(\xi) \leq |\xi|/R_m$ and $\mu(\xi) \geq \tau$) were well-known (see [8]). The fundamental contents of this theorem are the inverse type of assertions ((ii) → (i), $\Delta(\xi) \geq |\xi|/R_m$, $\mu(\xi) \leq \tau$) .

From this result immediately follows:

Corollary. *Let* (1) *be a formal power series. Then* $R_m > R_0$ (> 0) *if and only if* $P_m \neq \emptyset$. *Moreover,*

$$R_m = \frac{|\xi|}{\Delta(\xi)} \quad , \quad \forall\, \xi \in P_m \quad ;$$

P_m *is the set of poles of* f *in* D_m ; *for each* $\xi \in P_m$, $\mu(\xi)$ *is the order of multiplicity of this pole; and* $\sum_{\xi \in \mathbb{C}^*} \mu(\xi)$ *is equal to the number of poles of* f *in* D_m *(counting multiplicities).*

Also in [10] the following conjectures were announced:

C-2. *If* $\xi \in L_m$, *then* $R_{m-1} \geq |\xi|$ *and* ξ *is a singular point of* f .

C-3. $R_m = \frac{|\xi|}{\Delta(\xi)}$ <u>for all</u> $\xi \in L_m$.

C-4. <u>If</u> $0 \neq \xi \in D_m$, <u>then</u> $\lambda(\xi) = \mu(\xi)$.

Suppose $\lambda(\xi) = \lambda$. Then, by definition, we have that $\lim |\xi_{n,\lambda} - \xi| = 0$ and $\overline{\lim} |\xi_{n,\lambda+1} - \xi| > 0$. V. I. Buslaev [6] considered the case when ξ is such that, instead of $\overline{\lim} |\xi_{n,\lambda+1} - \xi| > 0$, the stronger condition $\underline{\lim} |\xi_{n,\lambda+1} - \xi| > 0$ is satisfied, and proved that then, in fact, $R_m = |\xi|/\Delta(\xi)$ and $\lambda(\xi) = \mu(\xi)$.

Using a different method, in a recent paper [21], S. P. Suetin proved the following:

<u>Theorem</u>. <u>Let</u> $\xi \in \mathbb{C}^*$. <u>If</u> $\lambda(\xi) \geq 1$ <u>then</u> $R_m \geq |\xi|$. <u>Moreover, if</u> $R_m > |\xi|$, <u>then</u> $\lambda(\xi) = \mu(\xi) = \tau$, <u>where</u> τ <u>is the multiplicity of the pole</u> ξ <u>of</u> f .

This result settles all the questions in C-2, 3, 4, except, the assertion in C-2, that ξ is a singularity of f . In fact:

a) Suppose $R_{m-1} < |\xi|$ and $\lambda(\xi) \geq 1$. Then, using Suetin's result, we have that $R_{m-1} < R_m$. But this can be possible only if in $\overline{D}_{m-1}$ there are exactly m poles of f . From Montessus' theorem we know that all the poles of $\{\pi_{n,m}\}$, $n \in \mathbb{N}$, must converge to points in $\overline{D}_{m-1}$ which contradicts the fact that $\lambda(\xi) \geq 1$. So, necessarily, we must have $R_{m-1} \geq |\xi|$.

b) If $\xi \in P_m (\subset L_m)$, then $R_m = \frac{|\xi|}{\Delta(\xi)}$ according to Gončar's result. So C-3 reduces to the proof that if $\lambda(\xi) \geq 1$ and $\mu(\xi) = 0$ then $R_m = |\xi|$, (because $\mu(\xi) = 0$ if and only if $\Delta(\xi) = 1$) . Suetin's result yields this because $R_m \geq |\xi|$ but $R_m > |\xi|$ is not possible since then $\lambda(\xi) = \mu(\xi)$.

c) The second part of Suetin's theorem is precisely C-4.

(V). Let $\pi_{n,m} = \frac{P_{n,m}}{Q_{n,m}}$, where $P_{n,m} = p$, $Q_{n,m} = q$ satisfy (2), have no common zeros in $\mathbb{C}^*$ and $Q_{n,m}$ is normalized the following way:

$$Q_{n,m}(z) = \prod_{|\xi_{n,j}| \leq R} (z - \xi_{n,j}) \prod_{|\xi_{n,j}| > R} \left(\frac{z}{\xi_{n,j}} - 1\right) ,$$

where R is a positive real number properly selected, according to the problem we wish to solve.

From the definition of $P_{n,m}$ and $Q_{n,m}$ it follows that

$$(P_{n+1,m}Q_{n,m} - P_{n,m}Q_{n+1,m})(z) = A_{n,m}z^{n+m+1} ,$$

so we have that

$$(\pi_{n+1,m} - \pi_{n,m})(z) = \frac{A_{n,m}z^{n+m+1}}{(Q_{n+1,m}Q_{n,m})(z)} .$$

From this formula, it is obvious that the convergence of $\{\pi_{n,m}\}$, $n \in \mathbb{N}$, is equivalent to the convergence of the series

$$\pi_{n_0,m}(z) + \sum_{n \geq n_0} \frac{A_{n,m}z^{n+m+1}}{(Q_{n+1,m}Q_{n,m})(z)} .$$

In [10], Gončar proved that

$$(5) \qquad \frac{1}{R_m} = \overline{\lim_{n \to \infty}} |A_{n,m}|^{1/n} ,$$

(of course (5) does not depend on the choice of R , $0 < R < \infty$) . This formula is of particular interest because, as is the case with all the results described in part IV of this section, it only depends on the nature of the m-th row (while Hadamard's formula essentially considers two consecutive rows).

In a very recent paper [15], Prohorov, Suetin and Vavilov have shown that if $R_m^n A_{n,m} \not\to 0$, then the second part of C-2 is also true (that is, ξ is a singularity of f) .

(VI). Since $\mu(\xi) \leq \lambda(\xi)$, evidently C-4 is equivalent (see [8], Theorem 2 and Lemma 1) to the proof that whenever $0 \neq \xi \in D_m$ then there exist a disk $B(\xi,r) = \{z : |z - \xi| < r\}$ and $\Lambda \subset \mathbb{N}$ such that

$$Q'_{n,m}\pi_{n,m} \rightrightarrows (z-\xi)^{\mu(\xi)} \cdot f(z) , \quad K \subset B(\xi,r) , \quad n \in \Lambda ,$$

where

$Q'_{n,m}(z) = (z - \xi_{n,1})(z - \xi_{n,2}) \ldots (z - \xi_{n,\mu(\xi)})$, $(Q'_{n,m} \equiv 1$ if $\mu(\xi) = 0)$.

In [7], Buslaev, Gončar, and Suetin proved a similar statement for $\xi = 0$. More precisely:

Theorem. Let (1) represent on analytic function in a neighborhood of $z = 0$. Then, there exists a subsequence $\{\pi_{n,m}\}$, $n \in \Lambda \subset \mathbb{N}$, such that

$$\pi_{n,m} \rightrightarrows f, \; K \subset \{z : |z| < CR_m\} \setminus \{z : f(z) = \infty\} , \quad n \in \Lambda ,$$

where $0 < C < 1$ is a constant which depends only on m .

That $C < 1$ in the class of all such functions is a consequence of a simple example given in [7]. In fact, take

$$f(z) = \frac{1 + 2^{1/3} \cdot z}{1 - z^3} .$$

Then, for $m = 2$ and any $n \in \mathbb{N}$ the poles of $\pi_{n,2}$ are easy to compute and you can see that $\pi_{n,2}$ always has at least one pole in $|z| < 1/2^{1/3}$, while of course, $1/2^{1/3} < 1 = R_2$. Similar examples have been constructed by Buslaev for any $m > 1$. The last theorem was used to prove that the conjecture formulated by G. A. Baker and P. Graves-Morris (see [3]) about the existence of a subsequence $\{\pi_{n,m}\}$, $n \in \Lambda \subset \mathbb{N}$, such that $\pi_{n,m} \rightrightarrows f$, $K \subset D'_m$, $n \in \Lambda$, is true if $R_m = \infty$. On the other hand, Buslaev's examples show that it fails if $R_m < +\infty$ for any $m > 1$. In this case what is true (see [7]) is that there exists a subsequence $\{\pi_{n,m}\}$, $n \in \Lambda \subset \mathbb{N}$, such that $\pi_{n,m} \rightrightarrows f$, $n \in \Lambda$, $K \subset D'_m \setminus S$, where S is a finite set with no more than $m - m' - 1$ points and m' is the number of poles of f in D_m . Buslaev's examples also show that the estimate for the number of points in S cannot be improved.

3. Generalized Padé Tables

(I). Let's consider a general interpolation scheme which generalizes the classical Padé approximants. The corresponding generalized Padé table is constructed in the following way. Let $\alpha = \{\alpha_{n,k}\}$, $k = 1,2,\dots,n$, $n \in \mathbb{N}$, be a system of interpolation points and $\beta = \{\beta_{n,k}\}$, $k = 1,2,\dots,n$, $n \in \mathbb{N}$, a system of fixed poles. It's supposed that $\alpha \cap \beta = \emptyset$. Moreover, it turns out to be convenient to assume that $\alpha \subset E$, $\beta \subset F$, where E and F are compact sets of $\overline{\mathbb{C}}$, $E \cap F = \emptyset$ and $0 \notin F$. Let's put

$$a_n(z) = \prod_{k=1}^{n} (z - \alpha_{n,k}) \quad , \quad b_n(z) = \prod_{k=1}^{n} \left(1 - \frac{z}{\beta_{n,k}}\right) \quad , \quad n \in \mathbb{N} .$$

Suppose that f is analytic on E $(f \in H(E))$; that is, there exists an open set U such that $E \subset U$ and $f \in H(U)$. It's easy to prove that there exist polynomials p,q $(q \not\equiv 0)$ such that (compare with (2))

$$\text{(6)} \qquad \deg p \leq n \quad , \quad \deg q \leq m \quad , \quad \frac{q\, b_{n-m} f - p}{a_{n+m+1}} \in H(E)$$

((n,m) is a fixed pair of natural numbers; whenever $n < m$, we take $b_{n-m} = b_{m-n}^{-1}$). Relations (6) define for each (n,m) fixed, a unique rational function

$$r_{n,m} = \frac{p}{q \cdot b_{n-m}} \quad ,$$

which in the sequel we will refer to as the (α,β) - Padé approximant of type (n,m) with respect to f. When $\alpha_{n,k} \equiv 0$, $\beta_{n,k} \equiv \infty$ this table reduces to the classical Padé table (of series (1) with $R_0 > 0$). When $\beta_{n,k} \equiv \infty$ we obtain the so-called multipoint Padé approximants. If $\alpha_{n,k} = \alpha_k$ and $\beta_{n,k} = \beta_k$, $n \in \mathbb{N}$, a Newton-type table is obtained. The definition of (α,β) - Padé table can be extended to the case when f is given as a set of formal interpolation data (for details, see [10]).

(II). Let $f \in H(E)$ and let's consider the m-th row

$$\text{(7)} \qquad r_{n,m} \quad , \quad m \geq 0 \ \text{fixed} \quad , \quad n \in \mathbb{N} \ ,$$

of the (α,β) - Padé table associated with f. When $m = 0$ all the poles of $r_{n,0}$ are fixed and convergence results have been well studied (see [22] and [1]). In this case, asymptotic conditions are imposed on the tables (α,β) which determine the geometry of the region of convergence as well as the rate with which the convergence takes place. In the general case, for sequences of type (7), if we wish to study convergence properties and the behavior of the free poles (zeros of q) it is natural, all the more, to require such asymptotic conditions.

Let E, F be a pair of regular compact sets which satisfy the conditions above and $G = \overline{\mathbb{C}} \setminus (E \cup F)$. The pair (E,F) is called a <u>regular condensor</u>. Let w be a continuous function in $\overline{\mathbb{C}}$ and harmonic in G such that $w|_E \equiv 0$ and $w|_F \equiv 1$. We put $\Phi = \exp(w/c)$, where $c = c(E,F)$ is the capacity of (E,F). That is,

$$c = \frac{1}{2\pi} \int_{\Gamma} \frac{\partial w}{\partial n}\, ds \quad ,$$

where Γ is any smooth curve contained in G which separates E from F, $\partial/\partial n$ the normal derivative to Γ in the direction from E to F, ds is the arc element. (When G is a doubly connected region, then $R = \exp(1/c)$ is the Riemann module of G.) We set $E_\rho = \{z : \Phi(z) < \rho\}$, $1 < \rho < \exp(1/c)$.

There exist tables α, β $(\alpha \subset E$, $\beta \subset F)$ such that

$$(8) \qquad \left| \frac{a_n(z)}{b_n(z)} \right|^{1/n} \rightrightarrows \lambda\Phi(z) \ , \quad K \subset G \ , \quad n \in \mathbb{N} \ ,$$

where λ is a certain constant. The set of all pairs (α,β) which satisfy (8) will be denoted by $W(E,F)$; the subset of $W(E,F)$ consisting of tables (α,β) of Newton-type will be denoted by $N(E,F)$. Different constructions for such tables may be found in [22] and [1].

We will now let $D_m = E_{\rho_m}$, the greatest canonical region in which f admits a meromorphic extension with no more than m poles: $\Delta(\xi)$, $\mu(\xi)$, $\lambda(\xi)$ are also defined in a fashion analogous to the definitions in § 2, part III. Direct results for multipoint Padé tables were obtained by E. B. Saff [17] and for (α,β) type by A. A. Gončar [8]. All the questions examined in §2 can be reconsidered in this more general setting. In the inverse direction, when $(\alpha,\beta) \in N(E,F)$, the first result was obtained by R. K. Kovačeva [13] for the case when $\sum_{\xi \in G} \mu(\xi) = m$. It is analogous to Gončar's result stated at the end of part (I), §2. Some other results formulated in parts IV and V, §2, also hold when $(\alpha,\beta) \in N(E,F)$ (see [10], [6]).

S. P. Suetin has also considered analogous problems for rational approximants relative to Faber series and expansions with respect to orthogonal polynomials (for direct results see [18] and [20], inverse type are in [19]). Similar problems may be posed for best rational approximants.

4. Main Diagonal of the Padé Table

(I). Let's consider the main diagonal of the Padé table associated with a formal power series of type (1):

$$(9) \qquad \pi_n = \pi_{n,n} \quad , \quad n \in \mathbb{N} \quad .$$

In the following, we will suppose additionally that for each $n \in \mathbb{N}$ we have

$$f(z) - \pi_{n,n}(z) = A_n z^{2n+1} + \ldots , \qquad n \in \mathbb{N} \quad .$$

If sequence (9) converges uniformly on each compact set of some disk $U_R = \{z : |z| < R\}$, $R > 0$, then obviously it tends to a function $\tilde{f}$ analytic in U_R whose expansion in a Taylor series about $z = 0$ is none other than (1). Moreover, series (1) is also convergent in U_R and $f(z) = \tilde{f}(z)$, $z \in U_R$.

However, the opposite is not true; f may be an analytic function in U_R such that the poles of sequence (9) have limit points in U_R . An essential question is if there is any other difficulty for the convergence of sequence (9) to f . In other words, if $f \in H(U_R)$ and for each $n \in \mathbb{N}$, $\pi_n \in H(U_R)$, does this imply that sequence (9) converges to f uniformly on each compact set contained in U_R ? In this form, the problem was considered in [2, p. 183]. The best of the results announced there states that, under the above, conditions, it follows that $\pi_n \rightrightarrows f$, $K \subset U_{cR}$, $n \in \mathbb{N}$, where $c = 1/\sqrt{3} = 0.577\ldots$. In the past few years, many attempts have been made in order to prove that $c = 1$. A similar question, in terms of subsequences of (9), has a negative answer. E. A. Rahmanov [16] gave an example of an analytic function in $|z| < 1$ with a subsequence $\{\pi_n\}$, $n \in \Lambda \subset \mathbb{N}$, whose poles are in $|z| \geq 1$ and such that the subsequence is unbounded in any neighborhood of $z = 4/5$.

A. A. Gončar proved in [11], using potential theory, that if for all $n \in \mathbb{N}$, $\pi_n \in H(U_R)$ $(\{\pi_n\} \subset H(U_R))$, then in fact (9) converges uniformly on each compact set contained in U_R $(\{\pi_n\} \subset C(U_R))$. As a result of this, $f \in H(U_R)$ and $c = 1$. It should be underlined that apriori no assumption is made about (1) being an analytic function in any neighborhood whatsoever of $z = 0$. In other words, he proved that

$$\{\pi_n\} \subset H(U_R) \rightarrow \{\pi_n\} \subset C(U_R) \rightarrow f \in H(U_R) \quad .$$

These relations enclose a general principle, which is true for a wide class of regions, and expresses an inverse type of problem in the theory of Padé approximants.

(II). Let G_∞ denote the class of all regions ω such that

(10) $$\infty \in \omega = D \setminus E \quad ,$$

where D is a region, and E is a closed subset with respect to D which satisfies

(11) $$\partial D \subset \partial \tilde{D} \quad , \quad \mathrm{cap}(E) = 0 \quad ,$$

where $\tilde{D}$ denotes the complement to the convex hull of the boundary of $D(\partial D)$ and $\mathrm{cap}(\cdot)$ denotes the logarithmic capacity of E. Finally, let G_0 be the class of all regions Ω such that $\omega = \{z : \frac{1}{z} \in \Omega\} \in G_\infty$.

In [11], Gončar proved the following:

<u>Theorem</u>. <u>If</u> $\Omega \in G_0$ <u>then</u>

$$\{\pi_n\} \subset H(\Omega) \rightarrow \{\pi_n\} \subset C(\Omega) \rightarrow f \in H(\Omega) \quad .$$

Obviously, in the theorem, instead of $\{\pi_n\} \subset H(\Omega)$, it is sufficient to assume that for each compact $K \subset \Omega$ there exists an $n(K)$ such that if $n > n(K)$, then π_n has no pole on K.

Because of the invariant properties of the diagonal of the Padé table with respect to bilinear transforms (in particular, $z \to \frac{1}{z}$), a similar result holds for regions of type G_∞ and formal Taylor series expansions at $z = \infty$.

To G_0 and/or G_∞ belong, for instance, the regions:

$U_R \setminus E$, $\operatorname{cap}(E) = 0$; $\bar{\mathbb{C}} \setminus E$, $\operatorname{cap}(E) = 0$ (E is a compact subset of $\mathbb{C}$) ; $\bar{\mathbb{C}} \setminus S$, where S is a compact subset of $\mathbb{R}$; and $\mathbb{C} \setminus [c,+\infty[$, $c > 0$.

(III). It is well-known that there is a close relationship between diagonal Padé approximants and continued fractions. Suppose we consider a Chebyshev type continued fraction

$$1 + \frac{a_1|}{|z + b_1} + \frac{a_2|}{|z + b_2} + \ldots + \frac{a_n|}{|z + b_n} + \ldots$$

($\{a_n\}$, $\{b_n\}$ are arbitrary sequences of complex numbers); or that continued fraction obtained from the above after the change of variables $z \to \frac{1}{z}$; that is

$$1 + \frac{a_1 z|}{|1 + b_1 z} + \frac{a_2 z^2|}{|1 + b_2 z} + \ldots + \frac{a_n z^2|}{|1 + b_n z} + \ldots .$$

The above result essentially means that to prove the uniform convergence of one of these continued fractions on each compact set contained in a region of type G_∞ , respectively G_0 , it is sufficient to show that its n-th partial fraction is analytical on each compact set of such a region for all sufficiently large n .

Similar results are also true for continued fractions of type

$$1 + \frac{a_1 z|}{|1} + \frac{a_2 z|}{|1} + \ldots + \frac{a_n z|}{|1} + \ldots$$

(see [11], Theorem 4).

(IV). For multipoint type Padé approximants similar problems have not yet been studied.

Another result concerning the analytic extension of a formal power series in terms of the asymptotic behavior of the poles of the main diagonal appears in a paper of A. A. Gončar and K. N. Lungu [9].

References

1. T. Bagby, On interpolating by rational functions, Duke Math. J, 36, 1, (1969), 95-104.

2. G. A. Baker Jr., Essentials of Padé Approximants, Academic Press, New York, (1975).

3. G. A. Baker Jr., P. R. Graves-Morris, Padé Approximants, Part I: Basic Theory. Enc. of Math. and Appl., v. 13, Addison-Wesley Pub. Co., (1981).

4. R. de Montessus de Ballore, Sur les fractions continues algébrique, Bull. Soc. Math. France, 30(1902), 28-36.

5. L. Bieberbach, Analytische Fortsetzung, Springer-Verlag, Berlin - Gottingen - Heidelberg, (1955).

6. V. I. Buslaev, On the poles of the m-th row of the Padé table, Mat. Sb. 117 (159), 4(1982), 435-441; Eng. transl. in Math U.S.S.R. Sb, 45 (1983).

7. V. I. Buslaev, A. A. Gončar, S. P. Suetin, On the convergence of subsequences of the m-th row of the Padé table, Mat. Sb. 120(162), 4(1983), 540-545.

8. A. A. Gončar, On the convergence of generalized Padé approximants of meromorphic functions, Mat. Sb., 98(140), 4(12), (1975), 564-577; Eng. transl. in Math U.S.S.R. Sb., 27(1975).

9. A. A. Gončar, K. N. Lungu, The poles of diagonal Padé approximants and the analytic extension of functions, Mat. Sb. 111(153), 2(1980), 119-132; Eng. transl. in Math U.S.S.R. Sb., 39(1981).

10. A. A. Gončar, The poles of the rows of the Padé table and the meromorphic extension of the functions, Mat. Sb., 115(157), 4(8), (1981), 590-613, Eng. transl. in Math U.S.S.R. Sb., 43(1982).

11. A. A. Gončar, On the uniform convergence of diagonal Padé approxmants, Mat. Sb., 118(160), 4(8), (1982), 535-556; Eng. transl. in Math. U.S.S.R. Sb., 42(1983).

12. J. Hadamard, Essai sur l'étude des fractions données par leur développement de Taylor, J. Math. Pures et Appl., (4), 8(1892), 101-186.

13. R. K. Kovačeva, Generalized Padé approximants and meromorphic continuation of functions, Mat. Sb., 109(151), 3(1979), 365-377; Eng. transl. in Math U.S.S.R. Sb., 37, 3(1980).

14. G. López, V. A. Prohorov, V. V. Vavilov, On an inverse problem for the rows of a Padé table, Mat. Sb., 110(152), 1(9), (1979), 117-127; Eng. transl. in Math U.S.S.R. Sb., 38, (1980).

15. V. A. Prohorov, S. P. Suetin, V. V. Vavilov, Poles of the m-th row of the Padé table and the singular points of functions, Mat. Sb. 122(164), 4(1983).

16. E. A. Rahmanov, On the convergence of Padé approximants in classes of holomorphic functions, Mat. Sb., 112(154), (1980), 162-169; Eng. transl. in Math. U.S.S.R. Sb., 40, (1981).

17. E. B. Saff, An extension of Montessus de Ballore's theorem on the convergence of interpolating rational functions, J. Approx. Theory 6, (1972), 63-67.

18. S. P. Suetin, On the convergence of rational approximants of polynomial expansions in the regions of meromorphicity of a given function, Mat. Sb., 105(147), 3, (1978), 413-430, Eng. transl. in Math. U.S.S.R. Sb., 33, (1979).

19. S. P. Suetin, Inverse problems for generalized Padé approximants, Mat. Sb., 109(151), (1979), 629-646; Eng. transl. in Math. U.S.S.R. Sb., 37(1980).

20. S. P. Suetin, On the theorem of Montessus de Ballore for rational approximants of orthogonal expansions, Mat. Sb. 114(156), 3, (1981), 451-464; Eng. transl. in Math. U.S.S.R. Sb., 42(1982).

21. S. P. Suetin, On the poles of the m-th row of the Padé table, Mat. Sb., 120(162), 4(1983), 500-504.

22. J. L. Walsh, Interpolation and Approximation by Rational Functions in the Complex Domain, 5th ed., Coll. Publ., v. 20, A.M.S., Providence, (1969).

SOME PROPERTIES AND APPLICATIONS OF CHEBYSHEV POLYNOMIAL AND RATIONAL APPROXIMATION

J C Mason
Department of Mathematics and Ballistics
Royal Military College of Science
Shrivenham
SWINDON
Wiltshire
England

Abstract. A number of key properties and applications of real and complex Chebyshev polynomials of the first and second kinds are here reviewed. First, in the overall context of L_p norms ($1\leq p\leq\infty$), there is a review of the orthogonality and minimality properties of Chebyshev polynomials, the best and near-best approximation properties of Chebyshev series expansions and Chebyshev interpolating polynomials, and the links between Chebyshev series and Fourier and Laurent series. Second, there is a brief discussion of the applications of Chebyshev polynomials to Chebyshev-Padé-Laurent approximation, Chebyshev rational interpolation, Clenshaw-Curtis integration, and Chebyshev methods for integral and differential equations. Several new or unpublished ideas are introduced in these areas.

1. PROPERTIES OF CHEBYSHEV POLYNOMIALS

1.1 Fourier and Laurent Series

The Chebyshev polynomials $T_n(x)$ and $U_n(x)$ of degree n of the first and second kinds, respectively, appropriate to the range [-1,1] of x, may be conveniently defined in terms of a transformation from x to θ as

$$T_n(x) = \cos n\theta, \quad U_n(x) = \sin(n+1)\theta \,/\, \sin\theta, \tag{1}$$

where $x = \cos\theta$.

This definition provides an immediate link with Fourier series. Suppose that F_n denotes the Fourier series projection of order n on $[-\pi,\pi]$, and that G_n and H_n respectively denote the first and second kind Chebyshev series projections of degree n on [-1,1]. Suppose further that a weighted second kind series projection H_n^* is defined by the relation

$$(H_n^* \; f^*)\;(x) = \sqrt{(1-x^2)}\;(H_n \; f)\;(x) \qquad (2)$$

where $f^*(x) = \sqrt{(1-x^2)}\; f(x)$.

Then it follows from (1) (compare [1]) that

$$(G_n \; f)\;(x) = (F_n \; g)\;(\theta) \quad \text{where } g(\theta) = f(\cos\theta) = f\,(x) \qquad (3)$$

Similarly $\sqrt{(1-x^2)}\;(H_{n-1}\; f)\;(x) = (H_{n-1}^* \; f^*)\;(x) = (F_n \; h)\;(\theta)$ (4)

where $h(\theta) = \sin\theta\; f(\cos\theta) = \sqrt{(1-x^2)}\; f(x) = f^*(x)$.

The half-range Fourier series in (3) and (4) are, respectively, Fourier cosine series and Fourier sine series.

In the case of a complex variable, it is convenient to define Chebyshev polynomials $T_n(z)$ and $U_n(z)$ of degree n of first and second kinds appropriate to an elliptical annulus $1 \leqslant \rho_1 \leqslant |z + \sqrt{(z^2-1)}| \leqslant \rho_2$ as follows

$$T_n(z) = \tfrac{1}{2}(w^n + w^{-n}), \; \sqrt{(z^2-1)}\,U_{n-1}(z) = \tfrac{1}{2}(w^n - w^{-n}), \qquad (5)$$

where $z = \tfrac{1}{2}(w+w^{-1})$ and $w = z+\sqrt{(z^2-1)}$.

From this definition an immediate link may be established with Laurent series on the circular annulus $\rho^{-1} \leqslant |w| \leqslant \rho$. $(\rho > 1)$.

Suppose that B_{mn} denotes the Laurent series projection of order m in z^{-1} and order n in z, and that G_n and H_n now denote first and second kind complex Chebyshev series projections of degree n on the elliptical domain E_ρ: $|z+\sqrt{(z^2-1)}| \leqslant \rho > 1$.

Suppose further that H_n is now the weighted projection given by

$$(H_n^* \; f^*)\;(z) = \sqrt{(z^2-1)}\;(H_n \; f)\;(z) \qquad (6)$$

where $f^*(z) = \sqrt{(z^2-1)}\; f(z)$.

Then it follows from (5) (see [1] for details) that

$$(G_n \; f)\;(z) = (B_{nn} \; g)\;(w) \qquad (7)$$

where $g(w) = g(z+\sqrt{(z^2-1)}) = f(z)$.

Similarly $\sqrt{(z^2-1)}\;(H_{n-1}\; f)\;(z) = (H_{n-1}^* \; f^*)\;(z) = (B_{nn} h)\,(w)$ (8)

where $h(w) = \frac{1}{2}(w-w^{-1}) \quad f(\frac{1}{2}(w+w^{-1})) = \sqrt{(z^2-1)}\ f(z) = f^*(z)$. Here $g(w)$ and $h(w)$ are to be interpreted as "even" and "odd" functions, satisfying $g(w) = g(w^{-1})$ and $h(w) = -h(w^{-1})$ respectively, and to ensure this we assume that in each case $f(z)$ is in the class $A(E_\rho)$ of functions analytic in the interior of E_ρ and continuous on its closure. We shall later also consider restrictions of this class to the elliptical contour ξ_r: $|z+\sqrt{(z^2-1)}| = r>1$.

For functions in $A(N_{\rho 1 \rho 2})$, where $N_{\rho 1 \rho 2}$ denotes the elliptical annulus $1<\rho_1 \leq |z+\sqrt{(z^2-1)}| \leq \rho_2$, it is necessary to consider the generalised Chebyshev series projection of order n (see [2])

$$J_{nn} = G_n + H^*_{n-1}, \tag{9}$$

which projects into a combined series in $\{T_k(z)\}$ and $\{\sqrt{(z^2-1)}U_{k-1}(z)\}$. In this case we deduce that

$$(J_{nn}\ f)\ (z) = (B_{nn}\ g)\ (w) \tag{10}$$

where $g(w) = g(z+\sqrt{(z^2-1)}) = f(z)$.

1.2 Discrete Orthogonality and Chebyshev Interpolation

The first kind Chebyshev polynomials satisfy the discrete relation (see [3])

$$n^{-1} \sum_{k=1}^{n} T_i(\alpha_{kn})\ T_j(\alpha_{kn}) = \begin{cases} 0 & i \neq j \\ 1 & i=j=0 \\ \frac{1}{2} & i=j>0 \end{cases} \tag{11}$$

where $\alpha_{kn} = \cos \frac{(2k-1)\pi}{2n}$ are the n zeros of $T_n(x)$.

It immediately follows that the projection G^I_{n-1} on the polynomial of degree n-1 interpolating at the zero of $T_n(x)$ is given by

$$(G^I_{n-1}\ f)\ (x) = \sum_{j=0}^{n-1}{}' c_j\ T_j\ (x) \tag{12}$$

where $c_j = \frac{2}{n} \sum_{k-1}^{n} f(\alpha_{kn})\ T_j(\alpha_{kn})$

and where the dash denotes that the first term is halved.

The second kind polynomials are also discretely orthogonal in the sense that

$$(n+1)^{-1} \sum_{k=1}^{n} (1-\beta_{kn}^2)\ U_i(\beta_{kn})\ U_j(\beta_{kn}) = \begin{cases} 0 & i \neq j \\ \frac{1}{2} & i=j \end{cases} \tag{13}$$

where $\beta_{kn} = \cos \frac{k\pi}{n+1}$ are the n zeros of $U_n(x)$.

Hence the projection H^I_{n-1} on the polynomial of degree n-1 interpolating at the zeros of $U_n(x)$ is given by

$$(H^I_{n-1} f)(x) = \sum_{j=1}^{n} d_j U_{j-1} \tag{14}$$

where $d_j = \frac{2}{n+1} \sum_{k=1}^{n} (1-\beta^2_{kn}) f(\beta_{kn}) U_{j-1}(\beta_{kn})$.

Again this is a fast algorithm.

1.3 L_p Orthogonality and Minimality

It follows from the definitions (1) that $\{T_k(x)\}$ and $\{U_k(x)\}$ are orthogonal polynomial systems on [-1,1] with respect to the weights $(1-x^2)^{-\frac{1}{2}}$ and $(1-x^2)^{\frac{1}{2}}$, respectively. In fact these standard results generalise into less well-known results expressible in the context of L_p norms. A full discussion of this topic is given by Atkinson and Mason [4], and we only summarise results and give some indication of proofs.

An L_p norm on a domain S with measure $d\upsilon$ is defined by

$$\|f\|_p = \left\{\int_S w_p \, |f|^p \, d\upsilon\right\}^{1/p} \quad (1 \leqslant p \leqslant \infty) \tag{15}$$

where a weight function w_p is defined for each of the Chebyshev polynomials T_k and U_k.

This w_p varies with p and with the choice of S and $d\upsilon$ and the variables x or z. (Norms on either a contour or a domain are covered in the complex case).

For the system $\{T_k\}$; we define

$$w_p = \begin{cases} (1-x^2)^{-\frac{1}{2}} & \text{for } S = [-1,1], \; d\upsilon = dx \\ |z^2-1|^{-\frac{1}{2}} & \text{for } S = \xi_r \quad, \; d\upsilon = |dz| \\ |z^2-1|^{-1} & \text{for } S = E_\rho \quad, \; d\upsilon = dS \\ & \quad \text{or } S = N_{\rho 1 \rho 2} \; , \; d\upsilon = dS \end{cases} \tag{16}$$

For the system $\{U_k\}$*;* we define

$$w_p = \begin{cases} (1-x^2)^{(p-1)/2} & \text{for } S = [-1,1], \quad d\upsilon = dx \\ |z^2-1|^{(p-1)/2} & \text{for } S = \xi_r, \quad d\upsilon = |dz| \\ |z^2-1|^{(p-2)/2} & \text{for } S = E_\rho \text{ or } N_{\rho 1 \rho 2}, d\upsilon = dS \end{cases} \tag{17}$$

We note that, on the elliptical-type domains E_ρ and $N_{\rho 1\rho 2}$ an element of area is

$$dS = r^{-1} \, |z^2-1|^{\frac{1}{2}} \, dr|dz| \tag{18}$$

It follows that L_p norms on E_ρ and $N_{\rho 1\rho 2}$ (abbreviated to E and N) are linked to the L_p norm on the contour ξ_r (abbreviated to ξ) by

$$(\| f \|_p^N)^p = \int_{\rho 1}^{\rho 2} r^{-1} (\| f \|_p^\xi)^p \, dr, \quad (\| f \|_p^E)^p = \int_1^\rho r^{-1} (\| f \|_p^\xi)^p \, dr. \tag{19}$$

Property 1 For $\phi_k = T_k$ or U_k, with w_p defined by (16), (17), and all p

$$I_{kn} = \int_S w_p \, \phi_k \phi_n |\phi_n|^{p-2} \, d\upsilon = 0 \quad (k = 0, 1, \ldots, n-1). \tag{20}$$

Proof Consider first $S = \xi_r$. By substituting $z = \frac{1}{2}(w+w^{-1})$ and $w = \rho e^{i\theta}$, the result follows by binomial expansion and orthogonality arguments (see [4]). The cases S = N and S = E may be obtained by integrating (20) using (18), and the case S = [-1,1] may be obtained from (20) in the limit as $\rho \to 1$. ■

We note that Property 1 reduces to standard L_2 orthogonality when p = 2, and that the real case p = 2s+2 (s integer) (in which the modulus signs may be deleted in (20)) is termed s-orthogonality and has been discussed in a wider context by Ossicini and Rosati [5].

Property 2 The Chebyshev polynomials T_n and U_n have the property of minimising $\| \phi_n \|_p$ over all polynomials ϕ_n with the same leading coefficients (2^{1-n} and 2^{-n} respectively), where $\| \cdot \|_p$ is defined by (15) and w_p, S, and $d\upsilon$ are given by (16) and (17) respectively.

Proof A sufficient condition for the minimisation of $\| \phi_n \|_p$ is precisely (20) (see Rice [6]). ■

We note that for $p = \infty$ Property 2 is the classical minimax property, and that for p = 1 we have the well-known property that U_n is minimal in an (unweighted) L_1 norm.

1.4 Best L_p Approximation By Lacunary Series

There are special functions whose partial Chebyshev series expansions are best L_p approximations. One such class is given by Atkinson and Mason [4]:

Property 3. The Lacunary Series

$$f(x) = \sum_{k=1}^{\infty} a^k \phi_{b^k}(x)$$

where ϕ_r is T_r or U_{r-1}, b is an integer, a is real, and $|ab| < \frac{1}{2}$, has the property that every partial sum is a best approximation in L_p (defined by (15)) $(1 \leq p \leq \infty)$.

Proof In the case of U_{r-1}, for example, the error takes the form (for some N, with $x = \cos\theta$):

$$\frac{1}{\sin\theta} \sum_{k=N}^{\infty} a^k \sin b^k\theta = \frac{1}{\sin\theta} a^N \sin b^N\theta. \quad (1+u)$$

where $u = \sum_{k=N+1}^{\infty} a^{k-N} \frac{\sin b^k\theta}{\sin b^N\theta}$ and $|u| < 1$.

Since u is a uniformly convergent cosine series, it is possible to construct a proof along similar lines to those of Properties 1 and 2. ∎

We remark that the conditions on a and b can be significantly relaxed in the case of T_r, and we note that the results for $p = \infty$ and $p = 1$ have been discussed by Freilich and Mason [7].

1.5 Near-Best L_p Approximations and Minimal Projections by Series Expansions

The Chebyshev series projections G_n, H^*_{n-1}, J_{nn} are shown by Mason [2] to yield near-best approximations, in the sense that, for all f in a specified normed function space,

$$\|f-Pf\| \leq (1+\|P\|) \quad \|f-f_B\| \qquad (21)$$

where f_B is a best approximation, P is the relevant projection, and $\|P\|$ is small for realistic values of n (in fact $\|P\| \leq 4$ for $n \leq 500$). More specifically the following results are given in [2], and the reader is referred there for proofs.

Property 4 Define

$$\sigma_n^{(p)} = \lambda_n^{|2p^{-1}-1|}, \text{ where } \lambda_n = \frac{1}{\pi}\int_0^n \frac{\sin\,(n+\frac{1}{2})\theta}{\sin\,\frac{1}{2}\,\theta} \sim \frac{4}{\pi^2}\log n,$$

(i) On $L_p\,[-1,1]$, $\|G_n\|_{L_p} \leqslant \sigma_n^{(p)}$

(ii) On $\{f(x) = \sqrt{(z^2-1)}\,F(x)\,|\,F \text{ in } L_p\,[-1,1],\ z \text{ on } \xi_1\}$,

$\|H^*_{n-1}\|_{L_p} \leqslant \sigma_n^{(p)}$

(iii) On $A(E\rho)$, $\|G_n\|_{L_p} \leqslant \sigma_n^{(p)}$

(iv) On $\{f(z) = \sqrt{(z^2-1)}\,F(z)\,|\,F \text{ in } A(E_\rho)\}$, $\|H^*_{n-1}\|_{L_p} \leqslant \sigma_n^{(p)}$

(v) On $A(N_{\rho 1 \rho 2})$, $\|J_{nn}\|_{L_p} \leqslant \sigma_n^{(p)}$

Now it is well-known that the Fourier series projection F_n is a minimal projection (see Chalmers and Mason [8] for a general L_p discussion). By identifying F_n with B_{nn} and J_{nn} according to relation (10) on $|w| = |z+\sqrt{(z^2-1)}| = \rho$, we immediately deduce the following result for the generalised Chebyshev projection J_{nn} restricted to the elliptical contour.

Property 5 J_{nn} is a minimal projection on $L_p\,(\xi_r)$.

However, it does not hence follow that G_n and H^*_{n-1} are minimal projections, just as it does not follow that Fourier cosine series or Fourer sine series projections are minimal.

We finally note that it is relatively inexpensive to calculate Chebyshev series coefficients. In the real case these are just Fourier coefficients of a related function (see §1.1), and so the Fast Fourier Transform is immediately applicable. In the complex case, Chebyshev series coefficients are also Laurent series coefficients, and again Fast Fourier Transform methods may be exploited (see [9]).

1.6 Near-Best Approximation By Interpolation

Efficient algorithms were given in §1.2 for interpolating a function at Chebyshev zeros. In fact the resulting approximations are near-best or best in the limited sense of the following classical results. The first is discussed in [10][11] and the second in [6].

Property 6 $\| G^{I}_{n-1} \|_{L_\infty} \sim \frac{2}{\pi} \log n$ on $[-1,1]$,

and $\| G^{I}_{n-1} \| \leqslant 5$ for $n \leqslant 500$.

Property 7 $H^{I}_{n-1} f$ is a best L_1 approximation to f on $C[-1,1]$, if $f - H^{I}_{n-1} f$ has no other zeros apart from those of U_n.

Analogous results are not known in the complex case, although (in L_∞) good practical choices of nodes for interpolation have been proposed for elliptical domains and annuli. Specifically for f in $A(E_\rho)$, Geddes [12] advocates the images of the zero on $|w| = \rho$ of

$$w^n = i\rho^n, \quad (\text{under } z = \tfrac{1}{2}(w+w^{-1})) \tag{22}$$

and for f in $A(N_{\rho_1\rho_2})$ Mason [13] proposes that the (Laurent-type) polynomial

$$\sum_{k=-n}^{n-1} a_k w^k$$

should be interpolated at the images of the zeros on $|w| = \rho_1, \rho_2$ of

$$w^n = i(\rho_2)^n, \quad w^n = -i(\rho_1)^n \tag{23}$$

For both sets of points (22) and (23), numerical values of the projection norm have been calculated ([12] and [13]) and appear to increase logarithmically with n. Note also that the choices of points (22) and (23) are appropriate in the limiting cases $\rho \to 1, \rho_1 \to 1, \rho_1 \to \rho_2$.

2. RATIONAL APPROXIMATION

Chebyshev methods of rational approximation, by analogy with polynomial methods, are generally based either on series expansions or on interpolation criteria, and we shall discuss one method of each type.

2.1 Chebyshev-Padé-Laurent Approximants

The derivation of Chebyshev-Padé rational approximants from Chebyshev series expansions has been studied by Gragg and Johnson [14] and Chisholm and Common [15], using different approaches. We briefly give a new extension of Gragg's treatment to cover generalised Chebyshev series, thus reproducing Chisholm and Common's generalised approximants.

Suppose that g(w) has the Laurent expansion

$$g(w) \sim \sum_{k=-\infty}^{\infty} c_k w^k = \sum_{k=0}^{\infty}{}' c_k w^k + \sum_{k=0}^{\infty}{}' c_{-k} w^{-k} \tag{24}$$

on the circular annulus $C_{\rho_1\rho_2}$: $1 \leqslant \rho_1 \leqslant |w| \leqslant \rho_2$.

Then a Padé-Laurent approximant may be defined by

$$g_{pqrs}(w) = \frac{\sum_{k=0}^{p} a_k w^k}{\sum_{k=0}^{q} b_k w^k} + \frac{\sum_{k=0}^{r} a_k^* w^{-k}}{\sum_{k=0}^{s} b_k^* w^{-k}} = \frac{A_p(w)}{B_q(w)} + \frac{A_r^*(w^{-1})}{B_s^*(w^{-1})} \tag{25}$$

$$= g(w) + \mathcal{O}(w^n) + \mathcal{O}((w^{-1})^m)$$

where $n = p + q + 1$, $m = r + s + 1$, and $\{a_k\}$, $\{b_k\}$, $\{a_k^*\}$, $\{b_k^*\}$ are obtained by standard Padé algorithms from the two parts of the expansion (24).

In particular, for $p = q$, $r = s$,

$$g_{pprr}(w) = [A_p(w)\, B_r^*(w^{-1}) + B_p(w)\, A_r^*(w^{-1})]/[B_p(w)\, B_r(w^{-1})]$$

$$= \sum_{k=-r}^{p} e_k w^k \Big/ \sum_{k=-r}^{p} f_k w^k \quad \text{say} \tag{26}$$

setting $g(w) = f(z)$ under $w = z+\sqrt{(z^2-1)}$,

with
$$w^{\pm k} = T_k(z) \pm \sqrt{(z^2-1)}\, U_{k-1}(z), \tag{27}$$

we obtain a generalised Chebyshev-Padé approximant on $N_{\rho_1\rho_2}$ of the form (26) with w^k, w^{-k} replaced by (27). This may in turn be written in the form

$$\frac{\sum_{k=0}^{t} e_k^{(1)} T_k(z) + \sqrt{(z^2-1)} \sum_{k=1}^{t} e_k^{(2)} U_{k-1}(z)}{\sum_{k=0}^{t} f_k^{(1)} T_k(z) + \sqrt{(z^2-1)} \sum_{k=1}^{t} f_k^{(2)} U_{k-1}(z)} \tag{28}$$

where $e_k^{(1)}$, $e_k^{(2)}$, $f_k^{(1)}$, $f_k^{(2)}$ are appropriately determined coefficients, and where $t = \max(p,r)$.

Note that for $p=r$ (28) reduces to a first kind approximant if $f(z)$ is in $A(E_\rho)$, since $e_k^{(2)} = f_k^{(2)} = 0$. Also if $f(z)$ is $\sqrt{(z^2-1)}$ times a function in $A(E_\rho)$, then (28) reduces to $\sqrt{(z^2-1)}$ times the ratio of a second kind sum and a first kind sum, since $e_k^{(1)} = f_k^{(2)} = 0$.

2.2 Chebyshev Rational Interpolation

If the form of rational approximation (25) is adopted on $C_{\rho_1\rho_2}$, but with $p + q + 1 = r + s = n$ (say), a new scheme of interpolation may be based on the images of the 2n points (23). Under the mapping $w = z + \sqrt{(z^2-1)}$, this approach corresponds to generalised Chebyshev rational approximation on $N_{\rho_1\rho_2}$. It is anticipated that, for analytic functions, this will normally lead to approximations somewhat better than those obtained by polynomial interpolation at the same number of nodes, especially for $p \simeq q$ and $r \simeq s$. Indeed, we would expect to obtain results comparable with those obtained by Chebyshev-Padé-Laurent approximants of the same form.

However, for non-analytic functions, results can be very unsatisfactory. Indeed, to illustrate this, we have obtained Chebyshev rational interpolants of form

$$f_{pq}(x) = \frac{a_0 + a_1 x + \ldots + a_p x^p}{1 + b_1 x + \ldots + b_p x^p}$$

to the C^1 function $f(x) = \begin{cases} \sin \pi x & -1 \leqslant x \leqslant 0 \\ x & 0 \leqslant x \leqslant 1 \end{cases}$

based on interpolation at the zeros of $T_{2p+1}(x)$.

Unacceptable results were obtained for $p = 1, 2, 3$, which were briefly as follows ($\|\epsilon\| = \max |f(x) - f_{pq}(x)|$):

$p = 1$: $\|\epsilon\| \simeq 0.7$

$p = 2$: f_{pq} has a pole near $x = -0.75$

$p = 3$: f_{pq} has a pole near $x = -0.45$.

On the other hand good results were obtained for a similar C^1 function

$$f(x) = \begin{cases} \sin x & -1 \leqslant x \leqslant 0 \\ x & 0 \leqslant x \leqslant 1, \end{cases}$$

namely

$$p = 1: \quad \|\epsilon\| \simeq 0.03$$
$$p = 2: \quad \|\epsilon\| \simeq 0.007$$
$$p = 3: \quad \|\epsilon\| \simeq 0.001.$$

3. INTEGRALS AND INTEGRAL EQUATIONS

3.1 Integrals

A well-established integration method is that of Clenshaw and Curtis [16], in which the integral is replaced by either a (truncated) Chebyshev series or a Chebyshev interpolating polynomial before being integrated. This has obvious advantages for indefinite integration, and it is also a rather competitive method for definite integration provided it is viewed as a product integration rule with appropriate weights and abscissae (see Sloan and Smith [17] , where convergence is also discussed for definite integration). Let us now simply point out that an elementary approximation theory argument (extending a discussion in [18]) may be used to ensure uniform convergence of indefinite integration as follows.

Consider Filippi's variant [19] of the method in which we determine

$$g(x) = \int_{-1}^{x} f(x)\,dx \qquad (-1 \leqslant x \leqslant 1) \tag{29}$$

by integrating, in place of f(x), the polynomial

$$f_{n-1}(x) = \sum_{k=1}^{n} b_k\, U_{k-1}(x)$$

where f_{n-1} is the partial sum of degree n-1 of the expansion of f(x) in $\{U_{k-1}(x)\}$. (In practice an interpolating polynomial at U_n zeros gives very similar results but is less easy to analyse). Then, since $T_k'(x) = kU_{k-1}(x)$, g(x) is replaced by

$$g_n(x) = \int_{-1}^{x} f_{n-1}(x)\,dx = \sum_{k=1}^{n} a_k\, k^{-1}\,[T_k(x) - T_k(-1)] \tag{30}$$

From (29) and (30),

$$g - g_n = \int_{-1}^{x} (f - f_{n-1})\,dx$$

and

$$\| g - g_n \|_\infty \leqslant \int_{-1}^{1} |f - f_{n-1}| \, dx = \| f - f_{n-1} \|_1 . \tag{31}$$

(Here the norms are traditional unweighted ones).

Hence, g_n converges uniformly to g provided f_{n-1} converges in L_1 to f, and the latter is guaranteed if f is L_2-integrable (see [18]).

<u>Property 8</u> <u>If f is L_2-integrable, then the integration method (30) converges uniformly.</u> [This result first appeared in [18]]

Now consider the analogous method for determining

$$h(x) = \int_{-1}^{x} f(x) \, (1-x^2)^{-\frac{1}{2}} \, dx \qquad (-1 \leqslant x \leqslant 1). \tag{32}$$

Suppose f(x) is approximated by the polynomial

$$f_n(x) = \sum_{k=0}^{n} a_k \, T_k(x),$$

namely the partial sum of the $\{T_k\}$ expansion of f, and then integrated. Then

$$\int_{-1}^{x} T_k(x) \, (1-x^2)^{-\frac{1}{2}} \, dx = \begin{cases} k^{-1} \, (1-x^2)^{\frac{1}{2}} \, U_{k-1}(x) & k \geqslant 1 \\ \pi - \cos^{-1}x & k = 0 \end{cases}$$

and so h(x) is replaced by

$$\begin{aligned} h_{n-1}(x) &= \int_{-1}^{x} f_n(x) \, (1-x^2)^{-\frac{1}{2}} \, dx \\ &= a_0(\pi - \cos^{-1}x) + \sum_{k=1}^{n} a_k k^{-1} \, (1-x^2)^{\frac{1}{2}} \, U_{k-1}(x). \end{aligned} \tag{33}$$

From (32) and (33),

$$h - h_{n-1} = \int_{-1}^{x} w(f - f_n) \, dx, \quad \text{where } w = (1-x^2)^{-\frac{1}{2}} .$$

Thus $\| h - h_{n-1} \|_\infty \leqslant \| w(f - f_n) \|_1$. (34)

<u>Lemma</u> <u>If f is L_2-integrable on [-1,1], then its partial expansion $f_n = G_n f$ in $\{T_k\}$ converges in the L_1 norm weighted by $w = (1-x^2)^{-\frac{1}{2}}$.</u>

<u>Proof</u> Let $x = \cos\theta$, then $u(\theta) = f(\cos\theta)$ is L_2-integrable and hence its partial Fourier cosine series $F_n u$ of order n converges in L_2 (as $n \to \infty$).

Hence $F_n u$ also converges in the (less strict) L_1 norm and so

$$\int_0^\pi |u(\theta) - (F_n u)(\theta)| \, d\theta = \int_{-1}^1 |f(x) - f_n(x)| \frac{dx}{(1-x^2)^{\frac{1}{2}}} \to 0 \quad .$$

<u>Property 9</u> <u>If f is L_2-integrable, then the integration method (33) converges uniformly.</u> [We believe this result to be new.]

<u>Proof</u> This follows immediately from (34) and the Lemma.

3.2 <u>Integral Equations</u>

Chebyshev polynomials have also been used to solve Fredholm integral equations with kernels of two rather different kinds. First, for nicely behaved kernels, an early method of Elliott [20] is attractively simple, since it is based on polynomial approximations to kernels as well as to solutions. We make no further comment, other than to note that the method is essentially a Chebyshev series rather than Chebyshev interpolation procedure. Second, Chebyshev polynomials happen to be naturally related to certain integral equations with Hilbert-type kernels for which they may be used to great advantage. In this context we summarise a standard elasticity problem, given, for example, by Gladwell and England [21], and we note that Razali and Thomas [22] have recently used similar ideas for related integral equation problems.

Suppose f(x) and g(x) are linked by Hilbert's integral equation

$$\pi^{-1} \int_{-1}^1 \frac{f(t)}{t-x} \, dt = g(x) \qquad |x| \leqslant 1 \quad . \tag{35}$$

Then Gladwell and England note that a complex potential $\Phi(z)$ exists, analytic in the z-plane cut along [-1,1] and $O(z^{-1})$ as $z \to \infty$, such that

$$\Phi^+(x) + \Phi^-(x) = -i\, g(x) \qquad \text{on } [-1,1]$$

$$\Phi^+(x) - \Phi^-(x) = f(x) \qquad \text{on } [-1,1] \quad .$$

<u>Property 10</u> <u>For the integral equation (35);</u>

(i) <u>if</u> $g(x) = \sum_{k=1}^{n} b_k U_{k-1}(x)$ <u>then</u> $f(x) = (1-x^2)^{-\frac{1}{2}} \sum_{k=0}^{n} b_k T_k(x)$

<u>where b_o is arbitrary,</u>

(ii) <u>if</u> $g(x) = \sum_{k=0}^{n} a_k T_k(x)$ <u>then</u> $f(x) = -(1-x^2)^{\frac{1}{2}} \sum_{k=1}^{n} a_k U_{k-1}(x)$

provided that $a_o = 0$ and that the solution is required to vanish at ± 1.

The reader will note the close similarity between Property 10 and the integration methods of §3.1.

4. ORDINARY DIFFERENTIAL EQUATIONS

4.1 Polynomial Methods

There has been extensive research into the use of Chebyshev polynomials in ordinary differential equations, and one of the earliest methods, the tau method of Lanczos [23], is effectively based on the elimination of as many leading terms as possible in the Chebyshev series expansion of the error in the differential equation. Thus, on [-1,1], for the equation

$$y' + y = 0, \quad y(0) = 1, \tag{36}$$

the tau method determines

$$y_n = a_o + a_1 x + \ldots + a_n x^n \tag{37}$$

to satisfy

$$y_n' + y_n = \tau\, T_n(x) \ , \ y_n(0) = 1 \ . \tag{38}$$

In general on [-1,1] for the equation $L(y) = 0$, one solves

$$L(y) = \tau_1\, T_{n-m+1}(x) + \tau_2\, T_{n-m+2}(x) + \ldots + \tau_r\, T_{n-m+r} \tag{39}$$

subject to m boundary conditions, where r is chosen to yield a well-determined algebraic system on equating coefficients in (39). This method has been developed to a high level of sophistication in the publications of E L Ortiz (see [24], for example).

It is worth noting that the tau method is in fact equivalent to a Galerkin method, since (39) implies that

$$\int L(y_n) \,.\, T_k(x)\ (1-x^2)^{-\frac{1}{2}}\ dx = 0 \qquad (k = 0, 1, \ldots, n-m) \ .$$

Indeed Urabe [25] developed Chebyshev methods from precisely this point of view. There is also an important method due to Clenshaw [26], which leads to similar results, but which is based on a more stable Chebyshev series representation of y_n in place of (37).

A more general "selected points" approach was also proposed by Lanczos [23], in which $L(y_n)$ is set to zero at the zeros of an appropriate Chebyshev polynomial. In fact identical results are obtained by the selected points and tau methods if, as Lanczos suggests, (39) is replaced by a form which is more convenient (and almost equivalent in practice), namely

$$L(y_n) = T_{n-m+1}(x)\ (\tau_1 + \tau_2 x + \ldots + \tau_r x^{r-1}).$$

Chebyshev methods for nonlinear differential equations, based essentially on the selected points method, are described by Wright [27] and Clenshaw [28]. Typically Newton's method is used to solve the resulting nonlinear algebraic system, solutions being obtained for successively higher degrees n.

4.2 Rational Methods

It can also be worthwhile to generalise the selected points method of §4.1 to determine rational approximations of form

$$y \simeq y_{pq} = \frac{a_0 + a_1 x + a_2 x^2 + \ldots + a_p x^p}{1 + b_1 x + b_2 x^2 + \ldots + b_q x^q} = \frac{A_p(x)}{B_q(x)} \tag{40}$$

to solve on [-1,1] a (nonlinear) differential equation

$$E(y) = 0, \quad \text{subject to m boundary conditions.} \tag{41}$$

Indeed precisely this (unpublished) idea was proposed by us in [29], and termed a "rational tau method". It consists of solving for $a_0, \ldots, a_p, b_1, \ldots, b_q$, subject to the boundary conditions on y_{pq}, the nonlinear algebraic system

$$E(y_{pq}) = 0, \quad \text{at the zeros of } T_{p+q-m+1}(x). \tag{42}$$

Several examples were discussed in [29], and the rational tau approximant was found to be consistently more accurate than the corresponding Padé approximant, which was typically used as an initial guess for Newton's method in solving (42). For example, consider

$$(1-x)\ y' + \tfrac{1}{2}y = 0\ ,\ y(0) = 1, \tag{43}$$

which has the true solution $(1-x)^{\frac{1}{2}}$, and determine a [p,p] approximant on [0,1] of form (40) by setting $a_0 = 1$ and solving

$$(1-x)\ y'_{pp} + \tfrac{1}{2}y_{pp} = \tau\ T^*_{2p}(x)\ .\ [B_p(x)]^{-2}. \tag{44}$$

Multiplying through by $(B_p)^2$ and equating coefficients of powers of x, 2p+1 nonlinear equations are obtained for $a_1, \ldots, b_p, \tau$. (More simply and generally the left-hand side of (44) may be set to zero at the 2p zeros of T^*_{2p}). These equations may be solved by Newton's method, using Padé coefficients as first guesses. For p = 2, 6 iterations sufficed to give full convergence of $a_1, \ldots, b_2, \tau$ to the values:

$$a_1 = -1.572237 \qquad b_1 = -1.072634 \qquad \tau = -0.000396$$

$$a_2 = 0.579987 \qquad b_2 = 0.174911$$

Rigorous error analysis is possible in this simple example. Writing ϵ for $y-y_{pp}$ and subtracting (44) from (43):

$$(1-x)\ \epsilon' + \tfrac{1}{2}\epsilon \ = -\tau\ T^*_{2p}(x)\ (B_p(x))^{-2}.$$

Since $a_0 = 1$, it follows that the maximum error in [0,1] occurs at x = 1 or where $\epsilon' = 0$. In either case $(1-x)\epsilon'$ is zero and hence, at a maximum,

$$|\epsilon| \leqslant 2|\tau|\ .\ |B_p(x)|^{-2}\quad .$$

It is readily seen that $|B_2(x)| \geqslant |B_2(1)| = 0.102$, and hence

$$\|\epsilon\|_\infty \leqslant 0.076.$$

This bound is in fact attained at x = 1. In contrast the corresponding [2,2] Padé approximant has a maximum error of about 0.2.

For nonlinear boundary value problems, in which Padé methods are not available, a continuation (or "embedding" or "variation of parameters") method may be used to treat the nonlinear equations (42) by Newton's method. In this method a sequence of boundary value problems is solved, starting from a straightforward problem, such as a boundary value problem corresponding to a chosen initial value problem. In [30] this procedure was successfully implemented to solve the Thomas-Fermi problem

$$x(y'')^2 - y^3 = 0,\ y(0) = 1,\ y'(X) = y(X)/X$$

for a range [3,8] of X. Using p = q = 5 in (40), solutions were obtained correct to 4 decimals, and indeed, by generating rational approximations in X to each of the coefficients $a_1, \ldots, a_5, b_1, \ldots, b_5$, a two-dimensional rational approximation was obtained which generated the whole family of solutions.

The rational tau method typically gives an approximation which is comparable with that which would be obtained by Chebyshev rational interpolation to the true solution (§2.2). This is illustrated in Table 1, where the respective coefficients are given in the approximation of e^x on [0,1] for $p = q = 2$. The tau method is here based on the equation $y' = y$ with $y(0) = 1$.

Table 1. Coefficients in Rational Approximants to e^x

	Rational Tau ($y'=y$)	Rational Interpolation ($y=e^x$)
a_0	1	1.00000273
a_1	0.54134796	0.54115857
a_2	0.10737636	0.10755997
b_1	-0.45859740	-0.45870610
b_2	0.06512693	0.06523842
$\|\epsilon\|_\infty$	0.0000122	0.0000074

5. PARTIAL DIFFERENTIAL EQUATIONS

Two quite distinct types of Chebyshev methods have been studied for two-dimensional partial differential equations. "One-dimensional methods" are based on the use of polynomials in one variable together with some kind of discretisation in the second variable. "Two-dimensional methods" make use of polynomials in both independent variables.

5.1 One-Dimensional Methods

These methods have now been quite widely and successfully developed, especially for time-dependent problems. Typically an approach like the "method of lines" is adopted to reduce the problem to a system of ordinary differential equations, which are solved by the types of (univariate) Chebyshev polynomial methods discussed in §4 above. Neat algorithms are possible, often based on vector representation of a discrete set of function values or polynomial coefficients, and robust and efficient software is now available for certain classes of problems. The reader is referred for full details to the work of Berzins, Dew, Knibb, and Scraton in England (see for example [31], [32]) and Gottlieb and Orszag in the USA (see for example [33], [34]), which has developed over a number of years from the early methods of Elliott [35], Mason [36], and Fox and Parker [37].

5.2 Two-Dimensional Methods

There has been far less successful activity in this area, probably because of the difficulty of dealing with boundary conditions for elliptic-type problems, especially when boundaries are curved. Indeed the standard type of two-dimensional approximation

$$p_{nn}(x,y) = \sum_{i=1}^{n} \sum_{j=1}^{n} c_{ij} \, x^{i-1} \, y^{j-1} \tag{45}$$

is generally most appropriate to rectangular domains. An early method of ours [38] uses vector product Chebyshev interpolation in the differential equation, based on a form such as

$$u(x,y) \simeq \phi(x,y) \, p_{nn}(x,y) + \psi(x,y) \tag{46}$$

where p_{nn} is the undetermined polynomial (45), and ϕ and ψ are chosen for convenience. Although the interpolation is performed over a rectangular domain, larger than the domain of the problem, good results have been obtained for a variety of problems with curved boundaries. However, the method involves solving a full linear algebraic system, and thus requires $O(N^3)$ operations, where $N = n^2$ is the total number of parameters.

More recently Haidvogel and Zang [39] introduced an attractive Chebyshev series method, which is basically a two-dimensional version of Clenshaw's series method for ordinary differential equations [28]. The method involves just $O(N^{3/2})$ operations, but is only applicable to the Poisson equation.

Recently Mason [40] has proposed an (unpublished) "vector tau method" for linear partial differential equations, which is designed for equations that take a simple form in one variable but can take a more complicated form in the other variable. Such a problem might be (for general $K(y)$ and $f(x,y)$)

$$L(u) = u_{xx} + K(y) \, u_{yy} + f(x,y) = 0 \; . \tag{47}$$

As a particular example, consider the torsion equation, namely (47) with $K(y) = 1$, $f(x,y) = 2$. In this case we adopt the form (46) with $\phi = (x^2-1)\,(y^2-1)$ and $\psi = 0$, on the square $-1 \leq x,y \leq 1$, and we replace $p_{nn}(x,y)$ by $p_{nn}(x^2,y^2)$ (for symmetry). Then the method is a combination of a tau method in one variable and a selected points method in the other variable.

Specifically we solve

$$L(p_{nn}) = \tau_i \, T_n(x^2) \quad \text{at } y = y_i \tag{48}$$

where $\{y_i\}$ are the n zeros of $T_n(y^2)$. (More τ terms may be used on the right-hand side if the differential equation requires it). By writing $\underline{\tau}$ for the vector $\{\tau_i\}$ and $\underline{c}_j$ for the vector of coefficients c_{ij} (j fixed), equation (48) may readily be expressed in matrix form (see [40] for details) as

$$\begin{aligned}
\underline{P}_2 \, \underline{c}_2 + \underline{Q}_1 \, \underline{c}_1 &= t_1 \, \underline{\tau} - \underline{2} \\
\underline{P}_{i+2} \, \underline{c}_{i+2} + \underline{Q}_{i+1} \, \underline{c}_{i+1} + \underline{R}_i \, \underline{c}_i &= t_{i+1} \, \underline{\tau} \qquad (i=1,\ldots,n-2) \\
\underline{Q}_n \, \underline{c}_n + \underline{R}_{n-1} \, \underline{c}_{n-1} &= t_n \, \underline{\tau} \\
\underline{R}_n \, \underline{c}_n &= t_{n+1} \, \underline{\tau} \quad .
\end{aligned} \tag{49}$$

Here the capital letters are known matrices, t_i is the coefficient of x^{i-1} in $T_n(x)$, and $\underline{2}$ is a vector of 2's.

Clearly the sparse vector system (49) is of precisely the form encountered in the traditional Lanczos tau method [23], except that matrices and vectors appear in place of scalars. Indeed the equations may be solved backwards expressing each $\underline{c}_i$ in terms of $\underline{\tau}$, until $\underline{\tau}$ is explicitly determined from the first equation. The total number of arithmetic operations is then $O(N^2)$.

The tau method can also be implemented in a tensor product form for a simple equation such as the Poisson equation. For example the torsion problem

$$u_{xx} + u_{yy} + 2 = 0 \text{ subject to } u = 0 \text{ on } x = \pm a, y = \pm b \tag{50}$$

can be solved approximately by the new technique of choosing

$$u^* = u_{nn} = (x^2-a^2)\,(y^2-b^2) \sum_{i=1}^{n} \sum_{j=1}^{n} c_{ij} \, x^{2i-2} \, y^{2j-2}$$

and satisfying

$$\begin{aligned}
u^*_{xx} + u^*_{yy} + 2 = {} & T^*_n(x^2)\,(\tau_1+\tau_2 y^2+\ldots+\tau_n y^{2n-2}) \\
& + T^*_n(y^2)\,(\sigma_1+\sigma_2 x^2+\ldots+\sigma_n x^{2n-2})
\end{aligned} \tag{51}$$

where $\tau_1,\ldots,\tau_n$, $\sigma_1,\ldots,\sigma_n$ are undetermined (tau) parameters. Indeed for a = b = 1 and n = 2, we deduce that

$$c_{11} = 3/5,\ c_{12} = c_{21} = 0,\ c_{22} = 4/5;\ \tau_1 = \sigma_1 = -1/5,\ \tau_2 = \sigma_2 = 6/5$$

and thus

$$u^*_{xx} + u^*_{yy} + 2 = T^*_2(x^2)\ (6y^2-1)/5 + T^*_2(y^2)\ (6x^2-1)/5. \quad (52)$$

Clearly the right-hand side of (51) is zero at the vector product of zeros of $T^*_n(x^2)$ and $T^*_n(y^2)$, and so the method must give identical results to the collocation method of Mason [38]. However, it is easily seen that the solution of (51) (expressing the c_{ij} in terms of the τ_i and σ_i by backward substitution, etc) only involves $O(N^{\frac{3}{2}})$ operations, and so the method has comparable efficiency to Haidvogel and Zang's method.

The error in replacing (50) by (52) is bound to be large at the corners $(\pm a, \pm b)$, since $u_{xx} + u_{yy} = 0$ on the boundaries, and this is confirmed in the bound 2 for the right-hand side of (52).

Clearly, our above new tau methods would need to be developed into piecewise methods in order to provide robust procedures and indeed Gottlieb and Orszag [34] have already taken this approach for time-dependent problems. Some progress in this general area has been made by L M Delves and co-workers (see [41] for example). Their "global element methods", which use bivariate polynomial approximations, exploit special properties of orthogonal polynomials to speed up computations.

References

[1] J C Mason, Minimal projections and near-best approximations by multivariate polynomial expansion and interpolation. In: "Multivariate Approximation II" (W Schempp and K Zeller, Eds) Birkhäuser Verlag, Basel, 1982 (pp 241-254).

[2] J C Mason, Near-best L_p approximations by real and complex Chebyshev series. IMA J. Numer. Anal. 3 (1983), 493-504.

[3] E W Cheney, "Introduction to Approximation Theory", McGraw Hill, New York, 1966.

[4] F V Atkinson and J C Mason, Minimal L_p properties of Chebyshev polynomials and series. (1984) (in preparation).

[5] A Ossicini and F Rosati, Bolletino Unione Math. Ital, (4) 11 (1975), 224-237.

[6] J R Rice, "The Approximation of Functions (Vol I)", Addison Wesley, 1964.

[7] J H Freilich and J C Mason, Best and near-best L_1 approximations by Fourier series and Chebyshev series. J of Approx. Th. 4 (1971) 183-193.

[8] B L Chalmers and J C Mason, Minimal L_p projections by Fourier, Taylor, and Laurent series. J. of Approx, Th. 40 (1984) (in press).

[9] K O Geddes and J C Mason, Polynomial approximation by projections on the unit circle. SIAM J. Numer. Anal. 12 (1975), 111-120.

[10] H Ehlich and K Zeller, Auswertung der Normen von Interpolations - operatoren. Math. Annalen 164 (1966), 105-112.

[11] M J D Powell, On the maximum errors of polynomial approximations defined by interpolation and least squares criteria. Computer J. 9 (1967), 404-407.

[12] K O Geddes, Near-minimax polynomial approximation in an elliptical region. SIAM J. Numer. Anal. 15 (1978), 1225-1233.

[13] J C Mason, Near-minimax interpolation by a polynomial in z and z^{-1} on a circular annulus. IMA J. Numer. Anal. 1 (1981), 359-367.

[14] W B Gragg and G D Johnson, The Laurent-Padé table. In "Information processing 74" (Proc. IFIP Congress 74), North Holland, Amsterdam, 1974, pp 632-637.

[15] J S R Chisholm and A K Common, Generalisations of Padé approximation for Chebyshev and Fourier series. Proc. 1979 International Christoffel Symposium (1980), pp 212-231.

[16] C W Clenshaw and A R Curtis, A method for numerical integration on an automatic computer. Numer. Mathematik 2 (1960), 197-205.

[17] I Sloan and W E Smith, Properties of interpolating product integration rules. SIAM J. Numer. Anal. 19 (1982), 427-442.

[18] J C Mason, Orthogonal polynomial approximation methods in numerical analysis. In: "Approximation Theory", A Talbot (Ed), Academic Press, London, 1970.

[19] S Filippi, Angenäherte Tschebyscheff - Approximation einer Stammfunktion - eine Modifikation des Verfahrens von Clenshaw und Curtis. Numer. Mathematik 6 (1964), 320-328.

[20] D Elliott, A Chebyshev series method for the numerical solution of Fredholm integral equations. Computer J. 6 (1963), 102-111.

[21] G M L Gladwell and A H England, Orthogonal polynomial solutions to some mixed boundary-value problems in elasticity theory. Q.J. Mech. Appl. Math. 30 (2), (1977), 175-185.

[22] M R Razali and K S Thomas, Singular integral equations and mixed boundary value problems for harmonic functions. In: "Treatment of Integral Equations by Numerical Methods", C T H Baker and G F Miller (Eds), Academic Press, London, 1982, pp 387-396.

[23] C Lanczos, "Applied Analysis", Prentice Hall, 1956.

[24] J H Freilich and E L Ortiz, Numerical solution of systems of ordinary differential equations with the Tau method: an error analysis. Maths of Comp. 39 (1982), 467-479.

[25] M Urabe, Galerkin's procedure for non-linear periodic systems and its extension to multi-point boundary value problems for general non-linear systems. In: "Numerical Solution of Non-linear Differential Equations" D Greenspan(Ed), Wiley, 1966, pp 297-327.

[26] C W Clenshaw, The numerical solution of linear differential equations in Chebyshev series. Proc. Camb. Phil. Soc. 53 (1957), 134-149.

[27] K Wright, Chebyshev collocation methods for ordinary differential equations. Comp. J. 6 (1964), 358-363.

[28] C W Clenshaw, The solution of van der Pol's equation in Chebyshev series. In: "Numerical Solution of Nonlinear Differential Equations". D Greenspan (Ed), Wiley, 1966, pp 55-63.

[29] J C Mason, Some new approximations for the solution of differential equations. D. Phil. Thesis, Oxford, 1965.

[30] J C Mason, Formulae for approximate solutions of the Thomas Fermi equation. Proc. Phys. Soc. 89 (1966), 772-774.

[31] D Knibb and R E Scraton, A note on the numerical solution of non-linear parabolic equations in Chebyshev series. Int. J. Comput. Math. 7 (1979), 217-225.

[32] M Berzins and P M Dew, A generalised Chebyshev method for non-linear parabolic equations in one space variable. IMA J. Numer. Anal. 1 (1981), 469-487.

[33] D Gottlieb, The stability of pseudospectral - Chebyshev methods. Maths of Comp. 36 (1981), 107-118.

[34] D Gottlieb and S A Orszag, "Numerical Analysis of Spectral Methods: Theory and Applications", SIAM Publications, Philadelphia 1977.

[35] D Elliott, A method for the numerical integration of the one-dimensional heat equation using Chebyshev series. Proc. Camb. Phil. Soc. 57 (1961), 823-832.

[36] J C Mason, A Chebyshev method for the numerical solution of the one-dimensional heat equation. Proc. 22nd ACM Nat. Conf. Thompson Book Co., Washington DC., (1967), 115-124.

[37] L Fox and I B Parker, "Chebyshev Polynomials in Numerical Analysis", Oxford University Press, 1968.

[38] J C Mason, Chebyshev polynomial approximations for the L-membrane eigenvalue problem. SIAM J. Appl. Math. 15 (1967), 172-186.

[39] D B Haidvogel and T Zang, The accurate solution of Poisson's equation by expansion in Chebyshev polynomials. J. Comp. Phys. 30 (1979), 167-180.

[40] J C Mason, The vector Chebyshev tau method - a new fast method for linear partial differential equations. RMCS Report 79/3 (1979), 25pp.

[41] L M Delves and C A Hall, An implicit matching principle for global element methods. JIMA. 23 (1979), 223-224.

POLYNOMIAL, SINC AND RATIONAL FUNCTION METHODS FOR APPROXIMATING ANALYTIC FUNCTIONS*

Frank Stenger*
Department of Mathematics
University of Utah
Salt Lake City, Utah 84112

Abstract. This paper presents practically useful constructive linear methods of approximation of analytic functions by polynomials, sinc functions and rational functions. Spaces of functions of the type frequently encountered in applications are described for approximation by each method. Within these spaces, the rate of convergence of each approximation is nearly optimal.

1. Introduction and Summary

There are relatively few basic tools of approximation of functions in applications. They are the polynomials, the Fourier polynomials, the sinc functions [6], and the rational functions. Occasionally other special functions such as Bessel functions, Legendre functions, and more general hypergeometric functions play a role, although the range of applicability of these is very limited compared with that of the above underlined classes.

Although splines are special cases of polynomials, we mention them as a separate entity. They are flexible and powerful tools for the approximation of data in one dimension, and in more than one dimension they are often used as finite elements for obtaining approximate solutions to differential and integral equations.

*Work supported by U.S. Army Research Contract No. DAAG29 83 K 0012.

Fig. 1.1 illustrates some frequently occurring spaces of functions encountered in the approximate solution of practical problems. For sake of simplicity we have restricted our illustration to special one-dimensional cases to better illustrate the basic differences.

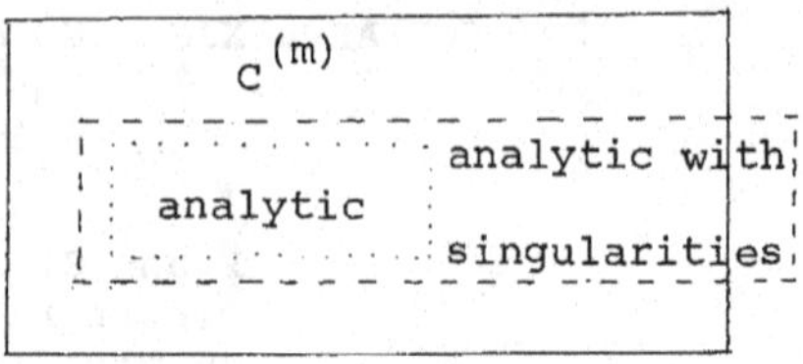

Figure 1.1
Typical Spaces of Functions

In a Sobolev space setting, splines are an optimal basis in $C^{(m)}[a,b]$, or slight and convenient variations of this space. (One such variation consists of the family of all functions f such that $f^{(m)}$ belongs to $C[a,b]$ for $k = 0, 1, \ldots, m-1$, and $f^{(m)}$ belongs to $L^2[a,b]$.) The flexibility of splines together with the convenient mathematical Sobolev space setting is what makes spline (and finite element, in more than one dimension) approximation so popular.

There is a price to be paid for this convenience, however, based on my own experience. In nearly all function approximation problems in applications the functions to be approximated are <u>piecewise analytic</u>, and may have <u>isolated singularities</u>. In the cases when singularities are absent, the space $C^{(m)}$ is too big, and the resulting approximations converge so slowly that it is usually impossible to compute solutions of three dimensional problems via finite element or finite difference methods. In the case when singularities are present, the function to be approximated may no longer be in $C^{(m)}$, and furthermore, the rate of convergence of polynomial or spline approximation is abruptly slowed to the point where it may be impossible to achieve a desired accuracy even in the one dimensional case.

	$C^{(m)}[a,b]$	Analytic No singularity on [a,b]	Analytic on (a,b), singularity at a or b
Polynomial	n^{-m}	exp(-cn)	n^{-c}
Spline	n^{-p} $p \le m$	n^{-p}	n^{-d} $q \le c$
Fourier Polynomial	n^{-m}	exp(-cn)	n^{-p}
Sinc Function	n^{-m}	$\exp(-cn^{1/2})$	$\exp(-cn^{1/2})$
Rational Function	n^{-m}	exp(-cn)	$\exp(-cn^{1/2})$

Figure 1.2 Order of Error of Approximation

Fig. 1.2 illustrates the order of error of approximation of functions in the spaces of functions in Fig. 1.1 by polynomials, spline polynomials, Fourier polynomials, sinc functions and rational functions. Given a function f in one of these spaces, the number n in Fig. 1.2 may be thought of as the number of evaluations of f required to obtain the approximation, and c, p and q denote positive constants. Notice the rapid exp(-cn) rate of convergence of the error of approximation of an analytic function without singularities on an interval by a polynomial or a rational function. This rapid rate is worthy of recognition and will be described more accurately in this paper. We shall also describe more accurately the rapid exp(-cn) rate of convergence achieved in the approximation of analytic functions that are also periodic. Notice also the drastic difference in the rate of convergence of the error to zero in the case of the absence compared with the case of the presence of singularities on an interval. Finally, the second and third entries in column 3 are for fixed degree splines at equispaced knots on [a,b]. It may be possible to achieve better rates of convergence by varying both the degree and the knot spacing.

Given a linear partial differential equation (PDE) such as, for example, and elliptic boundary value problem, and given that the coefficients of the PDE are analytic functions of one of the variables with the other variables held fixed, then the solution of the PDE also has this property. Mild singularities may occur on the boundary, however, particularly at corners of the region, or at points or curves on the boundary where the data has singularities, although the exact nature of these singularities is usually difficult to determine. It is in such instances that sinc or rational approximation is particularly powerful. Similar remarks apply to the approximation of solutions of integral equations.

In Sec. 2 of this paper we review practically useful properties of approximation of analytic functions by Fourier or ordinary polynomials. Our main tool here is the Laurent series from which all methods of approximation discussed in this paper are derived. In Sec. 3 of the paper we derive the Whittaker cardinal, or sinc expansion, along with some extensions to other intervals. Other formulas of approximation based on this formula may be found in [6]. In Sec. 4 of the paper we derive a family of rational functions which have interpolation and approximation properties very similar to those of the sinc functions.

2. Polynomial Approximation

The error of polynomial approximation of analytic functions can be conveniently studied via a Taylor or Laurent series representation of an analytic function. Let R denote the real line, C the extended complex plane, and Z the integers. Given a simply connected domain D in $C, \partial D$ will denote the boundary of D. We define $\int_D (\cdot)$ to be $\lim_{(n\to\infty)} \int_{C_n} (\cdot)$, where $\{C_n\}^\infty$, is a sequence of contours in D such that $C_n \to \partial D$ as $n \to \infty$, and such that whenever f is analytic in D then $\lim_{(n\to\infty)} \int_{C_n} |f(z)|dz = \inf \int_C |f(z)|dz$, where $\{C\}$ ranges over the set of all contours in D having ∂D as limit.

2.1 Approximation via the Taylor Polynomial

The Taylor polynomial is probably the simplest and most widely used method of approximation.

<u>Theorem 2.1:</u> Let $R > 0$, let f be analytic and bounded by M in the disc

$$(2.1) \quad D = \{z \in C: |z| < R\}$$

and let $0 < a < R$. Then for all $x \in [-a,a]$ and $n = 1,2,3,\ldots,$

$$(2.2) \quad |f(x) - \Sigma_{k=0}^{n-1} f^{(k)}(0)x^k/k!| \le (a/R)^n MR/(R-a).$$

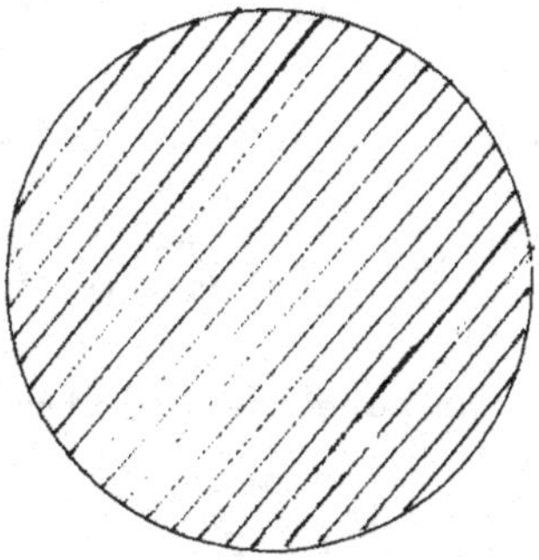

Figure 2.1
The Disc of Eq. (2.1)

<u>Proof:</u> If $-a \le x \le a$, then

$$(2.3) \quad f(x) - \Sigma_{k=0}^{n-1} f^{(k)}(0)x^k/k! = \frac{1}{2\pi i} \int_{|z|=R} \frac{f(z)}{(z-x)} \left(\frac{x}{z}\right)^n dz$$

The bound on the right hand side of (2.2) now follows if we note that $|f(z)| \le M$, $|x/z| \le a/R$, $|z-x| \ge R - a$, and $\int_{|z|=R} |dz| = 2\pi R$.

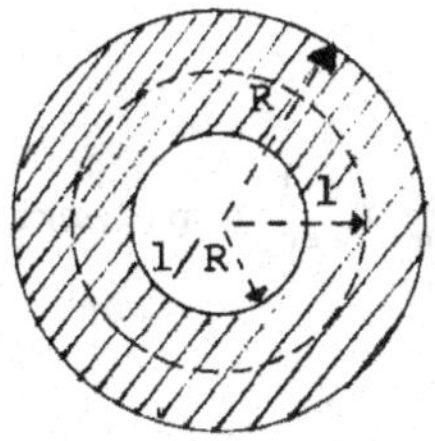

Figure 2.2
Annulus of Eq. (2.4)

2.2 Approximation via Fourier and Chebyshev Polynomials

Let $R > 1$ and let A_R denote the annulus

$$(2.4) \quad A = \{w \in C: 1/R < |w| < R\}.$$

<u>Theorem 2.2:</u> Let F be analytic and bounded by M in A_R. Let $1/R < r < R$, and let a_k be defined by

$$(2.5) \qquad a_k = \frac{1}{2\pi i} \int_{|z|=r} f(z) z^{-k-1}\, dz \quad .$$

Then, for $n = 1,2,3,\ldots,$ and $0 \le \theta \le 2\pi$,

$$(2.6) \qquad |F(e^{i\theta}) - \Sigma_{k=-n+1}^{n-1} a_k e^{ik\theta}| \le \frac{2MR}{R-1} R^{-n}.$$

Proof: We have the identity

$$(2.7) \qquad \begin{aligned} F(e^{i\theta}) - \Sigma_{k=-n+1}^{n-1} a_k e^{ik\theta} &= 1/(2\pi i) \int_{|z|=R} \frac{f(z)}{z-e^{i\theta}} \left(\frac{e^{i\theta}}{z}\right)^n dz \\ &- 1/(2\pi i) \int_{|z|=1/R} \frac{f(z)}{z-e^{i\theta}} \left(\frac{z}{e^{i\theta}}\right)^{n-1} dz. \end{aligned}$$

The bound (2.6) now follows by bounding the integrals on the right-hand side of (2.7).

If $m > 0$ and n are integers, one has the identity

$$(2.8) \qquad \frac{1}{m} \Sigma_{k=0}^{m-1} \exp(2\pi n k i/m) = \begin{cases} 1 & \text{if } n = \ell m,\ \ell \in Z \\ 0 & \text{otherwise} \end{cases}$$

which is basic to many discrete Fourier approximation procedures, including the fast Fourier transform method.

An important special subclass of functions of the class considered in Thm. 2.2 consists of those functions F that are analytic and uniformly bounded by M in A_R (Eq. (2.4)) for which we have $F(1/w) = F(w)$ for all w in A_R. Such functions F have the form

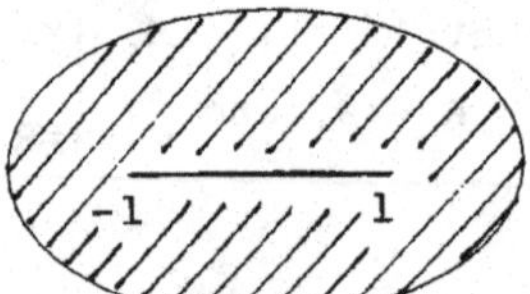

Figure 2.3.
Ellipse of Eq. (2.11)

$$(2.9) \qquad F(w) = a_0 + \Sigma_{k=1}^{\infty} a_k (w^k + w^{-k}).$$

Let us set

$$(2.10) \qquad z = (w + 1/w)/2$$

and let us define E_R by

(2.11) $$E_R = \{z \in C: z = (w + 1/w)/2, w \in A_R\}.$$

The transformation (2.10) maps the annulus A_R of Eq. (2.4) onto the ellipse E_R with foci at z = 1 and z = -1, and semi-major and semi-minor axes equal to (R + 1/R)/2 and (R - 1/R)/2 respectively. Under the transformation (2.10) we also have

(2.12) $$(w^n + 1/w^n)/2 = T_n(z)$$

where T denotes the Chebyshev polynomial. Defining f by f(z) = F(w) under the transformation (2.10), we have

(2.13) $$f(z) = a_0 + 2 \Sigma_{n=1}^{\infty} a_k T_k(z).$$

In Eq. (2.13) the a_k may also be defined by

(2.14) $$a_k = \frac{1}{\pi} \int_{-1}^{1} \frac{F(w)}{\sqrt{1-w^2}} dw, \qquad k = 0,1,2,3,\ldots .$$

Conversely, if f is analytic and uniformly bounded by M in E_R and if $a_0, a_1, a_2, \ldots,$ are defined by Eq. (2.14), then the sum (2.13) converges to f in E_R. Hence by Thm. 2.2 we have

Theorem 2.3: If f is analytic and uniformly bounded by M in E_R, and if a, k = 1,2,3,..., are defined by (2.14), then for all x in [-1,1],

(2.15) $$|f(x) - a_0 - \Sigma_{k=1}^{n-1} a_k T_k(x)| \le \frac{2MR}{R-1} R^{-n}.$$

We omit the practically more useful discrete analogue of Thm. 2.3 which results via application of the identities (2.8). We shall, however, state and prove the practically useful formula which results via Lagrange interpolation at the zeros of $T_n(x)$, and which can also be constructed using (2.8).

Theorem 2.4: If f is analytic and uniformly bounded by M in E_R

and if $x_k = \cos[(2k-1)\pi/(2n)]$, then for all $x \in [-1,1]$

$$(2.16) \qquad \left| f(x) - \Sigma_{k=1}^{n} \frac{F(x_n)T_n(x)}{(x-x_n)T_n'(x_n)} \right| \leq \frac{4M(R+R^{-1})}{[R+R^{-1}-2][R^n-R^{-n}]} .$$

Proof: The identity

$$(2.17) \qquad e_n(x) \equiv f(x) - \Sigma_{k=1}^{n} \frac{F(x_n)T_n(x)}{(x-x_n)T_n'(x_n)} = \frac{1}{2\pi i} \int_{\partial ER} \frac{f(z)T_n(x)}{(z-x)T_n(z)} dz$$

together with $|T_n(x)| \leq 1$, $|z - x| \geq (R+1/R)/2 - 1$, $|T_n(z)| \leq (R-1/R)/2$, $\int_{\partial ER} |dz| \leq 2\pi(R+1/R)$ yield (2.16).

Now let P_n denote the class of all polynomials of degree n, let f be an arbitrary continuous function on [-1,1], and let E_n and C_n be defined by

$$(2.18) \qquad \begin{aligned} E_n &= \inf_{p\in P} \sup_{-1<x<1} |f(x) - p(x)| \\ C_n(f) &= \sup_{-1<x<1} |e(x)| \end{aligned} ,$$

where e_n is defined by the first identity in (2.17). It may then be shown (Powell [3]) that

$$(2.19) \qquad \begin{aligned} C_n(f) &\leq 4E_n(f) \qquad \text{if} \quad n \leq 20 \\ C_n(f) &\leq 5E_n(f) \qquad \text{if} \quad n \leq 100. \end{aligned}$$

Moreover, (see [1]) the $O(R^{-n})$ bounds on the right hand sides of (2.5), (2.15), (2.16) are optimal, in the sense, e.g. for the case of (2.16), that there exists an f which is analytic and uniformly bounded by M in E_R such that

$$(2.20) \qquad E_n(f) \geq (c/n)R^{-n}$$

where c is a positive number. That is, the R in the above $O(R^{-n})$ bounds cannot be replaced by a larger number.

2.3. Singularities on the Interval of Approximation

The rate of convergence to zero of the error of polynomial approximation is severely slowed if one or more singularities is present on the interval of approximation.

Theorem 2.5 [1]: Let $0 < \alpha < 1$. Then there exists a constant $C > 0$ such that for $n = 1,2,3,\ldots$,

$$(2.21) \qquad \inf_{p\in P} \sup_{-1<x<1} |(1-x^2)^\alpha - p(x)| > Cn^{-\alpha}.$$

3. Sinc Function Approximations

The Whittaker cardinal function, or sinc function expansion may also be derived via Fourier series. Let $h > 0$, and let $g \in L^2(-\pi/h,\pi/h)$. Then, as is well known from Fourier series,

$$(3.1) \qquad g(t) = h\, \Sigma_{k\in\mathbb{Z}}\, g_k e^{ik\theta}$$

a.e. on $(-\pi/h,\pi/h)$, where

$$(3.2) \qquad g_k = \frac{1}{2\pi} \int_{-\pi/h}^{\pi/h} g(t) e^{-kht}\, dt.$$

Now let us extend the definition of g to R by

$$(3.3) \qquad G(t) = \begin{cases} g(t), & \text{if } -\pi/h < t < \pi/h \\ 0, & \text{if } t < -\pi/h, \text{ or if } t > \pi/h. \end{cases}$$

Let f denote the Fourier transform of G, that is,

$$(3.4) \qquad f(x) = \frac{1}{2\pi} \int_{\mathbb{R}} G(t) e^{-ixt}\, dt.$$

That is, by (3.3),

$$(3.5)\qquad f(x) = \frac{1}{2\pi}\int_{-\pi/h}^{\pi/h} g(t)e^{-ixt}\,dt.$$

Substituting into (3.5) the series (3.1), interchanging the order of integration and summation, and noting by (3.5) and (3.2) that $g_k = f(kh)$, we get

$$(3.6)\qquad f(x) = \Sigma_{k\in\mathbf{Z}}\, f(kh)S(k,h)(x) \quad ,$$

where the <u>sinc function</u> $S(k,h)$ is defined by

$$(3.7)\qquad S(k,h)(x) = \frac{\sin[(\pi/h)(x-kh)]}{(\pi/h)(x-kh)} \quad .$$

Given f defined on R let us set

$$(3.8)\qquad C(f,h)(x) = \Sigma_{k\in\mathbf{Z}}\, f(kh)S(k,h)(x).$$

This series is known as the Whittaker cardinal series representation of the function f. This series is also called the <u>sinc function expansion</u> of f, a terminology which originated in engineering literature.

Let $B(h)$ denote the family of entire functions f such that

$$(3.9)\qquad |f(z)| < A\, e^{\pi|z|/h}$$

for some constant A and for all $z \in C$, and such that $f \in L^2(R)$. Then, by the Paley-Wiener theorem, $f(z) = C(f,h)(z)$ for all $z \in C$.

Given f defined and continuous on R such that the series (3.8) converges, the resulting function $C(f,h)$ is in $B(h)$ and interpolates f at $z = kh$ for all $k \in Z$. Although $C(f,h)$ is not in general identically equal to f, there nevertheless exists a practically important class of functions such that whenever f is in this class then $C(f,h)$ is a very accurate approximation of f.

<u>Theorem 3.1</u>: Let $d > 0$, let f be analytic in the region

$$(3.10)\qquad D_d = \{z \in C: |Im(z)| < d\},$$

and let $N(f) < \infty$, where $N(f)$ is defined by

(3.11) $$N(f) = \int_{\partial D_d} |f(z)dz|.$$

Then for all $x \in R$,

(3.12) $$|f(x) - \Sigma_{k\in\mathbf{Z}} f(kh)S(k,h)(x)| \le N(f)/[2\,d\,\sinh(\pi d/h))],$$

and also

(3.13) $$|\int_R f(x)dx - h\,\Sigma_{k\in\mathbf{Z}} f(kh)| < e^{-\pi d/h}\,N(F)/[2\,\sinh(\pi d/h)].$$

Proof: Let $0 < y < d$, let n be a positive integer, and let $L_n = L_n(y)$ denote the boundary of the rectangular region $\{z = u+iv\colon u\in R, v\in R, |u| \le (n+1/2)h, |v| \le y\}$. Let $x \in R$ be fixed such that $|x| \le mh$, where $m < n$. Then

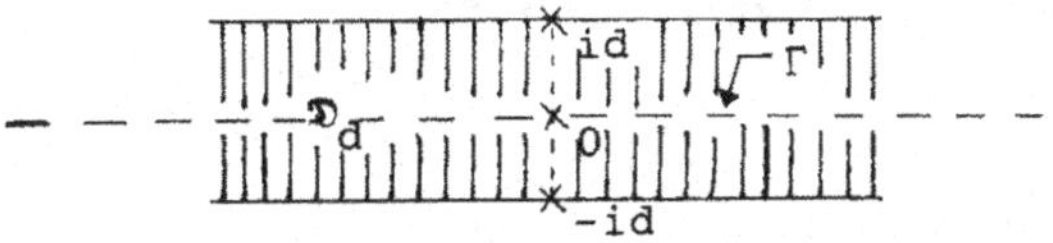

Figure 3.1
Rectangle D of Eq. (3.10)

(3.14) $$f(x) - \Sigma_{k\in\mathbf{Z}}\, f(kh)S(k,h)(x) = \frac{1}{2\pi i}\int_{L_n(y)} \frac{f(z)\,\sin(\pi x/h)}{(z-x)\sin(\pi z/h)}\,dz.$$

Now if z is on a horizontal segment of L_n then $|z - x| \ge d$, $|\sin(\pi z/h)| \ge \sin(\pi d/h)$, and since $x \in R$, $|\sin(\pi x/h| \le 1$. Hence the integral in (3.14) taken along the horizontal segments of L_n is bounded by the right hand side of (3.12). Along the vertical segments $|\sin(\pi z/h)| \ge 1$, while $|\sin(\pi x/h)| \le 1$. Hence the integrals along the vertical segments are bounded by the maximum of either $1/[(n+1/2)h-x]$ or $1/[(n+1/2)h+x]$ times $N(f)/(2\pi)$. Keeping x fixed and letting $n \to \infty$ yields (3.12). The proof of (3.13) follows by letting $n \to \infty$ in (3.14), integrating over R with respect to x, interchanging the order of integration in the double integral, and then bounding the resulting integral by proceeding similarly as for (3.14).

The functions $S(k,h)$ are replete with beautiful properties [6], such as being orthogonal over $\mathbb{R}$, yielding explicit expressions for their Fourier and Hilbert transforms, and yielding explicit function and derivative values at the points ℓh, $\ell \in \mathbb{Z}$. These combined with the explicit form of the coefficient of $S(k,h)$ in (3.8) and (3.12) give rise to a large family of explicit and accurate approximation formulas.

Thm. 3.1 can readily be extended to intervals other than $\mathbb{R}$ as follows.

<u>Theorem 3.2</u>: Let D be a simply connected domain in $\mathbb{C}$, with boundary points a and b where $b \neq a$. Let φ be a conformal mapping of D onto the region D_d defined in Eq. (3.10), let $\psi = \varphi^{-1}$ denote the inverse mapping, and set

$$
\text{(3.15)} \qquad \begin{aligned} \Gamma &= \{\psi(x) : x \in \mathbb{R}\} \\ z_k &= \psi(kh), \; k \in \mathbb{Z}. \end{aligned}
$$

Let $B(D)$ denote the family of all functions f that are analytic in D and such that $N(f) < \infty$, where

$$
\text{(3.16)} \qquad N(f) = \int_{\partial D} |f(z)dz|.
$$

Then for all $x \in \Gamma$,

$$
\text{(3.17)} \qquad \left|\frac{f(x)}{\varphi'(x)} - \sum_{k\in\mathbb{Z}} \frac{f(z_k)}{\varphi'(z_k)} S(k,h)\circ\varphi(x)\right| \le \frac{N(f)}{2\pi d \sinh(\pi d/h)},
$$

and also

$$
\text{(3.18)} \qquad \left|\int_\Gamma f(x)dx - \sum_{k\in\mathbb{Z}} \frac{f(z_k)}{\varphi'(z_k)}\right| \le \frac{e^{-\pi d/h} N(f)}{2 \sinh(\pi d/h)}.
$$

<u>Proof</u>: We set $z = \psi(w)$ and define F by $F(w) = f(\psi(w))\psi'(w) = f(z)/\varphi'(z)$. Then F satisfies the conditions of Thm. 3.1. Hence (3.17) is a consequence of (3.12). Eq. (3.14) thus has the equivalent form

$$(3.19)\qquad \frac{f(x)}{\varphi'(x)} - \Sigma_{k\in\mathbb{Z}} \frac{f(z_k)}{\varphi'(z_k)} S(k,h)\circ\varphi(x)$$

$$= \frac{1}{2\pi i} \int_{\partial D} \frac{f(z)\sin[\pi\varphi(x)/h]}{[\varphi(z)-\varphi(x)]\sin[\pi\varphi(z)/h]}\, dz.$$

Multiplying (3.19) by $\varphi'(x)$, integrating over Γ, and interchanging the order of integration in the resulting double integral on the right hand side, we get

$$(3.20)\qquad \int_\Gamma f(x)dx - \Sigma_{k\in\mathbb{Z}} \frac{f(z_k)}{\varphi'(z)}$$

$$= \frac{i}{2} \int_{\partial D} \frac{f(z)\exp[i\pi\varphi(z)/h \text{ sgn Im } \varphi(z)]}{\sin[\pi\varphi(z)/h]}\, dz.$$

Now if $z \in \partial D$ and $x \in \Gamma$, then $\varphi(x) \in R$, $\mathrm{Im}[\varphi(z)] = \pm d$, and

$$(3.21)\qquad |\sin[\pi\varphi(x)/h]/\sin[\pi\varphi(z)/h]| \le 1/\sinh[\pi d/h].$$

Hence the bounds on the right hand sides of (3.17) and (3.18) now follow readily from (3.19) and (3.20) respectively.

The representation (3.19) gives rise to many other approximation formulas. The practical use of these approximations depends on our being able to explicitly express the functions φ and ψ. Let us list some important examples of when this occurs.

Example 3.1: Let $\Gamma = (-1,1)$.
In this case we take

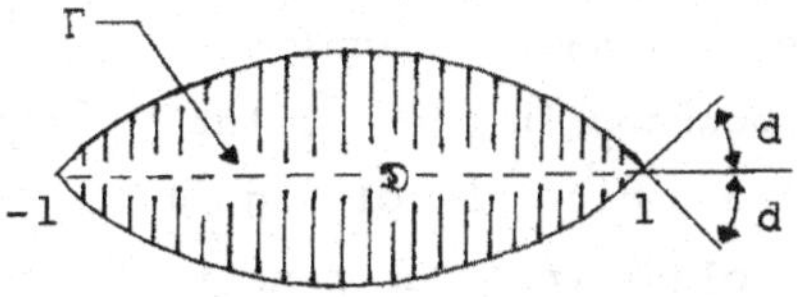

Figure 3.2
The "Eye" of Eq. (3.22).

$$(3.22)\qquad \begin{aligned} \varphi(x) &= \log[(1+x)/(1-x)] \\ \psi(x) &= (e^w-1)/(e^w+1) \\ z_k &= (e^{kh}-1)/(e^{kh}+1) \\ D &= \{z: |\arg[(1+z)/(1-z)]| < d\}. \end{aligned}$$

We shall give two examples for the case when $\Gamma = (0,\infty)$.

Example 3.2: Let $\Gamma = (0,\infty)$.
In this case we take

$$\varphi(x) = \log(x)$$

$$(3.23) \quad \psi(w) = e^{w}, \ z_k = e^{kh}$$

$$D = \{z \in C: |\arg(z)| < d\}.$$

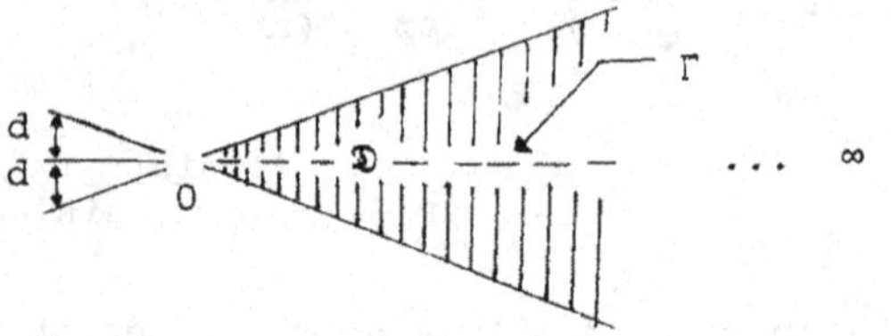

Figure 3.3
The Sector of Eq. (3.23).

This transformation is useful for functions f that are analytic and bounded in the sector of Fig. 3.3.

Example 3.3: Let $\Gamma = (0,\infty)$. In this case we take

$$\varphi(x) = \log[\sinh(x)]$$

$$\psi(w) = \log[e^{w}+(1+e^{2w})^{1/2}]$$

$$(3.24) \quad z_k = \log[e^{kh}+(1+e^{2kh})^{1/2}]$$

$$D = \{z \in C: |\arg[\sinh(z)]| < d\}.$$

This transformation is useful for the approximation of functions f that are analytic and bounded in the "bullet "-shaped region D of Fig. 3.4. This class includes functions which may have singularities at 0 and at ∞ and which may be oscillatory on $(0,\infty)$.

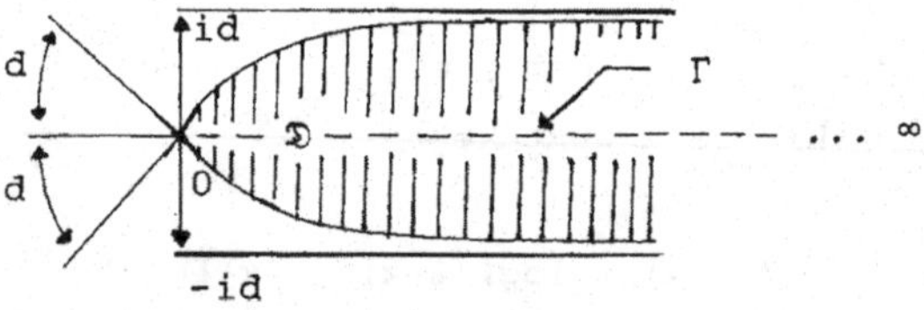

Figure 3.4
The "Bullet " of Eq. (3.24).

Finally, for practical considerations it is desirable to replace the infinite sums $\Sigma_{-\infty}^{\infty}$ in (3.12), (3.13), (3.17) and (3.18) by finite sums, Σ_{-N}^{N}, with N relatively small. To this end we make the additional assumption

(3.25) $|f(x)/\varphi'(x)| \leq A \exp[-\alpha|\varphi(x)|]$, $x \in \Gamma$,

where A and α are positive constants.

Example 3.4: Assumption (3.25) is equivalent to:

(3.26) $|f(x)| \leq A\, e^{-\alpha|x|}$ for the case of $\varphi(x) = x$;

(3.27) $|f(x)| \leq A(1 - x^2)^{\alpha-1}$ for the case of Ex. 3.1;

(3.28) $|f(x)| \leq Ax^{\alpha-1}/(1+x)^{2\alpha}$ for the case of Ex. 3.2;

(3.29) $|f(x)| \leq A[x/(1+x)]^{\alpha}e^{-\alpha x}$ for the case of Ex. 3.3.

Theorem 3.3: If f satisfies the assumptions of Thm. 3.2 and if f satisfies (3.25) on Γ, then:

(a) By taking $h = [\pi d/(\alpha N)]$, there exists a constant C such that for all $x \in \Gamma$,

(3.30) $$\left|\frac{f(x)}{\varphi'(x)} - \Sigma_{k=N}^{N} \frac{f(z_h)}{\varphi'(z)} S(k,h)\circ\varphi(x)\right| \leq CN^{1/2}\exp[-(\pi d\alpha N)^{1/2}];$$

(b) By taking $h = [2\pi d/(\alpha N)]^{1/2}$, there exists a constant C such that

(3.31) $$\left|\int_{\Gamma} f(x)dx - h\,\Sigma_{k=-N}^{N} \frac{f(z_h)}{\varphi'(z_h)}\right| \leq C\exp[-(2\pi d\alpha N)^{1/2}].$$

We remark that if the exact value of α in Thm. 3.3 is unknown, then by taking $h = (\gamma/N)^{1/2}$ where γ is a positive constant enables us to replace the right hand sides of (3.30) and (3.31) by $C\exp[-\delta N^{1/2}]$ where C and δ are positive constants.

Corollary 3.4: Let $\varphi'f$ be in $B(D)$, where $B(D)$ is defined as in Thm. 3.2, let N be a positive integer, let A and α be positive numbers, and let

(3.32) $|f(x)| \leq A \exp[-\alpha|\varphi(x)|]$

for all $x \in \Gamma$. By taking $h = [\pi d/(\alpha N)]^{1/2}$, there exists a constant C which is independent of N such that

$$(3.33) \qquad |f(x) - \Sigma_{k\in\mathbb{Z}} f(z_h) S(k,h)\circ\varphi(x)| \le CN^{1/2}\exp[-(\pi d\alpha N)^{1/2}].$$

Example 3.5: Let us approximate the function $f(x) = (1-x^2)^\alpha$ on [-1,1] by a linear combination of the functions $S(k,h)\circ\varphi(x)$, $k = -N, -N+1,\ldots,N$, where $0 < \alpha < 1$. In this case, by Ex. 3.1, we have $\varphi(x) = \log[(1+x)/(1-x)]$, so that $f(x)\varphi'(x) = 2(1-x^2)^{\alpha-1}$. Thus $\varphi' f$ is in B(D), (see Thm. 3.2) with $0 < d < \pi$. Hence, by Corollary 3.4 we have for all $x \in [-1,1]$,

$$(3.34) \qquad |(1-x^2)^\alpha - \Sigma_{k=-N}^{N} (1-z_k^2) S(k,h)\circ\varphi(x)| < CN^{1/2}\exp[-(\pi d\alpha N)^{1/2}] ,$$

where C is a constant. This rate of convergence is considerably better than that of polynomial approximation (compare Thm. 2.5).

Indeed, the rate of convergence (3.33) is optimal, in the sense that there does not exist a basis $\{u_{k,N}: k = -N,-N+1,\ldots,N$ and $N = 1,2,\ldots,\}$ such that

$$(3.35) \qquad \sup_{x\in\Gamma} \left|\frac{f(x)}{\varphi'(x)} - \Sigma_{k\in\mathbb{Z}} c_{k,N} u_{k,N}(x)\right| < C\exp[-\gamma N^{1/2}]$$

for all $N = 1,2,3,\ldots$, where c are constants, C and α are positive constants, and where $\gamma > (\pi d\alpha)^{1/2}$ for all f in the space of functions considered in Thm. 3.3.

4. Linear Rational Approximation

In this section we shall construct a family of rational functions which interpolate a function at the sinc interpolation points z (see Eqs. (3.15), (3.22), (3.23) and (3.24)). This will enable us to construct a rational function approximation scheme which is linear in the function to be approximated, with application to the same spaces and yielding roughly the same order of convergence as sinc approximation. We are thus able to identify practically important and fairly large spaces of functions for which we can expect rational approximation to be accurate. It is convenient to consider first the case of rational approximation over the interval $(0,\infty)$: we shall then describe a simple procedure for extending the results to the other intervals and regions discussed in Sec. 3, and indeed, to a more arbitrary contour Γ of the type defined in Eq. (3.15).

4.1 Rational Approximation Over $(0,\infty)$.

Let $0 < k < 1$, and let us use the following standard notation for elliptic functions.

$$u = u(k,w) = \int_0^w [(1-t^2)(1-k^2t^2)]^{-1/2}\, dt$$

$$\iff w = \mathrm{sn}(u,k)$$

$$k_1 = (1-k^2)^{1/2} \tag{4.1}$$

$$K = K(k) = u(k,1), \quad K_1 = K(k_1)$$

$$q = \exp[-\pi K_1/K], \quad q_1 = \exp[-\pi K/K_1].$$

Let $z \in \mathbb{C}$, let $0 < q < 1$, and let Φ be defined by

$$\Phi(z,q) = \frac{z-1}{z+1} \prod_{n=1}^{\infty} \frac{(z-q^n)(1/z-q^n)}{(z+q^n)(1/z+q^n)}. \tag{4.2}$$

We then prove the following.

<u>Lemma 4.1:</u> Let k and q be related as in Eq. (4.1), let $h = \log(1/q)$, and let $\Phi(z,q)$ be defined by Eq. (4.2). Then:

(i) Φ has the explicit expression

$$\Phi(z,q) = \left(\frac{1-k}{1+k}\right)^{1/2} \mathrm{sn}\left[\frac{(1+k)K}{\pi}\log(z); \frac{1-k}{1+k}\right]; \tag{4.3}$$

(ii) On $(0,\infty)$, Φ has the bound

(4.4) $$\sup_{0<z<\infty} |\Phi(z,q)| = [(1-k)/(1+k)]^{1/2} \leq 2\exp[-\pi^2/(2h)];$$

(iii) If $z = te^{i\theta}$, where $|z| = t$, $|\theta| = d$, $0 < t < \infty$, and where

(4.5) $$|\Phi(z,q)| = \exp\left[\frac{-\pi(\pi/2-d)}{h} - \Sigma_{m=1}^{\infty} \frac{\sinh[\pi^2 m(1-2d/\pi)/h]\cos[2m\pi \log(t)/h]}{m\cosh[\pi^2 m/h]}\right],$$

and so

(4.6) $$\exp[-\pi(\pi/2-d)/h-\epsilon] \leq |\Phi(z,q)| \leq \exp[-\pi(\pi/2-d)/h+\epsilon],$$

where

(4.7) $$\epsilon = \left|\frac{\exp[-2\pi d/h]}{1-\exp[-2\pi d/h]} - \frac{\exp[-2\pi(\pi-d)/h]}{1-\exp[-2\pi(\pi-d)/h]}\right|.$$

Proof: (i) The identity (4.3) follows if we set $z = e^{2iv}$ in (4.2), and then use the identities (16.37.1) to (16.37.4) as well as (16.36.3), (16.20.1) to (16.20.3), (16.14.1) and (16.14.2) of [4]. (ii) Since $\sup_{u\in\mathbb{R}}|sn(u,k)| = 1$ if $0 < k < 1$, the first identity in (4.4) follows from (4.3). Next, by [3, p. 378], if we set $2L = (1-k)/(1+k)$ we obtain the rapidly convergent positive-term series for q_1, namely

(4.8) $$q_1 = L + 2L^5 + 15L^9 + 150L^{13} + \ldots .$$

That is, $q_1 \geq L \geq (1-k)/[4(1+k)]$, so that $[(1-k)/(1+k)]^{1/2} \leq 2q_1^{1/2} = 2\exp[-\pi^2/(2h)]$ using the identity $\pi^2 = \log(q)\log(q_1)$. (iii) Let $g \in L^1(R) \cap C(R)$, and let G be defined by

(4.9) $$G(y) = \int_{\mathbb{Z}} g(x)e^{ixy}\,dx.$$

Then Poisson's summation formula states that

(4.10) $$\Sigma_{n\in\mathbb{Z}}\, g(nh) = (1/h)\,\Sigma_{m\in\mathbb{Z}}\, G(2\pi m/h), \quad h > 0.$$

Expanding $g(x) = \log|(e^x+z)/(e^x-z)|$ in powers of e^x/z if $x < \log|z|$ and in powers of ze^{-x} if $x > \log|z|$, substituting the result into the formula (4.9), and using (4.14) we get

$$(4.11) \qquad \Sigma_{n\in\mathbb{Z}} \log|(e^{nh}+z)/(e^{nh}-z)| - \pi(\pi/2-d)/h$$

$$= \Sigma_{m=1}^{\infty} \frac{\sinh[\pi^2 m(1-2d/\pi)/h]}{m \cosh[\pi^2 m/h]} \cos[2m\pi \log(t)/h]$$

which is just (4.5). Denoting the right hand side of Eq. (4.11) by u, we bound u be noting that the coefficients of the cosine terms are all of the same sign; replacing $m \cosh[\pi^2 m/h]$ by $(1/2)e^{\pi^2 m/h}$, we get $|u| \leq \varepsilon$, where ε is given in (4.7). The inequality (4.6) then follows from (4.5).

<u>Lemma 4.2</u>: Let $z = te^{i\theta}$, where $|z| = t$, let $d = |\theta|$, let ε be defined by (4.7), let N be a positive integer, let $h = \log(1/q)$, and let P be defined by

$$(4.13) \qquad P = \Pi_{m=-N}^{N} \left|\frac{z+q^m}{z-q^m}\right|.$$

(i) If $0 < d \leq \pi/2$, then

$$(4.14) \qquad P \leq \exp[\pi(\pi/2-d)/h + \varepsilon].$$

(ii) If $\pi/2 \leq d < \pi$, if $N^{1/2}q^N \leq t \leq N^{-1/2}q^{-N}$, and if $Q = \exp[\varepsilon+1/(4hN^{1/2})]$, then

$$(4.15) \qquad (1/Q)\exp[\pi(\pi/2-d)/h)/Q \leq P \leq Q \exp[\pi(\pi/2-d)/h]$$

(iii) If $d = \pi$, and if $N^{1/2}q^N \leq t \leq N^{-1/2}q^{-N}$, then

$$(4.16) \qquad P \leq 2 \exp[-\pi^2(1-N^{-1/2})/(2h)].$$

<u>Proof</u>: We have

$$(4.17) \qquad P = |\Phi(z,q)|W(-\infty,-N-1)W(N+1,\infty)$$

where

$$(4.18)\qquad W(m,n) = \Pi_{j=m}^{n} |z-q^j|/|z+q^j|.$$

(i) If $d \leq \pi/2$, then for $a > 0$, we have $|z-a|/|z+a| \leq 1$. Hence (4.14) follows from (4.17) and (4.6).

(ii) Upon replacing z by $t = |z|$, then t by $N^{1/2}q^N$, then $q^{n+1/2}/(1-q^{2n+1})$ by $1/[(2n+1)q^{n+1/2}]$, and finally using the identity $\Sigma[2/(2n+1)^2] = \pi^2/4$, we get

$$(4.19)\qquad \log[W(N+1,\infty)] = \mathrm{Re}\ \Sigma_{n=0}^{\infty}[2/(2n+1)]\ \Sigma_{j=N+1}^{\infty}\ (q^j/z)^{2n+1}$$

$$\leq \Sigma_{n=0}^{\infty}[2/(2n+1)][q^{N+1}/t]^{2n+1}/[1-q^{2n+1}]$$

$$\leq \Sigma_{n=0}^{\infty}[2/(2n+1)^2](q/N)^{n+1/2}$$

$$\leq (\pi^2/4)N^{-1/2}.$$

Similarly, $\log[W(-\infty,-N-1)]$ is also bounded by the extreme right hand side of (4.19). Combining these results with (4.17) yields (4.15).

(iii) Here we use (4.4) and proceed as in the proof of (ii) above.

<u>Lemma 4.3:</u> Let $0 < \alpha < 1$, let $0 < d \leq \pi/2$, and let D be defined by Eq. (3.23). Then for all $x > 0$,

$$(4.20)\qquad H(\alpha,d,x) \equiv \int_{\partial D} \frac{|z|^{\alpha-i}|1+z|^{1-2\alpha}}{|z-x|}\,dz$$

$$\leq C(\alpha,d)\begin{cases} x^{\alpha-1} & \text{if } 0 < x \leq 1 \\ x^{-\alpha} & \text{if } 1 \leq x < \infty \end{cases},$$

where $C(\alpha,d)$ is a constant depending only on α and d.

<u>Proof:</u> We omit the proof of this result, which can be found in [7].

<u>Theorem 4.4:</u> Let $0 < \alpha < 1$, let $0 < d \leq \pi/2$, and let f be analytic in the region D of Eq. (3.23), and for all $z \in D$, let

(4.21) $$|f(z)| \le A|z|^{\alpha}|1+z|^{-2\alpha},$$

where A is a constant. Let $B(z)$ be defined by

(4.22) $$B(z) = \frac{z}{1+z} \Pi_{j=-N}^{N} \frac{z-q^j}{z+q^j},$$

where N is a positive integer, and

(4.23) $$q = \exp[-\pi/(2\alpha N)^{1/2}].$$

Then for all $x \in (0,\infty)$,

(4.24) $$|f(x) - \Sigma_{j=-N}^{N} \frac{f(q^j)B(x)}{(x-q^j)B(q^j)}| \le CN^{\alpha/2} \exp[-d(2\alpha N)^{1/2}],$$

where C is a constant depending only on A, α, and d.

The following Corollary shows that the rational approximation method of Thm. 4.4 yields practically the same results as sinc approximation.

Corollary 4.5: Let the conditions of Thm. 4.4 be satisfied. If h and q are selected by the formula

(4.25) $$h = [\pi d/(\alpha N)]^{1/2}; \quad q = e^{-h},$$

then for all $x \in (0,\infty)$,

(4.26) $$|f(x) - \Sigma_{j=-N}^{N} f(q^j)S(j,h)\circ\log(x)| \le CN^{1/2} \exp[-(\pi d\alpha N)^{1/2}],$$

where C is a constant depending only on A, α, and d.

Proof of Thm. 4.4: The difference η between the absolute values on the left hand side of (4.24) is given by

(4.27) $$\eta = \frac{1}{2\pi i} \int_{\partial D} \frac{f(z)B(x)}{(z-x)B(z)} dz.$$

Hence, by Lemma 4.2 (i), we have

(4.28) $$|\eta| \le A/(2\pi)|B(x)|\exp[\pi(\pi/2-d)/h+\varepsilon]H(\alpha,d,x),$$

where $h = 1/[\pi(2\alpha N)^{1/2}]$. ε is defined by (4.7), and H by (4.20).

We now use Lemma 4.2 (iii) to bound $|B(x)|$ on the interval $J = \{x: q^N N^{1/2} \leq x \leq q^{-N} N^{-1/2}\}$. For $x \in R - J$, we have $|B(x)(1+x)/x| \leq 1$, and by (4.20)

$$(4.29) \qquad \frac{x}{1+x} H(\alpha,d,x) \leq C(\alpha,d) \cdot \begin{cases} x^{\alpha} & \text{if } 0 < x \leq 1 \\ x^{-\alpha} & \text{if } 1 \leq x < \infty. \end{cases}$$

Hence

$$(4.30) \qquad |\eta| \leq$$

$$\begin{cases} 2A \exp[(N^{-1/2}-1)\pi^2/(2h)+\pi^2/(2h)-\pi d/h+\varepsilon] C(\alpha,d) & \text{if } x \in J; \\ 2A \exp[\pi^2/(2h)-\pi d/h+\varepsilon] C(\alpha,d) N^{\alpha/2} q^{\alpha N} & \text{if } x \in R - J. \end{cases}$$

The choice of (4.23) for q now yields (4.24).

Proof of Corollary 4.5: By Ex. 3.2, $\varphi(z) = \log(z)$ is a conformal mapping of the region D onto the region D defined in Eq. (3.10). Since f is analytic in D, then by (4.21) $G = \varphi' f$ is clearly in $B(D)$ as defined in Thm. 3.2. Hence (4.26) follows from Corollary 3.4.

4.2 Rational Approximation over Other Intervals and Contours

The results of Thm. 4.4 and Corollary 4.5 make possible rational approximation over more general contours Γ as defined in Thm. 3.2. The procedure for doing this is very simple. Let φ, ψ, z_k, and Γ be as in Thm. 3.2, and define ρ by

$$(4.31) \qquad \rho(z) = \exp[\varphi(z)].$$

We next choose a positive integer N and define $B(z)$ by

$$(4.32) \qquad B(z) = \frac{\rho(z)}{1+\rho(z)} \Pi_{j=-N}^{N} \frac{\rho(z)-e^{jh}}{\rho(z)+e^{jh}}.$$

The proof of the following theorem is then easily carried out, using Thm. 4.4, Corollary 4.5 and Corollary 3.4.

Theorem 4.6: Let f be analytic in a simply connected region D, let φ, Γ, and z_k be defined as in Thm. 3.2, and let ρ and B be defined by (4.31) and (4.32) respectively. On Γ let f satisfy either the inequality

$$|f(x)| \leq A|\rho(x)|^{\alpha}|1+\rho(x)|^{-2\alpha}, \tag{4.33}$$

or else the inequality (3.32), where $A > 0$ and $0 < \alpha < 1$. Then, for all $x \in \Gamma$,

$$\left|f(x) - \Sigma_{k=-N}^{N} \frac{f(z_k)\rho(x)e^{kh} \quad \varphi'(z_k)}{[\rho(x)-e^{kh}]\varphi'(z_h)}\right| \leq CN^{\alpha/2} \exp[-d(2\alpha N)^{1/2}], \tag{4.34}$$

where C is a constant depending only on A, α, and d.

The extensions to the special regions of Sec. 3 is now immediate. We can approximate via rationals in $\rho(x)$ on Γ where:

$$\rho(x) = \begin{cases} (1+x)/(1-x), \text{ i.e., we get rationals} \\ \quad \text{in } x, \text{ on } [-1,1] \text{ --see Ex. 3.1;} \\ x \text{ on } (0,\infty) \text{--see Ex. 3.2 or sec 4.1;} \\ \sinh(x), \text{ i.e., we get rationals in } e^x \\ \quad \text{on } (0,\infty) \text{--see Ex. 3.3;} \\ e^x \text{ on } (-\infty,\infty). \end{cases} \tag{4.35}$$

References

[1] Bernstein, S. and C. de la Vallee Poussin, L'Approximation, Chelsea, N.Y. (1970).

[2] Burchard, H.G., and Höllig, W .G., N-Width and Entropy of H_p-Classes in $L_q(-1,1)$. To appear.

[3] Magnus, W., Oberhettinger, F. and Soni, R.P., Formulas and Theorems for the Special Functions of Mathematical Physics, Springer-Verlag, N.Y. (1956).

[4] National Bureau of Standards, Handbook of Mathematical Func-

tions, N.B.S. Applied Math. Series 55 (1964).

[5] Powell, M. J. D., Approximation Theory and Methods, Cambridge University Press (1981).

[6] Stenger, F., Numerical Methods Based on the Whittaker Cardinal or Sinc Functions, SIAM Rev. 22 (1981) 165-224.

[7] Stenger, F., Explicit, Nearly Optimal, Linear Rational Approximation with Preassigned Poles. Submitted for publication.

RATIONAL APPROXIMATION OF FRACTALS

Michael F. Barnsley, Stephen G. Demko
School of Mathematics
Georgia Institute of Technology
Atlanta, Georgia 30332

Abstract   Stationary distributions for certain Markov chains of inverse branches of rational maps are put forward as the basis of an approximation theory for fractals. Results on existence and on computability of moments are proved.

1. Introduction

Classical approximation theory concerns the description of smooth functions using approximating sets of smooth functions. It does not deal with those objects, broadly referred to as fractals, which possess features which cannot be fully resolved or simplified by magnification, such as coastlines, galaxies, and frost patterns on windows, cf. Mandelbrot [16]. However, the feasibility of modeling these objects is clear from the computer simulations of Mandelbrot, whose pictures of imagined landscapes and moonscapes are well known.

In this paper, we describe the moment theory of balanced measures, generalizing earlier work by Pitcher and Kinney [17], by Barnsley, Geronimo and Harrington [1-7] and by Bessis, Moussa and coworkers [8-11]. Not only do the introduced Markov chains allow one to build up a diverse class of measures, which carry local patterns to global ones, and interpolate and extrapolate formations from one scale to the next; but also, because of the explicit computability of their moments, they can be characterized in terms of discrete information sets such as may be obtained by experimental observation of a given structure. Conversely, using discrete data sets of the latter type and appropriate additional information it is possible to compute sequences of approximants for the given structure. The result can be an analytical description of the given object, good in a suitable sense on a number of different scales, available for the simulation of other properties.

We first present the theory together with a few illustrations.

We then illustrate an application, modeling the spectrum of a certain Schrödinger operator, of interest in statistical physics.

2. Theory

Let $\hat{\mathbb{C}}$ denote the Riemann sphere $\mathbb{C} \cup \{\infty\}$, and let $R_i: \hat{\mathbb{C}} \to \hat{\mathbb{C}}$ denote a rational transformation of degree $d(i) \geq 1$, for each $i \in \{1,2,\ldots,h\}$. We suppose $d(1) \geq 2$. Let $\{R_{ik}^{-1}: k = 1,2,\ldots,d(i)\}$ be a complete assignment of branches of the inverse of R_i. Let σ be a given probability measure, defined on the Borel subsets of $\hat{\mathbb{C}}$. Let $\{p(0), p(1), \ldots, p(h)\}$ be given probabilities, so that

$$\sum_{i=0}^{h} p(i) = 1, \; p(i) \geq 0 \text{ for each } i, \text{ and } p(1) > 0.$$

Then we consider the discrete time Markov chain

$$(1) \qquad P(z,A) = \sum_{i=1}^{h} \frac{p(i)}{d(i)} \sum_{k=1}^{d(i)} \delta_{R_{ik}^{-1}(z)}(A) + p(0)\sigma(A),$$

where $P(z,A)$ denotes the probability of transfer from $z \in \hat{\mathbb{C}}$ to a Borel set $A \subset \hat{\mathbb{C}}$, and $\delta_Y(A)$ takes value 1 if $Y \in A$ and value 0 if $Y \notin A$.

Our first concern is with the existence of a stationary probability measure for the Markov chain.

<u>Theorem 1</u>. There exists a probability measure μ such that

$$(2) \qquad \mu(A) = \int_{\hat{\mathbb{C}}} P(z,A)\,d\mu(z)$$

for all Borel subsets A of $\hat{\mathbb{C}}$.

<u>Proof</u>. A good reference for the framework of this proof is Dunford and Schwartz [14]. The linear operators, T_i, defined by

$$(T_i g)(z) = \frac{1}{d(i)} \sum_{j=1}^{d(i)} g(R_{ij}^{-1}(z))$$

map $C(\hat{\mathbb{C}})$ into itself. ($C(\hat{\mathbb{C}})$ is the Banach space of complex valued continuous functions on $\hat{\mathbb{C}}$.) Hence the adjoint operators T_i^* map the set of probability measures on $\hat{\mathbb{C}}$, $P(\hat{\mathbb{C}})$ into itself weak * continuously.

The affine map defined on $P(\mathbb{C})$ by

$$(3) \qquad S\nu = \sum_{i=1}^{h} p(i)T_i^*\nu + p(0)\sigma$$

is, therefore, a weak * continuous mapping of a weak * compact convex set into itself. By the Schauder fixed point theorem S has a fixed point, μ. Now it is straightforward to check that

$$(T_i^*\nu)(A) = \frac{1}{d(i)}\sum_{j=1}^{d(i)} \nu((R_{ij}^{-1})^{-1}(A)) = \frac{1}{d(i)}\sum_{j=1}^{d(i)} \int \delta_{R_{ij}^{-1}(z)}(A)\,d\nu(z)$$

holds for all Borel subsets A of $\hat{\mathbb{C}}$. Hence, for all Borel sets

$$\mu(A) = \sum_{i=1}^{h} \frac{p(i)}{d(i)} \int \delta_{R_{ij}^{-1}(z)}(A)\,d\mu(z) + p(0)\sigma(A) = \int P(z,A)\,d\mu(z).$$

We call such a measure μ a <u>balanced measure</u>. The existence and uniqueness of such a measure in the case where $p(1) = 1$ and $R_1(z)$ is a polynomial was first proved by Brolin [12]; he showed that μ is the electrostatic equilibrium measure supported on the Julia set for the polynomial $R_1(z)$. Existence in the case where $p(1) = 1$ and $R_1(z)$ is any rational transformation ($d(1) \geq 2$) was originally proved by Demko, using the above method, and independently by Mañè [15] who also proved uniqueness. The present results are an extensive generalization of earlier work, which preserve the rich structure of the original theory.

Let M denote the support of μ. When μ cannot be written as a finite sum of point masses (which is an exceptional case), $p(0) = 0$, and $d(i) \geq 2$ for $i = 1,2,\ldots,h$, then we call M a <u>probabilistic mixture of Julia sets</u>; it appears in general to consist of the union of the Julia sets for all finite words obtained by composing the rational functions $\{R_i(z)\}$. If some of the $d(i)$ equal unity, then we simply call M a <u>probabilistic mixture</u> and note that it may contain limit sets of Kleinian groups. If $p(0) \neq 0$ then we say M is a <u>probabilistic mixture with condensation</u>, σ being the condensing measure. M now consists of the support of σ together with all its preimages under rational inverse chains. The latter tend to draw suitably weighted smaller copies of the support of σ into the probabilistic mixture associated with $\{R_i\}$. A good description of this process in the context of condensed Julia sets is given by Barnsley, Geronimo, and Harrington [6].

In the following examples we give illustrations of what M can look like. It is important to realize that these correspond to very simple cases and are not representative of the diversity which can be obtained.

(i) $R_1(z) = (z^2-81)/9$, $p(1) = 1$, M = Julia set of $R_1(z)$ = the boundary of the region in the following sketch.

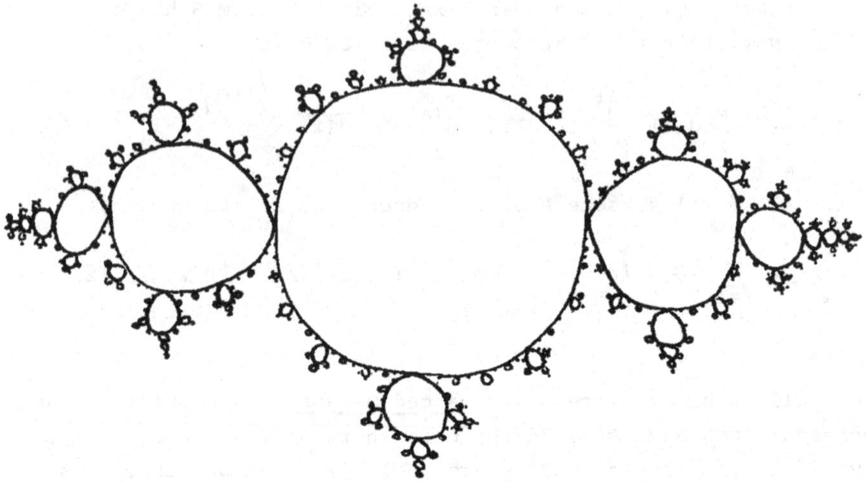

(ii) $R_1(z) = (z^2-81)/9$, $R_2(z) = z^2$, $p(1) = P(2) = 0.5$.

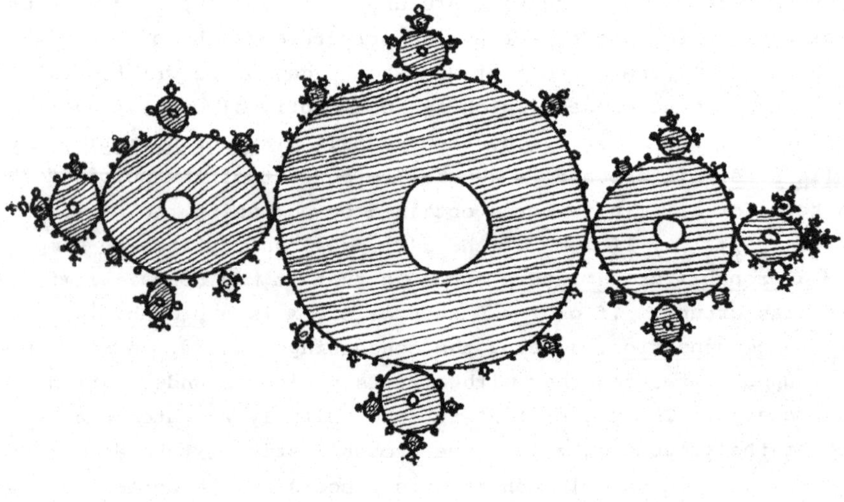

The shaded region represents M.

(iii) $R_1(z) = (81 \cdot z^2 - 0.1)/9$, $R_2(z) = (z^2-81)/9$, $p(1) = p(2) = 0.5$.

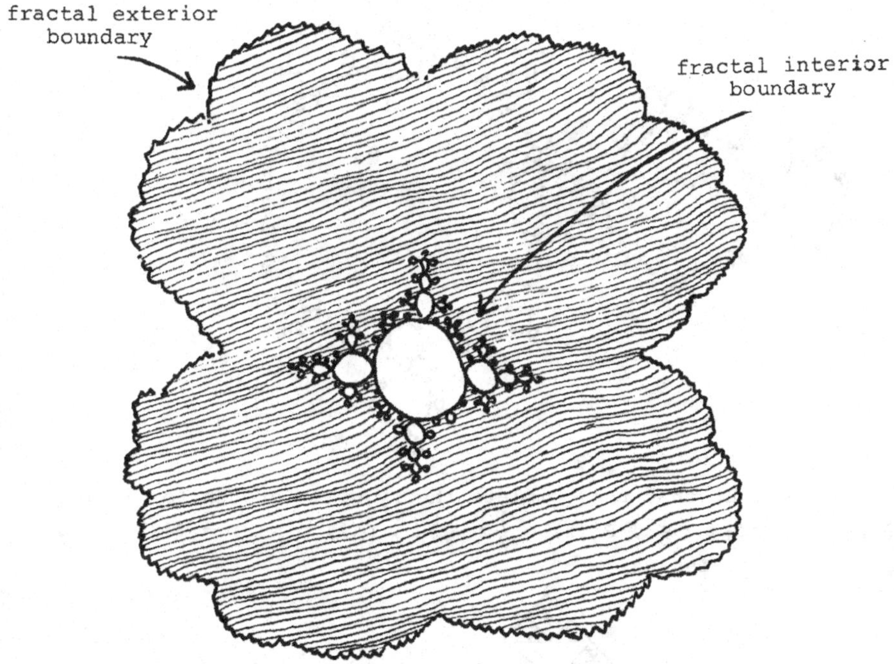

The shaded region represents M.

(iv) $R_1(z) = \frac{3}{4} \cdot z + \frac{1}{4 \cdot z^3}$, $p(1) = 1$. In this case M is a Julia set represented by the boundaries of the shaded components in the following photographs, which show a portion of M viewed on two scales. The fractal character is clear. (The shaded components actually represent the basins of attraction of the attractive fixed points ± 1, $\pm i$ of Newton's method applied to z^4-1. A Monroe EC8800 was used.)

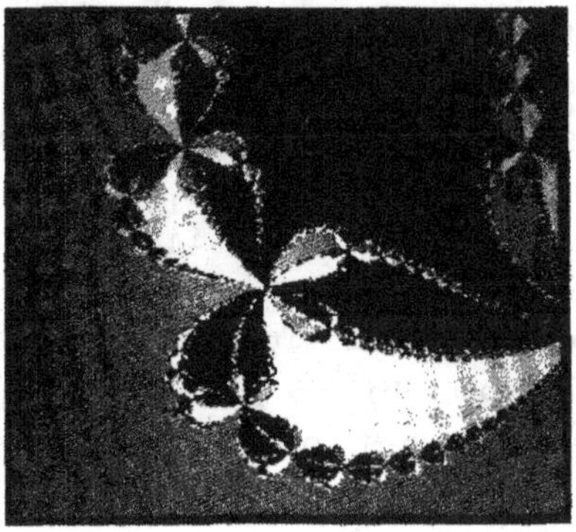

(v) $R_1(z) = \frac{1}{2}(z-2)^2 + 2$, $R_2(z) = z^2 + 4z + 2$, $p(1) = p(2) = 0.5$.

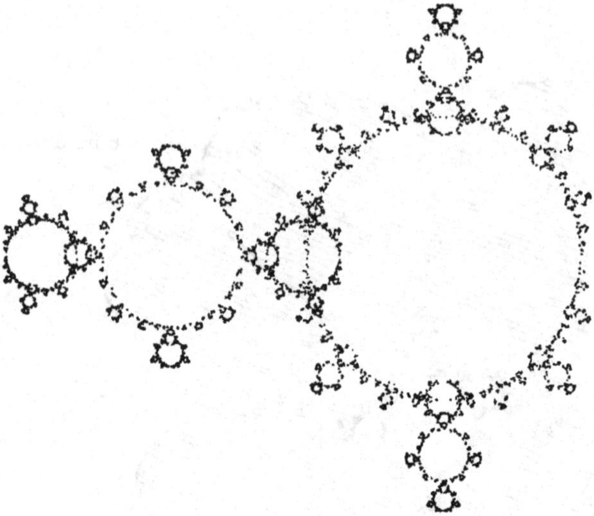

(vi) It turns out that the complete theory goes through for $R_1(z) = \frac{1}{2}z - \frac{1}{2}$, $R_2(z) = \frac{1}{2}z + \frac{1}{2}$, $R_3(z) = \frac{1}{2}z + \frac{1}{2}i$, $p(1) = p(2) = p(3) = 1/3$. The result is astonishing.

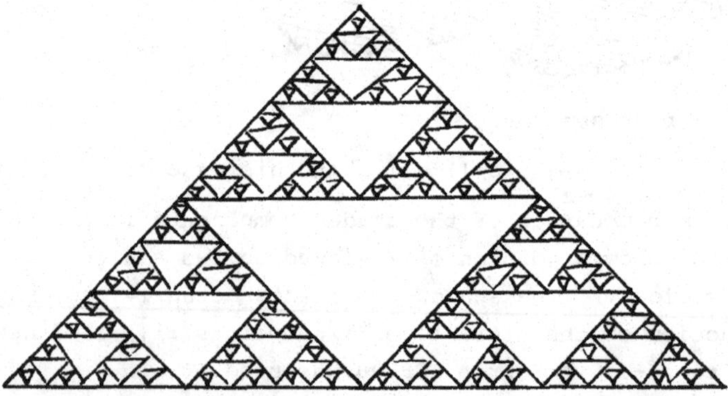

In general we expect a balanced measure μ to be an attractive fixed point of the operator S defined in (3). The specific construction of μ can then be thought of as follows. Start with almost any point $z_0 \in \mathbb{C}$ and choose recursively

$$z_{n+1} \in \{ \bigcup_{i,j} R_{ij}^{-1}(z_n)\} \cup \{\text{support of } \sigma\}, \quad \text{for } n = 0,1,2,\ldots,$$

where probability $p(i)/d(i)$ is attached to each of the numbers $R_{ij}^{-1}(z_n)$ and probability $p(0)\,d\sigma(z)$ is attached to z in the support of σ. Then we expect M to be the limit set, or probabilistic attractor, for the process, and

$$\mu(A) = \lim_{N\to\infty} \frac{|\{z_n | z_n \in A, n \le N\}|}{N}$$

whenever A is a Borel subset of $\hat{\mathbb{C}}$. Certainly this is true when $p(1) = 1$; it is also true when $p(0) + p(1) = 1$ and the support of σ does not contain an exceptional point for $R_1(z)$.

The "picture" of a balanced measure given below corresponds to $R_1(z) = (z^2-81)/9$ and $R_2(z) = z^2$ with probabilities 0.85 and 0.15 respectively. It consists of points $\{z_n\}_{n=10^2}^{n=10^4}$ computed as described above. The support of the measure is the one sketched in Ex. (ii) above.

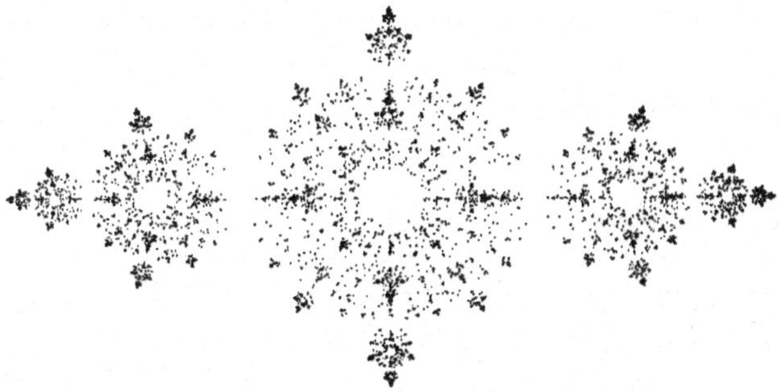

The analytical tool, apparently special to balanced measures, which extends a method for producing remarkable pictures into a constructive approximation theory, is the feasibility of characterizing such measures in terms of their moments. Here we generalize earlier results of Barnsley, Geronimo, and Harrington [1-6] and of Barnsley, Harrington [7] by showing that all of the moments $\int z^n d\mu$, $n = 0,1,2,\ldots$ can often be calculated explicitly recursively in terms of finite sets of parameters associated with the stochastic description of μ. In particular, the inverse procedure whereby μ is deduced from its moments and certain additional information, is available and can be applied to the problem of fractal approximation and reconstruction.

The key result which we use in the following theorems is that for $f \in L_1(\mu)$ we have from Theorem 1.

$$(4) \qquad \int_{\hat{\mathbb{C}}} f\, d\mu = \sum_{i=1}^{h} \frac{p(i)}{d(i)} \sum_{j=1}^{d(i)} \int_{\hat{\mathbb{C}}} f \circ R_{ij}^{-1} d\mu + p(0) \int_{\hat{\mathbb{C}}} f\, d\sigma .$$

Also, let c be a fixed point of a rational transformation $R_i: \hat{\mathbb{C}} \to \hat{\mathbb{C}}$. Define the expansion factor $E(i) = E(R_i, c)$ by

$$E(i) = \begin{cases} |R_i'(c)| & \text{if } c \neq \infty, \\ \lim_{z\to\infty} |z|/|R_i(z)| & \text{if } c = \infty. \end{cases}$$

Notice that

$$E(\psi R\psi^{-1}, \psi c) = E(R,c)$$

when ψ is a Möbius transformation. The following is not the strongest statement, but is relatively quickly formulated and proved.

<u>Theorem 2</u>. Let $c \in \hat{\mathbb{C}}$ be a fixed point of each $\{R_i: i = 1,2,\ldots,h\}$, $c \in \mathrm{supp}(\sigma)$, $E(1) > E(i)$ for $i \neq 1$, and g be a Möbius transformation such that $g(c) = \infty$. Then the moment integral

$$M_n = \int_{\hat{\mathbb{C}}} (g(z))^n d\mu$$

exists for positive integer n whenever

$$E(1) < \left(\sum_{i=1}^{h} \frac{p(i)}{d(i)}\right)^{-1/n}.$$

<u>Proof</u>. Without loss of generality, we take $c = 0$, so that $R_i(0) = 0$ and we can choose $g(z) = 1/z$. For $\Gamma > 0$ let $D(\Gamma)$ denote the disk $\{z: |z| \leq \Gamma\}$.

For simplicity let us first suppose each $E(i) > 0$, so that c is not a critical point of any R_i. Then let R_{i1}^{-1} denote the branch of the inverse of R_i such that $R_{i1}^{-1}(c) = c$. We choose $\Gamma > 0$ so small that

$$R_{11}^{-1}(D(\Gamma)) \subset R_{i1}^{-1}(D(\Gamma)) \quad \text{for } i = 2,3,\ldots,h,$$

$$(5) \qquad R_{11}^{-1}(D(\Gamma)) \cap R_{ij}^{-1}(\hat{\mathbb{C}}) = \emptyset \quad \text{for } j = 2,3,\ldots,d(i),$$

$$D(\Gamma) \in \mathrm{supp}(\sigma) = \emptyset,$$

and $D(\Gamma)$ contains no critical point of R_i, for $i = 1,2,\ldots,h$. Take $f(z) = \chi_{R_{11}^{-1}(D(\Gamma))}(z)$ (the characteristic function of $R_{11}^{-1}(D(\Gamma))$) in (4), to obtain

$$\mu(R_{11}^{-1}(D(\Gamma))) = \sum_{i=1}^{h} \frac{p(i)}{d(i)} \int_{\hat{\mathbb{C}}} \chi_{R_{11}^{-1}(D(\Gamma))}(R_{i1}^{-1}(z))\,d\mu(z) \tag{6}$$

$$\leq \left(\sum_{i=1}^{h} \frac{p(i)}{d(i)}\right)\mu(D(\Gamma)) = N^{-1}\mu(D(\Gamma))$$

where we define

$$N = \left(\sum_{i=1}^{h} p(i)/d(i)\right)^{-1}$$

Now choose K so that

$$N^{1/n} > K > \mathrm{Max}\{E(1),1\},$$

and choose $\Gamma > 0$ so small that

$$D(K^{-1}\Gamma) \subset R_{11}^{-1}(D(\Gamma)),$$

whence, using (5) as well,

$$\mu(D(K^{-k}\Gamma)) \leq \mu(R_{11}^{-k}(D(\Gamma))) \leq N^{-k}\mu(D(\Gamma))$$

for each positive integer k. We now have

$$\int_{D(\Gamma)} \frac{d\mu}{|z|^n} = \sum_{k=0}^{\infty} \int_{D(K^{-k}\Gamma)/D(K^{-k-1}\Gamma)} |z|^{-n} d\mu$$

$$\leq \sum_{k=0}^{\infty} \frac{1}{(K^{-k-1}\Gamma)^n}\, \mu(D(K^{-k}\Gamma))$$

$$\leq \mu(D(\Gamma))\left(\frac{K}{\Gamma}\right)^n \sum_{k=0}^{\infty} \left(\frac{K^n}{N}\right)^k < \infty.$$

Hence, whenever $E(1) < (\Sigma p(i)/d(i))^{-1/n}$, all of the moments $M_j = \int z^{-j} d\mu(z)$, for $j = 0,1,2,\ldots,n$, exist.

In the case where some of the E(i)'s equal zero, the above proof is modified as follows. Define an index set $I(i) = \{j \mid R_{ij}^{-1}(0) = 0\}$ for each $i = 2,3,\ldots,h$; and define the restricted inverse

$$\overline{R_{i1}^{-1}} = \bigcup_{j \in I(i)} R_{ij}^{-1}\,.$$

Then in (5) and (6) we replace R_{i1}^{-1} by $\overline{R_{i1}^{-1}}$ $i = 2,3,\ldots,h$, and we replace R_{ij}^{-1} for $j \neq 1$ by R_{ij}^{-1} for $j \notin I(i)$.

If all of the $E(i)$'s are less than unity, which occurs for example when $c = \infty$ and each R_i is a polynomial, then a similar result holds. Similarly to Theorem 1, one proves that there exists a balanced measure μ whose support does not include c. The existence of all of the integrals $\int |g(z)|^n d\mu$, $n = 0,1,2,\ldots$, follows at once.

We next consider the computation of $\int g(z)^n d\mu$. We will need the following result.

Lemma 1. Let

$$R(z) = \frac{z^d + a(1)z^{d-1} + \ldots + a(d)}{b(1)z^{d-1} + \ldots + b(d)}$$

where $a(i), b(i) \in \mathbb{C}$ and not all $b(i)$'s vanish. Let

$$S(n,z) = \sum_{i=1}^{d} (R_i^{-1}(z))^n$$

denote the nth symmetric function associated with $R(z)$. Then $\{S(n,z): n = 1,2,3,\ldots\}$ can be calculated explicitly recursively as follows. Let

$$c(k,z) = \begin{cases} a(k) - zb(k) & \text{for } k \in \{1,2,\ldots,d\}, \\ 0 & \text{for } k \notin \{1,2,\ldots,d\}. \end{cases}$$

Then

$$(7) \qquad S(n,z) = -nc(n,z) - \sum_{j=1}^{n-1} S(n-j,z)c_j(z).$$

$S(n,z)$ is a polynomial in z of degree at most n; and the coefficient of z^n is $b(1)^n$.

Proof. Consider the polynomial

$$w^d + c(1,z)w^{d-1} + \ldots + c(n,z) = 0,$$

whose roots are $R_i^{-1}(z)$. The formula (7) follows at once from Gaal [13]. The final statements in the Lemma follow by induction on n, starting from $s(1,z) = b(1)z - a(1)$.

Theorem 3. Let $c \in \hat{\mathbb{C}}$ be a fixed point of each R_i, with expansion factor $E(i)$, for $i = 1,2,\ldots,h$, and let g be a Möbius transformation such that $g(c) = \infty$. Suppose that, for some positive integer n, $g(z)^n \in L_1(\hat{\mathbb{C}},\mu)$ and

$$(8) \qquad \sum_{i=1}^{h} \frac{p(i)}{d(i)} E(i)^n < 1.$$

Let $M(\mu,k) = \int g(z)^k d\mu$ and $M(\sigma,k) = \int g(z)^k d\sigma$, $k = 0,1,2,\ldots,n$. Then $M(\mu,n)$ can be computed explicitly recursively in terms of $\{M(\mu,j): j = 0,1,\ldots,n-1\}$, $\{M(\sigma,k): k = 0,1,\ldots,n\}$, and the finite set of parameters $\{p(i),d(i),\text{coefficients of } R_i: i = 1,2,\ldots,h\}$ which characterize the Markov chain. The manner in which this can be done is given in the proof.

Proof. Without loss of generality we take $c = \infty$ and

$$R_i(z) = \frac{z^{d(i)} + a(i,1)z^{d(i)-1} + \ldots + a(i,d(i))}{b(i,1)z^{d(i)-1} + \ldots + b(i,d(i))},$$

so that

$$E(i) = |b(i,1)|, \quad i = 1,2,\ldots,h,$$

and we choose $g(z) = 1/z$. (Otherwise, use $g \circ R_i \circ g^{-1}$ in place of R_i and note that $\int z^n d\mu(g^{-1}(z)) = \int g(z)^n d\mu(z)$.)

Now choose $f(z) = z^n$ in (4) to obtain

$$(9) \qquad M(\mu,n) = \sum_{i=1}^{h} \frac{p(i)}{d(i)} \int S_i(n,z)d\mu + M(\sigma,n)$$

where $S_i(n,z)$ denotes the nth symmetric function associated with $R_i(z)$. We can write explicitly, via the recursion relation of Lemma 1,

$$S_i(n,z) = b(i,1)^n z^n + \sum_{j=0}^{n-1} \Gamma(i,n,j)z^j.$$

Hence (9) implies

$$(10) \qquad M(\mu,n)[1-\rho(n)] = \sum_{i=1}^{h} \frac{p(i)}{d(i)} \sum_{j=0}^{n-1} \Gamma(i,n,j)M(\mu,j) + M(\sigma,n)$$

where

$$\rho(n) = \sum_{i=1}^{h} \frac{p(i)}{d(i)} b(i,1)^n.$$

In particular

$$|\rho(n)| \leq \sum_{i=1}^{h} \frac{p(i)}{d(i)} E(i)^n < 1,$$

whence (10) can be solved for $M(\mu,n)$ as claimed.

In particular, if $1 \geq E(1) > E(i)$ for $i \neq 1$ or $E(j) < 1$ for all j, and $c \in \text{supp}(\sigma)$, then combining Theorem 1 (and the subsequent remark) with Theorem 2, we find that all of the moments $\int g(z)^n d\mu$ exist and can be calculated for $n = 0,1,2,\ldots$.

We next consider the uniqueness of balanced measures when the support is real.

Theorem 4. Let the support of the balanced measure μ be contained in $\mathbb{R} \cup \{\infty\}$; let $c = \infty$ be a fixed point of $R_i(z)$, with expansion factor $E(i)$, for $i = 1,2,\ldots,h$; and let $\infty \in \text{supp}(\sigma)$. If $E(1) = 1$ and $E(j) < 1$ for $j = 2,3,\ldots,h$, then the moment problem associated with the moment

$$M_n = \int_{-\infty}^{\infty} x^n d\mu(x) \qquad n = 0,1,2,\ldots$$

is determinate.

Proof. Let $A(\Gamma) = \{z \in \hat{\mathbb{C}}: |z| \geq \Gamma\}$. Then, much as in the proof of Theorem 2, one establishes that

$$\mu(R_{11}^{-1}(A(\Gamma))) \leq N^{-1}\mu(A(\Gamma)),$$

whenever Γ is sufficiently large, where

$$N = \left(\sum_{i=1}^{h} p(i)/d(i)\right)^{-1},$$

and where R_{11}^{-1} is the branch of R_1^{-1} which fixes ∞.

There are constants $\Gamma_0 > 0$ and $c > 0$ such that

$$|R_1^{-1}(z)| \leq \left(1 + \frac{c}{|z|}\right) \cdot |z| \qquad \text{for } |z| \geq \Gamma_0.$$

Let $\Gamma_n = nc/\varepsilon$ where ε is chosen small and positive, so that $\Gamma_n > \Gamma_0$ and $\exp(\varepsilon) < N$. Then

$$|R_{11}^{-1}(z)| \leq K_n|z| \quad \text{for } |z| \geq R_n,$$

where $K_n = (1 + \frac{\varepsilon}{n})$. It follows that

$$R_1^{-j}A(\Gamma_n) \supset A(K_n^j\Gamma_n) \quad \text{for } j = 0,1,2,\ldots .$$

We now have

$$\begin{aligned} |M_n| = |\int_{\hat{\mathbb{C}}} z^n d\mu(z)| &\leq \int_{|z|\leq\Gamma_n} |z|^n d\mu \\ &+ \sum_{j=0}^{\infty} \int_{A(K_n^{j+1}\Gamma_n)\backslash A(K_n^j\Gamma_n)} |z|^n d\mu \\ &\leq \Gamma_n^n + \sum_{j=0}^{\infty} (K_n^{n+1}\Gamma_n)^n \mu(A(K_n^j\Gamma_n)) \\ &\leq \Gamma_n^n + (\Gamma_n K_n)^n \sum_{j=0}^{\infty} (K_n^n/N)^j. \end{aligned}$$

Then noting that $K_n^n = (1+\varepsilon/n)^n < \exp(\varepsilon) < N$, we have

$$|M_n| \leq \frac{c^n n^n}{\varepsilon^n} \left(1 + \frac{N^2}{N-\exp(\varepsilon)}\right)$$

for $n = 0,1,2,\ldots$, where c and ε do not depend upon n. It follows that

$$\sum^{\infty} (M_{2n})^{-1/2n}$$

diverges, which is Carleman's condition for the moment problem to be determinate.

Corollary 1. Let $c = \infty$ be a fixed point of $R_i(z)$, with expansion factor $E(i)$, for $i = 1,2,\ldots,h$, and let $\infty \in \operatorname{supp}(\sigma)$. If $E(1) = 1$ and $E(j) < 1$ for $j = 2,3,\ldots,h$, there is at most one balanced measure whose support is contained in $\mathbb{R} \cup \{\infty\}$. (If $E(i) < 1$ for $i = 1,2,\ldots,h$ there is at most one balanced measure whose support is contained in $\mathbb{R}$.)

Proof. The first statement is immediate from Theorem 4, and the fact that the moments are unique by Theorem 3. Similarly, the second statement follows from uniqueness of the moments plus the determinacy of the moment problem when the support is real and bounded.

Example. We consider the reconstruction of the spectral density for a certain model Schrödinger equation. We have simplified the situation for brevity, but the example mimics the steps that were actually carried out [6] in an application of interest in statistical physics.

We start with an unknown measure in $\mathbb{R}$, for which we have the moments

$$M_0 = 1,\ M_1 = 1/2,\ M_2 = 9/4,\ ,\ M_3 = 1/2,\ M_4 = 57/8.$$

Also, by direct numerical computation, a portion of the measure is obtained. This is sketched below, where the heights of the line segments perpendicular to the x-axis give the spectral masses of the corresponding ordinates.

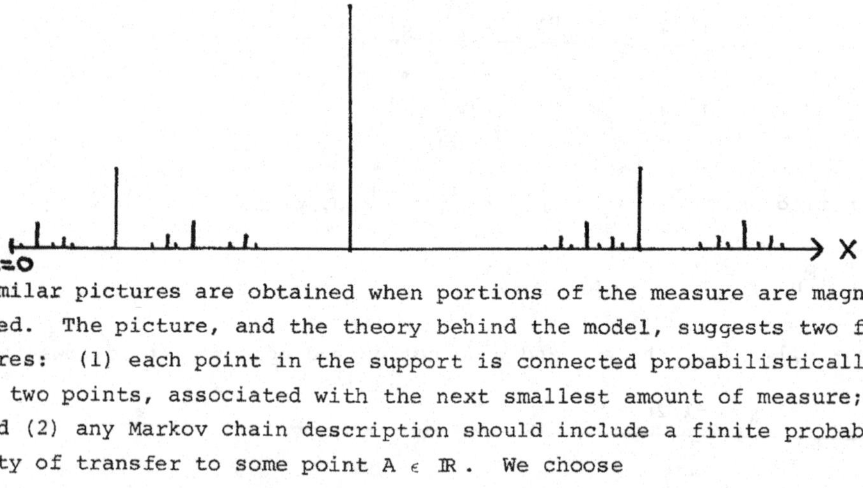

Similar pictures are obtained when portions of the measure are magnified. The picture, and the theory behind the model, suggests two features: (1) each point in the support is connected probabilistically to two points, associated with the next smallest amount of measure; and (2) any Markov chain description should include a finite probability of transfer to some point $A \in \mathbb{R}$. We choose

$$R(z) = z^2 - \lambda,\quad p(1) = p,$$
$$\sigma(z) = \delta_z(A),\quad p(0) = 1-p.$$

The moments of the corresponding balanced measure obey

$$M_0 = 1,\ M_1 = p(0)A,\ M_2 = p(1)(M_1 + \lambda) + p(0)A^2,$$
$$M_3 = p(0)A^3,\ M_4 = p(1)(M_2 + 2\lambda M_1 + \lambda^2).$$

Equating to the given moments and solving we find $p(0) = p(1) = 1/2$,

$\lambda = 3$, $A = 1$. Similarly, in the actual application, the correct fractal structure was exactly reconstructed.

ACKNOWLEDGEMENT

We thank Randy Nicklas for help in computing pictures of probabilistic mixtures.

References

1. M. F. Barnsley, J. S. Geronimo, A. N. Harrington, "Orthogonal polynomials associated with invariant measures on Julia sets," Bulletin A.M.S. 7 (1982), 381-384.

2. ———, "On the invariant sets of a family of quadratic maps," Comm. Math. Phys. (1983) 88, 479-501.

   ———, "Infinite dimensional Jacobi matrices associated with Julia sets," Proc. A.M.S. (1983) 88, 625-630.

3. ———, "Geometry electrostatic measure and orthogonal polynomials on Julia sets for polynomials," to appear Journal of Ergodic Theory and Dynamical Systems (1982).

4. ———, "Geometrical and electrical properties of some Julia sets," Ch. 1 of Quantum and Classical Models and Arithmetic Problems, (Dekker, 1984; edited by G. and D. Chudnovsky).

5. ———, "Some treelike Julia sets and Padé approximants," Letters in Math. Phys. 7 (1983), 279-286.

   ———, "Almost periodic operators associated with Julia sets," Preprint (Georgia Institute of Technology, 1983), submitted to Comm. Math. Phys.

6. ———, "Condensed Julia sets, with an application to a fractal lattice model Hamiltonian," submitted Trans. A.M.S. (1983).

7. M. F. Barnsley, A. N. Harrington, "Moments of balanced measures on Julia sets," to appear Trans. A.M.S. (1983).

8. D. Bessis, J. S. Geronimo, P. Moussa, "Mellin transforms associated with Julia sets and physical applications," Preprint (CEN-SACLAY, Paris, 1983).

9. D. Bessis, M. L. Mehta, P. Moussa, "Polynômes orthogonaux sur des ensembles de Cantor et iterations des transformations quadratiques," C. R. Acad. Sci. (Paris) 293 (1981), 705-708.

10. ———, "Orthogonal polynomials on a family of Cantor sets and the problem of iteration of quadratic maps," Letters in Math. Phys. 6 (1982), 123-140.

11. D. Bessis, P. Moussa, "Orthogonality properties of iterated polynomial mappings," Comm. Math. Phys. 88 (1983), 503-529.

12. H. Brolin, "Invariant sets under iteration of rational functions," Arkiv för Matematik 6 (1965), 103-144.

13. L. Gaal, Classical Galois Theory with Examples, Markham, Chicago, 1971.

14. N. Dunford, J. Schwartz, Linear Operators (Part I), John Wiley, 1957.

15. R. Mañè, "On the uniqueness of the maximizing measure for rational maps," Preprint (Rio de Janeiro, 1982).

16. B. Mandelbrot, The Fractal Geometry of Nature, (W. H. Freeman, San Francisco, 1982).

17. T. S. Pitcher, J. R. Kinney, "Some connections between ergodic theory and iteration of polynomials," Arkiv för Matematik 8 (1968), 25-32.

ON RATIONAL APPROXIMATION OF THE EXPONENTIAL AND THE SQUARE ROOT FUNCTION

Dietrich Braess
Institut für Mathematik
Ruhr-Universität, D-4630 Bochum, F.R. Germany

Abstract. Fifteen years ago Meinardus made a conjecture on the degree of the rational approximation of the function e^x on the interval $[-1,+1]$. The conjecture was recently proved via the approximation on the circle $|z| = \frac{1}{2}$ in the complex plane. The same method is now applied to the approximation of the square root function. Here we have a gap between the upper and the lower bound, the amount of which depends on the location of the branch point. To close the gap some folklore about Heron's method is collected and completed.

1. Introduction

In 1967 Meinardus [3,p.168] conjectured that the accuracy of the best approximation to e^x from R_{mn} on the interval $[-1,+1]$ is asymptotically

$$E_{mn}(e^x) = \frac{m!n!}{2^{m+n}(m+n)!(m+n+1)!}(1+o(1)) \text{ as } m+n \to \infty . \tag{1}$$

Here R_{mn} contains the rational functions of degree (m,n) and $C[-1,+1]$ is endowed with the uniform norm. To prove (1) we make use of a connection between the approximation on the interval $[-1,+1]$ and on the circle $|z| = \frac{1}{2}$ in the complex plane. To this end the (m,n)-th degree Padé approximant is modified such that the modulus of the error is almost constant on the circle.

The sharp result (1) is related to the fact that e^z is an entire function. This becomes obvious when the same method is applied to rational approximation of the square root function. To minimize the relative error the approximation with a weight function (see (19) below) is treated. For $m = n$ or $m = n+1$ we obtain

$$E_{mn}(\sqrt{x+a}) = \frac{4}{[2(\rho+\sqrt{\rho^2-1}-\lambda\rho^{-1})]^{n+m+1}}(1+o(1)) \quad \text{as } m\to\infty \tag{2}$$

whenever $a =: \frac{1}{2}(\rho+\rho^{-1}) \geq 3/2$. Here $|\lambda|\leq 1/2$. The formula provides upper and lower bounds which become close for large a, i.e., when the singularity $x = -a$ is far from the interval $[-1,+1]$.

2. Connection between the Approximation on an Interval and on a Circle

The crucial point in our analysis is Newman's trick [4] . It provides a connection between the approximation on an interval and on a circle. Specifically, this method can be successfully used for the study of the approximation of e^x and x^α.

Let p/q be an (m,n)-degree rational function. Given $x\in[-1,+1]$, put $z = x+iy$, where $x^2+y^2 = 1$. Then

$$u(x) := \frac{p(z)p(\bar{z})}{q(z)q(z)} \tag{3}$$

is again an (m,n)-degree rational function, i.e., the degree is not doubled when products of this special form are taken. To understand this, consider the product of two linear expressions:

$$(az+b)\cdot(a\bar{z}+b) = ab(z+\bar{z}) + a^2\bar{z}z + b^2$$

$$= 2abx + [a^2+b^2]. \tag{4}$$

Let $\alpha,\beta\in\mathbb{C}$. Then $\alpha\bar{\alpha}-\beta\bar{\beta} = 2\mathrm{Re}\,\bar{\alpha}(\alpha-\beta) - |\alpha-\beta|^2$. By applying this formula to the products

$$e^x = e^{z/2}\, e^{\bar{z}/2} \tag{5}$$

and (3) we obtain

$$e^x-u(x) = 2\,\mathrm{Re}\left\{e^{\bar{z}/2}\left(e^{z/2}-\frac{p(z)}{q(z)}\right)\right\}-\left|e^{z/2}-\frac{p(z)}{q(z)}\right|^2. \tag{6}$$

Henceforth only cases will be considered where the second term in (6) is small when compared with the first one. The following lemma deals with functions which admit a decomposition as in (5):

Lemma. Assume that f is a real analytic function in the unit disk $|z| < 1$ and that $qf-p$ with $p/q \in R_{mn}$ has n+m+1 zeros in the disk while q and f have none. Moreover, let $F(x) = f(\bar{z})f(z)$ where $|z| = 1$, $\mathrm{Re}\, z = x$. Then

$$2 \min_{|z|=1} |\bar{f}(f-\frac{p}{q})| \le E_{mn}(F)(1+o(1)) \le 2 \max_{|z|=1} |\bar{f}(f-\frac{p}{q})|. \tag{7}$$

Proof. The upper bound is clear, since a treatment of $f(z)f(\bar{z}) - p(z)p(\bar{z})/q(z)q(\bar{z})$ as in (6) shows that the right hand side of (7) equals $\|f-u\|$ apart from terms of higher order [4].

The lower bound [2] will be derived by using de la Vallée-Poussin's theorem. Note that

$$\mathrm{Re}\,\bar{f}\,(f-\frac{p}{q}) = \begin{cases} +|f(f-\frac{p}{q})| & \text{if } \arg \bar{f}(f-\frac{p}{q}) \equiv 0 (\mathrm{mod}\ 2\pi), \\ \\ -|f(f-\frac{p}{q})| & \text{if } \arg \bar{f}(f-\frac{p}{q}) \equiv \pi (\mathrm{mod}\ 2\pi). \end{cases} \tag{8}$$

By assumption $f^{-1}q^{-1}(qf-p)$ has m+n+1 zeros counting multiplicities but no pole in the unit disk. The winding number of this function is n+m+1. The argument of $f^{-1}q^{-1}(qf-p) = \bar{f}(f-p/q)/|f|^2$ is increased by $(m+n+1)2\pi$ when an entire circuit is performed. The argument is increased by $(m+n+1)\pi$ as z transverses the upper half of the circle. Since the function is real for $z = +1$ and $z = -1$, we get an alternant of the desired length m+n+2 and the lower bound follows from (8) and de la Vallée-Poussin's theorem. □

3. Proof of Meinardus' Conjecture

In order to apply the above lemma to e^x we need a rational approximation to $e^{z/2}$ on the unit circle. For convenience, we change the variable and consider the function $f(z) = e^z$ for $|z| = \frac{1}{2}$. With this instead of (7) we have

$$2 \min_{|z|=\frac{1}{2}} |e^{\bar{z}}(e^z-\frac{p}{q})| \le E_{mn}(e^x)(1+o(1)) \le 2 \max_{|z|=\frac{1}{2}} |e^{\bar{z}}(e^z-\frac{p}{q})| \tag{9}$$

Now the (m,n)-th degree Padé approximant [6] is inserted: $p(z) = \int_0^\infty (t+z)^m t^n e^{-t} dt$, $q(z) = \int_0^\infty t^m (t-z)^n e^{-t} dt$. From these representations one gets the well known expression for the remainder term:

$$e^z q(z) - p(z) = -\int_0^{-z} t^n (t+z)^m e^{-t} dt$$
$$= (-1)^n z^{m+n+1} \int_0^1 u^n (1-u)^m e^{uz} du.$$

Set $\alpha = (n+1)/(m+n+2)$. When we expand e^{uz} at $u = \alpha$, the linear term does not contribute to the integral and the (weighted) integral over the remainder is small when compared with the (weighted) integral over $e^{\alpha z}$:

$$e^z q(z) - p(z) = (-1)^n \frac{m!n!}{(m+n+1)!} z^{n+m+1} e^{\alpha z} (1+O(\frac{1}{m+n+3})) \text{ for } |z| \leq 1.$$

An asymptotic representation for q is already known from Perron [6,p.248]

$$q(z) = (m+n)!\, e^{-\frac{m}{m+n} z} (1+o(1)).$$

By collecting terms we get

$$e^z (e^z - \frac{p(z)}{q(z)}) = \frac{(-1)^m m!n!}{(m+n)!(m+n+1)!} z^{m+n+1} e^{\beta z} (1+o(1)) \quad (10)$$

with $\beta = 1+\alpha + n/(n+m) \leq 3$.

Now we observe that the right hand side of (10) has an absolute value which is nearly constant for $|z| = 1/2$. This phenomenon is denoted as *near circularity* [9]. A deviation from complete circularity is only caused by the term $e^{\beta z}$.

This deviation may be eliminated up to a o-term by a shift of the point at which the Padé approximation is evaluated. If we choose $z_0 = \beta/[4(m+n+1)]$, then

$$|e^{\beta z} (z-z_0)^{m+n+1}| = |z|^{m+n+1} (1+o(1)) \text{ for } |z| = \tfrac{1}{2}.$$

Indeed, $|z-z_0|^{m+n+1} = |2\bar{z}(z-z_0)|^{m+n+1} = |\frac{1}{2}(1-\beta\bar{z}/[m+n+1])|^{m+n+1} = 2^{-m-n-1} |e^{-\beta z}| (1+o(1))$.

Let z_0 be as above. Then $\tilde{p}(z)/\tilde{q}(z) := e^{z_0} p(z-z_0)/q(z-z_0)$ is the Padé approximant to e^z at z_0. From (9) and

$$e^z (e^z - \frac{\tilde{p}(z)}{\tilde{q}(z)}) = \frac{(-1)^n m!n!}{(m+n)!(m+n+1)!} (z-z_0)^{n+m+1} e^{\beta z} (1+o(1)),$$

it follows that Meinardus' conjecture (1) is true.

We mention that Trefethen [10] has used the same idea to prove Saff's conjecture [8] on the approximation of the exponential function in the disk $|z| \geq \rho$:

$$E_{mn}(e^z; |z| \leq \rho) = \frac{\rho^{m+n+1} m!\,n!}{(m+n)!\,(m+n+1)!}(1+o(1)) \text{ as } m+n \to \infty.$$

4. The Approximation of the Square Root Function

When applied to the square root function, Newman's trick yields an improvement which corresponds to the squaring of a certain characteristic parameter. First we will understand the meaning of this parameter.

Let $0 < x_1 < x_2$. Suppose that we know the best rational approximation of degree (m,n) to $\sqrt{x}$ on the interval $[x_1,x_2]$. Then a multiple is obviously the best approximation to $\sqrt{cx}$, $c > 0$, on this interval which in turn implies that we know the solution to $\sqrt{x}$ for the interval $[cx_1,cx_2]$. Consequently the solution mainly depends on the ratio

$$\frac{x_2}{x_1}. \tag{11}$$

Specifically, the accuracy is better, the smaller this parameter is. When we make a transformation and consider $\sqrt{\rho+x}$ with $\rho > 1$ on the unit interval $[-1,+1]$, then $x_1=\rho-1, x_2=\rho+1$ and

$$\frac{\rho+1}{\rho-1}$$

is the characteristic parameter.

For the decomposition of the square root function we recall a special case of (4):

$$\left.\begin{aligned}(\rho+z)(\rho+\bar{z}) &= \rho^2+1 + 2\rho x \\ &= 2\rho(a+x)\end{aligned}\right\} \quad \text{for } |z| = 1 \tag{12}$$

where $a = \frac{1}{2}(\rho+\rho^{-1})$. Therefore when setting $f(z) = \sqrt{\rho+z}$ the induced function is $F(x) = \text{const}\,\sqrt{a+x}$. The associated parameter (11) is squared:

$$\frac{a+1}{a-1} = \left(\frac{\rho+1}{\rho-1}\right)^2. \tag{13}$$

We note that ρ is the parameter of the ellipse with foci +1 and -1 in which $F(x) = (a+x)^{1/2}$ is an analytic function.

The "stair case" elements of the Padé table can be given in closed form [1]. More generally let $x_1, x_2, \ldots, x_k$ be positive numbers (which need not be distinct). Set

$$h(w) = \prod_{i=1}^{k} (\sqrt{x_i}+w).$$

Then by $p(x) = \frac{1}{2}\{h(\sqrt{x}) + h(-\sqrt{x})\}$, $q(x) = \frac{1}{2}x^{-\frac{1}{2}}\{h(\sqrt{x})-h(-\sqrt{x})\}$ a rational function is defined which interpolates $\sqrt{x}$ at $x_1, x_2, \ldots x_k$. Indeed we have

$$\sqrt{x} - \frac{p(x)}{q(x)} = - \frac{1}{q(x)h(\sqrt{x})} \prod_{i=1}^{k} (x_i - x). \tag{14}$$

Obviously, $p/q \in R_{m,m}$ if $k = 2m+1$ is odd, and $p/q \in R_{m,m-1}$ if $k = 2m$ is even.

Let $m=n$ or $m=n+1$ and $k=m+n+1$. We consider the (m,n)-th degree Padé approximant to $f(z) = (\rho+z)^{1/2}$. Since we expect that the point of the expansion will be shifted from the center of the circle $|z| = 1$, we give the expansion for the point $z_o = \rho_s - \rho$:

$$\begin{aligned} p(z) &= \frac{1}{2} \{(\sqrt{\rho_s} + \sqrt{\rho+z})^k + (\sqrt{\rho_s} - \sqrt{\rho+z})^k\}, \\ q(z) &= \frac{1}{2\sqrt{\rho+z}} \{(\sqrt{\rho_s} + \sqrt{\rho+z})^k + (\sqrt{\rho_s} - \sqrt{\rho+z})^k\}. \end{aligned} \tag{15}$$

The remainder term is easily determined to be

$$\sqrt{\rho+z} - \frac{p(z)}{q(z)} = - \frac{(\sqrt{\rho_s} - \sqrt{\rho+z})^k}{q(z)} = \frac{(\rho_s - \rho - z)^k}{(\sqrt{\rho_s} + \sqrt{\rho+z})^k q(z)}. \tag{16}$$

For $\rho, \rho_s > \frac{3}{2}$, $|\rho-\rho_s| < \frac{1}{4}$ and $|z| \leq 1$ the expression (16) becomes

$$2 \frac{(\rho_s-\rho-z)^k \sqrt{\rho+z}}{(\sqrt{\rho_s} + \sqrt{\rho+z})^{2k} - (\rho_s-\rho-z)^k} = 2 \frac{(\rho_s-\rho-z)^k \sqrt{\rho+z}}{(\sqrt{\rho_s} + \sqrt{\rho+z})^{2k}} (1+o(1)). \tag{17}$$

The situation is here less favorable than for the exponential function. For a moment let $\rho_s = \rho$. Then the deviation from circularity is governed by the denominator in (17) which is a multiple of $|1+\sqrt{1+z/\rho}|^{2k}$. The ratio of the largest value to the smallest value is not bounded as $k \to \infty$. It is not surprising that the deviation from circularity is less the farther the singularity $z = -\rho$ is from the circle $|z| = 1$.

Nevertheless the variation of the modulus of the remainder term can be reduced substantially by a shift of the point of the expansion. Specifically, we will choose $\rho_s = \sqrt{\rho^2-1}$. Then $\rho - \rho_s = (\rho + \rho_s)^{-1}$. Note that

$$(\sqrt{\rho_s} + \sqrt{\rho+z})^2 = 2(\rho_s + \rho + z) - \frac{1}{2} \frac{(\rho_s - \rho - z)^2}{\rho_s + \rho + z + 2\sqrt{\rho_s(\rho+z)}}.$$

Consequently, with ρ_s as specified we get

$$\left|\frac{2(\rho_s+\rho+z)}{(\sqrt{\rho_s}+\sqrt{\rho+z})^2}-1\right| \le \frac{1}{8\rho^2} \qquad \text{for } |z|\le 1, \rho \ge 2.$$

Finally we note that $\rho_s-\rho-z = -z(\rho+\rho_s)^{-1}(\rho_s+\rho+\bar{z})$ for $|z| = 1$. Hence, from (17) and the estimates above it follows that

$$|\bar{f}(f-\frac{p}{q})| = \frac{2F(x)}{[2(\rho+\sqrt{\rho^2-1}-\lambda\rho^{-1})]^k}, \quad |z| = 1, \ \rho \ge 2, \tag{18}$$

with some $|\lambda| \le \frac{1}{4}$.

Now we turn to the approximation problem with the weighted norm

$$\|f\| = \sup_{-1\le x\le +1} \frac{1}{\sqrt{x+a}} |f(x)|. \tag{19}$$

With this norm, the result (2) is a consequence of (18) and the lemma. The estimate (2) improves the upper bound in [1] .

When the accuracy for the uniform weight 1 is required, the result has to be multiplied by a factor which lies between $\sqrt{a-1}$ and $\sqrt{a+1}$.

The accuracy of the best polynomial approximation is well known:

$$E_{2n,o}(\sqrt{x+a}) \approx \text{const}\rho^{-2n}.$$

Therefore, rational approximation with the same number of coefficients is favorable by a factor of $[4-2\rho^{-2}]^{2n}$. This factor is larger than 3.93^{2n} whenever $a \ge 3$.

5. Heron's Algorithm and its Symmetrization

The gap which is left in (2) can actually be closed. The accuracy of the best approximation of degree $(2^\ell, 2^\ell-1)$ for $\ell=0,1,2,...$, can be computed recursively. From the recursion an estimate for the asymptotics can be determined. To this end we have to complete some folklore about Heron's method for the computation of square roots, which is possibly first noted in [7] (see also [5,11] and the literature cited there).

Given $\alpha > 0$, the sequence

$$u_{\ell+1} = \frac{1}{2}(u_\ell + \frac{\alpha}{u_\ell}), \quad \ell = 0,1,...$$

converges to $\sqrt{\alpha}$ for any $u_o > 0$. Analogously, if $u_\ell(x)$ is a rational approximation from $R_{m,m-1}$ to $\sqrt{x}$, then

$$u_{\ell+1} = \frac{1}{2}(u_\ell(x) + \frac{x}{u_\ell(x)}) \tag{20}$$

is a better approximating function from $R_{2m,2m-1}$. This is obvious from the identity [7]

$$\frac{u_{\ell+1}(x) - \sqrt{x}}{u_{\ell+1}(x) + \sqrt{x}} = \left(\frac{u_\ell(x) - \sqrt{x}}{u_\ell(x) + \sqrt{x}}\right)^2 . \tag{21}$$

Because of this squaring it is natural to define the distance

$$d(u,\sqrt{x}) = \max \left| \frac{u(x) - \sqrt{x}}{u(x) + \sqrt{x}} \right|, \tag{22}$$

and the minimal distance

$$e_{mn}(\sqrt{x+a}) = \inf\{d(u,\sqrt{x}) : u \in R_{mn}\}. \tag{23}$$

We note that the ratio

$$[u(x)-\sqrt{x}]/[u(x) + \sqrt{x}] \tag{24}$$

from (22) can be given in terms of the ratio $[u(x) - \sqrt{x}]/\sqrt{x}$ from (19). From this observation we conclude with some obvious arguments that a multiple of a solution for (23) provides a solution for the approximation problem from the preceding section. Specifically we get the correspondence

$$E_{mn}(\sqrt{x+a}) = \frac{2e_{mn}(\sqrt{x+a})}{1+e_{mn}^2(\sqrt{x+a})}, \quad e_{mn}(\sqrt{x+a}) = \frac{E_{mn}(\sqrt{x+a})}{1+\sqrt{1-E_{mn}^2(\sqrt{x+a})}} \tag{25}$$

Moreover, it follows that u is a solution of (23) if and only if the ratio (24) alternates m+n+1 times.

The best constant function for approximating $\sqrt{x}$ in the interval $[a-1,a+1]$ is now $\sqrt[4]{(a-1)(a+1)}$ and recalling (13) we have

$$\frac{a+1}{a-1} = \left(\frac{\rho+1}{\rho-1}\right)^2 = \left(\frac{1+e_{oo}}{1-e_{oo}}\right)^4$$

or

$$e_{oo} = \frac{1}{\rho + \sqrt{\rho^2-1}}, \quad E_{oo} = \frac{1}{\rho}. \tag{26}$$

Assume that u is optimal in $R_{m,m-1}$ and that there is an alternant of length 2m+1. Compute $u' \in R_{2m,2m-1}$ as in (20). Then

$$0 \le \frac{u'(x) - \sqrt{x}}{u'(x) + \sqrt{x}} \le e_{m,m-1}^2 .$$

and u' is a one-sided approximation. Rutishauser [7] observed already that a symmetrization improves the approximation. The factor $\lambda =$

$\max\{u'(x)/\sqrt{x}\} = (1-e^2_{m,m-1})/(1-e^2_{m,m-1})$ leads to a rational function $u" = \lambda^{-1/2}u' \in R_{2m,2m-1}$ with

$$-\rho \le \frac{u"(x) - \sqrt{x}}{u"(x) + \sqrt{x}} \le \rho \ , \tag{27}$$

where

$$\rho := \frac{e^2_{m,m-1}}{1+\sqrt{1-e^4_{m,m-1}}} = e_{2m,2m-1} \ ; \tag{28}$$

see also [7]. Here, the optimality of u" is a consequence of the following observation. The left hand side of (27) is an equality if x is a zero of $u(x)-\sqrt{x}$ and the right hand one is an equality if x belongs to the alternant of (24). Hence, $u"-\sqrt{x}$ alternates 4m+1 times in the same sense. This proves optimality and $\rho := e_{2m,2m-1}$.

Finally, e_{10} is obtained from e_{oo} by a formula analogous to (28). Therefore we can compute $e_{m,m-1}$ for m = 1,2,4,8,... recursively from (26) and (28). Moreover, $E_{2m,2m-1} = e^2_{m,m-1}$. From this and (28) it follows that the question from [5, p. 399]has to be answered in the negative way. The relations between the quantities treated above are depicted in the diagram on the following page. In particular, the relation $E^{-1}_{m,m-1} = (E^{1/2}_{2m,2m-1} + E^{-1/2}_{2m,2m-1})/2$ shows the connection with Gauß' arithmetic geometric mean. The quantity

$$r(\rho) = \lim_{m\to\infty} (4/E_{m,m-1}(\sqrt{x+a}))^{1/2m}$$

satisfies the functional equation

$$\log r(\rho) = 2 \log r(\tfrac{1}{2}(\rho^{1/2} + \rho^{-1/2})), \quad \rho \ge 1.$$

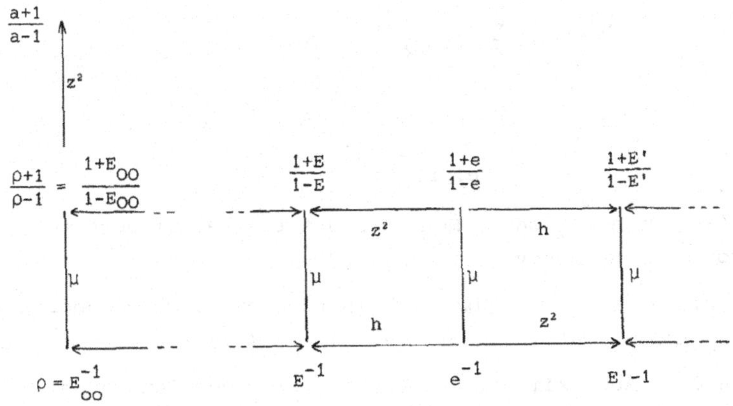

Relations between $E := E_{m,m-1}$, $e := e_{m,m-1}$ and $E' := E_{2m,2m-1}$

Here $\mu(z) = (z+1)/(z-1)$ and $h(z) = (z+z^{-1})/2$.

In order to get asymptotic estimates, we first observe that $e_{2m,2m-1} \geq \frac{1}{2} e^2_{m,m-1}$. On the other hand from $e_{2m,2m-1} \leq e_{m,m-1}$ and (28) it follows that

$$\frac{e_{2m,2m-1}}{1+\sqrt{1-e^2_{2m,2m-1}}} \leq \left(\frac{e_{m,m-1}}{1+\sqrt{1-e^2_{m,m-1}}}\right)^2 .$$

Knowing $e_{o,o}$ from (26) by successively squaring the upper and lower bounds we obtain

$$2(2\rho_1)^{-2m} \leq e_{m,m-1}(\sqrt{x+a}) \leq 2(2\rho_2)^{-2m}, \quad m = 1,2,4,8,... \tag{29}$$

where $\rho_1 = \rho + \sqrt{\rho^2-1}$ and $\rho_2 = 2\rho_1/(1+\sqrt{1-\rho_1^{-2}})$. Hence, we have $2\rho_2 < r(\rho) < 2\rho_1 < 4\rho$.

The denominators in (28) converge to 2 very rapidly. Therefore the upper and lower bounds become very close after a few steps. For $a = 3$, i.e. $\rho = 3 + \sqrt{8}$ we get e.g.

$$E_{o,o}(\sqrt{x+3}) = 4/(4\rho), \quad E_{1,o}(\sqrt{x+3}) = 4(3.97034\rho)^{-1}$$

$$4(3.97032\rho)^{-2m} \leq E_{m,m-1}(\sqrt{x+3}) \leq 4(3.97031\rho)^{-2m}, \quad m = 2,4,8,16,...$$

More generally, if $\rho > 1.01$ then $r(\rho)$ can be obtained with 8 correct digits by applying (28) four times at most. Moreover, $r(\rho) > 2\rho$ for $\rho > 1.01$. If, on the other hand, ρ is close to one, the asymptotic convergence rate can be obtained by adapting methods which are used for the estimate of $E_{n,n}(|x|)$, see e.g. [4]. One finally obtains

$$r(\rho) \approx \exp\{\frac{\pi^2}{2} / \log \frac{\rho+1}{\rho-1}\} \quad \text{as } \rho \to 1.$$

References

1. P. Borwein, On a method of Newman and a theorem of Bernstein. J. Approximation Theory 34,37-41(1982).
2. D. Braess, On the conjecture of Meinardus on rational approximation of e^x. J. Approximation Theory 36, 317-320(1982).
3. G. Meinardus, Approximation of Functions: Theory and Numerical Methods. Springer, New-York-Heidelberg 1967.

4. D.J. Newman, Approximation with Rational Functions. Regional Conference Series, No.41. Amer.Math.Soc.Providence,Rhode Island 1979.

5. I. Ninomiya, Best Rational Starting Approximations and Improved Newton Iteration for the Square Root. Math.Comp. 24,391-407(1970).

6. O. Perron, Die Lehre von den Kettenbrüchen II. Teubner, Stuttgart 1957.

7. H. Rutishauser, Betrachtungen zur Quadratwurzeliteration. Monatshefte Math. 67, 452-464(1963).

8. E. Saff, On the degree of best rational approximation to the exponential function. J. Approximation Theory 9,97-101(1973).

9. L.N. Trefethen, Near circularity of the error curve in complex Chebyshev approximation. J. Approximation Theory 31,344-367(1981).

10. L.N. Trefethen, The asymptotic accuracy of rational best approximation to e^z on a disk. J. Approximation Theory 40, 380-383 (1984).

11. K. Zeller, Newton-Čebyšev-Approximation. In "Iterationsverfahren, Numerische Mathematik, Approximationstheorie" (L. Collatz et al,Eds.) pp. 101-104, Birkhäuser, Basel 1970.

PADÉ-TYPE APPROXIMANTS AND LINEAR FUNCTIONAL TRANSFORMATIONS

Claude BREZINSKI and Jeannette VAN ISEGHEM
Laboratoire d'Analyse Numérique et d'Optimisation
Université de Lille I
59655 - Villeneuve d'Ascq Cedex
France

Abstract : Let $f(.) = \sum_{i=o}^{\infty} c_i g_i(.)$ be a series of functions and let $F(.) = \sum_{i=o}^{\infty} c_i h_i(.)$ be the series obtained by applying a linear functional transformation to f. It is shown that the Padé-type approximants of F can be deduced from that of f by application of the same functional transform. Some examples and applications are given. Convergence theorems are obtained. The particular case of the Laplace transform is studied in more detail.

1. Definitions and properties

Let f be a (formal) series of functions

$$f(t) = \sum_{i=o}^{\infty} c_i g_i(t) \quad ,$$

let G be the generating function of the sequence (g_i)

$$G(x, t) = \sum_{i=o}^{\infty} x^i g_i(t)$$

and let c be the linear functional on the space of polynomials defined by

$$c(x^i) = c_i \qquad i = 0, 1, \ldots .$$

Then, formally

$$f(t) = c(G(x, t))$$

where c acts on the variable x and t is a parameter.

Let v be an arbitrary polynomial of degree $k = k_1 + \ldots + k_n$: $v(x) = (x-x_1)^{k_1} \ldots (x-x_n)^{k_n}$ where the x_i's are distinct points in the complex plane. Let P(., t) be the Hermite interpolation polynomial of

$G(., t)$ at the zeros of v

$$P^{(j)}(x_i, t) = G^{(j)}(x_i, t) \quad i = 1, \ldots, n ; \quad j = 0, \ldots, k_i-1.$$

By definition $c(P(x, t))$ is called a Padé-type approximant (PTA) of the series f and, by analogy with the case of a power series, it is denoted by $(k-1/k)_f(t)$ although it is not necessarily a rational function v is called the generating polynomial of the approximant. The fundamental property of such Padé-type approximants is that $(k-1/k)(t) = f(t)+O(g_k(t))$.

If the x_i's are the zeros of the formal orthogonal polynomial of degree k with respect to the functional c, then $c(P(x, t))$ is the Padé approximant of f and is denoted by $[k-1/k]_f(t)$.

For a detailed exposition, and in particular for the construction of the whole table of approximants (p/q), see [1].

It is fundamental for a good understanding of the sequel to notice that, in general, $(k-1/k)_f$ as constructed above is different from the approximant $(k-1/k)_e$ where $e(t) = \sum_{i=0}^{\infty} d_i e_i(t)$ with $d_i=1$, $e_i(t)=c_i g_i(t)$ constructed from the same generating polynomial, since the linear functional and the generating function are not the same.

Now let L be a linear functional transform. We shall use the following notations which are not strictly correct but are in fact better because both variables clearly appear

$$h_i(p) = Lg_i(t)$$

$$F(p) = Lf(t) = \sum_{i=0}^{\infty} c_i h_i(p).$$

Let H be the generating function of the sequence $\{h_i\}$:

$$H(x, p) = \sum_{i=0}^{\infty} x^i h_i(p).$$

Obviously

$$H(x, p) = LG(x, t).$$

We set

$$Q(x, p) = LP(x, t).$$

Since L acts on P as a function of t then Q is a polynomial of degree at most $k-1$ in x and

$$Q^{(j)}(x_i, p) = LP^{(j)}(x_i, t) = LG^{(j)}(x_i, t) = H^{(j)}(x_i, p).$$

Thus $Q(., p)$ is the Hermite interpolation polynomial of $H(., p)$ at

$x_1, \ldots, x_n$ with the respective multiplicities $k_1, \ldots, k_n$ and we have

$$c(Q(x, p)) = c(LP(x, t)) = Lc(P(x, t))$$

since L acts on t and c on x. It means that

$$(k-1/k)_F(p) = L(k-1/k)_f(t).$$

Since, $\forall$ m and n, (m/n) is constructed from (k-1/k) we establish
Theorem 1 : Let $F(p) = Lf(t)$. Then, $\forall$ m, n $(m/n)_F(p) = L(m/n)_f(t)$ if both approximants are constructed from the same generating polynomial and the same linear functional c.

Let us comment on this theorem.

For this result to be valid it is essential, as we already pointed out above, that both series F and f have the same coefficients $\{c_i\}$. For example if

$$F(p) = f'(p) = \sum_{i=o}^{\infty} i\, c_i\, p^{i-1},$$

then $(m/n)_{f'}(t) = \frac{d}{dt}(m/n)_f(t)$ if the approximants of f' are constructed with the same c_i's and with $h_i(p) = i\, p^{i-1}$. The result is no longer true if the coefficients of f' are $\{i\, c_i\}$ and $h_i(p) = p^{i-1}$.

If v is the orthogonal polynomial of degree k with respect to c (or with respect to an adjacent functional) then the preceding result holds for Padé approximants since these orthogonal polynomials only depend on the functional c. Thus we have

$$[m/n]_F(p) = L[m/n]_f(t).$$

An obvious consequence of theorem 1 is that if we write

$$(k-1/k)_f(t) = \sum_{i=o}^{\infty} e_i\, g_i(t)$$

with $e_i = c_i$ for $i = 0, \ldots, m-1 \leq 2k-1$ then

$$(k-1/k)_F(p) = \sum_{i=o}^{\infty} e_i h_i(p).$$

When $g_i(t) = t^i$ this is the definition used by van Rossum [2] for Padé-approximants of an arbitrary series of functions, the practical problem being the summation of the defining series of (k-1/k).

Finally the inversion of the linear transform L can be performed in the same way. Starting from the expansion of F into a series of functions, replacing it by a PTA and inverting, gives an approximation of f. Such a method will be exemplified in the next section.

Let us now give an expression of the error for Padé-type approximants of any series of functions.

When $g_i(t) = t^i$ then

$$P(x, t) = \frac{1}{1-xt}\left(1 - t^k \frac{v(x)}{\tilde{v}(t)}\right)$$

with $\tilde{v}(t) = t^k v(t^{-1})$. Let L be defined by $L\, t^i = h_i(p)$, $i = 0, 1, \ldots$.

Applying L we get

$$Q(x, p) = H(x, p) - v(x)\, L \frac{t^k}{(1-xt)\tilde{v}(t)} .$$

Now, applying c, we obtain

$$F(p)-(k-1/k)_F(p) = c\left(v(x) L \frac{t^k}{(1-xt)\tilde{v}(t)}\right) = L\left(\frac{t^k}{\tilde{v}(t)}\, c\left(\frac{v(x)}{1-xt}\right)\right).$$

2. Padé-type approximants as partial sums of an interpolation series

Let f be analytic in a closed set C, and let $\Gamma=\partial C$ be the union of a finite number of non intersecting Jordan curves. Let $\{\beta_n\}$ be a sequence of points interior to C and let $\{\alpha_n\}$ be an arbitrary sequence of points distinct from the β's. We consider the interpolation series whose partial sums are [7, p. 188]

$$S_N(z) = \sum_{n=o}^{N} a_n \frac{(z-\beta_1)\ldots(z-\beta_n)}{(z-\alpha_1)\ldots(z-\alpha_n)},\quad a_n = \frac{1}{2i\pi}\int_\Gamma \frac{(\beta_{n+1}-\alpha_n)(t-\alpha_1)\ldots(t-\alpha_{n-1})}{(t-\beta_1)\ldots(t-\beta_{n+1})} f(t)\,dt,$$

$$a_o = f(\beta_1).$$

S_N interpolates f at the points β_i, $i = 1, \ldots, N+1$, in the sense of Hermite, and S_N is a (N/N+1) rational function. With the usual convention of omitting the term $(t-\alpha_i)$ or $(t-\beta_i)$ when α_i or β_i is infinite, S_N is the multipoint Padé-type approximant of f whose denominator is $(z-\alpha_1)\ldots(z-\alpha_N)$. In the particular case where all the β_i's coincide, S_N is the classical PTA or even the classical Padé approximant if the α_i's are conveniently chosen. For this reason the unknown function f will be considered in the sequel as expanded in a series of functions ϕ_n as in S_N above or corresponding to that type by a linear functional transformation.

3. Convergence

The convergence of the PTA of F can be studied, in some cases, from that of the PTA of f and conversely. Results in that direction have already been obtained by Van Rossum [5]. We shall now give some more results of that sort.

The first case we study is that of a general functional L between two spaces H and $\bar{H}$, where $H \subset \mathbb{C}^X$ is an Hilbert space with a complete orthonormal (CON) sequence (x_i^*) and a reproducing kernel $g : g(t, u) = \sum_i x_i^*(t)\, \overline{x_i^*(u)}$. We recall [3] :

Theorem 2 : $\forall f \in H$, the Fourier expansion $\sum_i (f, x_i^*)\, x_i^*$ of f converges to f on X. If every function in H is continuous, the convergence is uniform on every compact subset of X.

Let L be a linear functional : $H \to \bar{H}$, let δ_p be the Dirac function at the point p, let $y_i^* = Lx_i^*$ and let S_N be the partial sum of the Fourier expansion of f. The following result holds

Theorem 3 : $\forall f \in H$ we set $F = Lf$ and $T_N = LS_N$. Then $T_N = \sum_{i=0}^{N} (f, x_i^*) y_i^*$. The sequence $\{T_N\}$ converges to F for all p such that the norm of the linear operator $\delta_p \circ L$ is bounded. The convergence is uniform on every subset K where the norm of $\delta_p \circ L$ is uniformly bounded.

Proof :

$$(F-T_N)(p) = L(f-S_N)(t) = \delta_p \circ L(f-S_N).$$

Thus $|(F-T_N)(p)| \le ||\delta_p \circ L|| \cdot ||f-S_N||_H$ and the result follows. ■

Let us now consider the important case where L is an integral operator

$$Lf(p) = F(p) = \int_a^b K(p, t) f(t) dt, \quad H \subset L^2[a, b].$$

Our notations are as follows :

$\{w^{1/2}(t)\phi_n(t)\}_n$ is a complete orthonormal (CON) sequence in $L^2[a,b]$ with $\forall t \in]a, b[$, $w(t) > 0$.

$$f(t) = \sum_{n=0}^{\infty} b_n \phi_n(t) w^{1/2}(t) \quad \text{with } b_n = \int_a^b f(t)\phi_n(t) w^{1/2}(t) dt$$

$$F(p) = \sum_{n=0}^{\infty} a_n \Phi_n(p) \quad \text{with } \Phi_n(p) = \int_a^b K(p, t)\, \phi_n(t) w^{1/2}(t) dt.$$

The partial sums of the series F are the PTA of F.

The equalities between functions and series expansions are to be considered as formal. The nature of the convergence will be made more precise.

Theorem 4 : We assume that $f \in L^2[a, b]$ and that $\forall p$, $K(p, .) \in L^2[a,b]$. If $F(p) = \sum_{n=0}^{\infty} a_n \Phi_n(p)$ (pointwise convergence) then $a_n = b_n$ $(n \ge 0)$.

Proof :

$$F(p)-\sum_{n=o}^{N} b_n\Phi_n(p)=\int_a^b K(p,t)f(t)dt-\sum_{n=o}^{N}\int_a^b f(t)\phi_n(t)w^{1/2}(t)\Phi_n(p)dt.$$

$$|F(p)-\sum_{n=o}^{N} b_n\Phi_n(p)|^2\le\left|\int_a^b f^2(t)dt\right|\int_a^b|K(p,t)-\sum_{n=o}^{N}\phi_n(t)w^{1/2}(t)\Phi_n(p)|^2dt.$$

But $K(p,.)\in L^2[a,b]$ and $\{w^{1/2}\phi_n\}$ is a CON sequence. Thus

$$K(p,t)=\sum_{n=o}^{\infty}\alpha_n\phi_n(t)w^{1/2}(t) \text{ with } \alpha_n=\int_a^b K(p,t)\phi_n(t)w^{1/2}(t)dt=\Phi_n(p),$$

the last convergence being in the mean. The second integral in the right hand side of the inequality tends to zero when N goes to infinity and $\forall$ n, $a_n = b_n$.■

Corollary : Under the assumptions of theorem 4, the series expansion of f converges to f in the mean.

We must notice that the partial sums of F are the classical PTA and thus pointwise convergence results are known.

Theorem 5 : If the series of f converges to f in the mean, then the series of F converges to F, $\forall$ p such that $K(p,.)\in L^2[a,b]$. The convergence is uniform on every subset D such that $\sup_{p\in D}||K(p,.)||_{L^2}$ is bounded.

Proof :

$$F(p)-\sum_{n=o}^{N} a_n\Phi_n(p)=\int_a^b K(p,t)f(t)dt-\sum_{n=o}^{N} a_n\int_a^b K(p,t)\phi_n(t)w^{1/2}(t)dt.$$

$$|F(p)-\sum_{n=o}^{N} a_n\Phi_n(p)|^2\le\int_a^b|K(p,t)|^2dt\int_a^b|f(t)-\sum_{n=o}^{N} a_n\phi_n(t)w^{1/2}(t)|^2dt$$

and the result follows.■

Finally there is equivalence between pointwise convergence of the series F and convergence in the mean of the series f when f and K(p,.) are square integrable. If $H\subset L^2[a,b]$ has a reproducing kernel and if f is continuous, the convergence of the series f is uniform on compact subsets. The key assumption is that $f\in L^2[a,b]$ or, equivalently, that Σa_n^2 converges.

Let us now consider the problem of inverting L that is F is known, the unkown being f. We have to make the assumption that Σa_n^2 converges or any equivalent assumption easier to check. We set

$$\bar{F}(z)=\sum_{n=o}^{\infty} a_n z^n \quad , \quad F(p)=\sum_{n=o}^{\infty} a_n\Phi_n(p)$$

For example the assumption $\sum_{n=o}^{\infty}|a_n|<+\infty$ implies $\sum_{n=o}^{\infty} a_n^2<+\infty$. In some cases this assumption may be deduced from hypothesis on the function F.

Let us now look at a particular case where results on Σa_n have been proved. We set

$$\Phi_n(p)=\frac{(p-\lambda)^n}{(p+\lambda)^{n+1}}\sqrt{2\lambda},$$ then $T_N(F)$ is a classical PTA with a(N+1)th order pole at $p=-\lambda$ and with the change of variable $z=(p-\lambda)/(p+\lambda)$ we get $\sqrt{2\lambda}\,\bar{F}(z)=(p+\lambda)F(p)$.

The following result can be deduced [2]:

Theorem 6 : If $(p+\lambda)F(p)$ is analytic at infinity, the series of $\bar{F}$ converges for $|z| < R$ and $R > 1$.

In that case $\sum |a_n|$ converges which can be false if $R = 1$ even if the convergence of the series $\bar{F}$ is uniform on the closed unit disc.

The convergence of F is uniform on every compact subset of Ω where $\Omega = \chi(D_R)$, D_R being the disc of radius R and χ being defined by $\chi(p) = (p-\lambda)/(p+\lambda)$.

If L is the Laplace transform, a direct proof [2] exists for

Theorem 7 : If $(p+\lambda)F(p)$ is analytic at infinity, the series f converges uniformly on compact subsets of $[0, \infty)$.

4. Examples

To end the paper let us give two examples about the Laplace transform. In the first one Laguerre's polynomials are used.

Example 1 : We set

$$\phi_n(t)\, w^{1/2}(t) = \sqrt{2\lambda}\, e^{-\lambda t}\, L_n(2\lambda t).$$

$(\phi_n w^{1/2})$ is a CON sequence in $L^2[0, \infty)$. We have

$$\Phi_n(p)=\sqrt{2\lambda}\int_0^\infty e^{-pt}e^{-\lambda t}L_n(2\lambda t)\,dt=\sqrt{2\lambda}\sum_{k=0}^{n}\binom{n}{k}\frac{(-2\lambda)^k}{k!}\int_0^\infty e^{-(p+\lambda)t}t^k dt$$

$$=\sqrt{2\lambda}\sum_{k=0}^{n}\binom{n}{k}\frac{(-2\lambda)^k}{(p+\lambda)^{k+1}} = \sqrt{2\lambda}\,\frac{(p-\lambda)^n}{(p+\lambda)^{n+1}}.$$

We just have to check that a_n, defined by the series expansion of f, and $\bar{a}_n$ defined by $\sum_{n=0}^{N} \bar{a}_n\Phi_n(p) = (N/N+1)_F(p)$, are the same. We get

$$F(p)-\sum_{n=0}^{N}\bar{a}_n\sqrt{2\lambda}\frac{(p-\lambda)^n}{(p+\lambda)^{n+1}}=O((p-\lambda)^{N+1}) \quad \text{thus } \bar{a}_n\sqrt{2\lambda}=\frac{1}{n!}D^n(F(p)(p+\lambda)^{n+1})_{p=\lambda}.$$

$$D^n(F(p)(p+\lambda)^{n+1})_{p=\lambda}=\sum_{k=0}^{n}\binom{n}{k}D^kF(p)_{p=\lambda}\,D^{n-k}((p+\lambda)^{n+1})_{p=\lambda}$$

$$= \sum_{k=o}^{n} \binom{n}{k} \frac{n!}{k!} (2\lambda)^{k+1} \int_o^\infty e^{-\lambda t} f(t)(-t)^k dt.$$

$$\bar{a}_n = \sqrt{2\lambda} \int_o^\infty e^{-\lambda t} f(t)\ L_n(2\lambda t)dt = a_n.$$

The method can be numerically improved by a process which in fact constructs generalized PTA [4].

Example 2 : We shall now use Legendre's polynomials for inverting the Laplace transform [6]. Let us remark that the Legendre's polynomials P_k are orthogonals on $[0, \infty)$ with a change of variable, and that their Laplace transform are very particular rational functions :

$$P_n(x) = \sum_{k=o}^{n} \frac{(n+k)!}{(n-k)!(k!)^2} (-1)^k e^{-kt}, \quad x = 1-2e^{-t},$$

$$\Phi_n(p) = \int_o^\infty e^{-pt} e^{-t} P_n(1-2e^{-t})dt = \sum_{k=o}^{n} \frac{(-1)^k (n+k)!}{(n-k)!(k!)^2} \frac{1}{p+k+1} = \frac{p(p-1)\dots(p-n+1)}{(p+1)\dots(p+n+1)}.$$

As in section 3, we set

$$(N/N+1)_F(p) = \sum_{o}^{N} a_n \Phi_n(p), \quad a_n = \frac{1}{2i\pi} \int_\Gamma \frac{(2n+1)(t+1)\dots(t+n+1)}{t(t-1)\dots(t-n+1)} F(t)dt.$$

By the residue theorem we get :

$$a_n = (2n+1) \sum_{k=o}^{n} \frac{(-1)^k (n+k)!}{(n-k)!(k!)^2} F(k+1) = (2n+1) \sum_{k=o}^{n} \frac{(-1)^k (n+k)!}{(n-k)!(k!)^2} \int_o^\infty e^{-(k+1)t} f(t)dt$$

$$a_n = (2n+1) \int_o^\infty f(t)\ e^{-t}\ P_n(1-2e^{-t})\ dt$$

which is exactly the coefficient of the Fourier expansion of f :

$$f(t) = e^{-t} \sum_{k=o}^{\infty} a_n P_n (1-2e^{-t})$$

and

$$S_N(t) = e^{-t} \sum_{o}^{N} a_n P_n (1-2e^{-t}).$$

References

1. Brezinski, C., Padé type approximation and general orthogonal polynomials, Birkhäuser Verlag, Basel 1980.

2. Brochet, P., Contribution à l'inversion numérique de la transformée

de Laplace ..., Thèse 3e cycle, Université de Lille I, 1983.

3. Duc-Jacquet, M., Espaces hilbertiens à noyaux reproduisants, Lecture Notes, Université de Grenoble, 1979.

4. Iseghem, J. van, Applications des approximants de type Padé, Thèse 3e cycle, Université de Lille I, 1983.

5. Rossum, H. van, Generalized Padé approximants, in "Approximation Theory III", E.W. Cheney ed., Academic Press, New-York, 1980.

6. Sneddon, I.N., The use of integral transforms, Tata Mc Graw-Hill, New Delhi, 1974.

7. Walsh, J.L., Interpolation and approximation by rational functions in the complex domain, Amer. Math. Soc. Colloqium Publ. XX, Providence, 1969.

CONTINUED FRACTION SOLUTION OF THE GENERAL RICCATI EQUATION

J. S. R. Chisholm
Mathematical Institute
University of Kent
Canterbury, Kent
ENGLAND

Abstract The general Riccati equation is reduced to the standard form $z'(x) = b_0(x) - z^2(x)$. Successive iterations of a continued fraction solution of this equation are given in terms of a sequence $\{b_r(x);\ r=1,2,\ldots\}$ of functions which replace $b_0(x)$ in the standard form, and which are defined in terms of $b_0(x)$ and its derivatives.

1. Introduction

In 1933, McVittie [1] formulated a problem in general relativity in terms of an equation of the general form (primes denoting differentiation with respect to x)

$$y'' + (E_0+E_1y)y' + F_0 + F_1y + F_2y^2 + F_3y^3 = 0 , \qquad (1.1)$$

where $\{E_r;\ r=1,2\}$ and $\{F_r;\ r=1,\ldots,4\}$ were constants expressible in terms of two constants of integration a,b. He looked for solutions of (1.1) which were also solutions of a Riccati equation

$$y' = a_0 + a_1y + a_2y^2. \qquad (1.2)$$

This investigation has lasted for half a century; in a recent paper [2], McVittie has given explicit one-parameter solutions of (1.1): some are "Riccati solutions", satisfying an equation of form (1.2), provided that the constants a,b satisfy one of several relations; others are "non-Riccati" solutions, related to a more complicated first integral.

A different approach to second order non-linear equations is to form Padé approximants from perturbation series solutions. This method has proved very useful for soliton equations [3,4]. However, there are classes of solutions of interesting non-linear equations [5] which are not derivable by the simple Padé method; one part of our present investigation is to discover whether these solutions can be derived as Hermite-Padé approximants of some kind.

In some simple examples, when exact solutions can be derived as Padé approximants from the perturbation series, some of these solutions are "Riccati solutions", while others are not. It is a major problem to relate three different approaches to second-order non-linear equations:

(a) The intuitive approach, in which classes of solutions are found to a considerable extent by experiment. Solutions have been found [6] to some equations outside the class (1.1).

(b) The Riccati equation approach, which is being generalised by taking $\{E_r\}$ and $\{F_r\}$ in (1.1) to be functions of x. We find that there are solutions of Riccati form provided that the six functions $\{E_r, F_r\}$ satisfy a single condition.

(c) The Padé approach, which could be generalised by using Hermite-Padé approximation. We also hope that these methods can be used to approximate solutions of equations which do not have classes of solutions in closed form.

A first step towards relating approaches (b) and (c) is to find Padé-type solutions of Riccati equations (1.2). It is known [7,8,9,10,11] that certain classes of Riccati equations have solutions of continued fraction form, when the coefficients $\{a_r;\ r=0,1,2\}$ in (1.2) are polynomials in x. The method of finding a solution given in the next section is valid for general sets of functions $\{a_r(x)\}$.

2. The Continued Fraction Solution

First, we reduce the equation

$$y' = a_0(x) + a_1(x)y + a_2(x)y^2 \tag{2.1}$$

to standard form by changes of unknown. Defining

$$v(x) = -a_2(x)\, y(x) \, , \tag{2.2}$$

(2.1) becomes

$$v' = A_0 + A_1 v - v^2 , \qquad (2.3)$$

where

$$A_0 = -a_0 a_2, \qquad A_1 = a_1 + (a_2'/a_2). \qquad (2.4)$$

The linear term in (2.3) can be eliminated by putting

$$z = v - \tfrac{1}{2} A_1$$

$$= - a_2 y - \tfrac{1}{2} a_1 - (a_2'/2a_2) . \qquad (2.5)$$

Then if

$$b_0(x) = A_0 + \tfrac{1}{4} A_1{}^2 - \tfrac{1}{2} A_1' , \qquad (2.6)$$

(2.3) is equivalent to tne "standard form"

$$z' = b_0(x) - z^2. \qquad (2.7)$$

First we study a restricted form of continued fraction iteration, which is later generalised. The iteration is started by replacing the unknown $z(x)$ by a function $u(x)$, where

$$z(x) = \frac{\beta(x)}{1 + u(x)} , \qquad (2.8)$$

and $\beta(x)$ is a function to be chosen. Differentiating (2.8) gives

$$z' = \frac{\beta'(1+u) - \beta u'}{(1+u)^2} . \qquad (2.9)$$

Substituting from (2.8) and (2.9) into (2.7) gives

$$\beta'(1+u) - \beta u' = b_0(1+u)^2 - \beta^2 ;$$

multiplying by $\beta^{-2} b_0$ and re-arranging terms, this can be written

$$\left(\frac{b_0 u}{\beta}\right)' = \frac{b_0}{\beta^2} (\beta' + \beta^2 - b_0) + \frac{b_0 u}{\beta} \left(\frac{b_0'}{b_0} - \frac{2b_0}{\beta}\right) - \left(\frac{b_0 u}{\beta}\right)^2 . \qquad (2.10)$$

We are still free to choose $\beta(x)$. The simplest choice is

$$\beta = \beta_0 \equiv 2b_0^2/b_0' , \tag{2.11}$$

which makes the linear term in (2.10) vanish. This choice is invalid if b_0 is constant.

Then if we define

$$w = b_0 u/\beta = b_0' u/2b_0 , \tag{2.12}$$

(2.10) becomes

$$w' = b_1(x) - w^2 , \tag{2.13}$$

where

$$b_1 = b_0(\beta_0' + \beta_0^2 - b_0)/\beta_0^2 ;$$

using (2.11), this gives

$$b_1 = (4b_0^4 + 3b_0 b_0'^2 - 2b_0^2 b_0'')/4b_0^3 . \tag{2.14}$$

Substituting from (2.11) and (2.12) into (2.8) gives

$$z = \frac{2b_0^2}{b_0' + 2b_0 w} . \tag{2.15}$$

Since (2.13) is of the same form as (2.7), (2.15) can be regarded as the first iteration in developing a continued fraction solution of (2.7). Thus, if we define iteratively

$$\beta_r = 2b_r^2/b_r' \tag{2.16}$$

and

$$b_{r+1} = b_r(\beta_r' + \beta_r^2 - b_r)/\alpha_r^2 \tag{2.17}$$

for $r=1,2,3,\ldots$, then the relation (2.15) can be iterated to give the continued fraction solution

$$z = \frac{2b_0^2}{b_0'+}\ \frac{4b_0 b_1^2}{b_1'+}\ \frac{4b_1 b_2^2}{b_2'+} \cdots . \tag{2.18}$$

This solution will fail at the rth iteration if b_r is constant. The whole question of the validity and convergence of (2.18) remains to be investigated.

There is no constant of integration in (2.18). However, it appears that in (2.8) we have an arbitrary *function* $\beta(x)$. If we keep $\beta(x)$ arbitrary, we find that both (2.14) and (2.15) are unchanged. Hence the solution (2.18) is independent of the choice of $\beta(x)$, and contains no arbitrary constant.

In order to try to introduce an arbitrary constant into the solution of (2.7), the transformation (2.8) is generalised to

$$z(x) = \alpha + \frac{\beta(x)}{\gamma(x) + w(x)}, \tag{2.19}$$

where the constant α and functions $\beta(x)$, $\gamma(x)$ are to be chosen; then

$$z' = \frac{\beta'(\gamma+w) - \beta(\gamma'+w')}{(w+\gamma)^2}. \tag{2.20}$$

Substituting into (2.7) gives

$$\begin{aligned} \beta w' &= [\beta'\gamma-\beta\gamma'+(\alpha\gamma+\beta)^2-b_0\gamma^2] \\ &\quad + w[\beta'+2\alpha(\alpha\gamma+\beta)-2b_0\gamma] \\ &\quad + w^2[\alpha^2-b_0]. \end{aligned} \tag{2.21}$$

In order to reduce this equation to the form (2.7), we choose

$$\beta' + 2\alpha(\alpha\gamma+\beta) - 2b_0\gamma = 0$$

or

$$\gamma = \frac{\beta' + 2\alpha\beta}{2(b_0-\alpha^2)}, \tag{2.22}$$

and

$$\beta = b_0 - \alpha^2. \tag{2.23}$$

With this choice,

$$\gamma = \alpha + \tfrac{1}{2}\, b_0'(b_0^2-\alpha^2)^{-1} = \alpha + \tfrac{1}{2}\,\beta^{-1}\beta'. \tag{2.24}$$

Equation (2.21) then becomes

$$w' = b_1(x) - w^2 , \tag{2.25}$$

where

$$b_1 = \beta^{-1}[\beta'\gamma - \beta\gamma' + (\alpha\gamma+\beta)^2 - b_0\gamma^2]$$

$$= \beta^{-1}\beta'\gamma - \gamma' - \gamma^2 + 2\alpha\gamma + \beta.$$

Using (2.23) and (2.24), this gives

$$b_1 = b_0 + \frac{\alpha b_0'}{b_0-\alpha^2} + \frac{3b_0'^2}{4(b_0-\alpha^2)^2} - \frac{b_0''}{2(b_0-\alpha^2)} ; \tag{2.26}$$

when $\alpha=0$, this reduces to (2.14).

As before, we define $\{b_r;\ r=1,2,\ldots\}$ by generalising (2.26):

$$b_{r+1} = b_r + \frac{\alpha b_r'}{b_r-\alpha^2} + \frac{3b_r'^2}{4(b_r-\alpha^2)^2} - \frac{b_r''}{2(b_0-\alpha^2)} . \tag{2.27}$$

Then using the generalisations of (2.23) and (2.24), the iteration of (2.19) gives

$$z = \alpha + \frac{b_0-\alpha^2}{\frac{1}{2}b_0'(b_0-\alpha^2)^{-1}+\alpha+w}$$

$$= \alpha + \frac{b_0-\alpha^2}{\frac{1}{2}b_0'(b_0-\alpha^2)^{-1}+2\alpha+}\ \frac{b_1-\alpha^2}{\frac{1}{2}b_1'(b_1-\alpha^2)+2\alpha+}\ \cdots . \tag{2.28}$$

This generalisation of (2.17) and (2.18) appears to contain the arbitrary constant α. Certainly successive approximants to (2.28) vary with α. What has not been proved is that, when the continued fraction converges, its limit depends upon α. That this is not obvious can be seen by examining the solution when the function b_0 is constant, so that $b_0'=b_0''=0$. Then (2.26) and (2.27) give

$$b_r = b_0 \quad (r=1,2,\ldots) ,$$

and (2.28) becomes

$$z = \alpha + \frac{b_0-\alpha^2}{2\alpha+}\ \frac{b_0-\alpha^2}{2\alpha+}\ \cdots . \tag{2.29}$$

Thus z satisfies

$$z = \alpha + \frac{b_0 - \alpha^2}{\alpha + z}$$

or

$$z^2 = b_0 , \tag{2.30}$$

which gives solutions of (2.7) which are independent of α. It may be, however, that the solution (2.28) is α-dependent in general.

I have not yet investigated the convergence properties of the continued fraction (2.28). This and the question of α-dependence are problems which need to be studied.

3. Acknowledgments

In 1972, Ellis [12] discovered a continued fraction solution to the general Riccati equation; his definition of the iterative process involved successive integrations, rather than successive differentiations as in (2.27). The two solutions are therefore of very different form. I wish to thank Professor Homer Ellis for a discussion, and I am also grateful to Professor Arne Magnus, Professor Marcel de Bruin, Dr. Arieh Iserles and Dr. Peter Graves-Morris for their comments. My chief indebtedness is to Professor Philip Burt and Professor George McVittie for many discussions and communications in an ongoing collaboration which motivated this particular piece of work. I also thank Mrs. Sandra Bateman for her careful preparation of the paper, and those who have arranged this conference for their excellent work. Finally, I am pleased to acknowledge the assistance of a N.A.T.O. grant in support of this work.

References

1. G. C. McVittie, "The Mass-particle in an Expanding Universe", Mon.Not.Roy.Ast.Soc. 93, 325 (1933).

2. G. C. McVittie, "Elliptic Functions in Spherically Symmetric Solutions of Einstein's Equations", Ann.Inst.Henri Poincaré 40, 3, 231 (1984).

3. C. Liverani and G. Turchetti, "Existence and Asymptotic Behaviour of Padé Approximants to Korteweg-de-Vries Multisoliton Solutions", J.Math.Phys. 24, 1, 53 (1983).

4. F. Lambert and Musette, "Solitons from a Direct Point of View II", preprint VUB/TF/83/06, Vrije Universiteit Brussel (1983).

5. P. B. Burt, "Quantum Mechanics and Non-linear Waves" (Harwood Academic, 1982).

6. P. B. Burt, ibid., pp. 111-113.

7. E. Laguerre, "Sur la réduction en fractions continues d'une fraction qui satisfait à un equation differentielle linéare du premier ordre dont les coefficients sont rationels", J. de Math, Pures et Appliqués 1, 135-165 (1885).

8. E. P. Merkes and W. T. Scott, "Continued Fraction Solutions of the Riccati Equation", J.Math.Analysis and Applic. 4, 309 (1962).

9. Wyman Fair, "Padé Approximation to the solution of the Riccati Equation", Math. Comp. 18, 627 (1964).

10. G. A. Baker, Jr. and P. R. Graves-Morris, "Padé Approximants", part II, pp. 162-165 (Addison-Wesley, 1981).

11. A. N. Stokes, "Continued Fraction Solutions of the Riccati Equation", Bull. Austral. Math. Soc. 25, 207 (1982).

12. H. G. Ellis, "Continued Fraction Solutions of the General Riccati Differential Equation", Rocky Mountain J. Math. 4, 2, 353 (1974).

ORDER STARS, CONTRACTIVITY AND A PICK-TYPE THEOREM

Arieh Iserles
King's College
University of Cambridge
Cambridge CB4 1LE
England

Abstract. Given a function f that is analytic in the complex domain V and such that $|f|\equiv 1$ along ∂V (with the possible exception of essential singularities) we examine analytic approximations R to f that are contractions in $\mathrm{cl}V$. By applying the theory of order stars we demonstrate that the nature of essential singularities and zeros of f imposes surprisingly severe upper bounds on the degree of interpolation by a contractive approximation R. It is proved that, subject to V being conformal to the unit disk, contractive interpolations that satisfy the given bounds are attained by rational functions. Finally, we apply our theory to prove a version of the classical Pick theorem that is valid in every complex domain.

1. Introduction

Let V be a complex domain whose boundary ∂V is composed out of a finite number of Jordan curves and f a complex function such that a) $f\not\equiv$ constant is analytic in a complex domain U, $\mathrm{cl}V\subset U$, with the possible exception of a finite number of essential singularities along ∂V; b) $|f|\equiv 1$ along ∂V away from possible essential singularities and c) if $\tilde{z}\in\partial V$ is an essential singularity of f then for every sufficiently small $\varepsilon>0$ it is true that $|f(z)|<1$ for every z in $V\cap\{|z-\tilde{z}|<\varepsilon\}$. It follows that $|f(z)|<1$ for every $z\in V$. The function f is interpolated in $\mathrm{cl}V$ by an analytic function R, $R\not\equiv f$. It is often important that the interpolation R preserves the property that $|f(z)|<1$ in V. Moreover, one is interested in attaining the best degree of interpolation by a function R from a certain class, e.g. a rational function with a fixed number of zeros and poles. It comes as no surprise that the qualitative (contractivity) and quantitative (degree of interpolation) requirements may be in conflict. Hence it makes sense to ask how well can we interpolate by a contraction. The present paper gives a partial answer to this question.

Important examples of the given framework occur when the stability of certain methods for the numerical solution of differential equations is investigated by approximation-theoretic methods. For example, A-stability of one-step (and certain multi-step) methods for the numerical solution of ordinary differential equations is equivalent to $|R(z)|<1$, $z\in\mathbb{C}$, $\mathrm{Re}\, z<0$, where R is a rational function that depends on the numerical method, and interpolates the exponential [6]. In this case $f(z)=\exp(z)$ and $V=\{z\in\mathbb{C} : \mathrm{Re}\, z<0\}$.

The theory of order stars, initiated by Wanner, Hairer and Nørsett [10], has been applied with great effect to this and similar problems [3,4,5,7]. In Section 2 we sketch elements of this theory that are relevant to the present work. They are applied in Section 3 to derive upper bounds on the degree of interpolation by contractions. In Section 4 we demonstrate that in the important case of V being smoothly conformal to the complex unit disk some upper bounds of Section 3 can actually be attained by mappings of rational functions. Finally, in Section 5 we use our theory to prove a generalisation of the classical Pick theorem.

2. Order stars

Let $\sigma(z):=R(z)/f(z)$, $z\in \mathrm{cl}V$. We set

$A:=\{z\in \mathrm{cl}V : |\sigma(z)|>1\}$,

$D:=\{z\in \mathrm{cl}V : |\sigma(z)|<1\}$,

and call the pair {A,D} the order star of (f,R). The following four propositions follow by standard methods of complex analysis [1,10]:

Proposition 1: If

a) $z_o\in \mathrm{cl}V$;

b) $f(z)=d(z-z_o)^k+O(|z-z_o|^{k+1})$, $d\neq 0$ (z_o is a zero of f of multiplicity $k\geq 0$); and

c) $R(z)=f(z)+c(z-z_o)^p+O(|z-z_o|^{p+1})$, $c\neq 0$ (z_o is an interpolation point of multiplicity p), where $p\geq\max\{1, k+1\}$

then $z_o\in\partial A$ and it is adjoined by exactly p-k sectors of A, separated by p-k sectors of D, each of the angle $\pi/(p-k)$.

Proof: The desired result is a consequence of

$$\sigma(z) = 1 + \frac{c}{d}(z-z_o)^{p-k} + O(|z-z_o|^{p-k+1}). \qquad \square$$

We say that the function R is a V-contraction if it is analytic in $c\ell V$ and $|R(z)| \leq 1$ there.

Proposition 2: R is a V-contraction if and only if it is analytic in $c\ell V$ and $A \cap c\ell V = \phi$.

Proof: Follows from the definition of the order star by the maximal modulus principle. $\square$

We call the connected components of A A-regions. An A-region is said to be of multiplicity m if its directed boundary passes through exactly m interpolation points (i.e. zeros of the equation $R(z)=f(z)$), which need not be distinct. Similar terminology is used for connected components of D.

Proposition 3: If no essential singularities are present on a boundary of an A-region (D-region) then its multiplicity m equals the number of poles (zeros) of σ inside the region. Further, $m \geq 1$.

Proof: It follows from the Cauchy-Riemann equations in polar coordinates that $\arg \sigma$ decreases strictly monotonically along the positively oriented boundary of an A-region. The proposition now follows from the argument principle. $\square$

It is a straightforward consequence of the Picard theorem that every essential singularity of f along ∂V lies on ∂A.

Proposition 4: If $\tilde{z} \in \partial V$ is an essential singularity of f then it may be approached by at most one sector of A from within V.

Proof: Follows by asymptotic analysis, because $|f(z)|<1$ inside V. $\square$

Figure 1 depicts an approximation

$$R(z) = -\frac{1}{4}(\sqrt{6}+\sqrt{2})\,\frac{z-\frac{1}{2}(\sqrt{6}-\sqrt{2})}{1-\frac{1}{2}(\sqrt{6}-\sqrt{2})z}$$

to

$$f(z) = \left(\frac{z-\frac{\sqrt{2}}{2}}{1-\frac{\sqrt{2}}{2}z}\right)^2 .$$

Hence V is the open complex unit disk. There is an interpolation point of degree 2 at the origin and it can be deduced at once from Proposition 3 that there are no additional interpolation points in $c\ell V$. Note that $|R(z)| \leq \frac{1}{4}(\sqrt{6}+\sqrt{2}) \approx 0.965925826$ in $c\ell V$ and R is indeed a V-contraction.

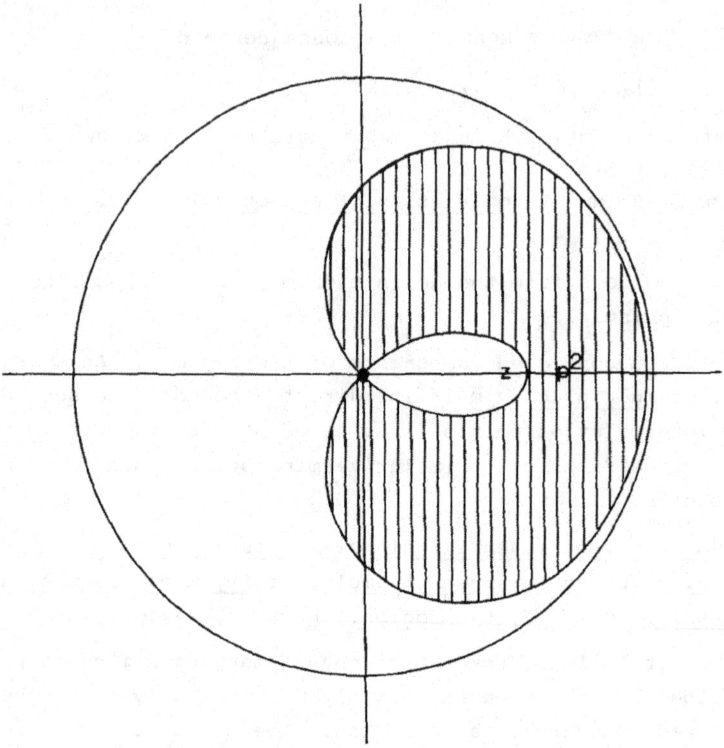

Figure 1: A zero of σ is denoted by "z", whilst "p^2" denotes a double pole of that function.

3. Upper bounds on the degree of interpolation

We say that $\tilde{z}\in c\ell V$ is an interpolation point of degree p if $R(z) = f(z) + c(z-\tilde{z})^p + O(|z-\tilde{z}|^{p+1}),\ c \neq 0$.

Theorem 1: Let f be analytic in a complex domain U, $c\ell V \subset U$, with L zeros inside V and set

$$\alpha(\tilde{z}) := \begin{cases} 2\pi & : \tilde{z}\in V \\ \text{the (inner) angle of } \partial V \text{ at } \tilde{z}. & : \tilde{z}\in \partial V. \end{cases}$$

Then, subject to V-contractivity, the degree of interpolation p at $\tilde{z}$ may not exceed $(2L+1)\pi/\alpha(\tilde{z})$. Note that unless f is a constant, $L \geq 1$.

Proof: By Proposition 2 and V-contractivity no A-region may cross ∂V. Moreover, analyticity of R implies that σ has exactly L poles in V. Hence, by Proposition 3, the sum of multiplicities of all the

A-regions in V may not exceed L-k, where $k \geq 0$ is the multiplicity of a (possible) zero of f at $\tilde{z}$.

If $1 \leq p \leq k \leq L$ then $\tilde{z} \in V$, $\alpha(\tilde{z}) = 2\pi$ and, trivially, $p \leq (2L+1)\pi/\alpha(\tilde{z}) = L + \frac{1}{2}$. Otherwise $p \geq k+1$ and Proposition 1 is applicable. Hence $\tilde{z}$ is approached by p-k equi-angular sectors of A and p-k such sectors of D. If $\tilde{z} \in V$ then, since the number of sectors of A that adjoin $\tilde{z}$ may not exceed the sum of multiplicities of all the A-regions in V, it follows that $L-k \geq p-k$, hence $p \leq L$. If $\tilde{z} \in \partial V$ then k=0 and, given that q sectors of A approach $\tilde{z}$ inside V, equi-angularity gives

$$\frac{2q-1}{p}\pi \leq \alpha(\tilde{z}) \leq \frac{2q+1}{p}\pi .$$

Hence $p \leq (2q+1)\pi/\alpha(\tilde{z})$. The proof follows since, by Proposition 3, $q \leq L$. □

Similar arguments can be applied in a more general framework [2], yielding the following upper bounds for V-contractive interpolations:

Theorem 2: _If f is analytic in_ $c\ell V$ _then the sum of degrees of interpolation within_ V _may not exceed_ L.

Theorem 3: _If f possesses_ M _essential singularities along_ ∂V _then the degree of interpolation at_ $\tilde{z} \in c\ell V$ _may not exceed_ $(2(L+M+m)+1)\pi/\alpha(\tilde{z})$ _and the sum of degrees of interpolation within_ V _may not exceed_ L+M+m, _where_ m _is the number of zeros of_ R _in_ $c\ell V$.

In [10] Wanner, Hairer and Nørsett proved the first Ehle conjecture: the [m/n] Padé approximation to exp(z) is A-acceptable only if $m \leq n \leq m+2$. This result is now a straightforward consequence of Theorem 3: given $f(z) = \exp(z)$, $V = \{z \in \mathbb{C} : \mathrm{Re}\, z < 0\}$, with a single essential singularity at $\infty \in \partial V$, we maximise the degree of interpolation at the origin by a rational function from $\pi_{m/n}$. Hence M=1, L=0, $\alpha(0) = \pi$ and the upper bound of the last theorem yields $n \leq m+2$. Therefore the Wanner-Hairer-Nørsett theorem is true, $m \leq n$ being trivial.

It should be pointed out that it is possible to derive Theorem 2 (as well as Theorem 5 in Section 5) without order stars [9]: for example, let both f and R be analytic in V and continuous in $c\ell V$, $|f| \equiv 1$ on ∂V and $|R| < 1$ in V. Then, if f has L zeros in V, it follows by the Rouché theorem that f-R also has L zeros there. Hence, the number of interpolation points is bounded by L. However, this proof breaks down if either interpolation points or essential singularities are allowed along ∂V.

4. Satisfaction of the upper bounds

Let $V \neq ¢$ be simply connected with analytic boundary. Then it is possible to verify whether some of the upper bounds of Section 3 are attainable. This is done by using the Pick theorem [8]:

Theorem: Let $W = \{z \varepsilon ¢ : |z| < 1\}$. Given $z_o, \ldots, z_p \varepsilon W$ and complex numbers $\omega_o, \ldots, \omega_p$ there exists a unique q-tuple $(\kappa_1, \kappa_2, \ldots, \kappa_q)$, $q \leq p$, of numbers in W and a $\lambda \varepsilon ¢$ such that the scaled Blaschke product

$$g(z) = \lambda B_q(z; \kappa_1, \kappa_2, \ldots, \kappa_q) := \lambda \prod_{j=1}^{q} \frac{z - \kappa_j}{1 - \bar{\kappa}_j z}$$

satisfies $g(z_j) = \omega_j$, $0 \leq j \leq p$, and such that

$$|g(e^{i\theta})| \equiv |\lambda| < \max_{0 \leq \psi \leq 2\pi} |h(e^{i\psi})|$$

for any other function h that is analytic in W and satisfies there the interpolation conditions $h(z_j) = \omega_j$, $0 \leq j \leq p$.

Let us first assume that V=W. Then given f and simple interpolation points $z_o, \ldots, z_p \varepsilon W$, we set $\omega_j = f(z_j)$, $0 \leq j \leq p$. The following result is a straightforward consequence of the Pick theorem:

Theorem 4: If f is the Pick interpolant it cannot be interpolated by any different W-contraction. Otherwise the bound of Theorem 2 is satisfied for R that is the Pick interpolant for the underlying points $z_o, \ldots, z_p$.

Since the Pick theorem can be trivially extended to an arbitrary simply-connected domain with analytic boundary that is not the whole complex plane by a smooth conformal mapping, Theorem 4 is true within this framework. Furthermore, interpolation points of degree 2 or more can be dealt with by a limiting process (cf. Figure 1). Essential singularities along ∂V can be incorporated by using recent results of Scales [8]. Note, however, that our analysis is no longer true if interpolation points are allowed on the boundary of V.

5. An extension of the Pick theorem

The relationship between the Pick theorem and our results is much deeper than the last section may imply. Indeed, order stars lead to a Pick-type theorem in a general complex domain that need not be conformal to the unit disk:

Theorem 5: *Let* V *be an arbitrary complex domain,* $z_o,\ldots,z_n \varepsilon V$ *and* $\omega_o,\ldots,\omega_n \varepsilon ¢$. *If a function* g *is analytic in* $c\ell V$, *satisfies the interpolation conditions* $g(z_j)=\omega_j$, $0\le j\le n$, *has at most* n *zeros inside* V *and is of constant modulus along* ∂V *then*

$$||g||_{L_\infty} = \sup_{z\varepsilon V}|g(z)| < \sup_{z\varepsilon V}|h(z)| = ||h||_{L_\infty}$$

for any other analytic function h *such that* $h(z_j)=\omega_j$, $0\le j\le p$.

Proof: We set $\tilde{g} := g/||g||_{L_\infty}$, $\tilde{h} := h/||g||_{L_\infty}$, regard $\tilde{h}$ as an approximation to $\tilde{g}$ and consider the order star of $(\tilde{g},\tilde{h})$. The sum of degrees of interpolation is at least n+1, since $h(z_j) = g(z_j)$, $0\le j\le n$. Furthermore, $\tilde{g}$ has at most n zeros in V. Thus, by Theorem 2, $\tilde{h}$ cannot be a V-contraction and, indeed,

$$||h||_{L_\infty} = \max_{z\varepsilon\partial V}|h(z)| = \max_{z\varepsilon\partial V}\{||g||_{L_\infty}|\tilde{h}(z)|\} > ||g||_{L_\infty}. \qquad \square$$

Theorem 5 states nothing about the existence of such a function g for an arbitrary domain V - this is, indeed, an open problem. It is a happy coincidence that such an interpolation exists in the unit disk, in the form of a scaled Blaschke product.

The author is happy to acknowledge that the alternative proof of Theorems 2 and 5 (by using the Rouché theorem) is due to L.N. Trefethen from the Courant Institute of Mathematical Sciences.

References

1. Iserles, A., Order stars, approximations and finite differences I: the general theory of orders stars, Tech. Rep. DAMTP NA3 (1983), University of Cambridge, to appear in SIAM J. Math. Analysis.

2. Iserles, A., Order stars, approximations and finite differences II: theorems in approximation theory, Tech. Rep. NA9 (1983), University of Cambridge, to appear in SIAM J. Math. Analysis.

3. Iserles, A., Order stars, approximations and finite differences III: finite differences for $u_t=\omega u_{xx}$, Tech. Rep. NA11 (1983), University of Cambridge, to appear in SIAM J. Math. Analysis.

4. Iserles, A. and Powell, M.J.D., On the A-acceptability of rational approximations that interpolate the exponential function, IMA J. Num. Analysis 1 (1981), 241-251.

5. Iserles, A. and Strang, G., The optimal accuracy of difference schemes, Trans. Amer. Math. Soc. 277 (1983), 779-803.

6. Lambert, J.D., *Computational Methods in Ordinary Differential Equations*, J. Wiley, London, 1973.

7. Nørsett, S.P. and Wanner, G., The real pole sandwich for rational approximations and oscillation problems, BIT 19 (1979), 79-94.

8. Scales, W.A., Interpolation with meromorphic functions of minimal norm, Ph.D. dissertation, Univ. of Cal., San Diego (1982).

9. Trefethen, L.N., Personal communication.

10. Wanner, G., Hairer, E. and Nørsett, S.P., Order stars and stability theorems, BIT 18 (1973), 475-489.

BERNSTEIN AND MARKOV INEQUALITIES FOR CONSTRAINED POLYNOMIALS

Michael A. Lachance
Department of Mathematics
University of Michigan-Dearborn
Dearborn
Michigan 48128
U.S.A.

Abstract. Pointwise and uniform bounds are determined for the derivatives of real algebraic polynomials $p(x)$ which on the interval $[-1,1]$ satisfy $(1-x^2)^{\lambda/2}|p(x)| \leq 1$, λ a fixed positive integer. The pointwise bounds are investigated with regard to their sharpness while the uniform bounds are shown to be best possible in an asymptotic sense.

1. Introduction

We begin our discussion by stating the classical inequalities of S. Bernstein [1] and A. Markov [6].

Theorem 1. Let $p(x)$ *be a real algebraic polynomial of degree at most* n *and assume* $|p(x)| < 1$ *for* $-1 \leq x \leq 1$. *Then*

(1) $$|p'(x)| \leq n\,(1-x^2)^{-\frac{1}{2}}, \quad -1 < x < 1$$

(2) $$\|p'\| \leq n^2,$$

where $\|\cdot\|$ *denotes the supremum norm on the interval* $[-1,1]$.

These inequalities are best possible in the following sense. The bound in (1) is attained at certain points of the interval when $p(x) = T_n(x)$, the Chebyshev polynomial of the first kind of degree n, and is asymptotically sharp for each fixed x as n becomes infinite.

The polynomial $T_n(x)$ is also extreme for (2) since $T_n'(1) = n^2$.

In 1970 P. Turán posed the following curved majorant generalization of Theorem 1.

Problem. Let $\Phi(x)$ be a nonnegative function defined on $(-1,1)$ and let $p(x)$ denote a real polynomial of degree at most n. If

$$(3) \qquad |p(x)| \le \Phi(x), \ -1 < x < 1,$$

then what bounds may be given for $|p^{(k)}(x)|$, $k=0,1,\ldots,n$ in $[-1,1]$?

In partial response to this question several authors [7,8,9,10] have investigated majorants of the form $(1-x^2)^{\frac{1}{2}}$, $|x|$, and more generally $(1-x)^{\lambda/2}(1+x)^{\mu/2}$ where λ and μ are positive integers. In the present note we consider the case when

$$(4) \qquad \Phi(x) = (1-x^2)^{-\lambda/2}, \ \lambda = 1,2,3,\ldots \quad .$$

In particular, for the polynomial $p(x)$ of degree at most n satisfying (3), when $\Phi(x)$ takes the form (4), we will show that

$$|p^{(k)}(x)| \le [2(n+\lambda)]^k (1-x^2)^{-(\lambda+k)/2}, \ -1 < x < 1,$$

$$\|p^{(k)}\| \le [2(n+\lambda)]^k \binom{n+\lambda}{\lambda+k} \sim \frac{2^k}{(\lambda+k)!} n^{\lambda+2k} \quad ,$$

for each $k = 0,1,\ldots,n$.

The outline of the paper is as follows. In section 2 we state our main results and indicate to what extent they are sharp. We offer some applications of these results in section 3, and in section 4 we conclude with the proofs of our results. In the remainder of this section we introduce some necessary notation and summarize some known results which will be of importance in the sequel.

For each nonnegative integer n we denote by π_n the collection of real polynomials of degree at most n, and for each positive integer λ let

$$(5) \qquad \pi_n(\lambda) := \{p\epsilon\pi_n : (1-x^2)^{\lambda/2}|p(x)| \le 1 \text{ for } -1 \le x \le 1\}.$$

The class $\pi_n(\lambda)$ has been investigated previously in [3,4] and we refer the interested reader to those works for the details regarding the following results.

If for each pair of integers $n \geq 0$, $\lambda \geq 1$ we set

$$(6) \qquad E_n(\lambda) := \min \{\|(1-x^2)^{\lambda/2}(x^n - g(x))\| : g\varepsilon\pi_{n-1}\},$$

then it is known that there exists a unique monic extremal polynomial $Q_{n,\lambda}(x)$ of precise degree n for which $\|(1-x^2)^{\lambda/2}Q_{n,\lambda}(x)\| = E_n(\lambda)$. Normalizing $Q_{n,\lambda}(x)$ we obtain what we will call the constrained Chebyshev polynomial

$$(7) \qquad T_{n,\lambda}(x) := Q_{n,\lambda}(x)/E_n(\lambda).$$

The polynomial $T_{n,\lambda}(x)$ is characterized by the existence of a unique set of n+1 points $\xi_i = \xi_{n,i}(\lambda)$ satisfying

$$(8) \qquad -a_n(\lambda) \leq \xi_0 < \xi_1 < \xi_2 < \ldots < \xi_n \leq a_n(\lambda) := \{1-[\lambda/(n+\lambda)]^2\}^{\frac{1}{2}}$$

on which

$$(9) \qquad T_{n,\lambda}(\xi_i) = (-1)^{n-i}\,(1-\xi_i^2)^{-\lambda/2}, \quad i=0,1,\ldots,n.$$

By the symmetry of our function $\Phi(x)$ it is also known that $\xi_i = -\xi_{n-i}$, for each $i=0,1,\ldots,n$. This next result is a domination property of the constrained Chebyshev polynomials [3, Theorem 3.3]. Let $p\varepsilon\pi_n(\lambda)$. Then for $x < \xi_0$ and for $x > \xi_n$ we have

$$(10) \qquad |p(x)| \leq |T_{n,\lambda}(x)|.$$

Finally for the polynomial p(x) constrained by the zeros at each end-point of [-1,1] its absolute maximum is attained only in a smaller subinterval [3, Theorem 2.5]. Let $p\varepsilon\pi_n(\lambda)$. Then with $a_n(\lambda)$ defined in (8)

$$(11) \qquad \|(1-x^2)^{\lambda/2}p(x)\| = \max \{|(1-x^2)^{\lambda/2}p(x)| : |x| \leq a_n(\lambda)\}.$$

2. Main Results

Our first result is an analog of the Bernstein and Markov inequalities for the collection of polynomials $\pi_n(\lambda)$ defined in (5).

Theorem 2. For each pair of integers $n \geq 0$, $\lambda \geq 1$ let $p(x)$ be a polynomial in $\pi_n(\lambda)$. Then

$$(12) \qquad |p'(x)| \leq [2(n+\lambda)]\,(1-x^2)^{-(\lambda+1)/2}, \quad -1 < x < 1,$$

$$(13) \qquad \|p'\| \leq [2(n+\lambda)] \binom{n+\lambda}{\lambda+1} \sim \frac{2}{(\lambda+1)!}\, n^{\lambda+2}.$$

With regard to the sharpness of this theorem the author suspects that the constant 2 in (12) is not best possible. In an effort to ascertain the optimal constant we define

$$(14) \qquad \gamma_n(\lambda) := \max \left\{ \left\| \frac{(1-x^2)^{(\lambda+1)/2} p'(x)}{(n+\lambda)} \right\| : p \in \pi_n(\lambda) \right\}.$$

From (12) it follows that $\gamma_n(\lambda) \leq 2$. To determine lower bounds on $\gamma_n(\lambda)$ it seems natural to consider polynomials from $\pi_n(\lambda)$ which exhibit very rapid growth off the interval $[-1,1]$. A natural choice according to the domination property of (10) would be the constrained Chebyshev polynomials. Several of these can be determined explicitly and these are displayed below.

$$(15) \qquad T_{1,\lambda}(x) = [(1+\lambda)^{1+\lambda}/\lambda^{\lambda}]^{\frac{1}{2}} x,$$

$$(16) \qquad T_{n,1}(x) = U_n(x),$$

$$(17) \qquad T_{n,2}(x) = T_{n+2}(\cos[\pi/2(n+2)]x) \,/\, (x^2-1),$$

where $T_n(x)$ and $U_n(x)$ are the classical Chebyshev polynomials of the first and second kind, respectively. Using these extremals and another special case $T_{2,\lambda}$ we can generate the following lower bounds for $\gamma_n(\lambda)$.

	$\lambda = 1$	$\lambda = 2$	$\lambda = 3$	$\lambda = 4$	
$n = 1$	1.000	.886	.770	.705	...
$n = 2$	1.026	.946	.875	.815	...
$n = 3$	1.042	.988			
$n = 4$	1.049	1.010			
	.	.			
	.	.			
	.	.			

Table 1. Lower bounds for $\gamma_n(\lambda)$.

Since $\pi_n(\lambda) \subset \pi_{n+1}(\lambda)$ it follows that the sequence $(n+\lambda)\gamma_n(\lambda)$ is increasing for each fixed λ. In the special case when $n = 1$ it is easy to verify from (15) that

$$(18) \qquad \gamma_1(\lambda) = [(1+1/\lambda)^{\lambda}/(1+\lambda)]^{\frac{1}{2}} \to 0 \quad \text{as } \lambda \to \infty.$$

Furthermore taking advantage of the trigonometric representation of $U_n(x)$ when $\lambda = 1$ we can show that

$$(19) \qquad \overline{\lim_{n\to\infty}}\, \gamma_n(1) \geq 1.063\ 103\ 659.$$

One would expect that the constant in (19) would be the optimal value to replace 2 in inequality (12).

As for the bound in inequality (13) it is not in general best possible for fixed λ and n, but we will show that it does provide the correct order of growth asymptotically for λ fixed and n large. For this purpose we introduce the collection of ultraspherical polynomials [12, p. 80] P_n^{λ} which satisfy for real $\lambda > 0$

$$(20) \qquad \int_{-1}^{1} (1-x^2)^{\lambda-\frac{1}{2}}\, P_n^{\lambda}(x)\, P_m^{\lambda}(x)\, dx = 0, \ n \neq m,$$

$$(21) \qquad P_n^{\lambda}(1) = \binom{n+2\lambda-1}{n}.$$

For each fixed positive integer λ a corresponding sequence of suitably normalized ultraspherical polynomials will be contained in $\pi_n(\lambda)$ for each n, respectively, and will exhibit the order of growth specified in (13). Thus we define

$$(22) \qquad R_{n,\lambda}(x) := \frac{(\lambda-1)!}{(n+\lambda)^{\lambda-1}}\, P_n^{\lambda}(x), \ \lambda = 1,2,3,\ldots \quad .$$

We argue inductively that $R_{n,\lambda}(x)$ is an element of $\pi_n(\lambda)$. First, for $\lambda = 1$ and $n \geq 0$, $R_{n,1}(x) = U_n(x)$ which is clearly in $\pi_n(1)$. Next, for $\lambda \geq 1$ it follows from (22) and [12, p. 80, (4.7.14)] that

$$(23) \qquad \frac{d}{dx} R_{n,\lambda}(x) = 2(n+\lambda)\, R_{n-1,\lambda+1}(x), \ n \geq 1.$$

The first inequality of Theorem 2 implies that if $p\epsilon\pi_n(\lambda)$, then $p'/[2(n+\lambda)]\epsilon\pi_{n-1}(\lambda+1)$. This observation together with (23) implies that $R_{n-1,\lambda+1}(x)\epsilon\pi_{n-1}(\lambda+1)$. Thus for each pair of integers $n \geq 0$, $\lambda \geq 1$ we have

(24) $\quad R_{n,\lambda}(x) \in \pi_n(\lambda)$.

From [12, Theorem 7.33.1] we conclude that for fixed λ and n large

(25) $$\| R_{n,\lambda} \| = |R_{n,\lambda}(1)| = \frac{(\lambda-1)!}{(n+\lambda)^{\lambda-1}} \binom{n+2\lambda-1}{n} \sim \frac{(\lambda-1)!}{(2\lambda-1)!} n^{\lambda},$$

(26) $$\| \frac{d}{dx} R_{n,\lambda}(x) \| = 2(n+\lambda) \; \| R_{n-1,\lambda+1} \| \sim 2\frac{\lambda!}{(2\lambda+1)!} n^{\lambda+2} .$$

Therefore the sequence defined in (22) does show that the order of growth prescribed in (13) is asymptotically best possible.

Our next result extends the inequalities of Theorem 2 to all orders of derivates.

Theorem 3. *For a given pair of integers* $n \geq 0$, $\lambda \geq 1$ *let* $p(x)$ *be a polynomial in* $\pi_n(\lambda)$. *Then for each* $k = 0,1...,n$ *we have*

(27) $$|p^{(k)}(x)| \leq [2(n+\lambda)]^k \quad (1-x^2)^{-(\lambda+k)/2}, \; -1 < x < 1 ,$$

(28) $$\| p^{(k)} \| \leq [2(n+\lambda)]^k \binom{n+\lambda}{\lambda+k} \sim \frac{2^k}{(\lambda+k)!} n^{\lambda+2k} .$$

The sharpness of inequality (27) is subject to the same limitations as in inequality (12). However the sequence defined in (22) again shows that the order of growth in (28) asymptotically best possible. For the case $k = 0$, the inequality (28) generalizes a theorem of I. Schur [11, Theorem IV.].

3. Applications

In this section we present several applications of the results of section 2. The first theorem which we state is a simple consequence of the properties of the constrained Chebyshev polynomials in (7) together with Theorem 3.

Theorem 4. *For a given pair of integers* $n \geq 0$, $\lambda \geq 1$ *let* $p(x)$ *be a polynomial in* $\pi_n(\lambda)$. *Then for each* $k = 0,1,...,n$

(29) $\qquad \|p^{(k)}\| \leq [2(n+\lambda)]^k \; T_{n-k,\lambda+k}(1).$

Proof. The case when k = 0 could easily be deduced from previously known results. Let $k \geq 0$ be fixed. Then inequality (27) of Theorem 3 implies for $p \in \pi_n(\lambda)$ that $p^{(k)}/[2(n+\lambda)]^k \in \pi_{n-k}(\lambda+k)$. Consequently

(30) $$|p^{(k)}(x)| \leq \frac{[2(n+\lambda)]^k}{(1-x^2)^{(\lambda+k)/2}} \leq \frac{[2(n+\lambda)]^k}{(1-\xi^2)^{(\lambda+k)/2}} = [2(n+\lambda)]^k T_{n-k,\lambda+k}(\xi)$$

for $|x| \leq \xi := \xi_{n-k,n-k}(\lambda+k)$ as defined in (8) and the last equality following from (9). The domination property (10) of the constrained Chebyshev polynomials implies that

(31) $$|p^{(k)}(x)| \leq |T_{n-k,\lambda+k}(x)| . [2(n+\lambda)]^k$$

for $\xi \leq |x| \leq 1$. Since $|T_{n-k,\lambda+k}(x)|$ is increasing (decreasing) for $x > \xi$ (for $x < -\xi$) we conclude that

(32) $$|T_{n-k,\lambda+k}(\xi)| \leq |T_{n-k,\lambda+k}(x)| \leq T_{n-k,\lambda+k}(1)$$

for $\xi \leq |x| \leq 1$. The inequality (29) follows on combining (30), (31), and (32). □

For those instances where the constrained Chebyshev polynomials can be determined directly, (29) represents an improvement over (28). We next offer an extension of a theorem for ultraspherical polynomials which appears in [12, Theorem 7.33.2]. We state it as

Theorem 5. Let $0 < \lambda < 1$ and let $0 \leq \theta \leq \pi$. Then

(33) $$(\sin\theta)^\lambda |P_n^\lambda(\cos\theta)| < (n/2)^{\lambda-1}\,\Gamma(\lambda)^{-1}.$$

From (22) and (24) after making the substitution $x = \cos\theta$ we have

Theorem 6. Let λ be a positive integer and let $0 \leq \theta \leq \pi$. Then

(34) $$(\sin\theta)^\lambda\, |P_n^\lambda(\cos\theta)| \leq (n+\lambda)^{\lambda-1}\,\Gamma(\lambda)^{-1}.$$

Finally for the so called incomplete polynomials introduced by G. G. Lorentz [5] we have

Theorem 7. *For each pair of integers* $m \geq 0$, $s \geq 1$ *let* $q(t)$ *be a polynomial in* π_m. *If* $|t^s q(t)| \leq 1$ *for* $0 \leq t \leq 1$, *then*

(35) $$|t^s q'(t)| \leq 2(s+m)\ [t(1-t)]^{-\frac{1}{2}},\ 0 < t < 1,$$

(36) $$\max\ \{|q(t)|: 0 \leq t \leq 1\} \leq T_{2m,2s}(1).$$

The proof of this result is immediate after making the substitution $t = 1-x^2$ and applying (12) and (29) when $k = 0$. We remark that (36) implies in particular that for $q(t)$ as in Theorem 7,

(37) $$|q(0)| \leq T_{2m,2s}(1).$$

Previously in [2] it was shown that for such $q(t)$ with $m \geq 1$ and $s \geq 1$

(38) $$|q(0)| \leq (1+m/2s)^{2s}(1+2s/m)^m.$$

In the special case when $s = 1$, (37) appears to be an improvement over (38) (compare [11, Theorem III]):

(39) $$|q(0)| \leq (m+1)\ \cot(\pi/4(m+1)) < (1+m/2)^2\ (1+2/m)^m.$$

4. Proofs of Main Results

The proof of Theorem 2 depends entirely upon trigonometric substitutions and inequalities for trigonometric polynomials, hence the restriction of λ to positive integer values. The two principle lemmas which we will use can be found in Q. I. Rahman's very interesting paper [10]. The first of these we state without proof referring the interested reader to [10, Lemma 1] for details.

Lemma 1. *Let* $t(\theta)$ *be a real trigonometric polynomial of degree* n. *If* $|t(\theta)| \leq 1$, *for all* θ, *then*

(40) $$n^2\ [t(\theta)]^2 + [t'(\theta)]^2 \leq n^2.$$

In particular, we have

(41) $|t'(\theta)| \le n.$

Our next lemma extends slightly Lemma 2 of [10].

Lemma 2. For a given pair of integers $n \ge 0$, $\lambda > 1$ let $p(x)$ be a polynomial in $\pi_n(\lambda)$. Then for each $k = 0,1,\ldots,\lambda$ we have

(42) $$\binom{n+\lambda}{k}^{-1}\binom{\lambda}{k} p(x) \quad \varepsilon \ \pi_n(\lambda-k),$$

where we define $\pi_n(0) := \{p\varepsilon\pi_n \ : \ |p(x)| \le 1 \text{ for } -1 \le x \le 1\}$.

Proof. Let $\lambda \ge 1$ be fixed and let $k = 1$. Setting $x = \cos\theta$ we apply (40) to the trigonometric polynomial $t(\theta) := (\sin\theta)^\lambda\ p(\cos\theta)$ of degree $n+\lambda$ to obtain

(43) $$(n+\lambda)^2[\sin^\lambda\theta p(\cos\theta)]^2 + [\lambda\sin^{\lambda-1}\theta\cos\theta\ p(\cos\theta) - \sin^{\lambda+1}\theta\ p'(\cos\theta)]^2 \le (n+\lambda)^2.$$

Letting $x_o = \cos\theta_o$ denote a relative extreme point for $p(x)$ implies $p'(x_o) = 0$ and hence (43) becomes after simplification

(44) $$[\sin^{\lambda-1}\theta_o\ p(\cos\theta_o)]^2 \le \frac{(n+\lambda)^2}{[(n+\lambda)^2-\lambda^2]\sin^2\theta_o + \lambda^2} \le \frac{(n+\lambda)^2}{\lambda^2}.$$

From the definition of $\pi_n(\lambda)$ in (5) the conclusion of the lemma follows when $k = 1$. For $k > 1$ we simply iterate the above result for each new value of λ. □

Proof of Theorem 2. Let $n \ge 0$ and $\lambda \ge 1$ be fixed and assume $p(x)\ \varepsilon\pi_n(\lambda)$. Again we make the substitution $x = \cos\theta$ and this time apply (41) to the trigonometric polynomial $t(\theta) := (\sin\theta)^\lambda\ p(\cos\theta)$ of degree $n+\lambda$ to get

(45) $$|\lambda\sin^{\lambda-1}\theta\cos\theta\ p(\cos\theta) - \sin^{\lambda+1}\theta\ p'(\cos\theta)| \le n+\lambda.$$

From the triangle inequality and Lemma 2 when $k = 1$ we have

(46) $$\begin{aligned}|\sin^{\lambda+1}\theta\ p'(\cos\theta)| &\le (n+\lambda) + \lambda\sin^{\lambda-1}\theta\ |p(\cos\theta)\cos\theta| \\ &\le (n+\lambda) + \lambda[(n+\lambda)/\lambda]\ |\cos\theta| \\ &\le 2(n+\lambda).\end{aligned}$$

Recalling the substitution $x = \cos\theta$, inequality (12) follows immediately from this last inequality. Since (12) implies for $p \in \pi_n(\lambda)$ that $p'/[2(n+\lambda)] \in \pi_{n-1}(\lambda+1)$, we apply Lemma 2 to the class $\pi_{n-1}(\lambda+1)$ with $k = \lambda+1$ to obtain (13). □

Proof of Theorem 3. Inequalities (27) and (28) follow directly from Theorem 2 and Lemma 2 by repeatedly iterating the given constrained polynomial. □

References

1. S. Bernstein, Sur l'ordre de la meilleure approximation des fonctions continues par des polynômes de degré donné, Mem. Acad. Roy. Belg. (2) 4 (1912), 1-103.

2. J.H.B. Kemperman and G.G. Lorentz, Bounds for polynomials with applications, Neder. Akad. Wetensch. Proc. Ser. A. 82 (1979), 13-26.

3. M.A. Lachance, E.B. Saff, and R.S. Varga, Bounds for incomplete polynomials vanishing at both endpoints of an interval, in "Constructive Approaches to Mathematical Models" (C.V. Coffman and G.J. Fix, Eds.), 421-437, Academic Press, New York, 1979.

4. M.A. Lachance, E.B. Saff, and R.S. Varga, Inequalities for polynomials with a prescribed zero, Math. Zeit. 168 (1979), 105-116.

5. G.G. Lorentz, Approximation by incomplete polynomials (problems and results), in "Padé and Rational Approximation: Theory and Applications" (E.B. Saff and R.S. Varga, Eds.), 289-302, Academic Press, New York, 1977.

6. A. Markov, On a certain problem of D.I. Mendeleieff, Utcheniya Zapiski Imperatorskoi Akademii Nauk (Russia), 62 (1889), 1-24.

7. D.J. Newman and T.J. Rivlin, On polynomials with curved majorants, Can. J. Math., 34 (1982) 961-968.

8. R. Pierre and Q.I. Rahman, On a problem of Turán about polynomials, Proc. Amer. Math. Soc., 56 (1976), 231-238.

9. R. Pierre and Q.I. Rahman, On a problem of Turán about polynomials, II, Can. J. Math., 33 (1981), 701-733.

10. Q.I. Rahman, On a problem of Turán about polynomials with curved majorants, Trans. Amer. Math. Soc., 163 (1972), 447-518.

11. I. Schur, Über das maximum des absoluten betrages eines polynoms in einem gegebenen interval, Math. Zeit., 4 (1919), 271-287.

12. G. Szegö, Orthogonal Polynomials, Colloquium Publication, Vol. 23, 4th ed., Providence, Rhode Island, American Mathematical Society, 1978.

MULTIVARIATE INTERPOLATION

G. G. Lorentz* and R. A. Lorentz
Department of Mathematics, The University of Texas, Austin, Texas 78712
GMD, Schloss Birlinghoven, 5205 St. Augustin 1, FDR

Abstract. We consider interpolation of multivariate functions by algebraic polynomials in $\mathbb{R}^s$, $s \geq 2$. Since our methods and results do not depend on dimension $s \geq 2$, we restrict ourselves to bivariate interpolation, $s = 2$. Using methods of Birkhoff interpolation from [7], we characterize all regular interpolation matrices, and state some results about conditionally regular ones.

§1. General Introduction

Our interpolating polynomials will be either of a given total degree n, or of given coordinate degrees n_1, n_2. But more generally, they will be

$$P(x,y) = \sum_{(i,k)\in S} a_{i,k} x^i y^k , \tag{1.1}$$

where S is some fixed lower set (see §2) of lattice points i,k in $\mathbb{R}^2$.

A polynomial P interpolates f if the values of P and of some of its derivatives match the corresponding values or derivatives of f. In this sense, methods of multivariate interpolation proposed by Hakopian [3,4] and Kergin [6,9] are not true interpolation. Thus, in Hakopian interpolation, one matches the averages of f over some simplices by the corresponding averages of P; Kergin interpolation is still farther removed from the problems that we consider.

An interpolation method will be called regular, if it is uniquely solvable for P for any selection of the knots of interpolation.

It is clear why people have avoided multivariate interpolation: even Lagrange interpolation is not regular if the knots are chosen

*Supported in part by NSF Grant MCS8303353.

arbitrarily in an open subset of $\mathbb{R}^2$.

There are three ways out of this dilemma: (a) one can explore other non-Lagrangean, but regular methods of interpolation. This way is pursued in our §3, where we describe all regular interpolation matrices. One can, (b) restrict the choice of interpolation knots. For Lagrange interpolation in $\mathbb{R}^s$ this has been done with success by Chung and Yao [1], but it is not known how to proceed in general. The third way, (c) is to study conditionally regular interpolation methods. They are solvable not for all selection of knots, but only for most of them. Roughly speaking, they are uniquely solvable with probability 1. For methods of this type, if one has a concrete problem, and selects the interpolation knots at random, it will be extremely unlikely that the problem will be unsolvable. We hope to develop a corresponding theory in a forthcoming paper. Some of the results obtained are announced in §4.

At this stage, the difference between different dimensions $s \geq 2$ is immaterial. But the difference between $s = 2$ and $s = 1$ is enormous. Bivariate interpolation does not resemble univariate Lagrange-Hermite interpolation. It is much closer to non-Hermitian, that is, the Birkhoff interpolation. Definitions and methods of the latter (see for example the book [7]) seem to be quite essential here.

§2. Technical Introduction

A lower set S consists of a finite number of pairs (i,k) of non-negative integers, for which $(i',k') \leq (i,k)$ (that is, $i' \leq i$, $k' \leq k$) and $(i,k) \in S$ imply $(i',k') \in S$. A subset $B \subset S$ is an upper set if $(i',k') \geq (i,k)$ and $(i,k) \in B$ imply $(i',k') \in B$. One sees that complements of lower subsets of S are upper subsets and conversely. We call $n = \max_{(i,k)\in S} (i+k)$ the order of S.

In the following, a lower set S will be fixed. The corresponding polynomials $P(x,y)$, given by (1.1), form a linear space $\mathcal{P}_S$ of dimension $|S|$. For example, if S is the triangle $i+k \leq n$, this set has dimension $\frac{1}{2}(n+1)(n+2)$.

An interpolation matrix $E = (e_{q,i,k})$ is a set of 0's and 1's defined for $q = 1,\dots,m$ and $(i,k) \in S$. We use the notation

$\varepsilon_{i,k} = \sum_{q=1}^{m} e_{q,i,k}$, and sometimes decompose E into its "rows" E_q, $q = 1,\dots,m$: $E = E_1 \oplus \dots \oplus E_m$. Here, for each fixed q,

$E_q = (e_{q,i,k})_{(i,k)\in S}$. For a subset $A \subset S$, we use the notation

$$|E_A| = \sum_{\substack{(i,k)\in A \\ q=1,\dots,m}} e_{q,i,k} .$$

The order of a matrix E is the order of S . A matrix E is normal if $|E| := |E_S| = |S|$. It satisfies the <u>lower</u> (or <u>upper</u>) Pólya conditions if

(2.1) $\quad |E_A| \geq |A|$ for each lower set A

or

(2.2) $\quad |E_A| \leq |A|$ for each upper set A .

For normal matrices, the two sets of conditions are equivalent.

An interpolation problem is determined by a normal matrix E , a set $Z = \{z_1,\dots,z_m\}$ of disjoint interpolation knots $z_q = (x_q,y_q)$ in $\mathbb{R}^2$ and by a set $C = \{c_{q,i,k}\}$ of real numbers defined for $e_{q,i,k} = 1$. Then the problem is to find P from the conditions

(2.3) $$\frac{\partial^{i+k}}{\partial x^i \partial y^k} P(x_q,y_q) = c_{q,i,k} , \quad \text{for all } q,i,k \text{ with } e_{q,i,k} = 1 .$$

Equations (2.3) form a system of linear equations in the coefficients $a_{i,k}$ of (1.1). If E,Z are given, the problem (2.3) is solvable for all choices of C if and only if the determinant of the system (2.3),

(2.4) $\quad D(E,Z)$

is different from zero. Then the pair (E,Z) is called regular. The <u>matrix E is regular</u>, if (E,Z) is regular for all sets $Z \subset \mathbb{R}^2$.

Now $D(E,Z)$ is a polynomial of high degree in the $2m$ variables x_q,y_q . Either it is identically zero or it vanishes only on a set of measure zero in $\mathbb{R}^{2m}$ which is also of first category. In the first case E is singular. <u>In the second case we say that E is conditionally regular</u>.

The following two theorems are well-known in univariate interpolation.

Theorem 1. A normal matrix can be conditionally regular only if it satisfies the Pólya conditions.

Proof. Let (2.1) be violated for some lower set A. We consider a polynomial

$$(2.5) \qquad P(x,y) = \sum_{(i,k)\in A} a_{i,k} x^i y^k$$

and the interpolation problem (2.3) with $c_{q,i,k} \equiv 0$. The homogenous equations (2.3) for $(i,k) \in A$ are $|E_A|$ in number while the number of coefficients $a_{i,k}$ is $|A| < |E_A|$. So there exists a nontrivial polynomial (2.5) which satisfies the conditions. Moreover, if $(i',k') \in S\setminus A$ and $(i,k) \in A$ then either $i' > i$ or $k' > k$. In the first case $\partial^{i'}/\partial x^{i'}(x^i y^k) = 0$, in the second case $\partial^{k'}/\partial y^{k'}(x^i y^k) = 0$. Hence $\partial^{i'+k'}/\partial x^{i'}\partial y^{k'}(P) \equiv 0$ for $(i',k') \notin A$. It follows that P satisfies the homogenous conditions (2.3) for all selections of knots Z. The matrix E is singular. □

Unlike the univariate case, Pólya conditions are by no means sufficient for conditional regularity. One can take for example $m = 2$, $n = 2$, with the triangle $i + k \leq 2$ for S, and the matrix $E = E_1 \oplus E_2$, where

$$E_1 = E_2 = \begin{pmatrix} 0 & & \\ 1 & 0 & \\ 1 & 1 & 0 \end{pmatrix} .$$

Without loss of generality let $z_1 = (0,0)$, $z_2 = (x,y)$. One computes that $D(E,Z) \equiv 0$, and yet E satisfies (2.1) and (2.2).

We say that the restriction E_A of E to a subset $A \subset S$ which satisfies $|E_A| = |A|$ is regular if it is regular for all polynomials of the form (2.5) with $(i,k) \in A$.

Theorem 2. Let $S = A \cup B$ be a disjoint decomposition of S, with $|E_A| = |A|$ and $|E_B| = |B|$. If A is a lower and B an upper subset of S, then E is regular if and only if both E_A and E_B are regular.

Proof. (a) Let E_A, E_B be regular, let P be a polynomial orthogonal to E,Z. We decompose P in an obvious way, $P = P_A + P_B$.

Homogeneous equations (2.3) with $(i,k) \in B$ for P are identical with those for P_B, hence $P_B \equiv 0$. Then the regularity of E_A yields $P_A \equiv 0$.

(b) Let E be regular. Let P_A be a polynomial orthogonal to E_A, Z. Then it is also orthogonal to E, Z, hence $P_A \equiv 0$, and E_A is regular.

Let now a polynomial P_B be orthogonal to E_B, Z. Because of the regularity of E_A, we can find a polynomial P_A for which

$$\frac{\partial^{i+k} P_A}{\partial x^i \partial y^k}(x_q, y_q) = \frac{\partial^{i+k} P_B}{\partial x^i \partial y^k}(x_q, y_q) \quad \text{for} \quad (i,k) \in A \quad q = 1, \dots, m \quad .$$

Then $P_B - P_A$ will be orthogonal to E, Z, hence $P_B - P_A \equiv 0$, thus $P_B \equiv 0$. □

§3. Characterization of regular matrices

A normal matrix E is called an Abel matrix if for each $(i,k) \in S$, the equation $e_{q,i,k} = 1$ is satisfied for exactly one q. Alternatively, $\varepsilon_{i,k} = 1$ for all $(i,k) \in S$, or $|E_A| = |A|$ for each subset $A \subset S$. A special case is a "Taylor matrix", when $m = 1$ and all $e_{1,i,k} = 1$.

Theorem 3. An interpolation matrix E is regular if and only if it is an Abel matrix.

Proof. (a) The condition is sufficient. We prove this by induction on the order n of E. Let the statement be true for all matrices of order $< n$.

We decompose $S = A \cup B$, where B is the upper set consisting of all $(i,k) \in S$ with $i+k = n$, and A is its complement, and apply Theorem 2. Then $|E_B| = |B|$. The matrix B is regular, because equations (2.3) for a polynomial $P_B = \sum_{i+k=n} a_{i,k} x^i y^k$ are

$$\text{(3.1)} \qquad i!k!a_{i,k} = c_{q,i,k} \quad \text{if} \quad e_{q,i,k} = 1 \quad ,$$

and they determine P_B uniquely. And E_A is regular by the inductive assumption.

(b) The conditions are necessary. Let E be regular, we take the knots to be $z_q = (x_q,0)$, $q = 1,\dots,m$. For a polynomial P given by (1.1) we have

$$\frac{\partial^{i+k} P}{\partial x^i \partial y^k}(x_q,0) = k!\,\frac{d^i Q_k}{dx^i}(x_q) \quad ,$$

where for each fixed k , Q_k denotes the polynomial of one variable

$$(3.2) \qquad Q_k(x) = \sum_{i=0}^{n_k} a_{i,k}x^i \quad ,$$

and n_k stands for the largest i for which $(i,k) \in S$. Thus, the problem (2.3) reduces to the sequence $k = 0,1,\dots$ of one-dimensional problems

$$(3.3) \qquad k!\,Q_k^{(i)}(x_q) = c_{q,i,k} \quad \text{for} \quad e_{q,i,k} = 1 \ .$$

For a given k , the problem (3.3) can be solvable only if the number of equations is not more than $n_k + 1$. That is,

$$(3.4) \qquad \sum_{i=0}^{n_k} \varepsilon_{i,k} \le n_k + 1 \quad , \quad k = 0,1,\dots \ .$$

We define the lower sets

$$A_i = \{(j,\ell) : j \le i\} \cap S \quad , \quad B_k = \{(j,\ell) : \ell \le k\} \cap S$$

$$C_{i,k} = A_i \cup B_k$$

and the upper set $D_{i,k} = S \setminus C_{i,k}$.

Summing (3.4), we obtain $|E_{B_k}| \le |B_k|$. Because of the Pólya conditions, and by symmetry,

$$(3.5) \qquad |E_{A_i}| = |A_i| \quad , \quad |E_{B_k}| = |B_k| \ .$$

Now for the set $D_{i,k}$ we get

$$\begin{aligned} |E_{D_{i,k}}| &= |S| - \{|E_{A_i}| + |E_{B_k}| - |E_{C_{i,k}}|\} \\ &\ge |S| - \{|A_i| + |B_k| - |C_{i,k}|\} = |D_{i,k}| \ . \end{aligned}$$

By upper Pólya conditions,

$$(3.6)\qquad |E_{D_{i,k}}| = |D_{i,k}| \quad .$$

In particular, let (i,k) satisfy $i+k=n$. If $i,k>0$, then $D_{i-1,k-1}$ is a one point set consisting of (i,k) , and from (3.6), $\varepsilon_{i,k}=1$. If $i=n$, $k=0$, then $A_n \setminus A_{n-1}$ is a one point set consisting of $(n,0)$ and from (3.5), $\varepsilon_{n,0}=1$. Similarly if $i=0$, $k=n$. We obtain

$$(3.7)\qquad \varepsilon_{i,k} = 1 \quad \text{whenever} \quad i+k=n \quad .$$

Let B be the subset of S with $i+k=n$, then $|E_B| = |B|$. Therefore, for $S^* = S \setminus B$, the matrix $E^* = E_{S^*}$ is normal and has order $< n$. It must be regular, for otherwise there would exist a non-trivial polynomial $P \in P_{S^*}$ and a set of knots Z^* so that P is annihilated by E^*, Z^* . The derivatives of P of order n are automatically zero. Thus, P is also annihilated by E, Z^* , in contradiction to the regularity of E .

We can now apply induction with respect to n ; if E^* is an Abel matrix, so is also E . □

§4. Conditional regularity

We collect here, without proofs, some results which in many cases guarantee conditional regularity or singularity of interpolation matrices.

As in [8], the main tools are shifts and differentiation of the determinant (2.4). Let $E = E_1 \oplus \dots \oplus E_m$ be an interpolation matrix. Let $A_q := A(E_q)$, $q = 1,\dots,m$ be the support of E_q , let $B_q := S \setminus A_q$. A <u>shift</u> Λ of E_1 takes a one $e_{1,i,k} = 1$ into position $(i+1,k)$ or $(i,k+1)$; it is permitted if this position was occupied by zero. The shift Λ transforms E_1 into ΛE_1 and E into ΛE . A <u>multiple shift</u> Λ^* of orders (α,β) is a repeated application of $\alpha+\beta$ shifts, α of them to the right, and β upward. The existence of shifts which preserve the Pólya condition is far from obvious, but can be proved:

<u>Theorem</u> 4. <u>Let</u> E <u>be an interpolation matrix that satisfies</u>

the Pólya condition. If for some q,i,k, $\varepsilon_{i,k} > 1$ *and* $e_{q,i,k} = 1$, *then there exists a multiple shift* Λ^* *of this one in* E_q *for which* $\Lambda^* E$ *also satisfies the Pólya condition.*

Let $z = (x,y)$ be variable, $z_2,\dots,z_m$ constant, then the derivatives of the determinant (2.4) at $z = z_2$ are given by

$$(4.1)\qquad \left.\frac{\partial^{\alpha+\beta} D}{\partial x^\alpha \partial y^\beta}\right|_{z=z_2} = \sum D(\Lambda^* E, Z^*) \quad , \quad Z^* = (z_2,\dots,z_m) \quad ,$$

the sum extended to all shifts Λ^* of E_1 which take $A_1 = A(E_1)$ into some subset B^* of B_2, and for which $\Lambda^* E$ is a Pólya matrix.

Important are the cases when for given α, β there is only one such set, and when the terms of (4.1) do not cancel. This is the case when $\alpha = 0$ or $\beta = 0$ and often when E is a two row matrix. Then, by means of the Taylor expansion of $D(E,Z)$ into powers of $x - x_2$ and $y - y_2$, conditional regularity of E can be reduced to that of some matrix E^* with $m-1$ rows. This allows us to obtain the *decomposition theorem*:

Theorem 5 *Let* E *be a matrix* $E = \oplus \Sigma E_{j,\ell}$ *with rows* $E_{j,\ell}$, $j,\ell = 1,2,\dots$. *Let* $E_{j,\ell}$ *have support* $A_{j,\ell}$. *Then* E *is conditionally regular if* S *is an upward shift of the bands* A_ℓ, $\ell = 1,2,\dots$ *and each* A_ℓ *is the left shift of* $A_{j,\ell}$.

A band is bounded from above by a non-increasing curve, from below by a non-decreasing curve.

We denote by $S_{j,\ell}$ the shifts of the $A_{j,\ell}$ obtained in this way.

Special examples: One can take for the $S_{j,\ell}$ single points of S. This gives the conditional regularity of the Lagrange interpolation. One can take a decomposition of S - when S is a rectangle - into Bogner-Fix-Schmit rectangles (used in the finite element methods, see [2, p. 76]) with subsequent shifts.

For two row matrices $E = E_1 \oplus E_2$ one can apply theorems of measure theory which describe existence and uniqueness of plane sets with prescribed cross-functions.

For the set $A_1 = A(E_1)$ we define its y-cross-functions as follows. For each $k = 0,1,\dots,$ let $P(k)$ be the number of $(i,k) \in A_1$. We extend this to a function $P(y)$ for all $y \geq 0$, which is constant on intervals $[k,k+1)$, $k = 0,1,\dots$. Then $p(y)$ is the decreasing rearrangement of $P(y)$. In a similar way we define the x-cross functions $Q(x)$ and $q(x)$ of the set $B_2 = S \setminus A(E_2)$.

Theorem 6. *If* $A_1 < B_2$ *(which means that* $z_1 \in A_1$ *and* $z_2 \in B_2$ *imply* $z_1 < z_2$ *in the order of the plane), then* $E = E_1 \oplus E_2$ *is conditionally regular if* p *and* q *are inverses of each other:*
$p^{-1}(u) = q(u) \quad , \quad u \geq 0 \quad .$

References

[1] K. C. Chung and T. H. Yao, On lattices admitting unique Lagrange interpolations, SIAM J. Numer. Anal. 14 (1977), 735-743.

[2] Ciarlet, P.G., The finite element method for elliptic problems, North Holland, New York, 1978.

[3] H. A. Hakopian, Multivariate divided differences and multivariate interpolation of Lagrange and Hermite type, J. Approximation Theory 34 (1982), 286-305.

[4] H. Hakopian, Multivariate spline functions, B-spline basis and polynomial interpolations, SIAM J. Numer. Anal. 19 (1982), 510-517.

[5] S. Karlin and J. M. Karon, Poised and non-poised Hermite-Birkhoff interpolation, Indiana Univ. Math. J. 21 (1972), 1131-1170.

[6] P. Kergin, A natural interpolation of C^K functions, J. Approximation Theory 29 (1980), 278-293.

[7] G. G. Lorentz, K. Jetter and S. D. Riemenschneider, Birkhoff Interpolation, Encyclopedia of Mathematics and its Applications, vol. 19, Addison-Wesley, Reading, 1983.

[8] G. G. Lorentz and K. L. Zeller, Birkhoff interpolation problem: coalescence of rows, Arch. Math. 26 (1975), 189-192.

[9] C. A. Micchelli, A constructive approach to Kergin interpolation in R^k: Multivariate B-splines and Lagrange interpolation, Rocky Mountain J. Math. 10 (1980), 485-497.

THE STRONG UNIQUENESS CONSTANT IN COMPLEX APPROXIMATION

T. J. Rivlin
Mathematical Sciences Department
IBM Thomas J. Watson Research Center
Yorktown Heights, N. Y. 10598
U.S.A.

Abstract. Let f(x) belong to the set C(B) of continuous functions, possibly complex valued, on a compact subset, B, of a finite dimensional euclidean space. Let V be a finite dimensional subspace of C(B). $v_0 \in V$ is a strongly unique best approximation to f if (1) $\|f - v\| \geq \|f - v_0\| + \gamma \|v - v_0\|$ holds for some $\gamma(f, B, V) > 0$ and all $v \in V$. The largest γ ($= \gamma^*$) such that (1) holds is the strong uniqueness constant for f. The strong uniqueness constant has previously been determined in essentially one case, namely, the approximation of a monomial by lower degree polynomials in the real case. When $f(z) = z^n$, B is the closed unit disc in the complex plane and V is the set of polynomials of degree at most $n-1$, $\gamma^* = 1/n$. We show that this fact is a trivial consequence of a simple result of Szász (1917). Some related results and problems are also discussed.

1. Introduction

Suppose B is a compact set in the complex plane, $f \in C(B)$ and V is a finite dimensional subspace of C(B). $v_0 \in V$ is called a strongly unique best approximation to f (out of V on B) if there exists a positive constant γ ($= \gamma(f, B, V)$) such that

(1) $\|f-v\| \geq \|f - v_0\| + \gamma \|v - v_0\|$, all $v \in V$.

The largest γ ($= \gamma^*$) such that (1) holds is the strong uniqueness constant (s.u.c.) for f.

For $g \in C(B)$ put $E(g, B) = \{z \in B : |g(z)| = \|g\|\}$. Given $f \in C(B)$ and $v_0 \in V$, suppose

(2) $$\min_{\substack{v \in V \\ \|v\|=1}} \max_{z \in E(f-v_0,B)} \mathrm{Re}[(\mathrm{sgn}\, \overline{f(z) - v_0(z)})\, v(z)] =: c\,,$$

then v_0 is a strongly unique best approximation to f if, and, only if $c > 0$, and if $c > 0$ then $\gamma^* = c$. (Bartelt and McLaughlin [1]). Thus (2) provides a means of calculating γ^*. It is sometimes more convenient to calculate γ^* from the dual extremal problem. Namely, suppose

$$(3)\qquad \tilde{V} = \{v \in V : \mathrm{Re}[\mathrm{sgn}\,\overline{(f(z)-v_0(z))}\, v(z)] \le 1,\ z \in E\,(f - v_0, B)\}\ .$$

Then if

$$(4)\qquad b = \sup_{v \in \tilde{V}} \|v\|\ ,$$

v_0 is a strongly unique best approximation to f if, and only if, $b < \infty$, and if $b < \infty$

$$\gamma^* = \frac{1}{b}\ .$$

It is difficult to determine the s.u.c. explicitly in non-trivial problems. There is essentially only one such determination in the literature and it is in the case of real approximation. It is the following. Let $x : (x_1,...,x_r)$ be a point in $\mathbb{R}^r$, $r \geq 1$. Let i and n denote points in $\mathbb{R}^r$ with non-negative integer coordinates. Suppose $n \neq 0$ is given, let V be the space of real polynomials forming the span of $x^i := x_1^{i_1}...x_r^{i_r}$ where $0 \le i_j \le n_j$, $j = 1,...,r$ and $i \neq n$. Put $T_n(x) = T_{n_1}(x_1)...T_{n_r}(x_r)$. $v_0 = 0$ is the strongly unique best approximation to $T_n(x)$ out of V on $I^r := [-1, 1]^r$ and $\gamma^*(T_n, I^r, V) = (2^r \prod_{i=1}^{r} n_i - 1)^{-1}$. (Cline [2] for $r = 1$, Rivlin [6] for $r \geq 1$) . The calculation of a s.u.c. in a complex polynomial approximation problem is implicit in Szász [7] and, along similar lines, in Newman [4]. Our purpose here is to recall these results, exhibit the strong uniqueness constant and discuss some related issues.

In the case of rational approximation in the plane (See Gutknecht [3] for information about strong uniqueness in this case) the determination of strong uniqueness constants (defined just as in the linear subspace case) remains to be accomplished.

1. Determination of a Strong Uniqueness Constant

The result of Szász mentioned above (quoted in Pólya and Szegö [5; Abschn. VI, No. 61]) is the following:

Proposition. Let $p(z) = c_1 z + c_2 z^2 + ... + c_n z^n$ satisfy

$$(5)\qquad 1 - \mathrm{Re}\, p(e^{i\theta}) \geq 0\ , \qquad 0 \le \theta \le 2\pi$$

then

$$(6)\qquad \sum_{j=1}^{n} |c_j| \le n\,,$$

and equality holds in (6) for

$$(7)\qquad q(z) = -\frac{2}{n+1}\,(nz + (n-1)z^2 + \dots + 2z^{n-1} + z^n)\,.$$

We reproduce the simple proof here.

Proof. According to the well-known representation theorem of Fejér-Riesz (cf. Polyá and Szegö [5; Abschn. VI, No. 40]) (5) implies that

$$1 - \operatorname{Re} p(e^{i\theta}) = |\gamma_0 + \gamma_1 e^{i\theta} + \dots + \gamma_n e^{in\theta}|^2\,, \qquad 0 \le \theta \le 2\pi,$$

hence

$$1 = \sum_{j=0}^{n} |\gamma_j|^2 \quad \text{and} \quad -c_k = 2 \sum_{j=0}^{n-k} \bar{\gamma}_j\, \gamma_{j+k}\,.$$

Thus, putting $c_0 = 1$, we obtain

$$\sum_{k=0}^{n} |c_k| \le \sum_{j=0}^{n} |\gamma_j|^2 + 2\sum_{k=1}^{n} \sum_{j=0}^{n-k} |\gamma_j\, \gamma_{j+k}| = (\sum_{j=0}^{n} |\gamma_j|)^2 \le n+1,$$

establishing (6). Finally note that $0 \le |1 + e^{i\theta} + \dots + e^{in\theta}|^2$ $= n + 1 + 2\operatorname{Re}(ne^{i\theta} + (n-1)e^{i2\theta} \dots + e^{in\theta})$, hence, if $p = q$ (5) is satisfied and equality holds in (6).

Another proof of this proposition, which also relies on the Fejér-Riesz representation, is given in Newman [4].

Now consider the best approximation of z^n out of P_{n-1} on $\overline{D} : |z| \le 1$. Putting $f(z) = z^n$, $V = P_{n-1}$, $v_0 = 0$ and $B = \overline{D}$ in (3) we obtain

$$\begin{aligned} \tilde{V} &= \{p(z) = a_0 + \dots + a_{n-1}\, z^{n-1} : \operatorname{Re} \bar{z}^n\, p(z) \le 1,\ |z| = 1\} \\ &= \{p(z) = a_0 + \dots + a_{n-1}\, z^{n-1} : \operatorname{Re}(\bar{a}_{n-1} z + \dots + \bar{a}_0 z^n) \le 1,\ |z| = 1\}\,. \end{aligned}$$

But since $\|p\| \le |a_0| + \dots + |a_{n-1}|$ the Proposition implies that $b = n$ and hence

$$\gamma^*(z^n, \overline{D}, P_{n-1}) = \frac{1}{n}\,.$$

Thus we have the

Theorem. The zero function is the strongly unique best approximation to z^n out of P_{n-1} on $|z| \le 1$. The strong uniquenesss constant is $1/n$.

Corollary: If $p(z) = a_0 + \dots + a_n z^n$ then

$$\|p\| \ge |a_n| + \frac{1}{n}\|a_0 + \dots + a_{n-1}z^{n-1}\|.$$

Remarks.

Let $P_{n,k} = \{p \in P_n : p(z) = a_{n-k}z^{n-k} + \dots + a_n z^n\}$.

1. The zero function is the strongly unique best approximation to 1 out of $P_{n,n-1}$ on $|z| \le 1$ for every $n \ge 1$. The strong uniqueness constant is $1/n$.

2. Consider the best approximation to z^n on $|z| \le 1$ out of P_k, $0 \le k < n-1$. Choose $f = z^n$, $V = P_k$, $v_0 = 0$ and $B = \overline{D}$ in (3). Then

$$\begin{aligned}\tilde{V} &= \{p(z) = a_0 + \dots + a_k z^k : \operatorname{Re}\,(z^n(\bar{a}_0 + \bar{a}_1 z^{-1} + \dots + \bar{a}_k z^{-k})) \le 1,\ |z| = 1\}\\ &= \{a_0 + \dots + a_k z^k : \operatorname{Re}\,(c_{n-k}z^{n-k} + \dots + c_n z^n) \le 1,\ |z| = 1\}\end{aligned}$$

where $c_{n-j} = \bar{a}_j$, $j = 0,\dots,k$. Invoking the Fejér-Riesz representation yields

$$0 \le 1 - \operatorname{Re}\,(c_{n-k}e^{i(n-k)\theta} + \dots + c_n e^{in\theta}) = |\gamma_0 + \gamma_1 e^{i\theta} + \dots + \gamma_n e^{in\theta}|^2$$

with

$$1 = \sum_{j=0}^{n} |\gamma_j|^2 \quad ; \quad -c_\nu = 2\sum_{j=0}^{n-\nu} \bar{\gamma}_j\,\gamma_{j+\nu} \quad , \qquad \nu = n-k,\dots,n.$$

If now, following another idea used by Szász [7], we recall that $2ab \le a^2 + b^2$, we obtain

$$|c_\nu| \le 2\sum_{j=0}^{n-\nu} |\gamma_j|\,|\gamma_{j+\nu}| \le \sum_{j=0}^{n-\nu} (|\gamma_j|^2 + |\gamma_{j+\nu}|^2)$$

and hence

$$\sum_{\nu=n-k}^{n} |c_\nu| \le \sum_{\nu=0}^{n} \lambda(\nu, k, n)\,|\gamma_\nu|^2$$

where

$$\lambda(\nu, k, n) = (k+1-\nu)_+ + (\nu-(n-(k+1)))_+ \le k+1 \,. \tag{8}$$

Equality holds in (8) only if $\nu = 0, n$. Therefore either

$$\sum_{j=0}^{k} |a_j| = \sum_{\nu=n-k}^{n} |c_\nu| < k+1,$$

or $\gamma_1 = \gamma_2 = \ldots = \gamma_{n-1} = 0$ which implies that $a_1 = a_2 = \ldots = a_{k-1} = 0$ and $|a_n| = 2|\gamma_0|\,|\gamma_n| \leq |\gamma_0|^2 + |\gamma_n|^2 = 1$. Thus in view of (4) we are led to the following conclusion:

If $0 \leq k \leq n-1$, 0 is the strongly unique best approximation to z^n out of P_k on $|z| \leq 1$, and the strong uniqueness constant $\gamma^*_{n,k}$ satisfies

$$\gamma^*_{n,k} \geq \frac{1}{k+1}$$

with equality holding only for $k = 0, n-1$.

References

1. Bartelt, M. W., and H. W. McLaughlin, Characterizations of strong unicity in approximation theory, J. Approx. Theory 9 (1973), 255-266.
2. Cline, A. K., Lipschitz conditions on uniform approximation operators, J. Approx. Theory 8 (1973), 160-172.
3. Gutknecht, M. H., On complex rational approximation; Part I: The characterization problem, Computational Aspects of Complex Analysis (Eds. H. Werner, L. Wuytack, E. Ng, H. J. Bünger), D. Reidel, Dordrecht, Holland (1973), 79-101.
4. Newman, D. J., Polynomials and rational functions, Approximation Theory and Applications (Ed. Z. Ziegler), Academic Press, N. Y. (1981), 265-282.
5. Pólya, G., and G. Szegö, Aufgaben und Lehrsätze aus der Analysis, Vol. II, Dover Publications, N. Y., 1945.
6. Rivlin, T., The best strong uniqueness constant for a multivariate Chebyshev polynomial, J. Approx. Theory, to appear.
7. Szász, O., Über nichtnegative trigonometrische Polynome, Sitzungsb. d. Bayer. Akad. der Wiss. Math. - Phys. Kl., 1917, 307-320.

ON THE MINIMUM MODULI OF NORMALIZED POLYNOMIALS

S. Ruscheweyh and R. S. Varga*
Mathematisches Institut
Universität Würzburg
D-8700 Würzburg am Hubland
West Germany

Abstract. Consider any complex polynomial $p_n(z) = 1 + \sum_{j=1}^{n} a_j z^j$ which satisfies $\sum_{j=1}^{n} |a_j| = 1$, and let Γ_n denote the supremum of the minimum moduli on $|z| = 1$ of all such polynomials $p_n(z)$. We show that

$$1 - \frac{1}{n} \le \Gamma_n \le \sqrt{1 - \frac{1}{n}}, \quad \text{for all} \quad n \ge 1 .$$

If the coefficients of $p_n(z)$ are further restricted to be positive numbers and if $\tilde{\Gamma}_n$ denotes the analogous supremum of the minimum moduli on $|z| = 1$ of such polynomials, we similarly show that

$$1 - \frac{1}{n} \le \tilde{\Gamma}_n \le \sqrt{1 - \frac{3}{(2n+1)}}, \quad \text{for all} \quad n \ge 1 .$$

We also include some recent numerical experiments on the behavior of Γ_n, as well as some related conjectures.

1. Introduction

Consider any non-constant complex polynomial $p_n(z) = \sum_{j=0}^{n} a_j z^j$ with $p_n(0) \neq 0$, and normalize $p_n(z)$ so that

$$(1.1) \qquad p_n(z) = 1 + \sum_{j=1}^{n} a_j z^j, \quad \text{where} \quad \sum_{j=1}^{n} |a_j| \neq 0 .$$

---

*Research supported by the Air Force Office of Scientific Research, the Department of Energy, and the Alexander von Humboldt Stiftung.

By a well-known result of Cauchy (cf. Marden [4, p. 126]), if R (the Cauchy radius of $p_n(z)$) is the unique positive zero of

$$(1.2)\qquad 1 - |a_1|\cdot R - \dots - |a_n|\cdot R^n \quad ,$$

then each zero $\hat{z}$ of $p_n(z)$ of (1.1) satisfies $|\hat{z}| \geq R$. On further normalizing the Cauchy radius R of $p_n(z)$ to be unity, i.e., on assuming

$$(1.3)\qquad \sum_{j=1}^{n} |a_j| = 1 \quad ,$$

then any polynomial $p_n(z)$ in (1.1) which satisfies (1.3) evidently has no zeros in $|z| < 1$. (It may have zeros on $|z| = 1$, as the examples $1 + z^n$ show.)

Our interest is in the following problem, which is related to the recent study of global descent functions for determining zeros of polynomials (cf. Henrici [2,3] and Ruscheweyh [6]). Consider the set of normalized polynomials

$$(1.4)\qquad S_n := \{p_n(z) = 1 + \sum_{j=1}^{n} a_j z^j \; : \; \sum_{j=1}^{n} |a_j| = 1\} \; , \quad \text{for each } n \geq 1 \quad ,$$

and put

$$(1.5)\qquad m(p_n) := \min\{|p_n(e^{i\theta})| : \theta \text{ real}\} \; , \quad \text{for any } p_n \in S_n \quad .$$

Our main question is: <u>How large can</u> $m(p_n)$ <u>be on the set</u> S_n ? Thus, on setting

$$(1.6)\qquad \Gamma_n := \sup\{m(p_n) : p_n \in S_n\} \quad , \qquad (n \geq 1) \quad ,$$

our goal here is to establish rigorous upper and lower bounds for Γ_n , as a function of n . We also report on some numerical experiments, which in turn have inspired some related mathematical conjectures.

2. Upper and Lower Bounds for Γ_n

The following inequality, based on conformal mappings, was derived in Ruscheweyh [6]:

$$(2.1) \qquad m^2(p_n) + \sum_{j=1}^{n} |a_j|^2 \le 1 \ , \quad \text{for any} \quad p_n \in S_n \ .$$

As the Cauchy-Schwarz inequality applied to (1.3) gives that $\sum_{j=1}^{n} |a_j|^2 \ge \frac{1}{n}$, it follows from (2.1) that

$$(2.2) \qquad m(p_n) \le \sqrt{1 - \frac{1}{n}} \ , \quad \text{for any} \quad p_n \in S_n \ ,$$

which yields (cf. (1.6)) the upper bound

$$(2.3) \qquad \Gamma_n \le \sqrt{1 - \frac{1}{n}} \ , \quad \text{for all} \quad n \ge 1 \ .$$

To similarly derive a lower bound for Γ_n , consider the specific polynomial

$$(2.4) \qquad Q_n(z) := 1 + \frac{2}{n(n+1)} \sum_{k=1}^{n} (n+1-k) z^k \ ,$$

which is an element of S_n for each $n \ge 1$. Now, $Q_n(z)$ can also be expressed as

$$Q_n(z) = \frac{n(n+1) - 2n^2 z + (n-2)(n+1) z^2 + 2z^{n+2}}{n(n+1)(1-z)^2} \ ,$$

and evaluating the above expression for $z = -1$ gives

$$(2.5) \qquad Q_n(-1) = \begin{cases} 1 - \frac{1}{n+1} \ , & \text{for } n \text{ an even positive integer;} \\ 1 - \frac{1}{n} \ , & \text{for } n \text{ an odd positive integer .} \end{cases}$$

Next, on defining

$$(2.6) \qquad g_n(z) := \frac{1}{(n+1)} \sum_{k=0}^{n} (n+1-k) z^k \ , \qquad (n = 1,2,\ldots) \ ,$$

then $Q_n(z)$ and $g_n(z)$ are related through

$$(2.7) \qquad Q_n(z) = \frac{n-2}{n} + \frac{2}{n} g_n(z) \ .$$

Writing $g_n(z)$ of (2.6) as

$$(2.8)\qquad g_n(z) = \sum_{k=0}^{\infty} \alpha_k(n) z^k \ , \text{ where } \ \alpha_k(n) := \begin{cases} \dfrac{n+1-k}{n+1} \ , & k = 0,1,\ldots,n \ ; \\ 0 \ , & k \geq n+1 \ , \end{cases}$$

then the coefficients $\alpha_k(n)$ of $g_n(z)$ can be seen to be doubly monotonic, i.e.,

$$(2.9)\qquad \alpha_k(n) \geq 0 \ , \quad \text{and} \quad \alpha_k(n) - \alpha_{k+1}(n) \geq \alpha_{k+1}(n) - \alpha_{k+2}(n) \ ,$$

for all $k \geq 0$ and all $n \geq 1$. From a well-known result of Fejér [1], it follows that

$$(2.10)\qquad \operatorname{Re} g_n(z) \geq \frac{1}{2} \ , \quad \text{for all } |z| \leq 1 \ .$$

Consequently (cf. (2.7)),

$$(2.11)\qquad \operatorname{Re} Q_n(z) = \frac{n-2}{n} + \frac{2}{n} \operatorname{Re} g_n(z) \geq 1 - \frac{1}{n} \ , \quad \text{for all } |z| \leq 1 \ ,$$

which implies that

$$(2.12)\qquad m(Q_n) \geq 1 - \frac{1}{n} \ , \quad \text{for all } n \geq 1 \ .$$

(Note that because of (2.5), equality evidently holds in (2.12) for every odd positive integer n.) But, as $Q_n(z)$ is an element of S_n, (2.12) implies that

$$(2.13)\qquad 1 - \frac{1}{n} \leq \Gamma_n \ , \quad \text{for all } n \geq 1 \ .$$

Combining (2.13) and (2.3) then yields our first result of

Proposition 1. For each positive integer n,

$$(2.14)\qquad 1 - \frac{1}{n} \leq \Gamma_n \leq \sqrt{1 - \frac{1}{n}} \ .$$

Obviously, the bounds of (2.14) are tight for $n = 1$ and give $\Gamma_1 = 0$. For $n > 1$, there is however a gap between the upper and lower bounds in (2.14).

Because of the bounds of (2.14), it is reasonable to express Γ_n as

(2.15) $\quad \Gamma_n := 1 - \frac{\gamma_n}{n} \quad , \quad \text{for all} \quad n \geq 1 \quad ,$

so that from (2.14),

(2.16) $\quad n\left(1 - \sqrt{1 - \frac{1}{n}}\right) \leq \gamma_n \leq 1 \quad , \quad \text{for all} \quad n \geq 1 \quad .$

Now, the lower bound in (2.16) is strictly decreasing as a function of n and has the limit $1/2$, so that

(2.17) $\quad \frac{1}{2} < \gamma_n \leq 1 \quad , \quad \text{for all} \quad n \geq 1 \quad .$

Next, consider the following subset, $\tilde{S}_n$, of S_n :

(2.18) $\quad \tilde{S}_n := \{p_n(z) = 1 + \sum_{j=1}^{n} a_j z^j \ : \ p_n \in S_n \ \text{ and } \ a_j > 0 \text{ for all } 1 \leq j \leq n\}$

For $\tilde{S}_n$, we can associate the analogous quantity $\tilde{\Gamma}_n$:

(2.19) $\quad \tilde{\Gamma}_n := \sup\{m(p_n) \ : \ p_n \in \tilde{S}_n\} \quad , \quad (n \geq 1) \quad .$

Obviously, as $\tilde{S}_n \subset S_n$, then

(2.20) $\quad \tilde{\Gamma}_n \leq \Gamma_n \quad , \quad \text{for all} \quad n \geq 1 \quad .$

But, noting that $Q_n(z)$ of (2.4) is also an element of $\tilde{\Gamma}_n$, then from (2.12), we also deduce

(2.21) $\quad 1 - \frac{1}{n} \leq \tilde{\Gamma}_n \quad , \quad \text{for all} \quad n \geq 1 \quad .$

From (2.20) and (2.21), the upper and lower bounds of (2.14) apply equally well to $\tilde{\Gamma}_n$. But, for an improved upper bound for $\tilde{\Gamma}_n$, we establish

Proposition 2. For each positive integer n,

(2.22) $\quad 1 - \frac{1}{n} \leq \tilde{\Gamma}_n \leq \sqrt{1 - \frac{3}{(2n+1)}} \quad .$

Proof. To obtain the upper bound in (2.22), set

(2.23) $\quad M(p_n) := \max\{|p_n(e^{i\theta})| : \theta \text{ real}\} \quad , \quad \text{for any} \quad p_n \in S_n \quad .$

Note that $M(p_n) = 2$ for any $p_n \in \tilde{S}_n$, for each $n \geq 1$. Next, for each $p_n(z) = \sum_{j=0}^{n} a_j z^j$ (with $a_0 := 1$) in $\tilde{S}_n$, consider the real trigonometric polynomial

$$(2.24) \qquad T_n(\theta;p_n) := |p_n(e^{i\theta})|^2 - \sum_{j=0}^{n} a_j^2 ,$$

which has the explicit form (without constant term)

$$(2.25) \qquad T_n(\theta;p_n) = 2\sum_{k=1}^{n} \cos(k\theta) \sum_{j=0}^{n-k} a_j a_{j+k} .$$

Setting $\lambda_0 := \sum_{j=0}^{n} a_j^2$, it is evident from (2.24) that

$$(2.26) \qquad \begin{cases} \max\limits_{\theta \text{ real}} T_n(\theta;p_n) = M^2(p_n) - \lambda_0 = 4 - \lambda_0 , & \text{and} \\ -\min\limits_{\theta \text{ real}} T_n(\theta;p_n) = \lambda_0 - m^2(p_n) . \end{cases}$$

From Pólya-Szegö [5, p. 84, Exercise 58], it is known that

$$\max_{\theta} T_n(\theta;p_n) \leq -n \cdot \min_{\theta} T_n(\theta;p_n) ,$$

or equivalently, using (2.26),

$$(2.27) \qquad 4 + n \cdot m^2(p_n) \leq (n+1)\lambda_0 .$$

But since $\sum_{j=1}^{n} a_j^2 \leq 1 - m^2(p_n)$ from (2.1), then

$$\lambda_0 := 1 + \sum_{j=1}^{n} a_j^2 \leq 2 - m^2(p_n) .$$

Substituting the above in (2.27) then yields

$$(2.28) \qquad m^2(p_n) \leq \frac{2n-2}{2n+1} = 1 - \frac{3}{2n+1} , \quad \text{for all } p_n \in \tilde{S}_n ,$$

which gives the desired upper bound of (2.22). □

In analogy with (2.15), we similarly define

(2.29) $\tilde{\Gamma}_n := 1 - \frac{\tilde{\gamma}_n}{n}$, for all $n \geq 1$.

Thus, from (2.22),

(2.30) $n\left\{1 - \sqrt{1 - \frac{3}{2n+1}}\right\} \leq \gamma_n \leq 1$, for all $n \geq 1$,

which further yields

(2.31) $0.732213 \ldots = 3\left(1 - \sqrt{\frac{4}{7}}\right) \leq \tilde{\gamma}_n \leq 1$, for all $n \geq 1$,

as well as

(2.32) $\frac{3}{4} \leq \underline{\lim}_{n\to\infty} \tilde{\gamma}_n \leq \overline{\lim}_{n\to\infty} \tilde{\gamma}_n \leq 1$.

3. Computational Results

Intrigued by these numbers Γ_n , $\tilde{\Gamma}_n$, γ_n , and $\tilde{\gamma}_n$, we embarked on some numerical calculations to give further insight into their behavior as $n \to \infty$. First, we conjecture that $\Gamma_n = \tilde{\Gamma}_n$ for all $n \geq 1$, and we hope to establish this in a later work. Thus, our calculations (to be described below) were aimed at determining sharp lower bounds for $\tilde{\Gamma}_n$.

The idea in our calculations was to find an "extremal" $\tilde{p}_n(z)$ in $\tilde{S}_n$ such that

(3.1) $\tilde{p}_n(e^{i\theta}) = m(\tilde{p}_n)$ in precisely n distinct points $\{\theta_j\}_{j=1}^n$.

For convenience, all calculations were performed for n even, say $n = 2k$, $k \geq 1$. Then, for $p_{2k}(z) = 1 + \sum_{j=1}^{2k} a_j z^j$ in $\tilde{S}_{2k}$ (so that $a_j > 0$ for $j = 1,\ldots,2k$, and $a_1 + a_2 + \ldots + a_{2k} = 1$) , write

(3.2) $$\left|1 + \sum_{j=1}^{2k} a_j e^{ij\theta}\right|^2 = m^2(p_{2k}) + 2^{2k} a_{2k} \prod_{j=1}^{k} (\varepsilon_j + \cos\theta)^2 ,$$

with the objective of maximizing $m^2(p_{2k})$. Writing

$$(3.3)\qquad \left|1+\sum_{j=1}^{2k} a_j e^{ij\theta}\right|^2 = \sum_{j=0}^{2k} A_j \cos j\theta \ , \text{ and } \prod_{j=1}^{k} (\varepsilon_j + \cos\theta)^2 = \sum_{j=0}^{2k} B_j \cos j\theta \ ,$$

where $A_j = A_j(a_1,\ldots,a_{2k})$, and where $B_j = B_j(\varepsilon_1,\ldots,\varepsilon_k)$, then (3.2) and (3.3) give us that

$$(3.4)\qquad \begin{cases} m^2(p_{2k}) = A_0 - 2^{2k} a_{2k} B_0 \ , \\ A_j - 2^{2k} a_{2k} B_j = 0 \ , \quad j = 1,\ldots,2k-1 \ , \\ a_1 + a_2 + \ldots + a_{2k} = 1 \ . \end{cases}$$

We then formulate this as the following <u>Lagrange multiplier problem</u>. Consider

$$(3.5)\qquad F(a_1,\ldots,a_{2k},\varepsilon_1,\ldots,\varepsilon_k,\lambda_1,\ldots\lambda_{2k}) := (A_0 - 2^{2k}a_{2k}B_0) + \sum_{j=1}^{2k-1} \lambda_j (A_j - 2^{2k}a_{2k}B_j) + \lambda_{2k}(a_1 + \ldots + a_{2k} - 1),$$

subject to the conditions

$$(3.6)\qquad \frac{\partial F}{\partial a_j} = 0 \ , \ j = 1,\ldots,2k \ ; \quad \frac{\partial F}{\partial \varepsilon_j} = 0, \ j = 1,2,\ldots k \ ; \quad \frac{\partial F}{\partial \lambda_j} = 0, \ j = 1,2,\ldots,2k \ .$$

As a start-vector, we used the polynomials $Q_{2k}(z)$ of (2.4), to give the initial estimates for the coefficients $\{a_j\}_{j=1}^{2k}$. Similarly, the negative cosines of the points of local minima of $Q_{2k}(e^{i\theta})$ were used as initial estimates for $\{\varepsilon_j\}_{j=1}^{k}$, from which the associated constants $\{A_j\}_0^{2k}$ and $\{B_j\}_0^{2k}$ were determined. Then, solving the linear system $\{\frac{\partial F}{\partial a_j} = 0\}_{j=1}^{2k}$ (in the parameters $\lambda_1,\ldots,\lambda_{2k}$) determined our initial estimates for $\{\lambda_j\}_{j=1}^{2k}$. From this point, a standard nonlinear Newton procedure was used and this converged quadratically in all cases treated, thanks, no doubt, to the good start polynomials $Q_{2k}(z)$. These calculations were carried out by Timothy S. Norfolk using Richard Brent's MP (multiple precision) package on the VAX-11/780 in the Department of Mathematical Sciences at Kent State University.

Below, we give the converged values $\{\hat{\Gamma}_{2k}\}_{k=1}^{11}$ from our numerical experiments (rounded to twelve decimals), along with the associated numbers $\hat{\gamma}_{2k} := 2k(1-\hat{\Gamma}_{2k})$. (These constants $\hat{\Gamma}_{2k}$ are surely lower bounds for $\tilde{\Gamma}_{2k}$ and Γ_{2k}, but we conjecture that $\hat{\Gamma}_{2k} = \Gamma_{2k}$ in all cases below.)

k	$\hat{\Gamma}_{2k}$	$\hat{\gamma}_{2k}$
1	0.544 331 053 952	0.911 337 892 096
2	0.778 192 979 320	0.887 228 082 721
3	0.853 294 443 051	0.880 233 341 695
4	0.890 391 158 846	0.876 870 729 236
5	0.912 511 021 366	0.874 889 786 340
6	0.927 201 419 083	0.873 582 971 010
7	0.937 667 439 454	0.872 655 847 650
8	0.945 502 263 242	0.871 963 788 123
9	0.951 587 367 216	0.871 427 390 110
10	0.956 450 029 314	0.870 999 413 712
11	0.960 425 000 586	0.870 649 987 104

Table 1

Finally, it seems reasonable to suppose that constants μ_0, $\mu_1, \ldots$ exist, independent of $2k$, such that

(3.7) $$2k(1-\hat{\Gamma}_{2k}) = \mu_0 + \frac{\mu_1}{2k} + \frac{\mu_2}{(2k)^2} + \ldots \quad , \quad \text{for } k \to \infty .$$

Assuming (3.7), the Richardson extrapolation method (with $x_k = \frac{1}{2k}$) was then numerically applied to the column of numbers $\{\hat{\gamma}_{2k}\}_{k=1}^{11}$ of Table 1, to accelerate the convergence of these numbers. Numerically, this Richardson extrapolation converged very rapidly, so much so that we were led to our final two conjectures. We conjecture that

$$\lim_{k \to \infty} 2k(1 - \hat{\Gamma}_{2k}) \quad \text{exists, i.e.,}$$

(3.8) $$\lim_{k \to \infty} 2k(1 - \hat{\Gamma}_{2k}) = \mu_0 \quad ,$$

and we further conjecture that

(3.9) $$\mu_0 \doteq 0.867\ 189\ 051 \quad .$$

References

1. L. Fejér, "Trigonometrische Reihen und Potenzreihen mit mehrfach monotoner Koeffizientenfolge", Trans. Amer. Math. Soc. 39 (1936), 18-59.
2. P. Henrici, Applied and Computational Complex Analysis, volume 1, John Wiley, New York, 1974.
3. P. Henrici, "Methods of descent for polynomial equations", Computational Aspects of Complex Analysis (H. Werner, L. Wuytack, E. Ng, and H. J. Bünger, eds.), pp. 133-147. D. Reidel Publishing Co., Boston, 1983.
4. M. Marden, Geometry of Polynomials, Mathematical Surveys 3, American Mathematical Society, Providence, R.I., 1966.
5. G. Pólya and G. Szegö, Aufgaben und Lehrsatze aus der Analysis, volume II, Springer-Verlag, Berlin, 1954.
6. S. Ruscheweyh, "On a global descent method for polynomials", Numerische Mathematik (submitted).

ON THE BLOCK STRUCTURE OF THE LAURENT-PADÉ TABLE

A. BULTHEEL
Department of Computer Science
K.U. Leuven
Celestijnenlaan 200A,
B-3030 Leuven (Belgium)

Abstract. Some ideas on the block structure of a formal Laurent-Padé table will be given. The structure is derived from the block structure of the table of Toeplitz determinants which also defines the blocks of a classical Padé table.

1. Formal Laurent series

With a bi-infinite sequence of complex numbers $\{f_k\}_{-\infty}^{\infty}$ we associate a formal *Laurent series* (fls)

(1.1) $$F(z) = \sum_{-\infty}^{\infty} f_k z^k .$$

$F(z)$ will be fixed throughout this paper. With F we associate the series

(1.2) $$Z(z) = \frac{1}{2} f_0 + \sum_{1}^{\infty} f_k z^k \quad \text{and} \quad \hat{Z}(z) = \frac{1}{2} f_0 + \sum_{1}^{\infty} f_{-k} z^{-k} .$$

A *projection operator* $\Pi_{m:n}$ is defined by

(1.3) $$\Pi_{m:n} F(z) = \sum_{k=m}^{n} f_k z^k .$$

A *Laurent polynomial* is an expression like in the right hand side of (1.3) with m and n finite. Its *degree* is $\max\{|c| \mid f_c \neq 0\}$. The notations O_+ and O_- are introduced via

(1.4) $$F(z) = O_+(z^m) \quad \text{iff} \quad \Pi_{-\infty:m-1} F(z) = 0 ,$$
$$F(z) = O_-(z^n) \quad \text{iff} \quad \Pi_{n+1:\infty} F(z) = 0 .$$

$\text{Sup}\,\{m \mid F(z) = O_+(z^m)\}$ is called the *+ order* of $F(z)$ and we denote it as $\text{ord}_+(F)$. Similarly $\text{Inf}\{n \mid F(z) = O_-(z^n)\}$ is called the

- *order* of $F(z)$ and we denote it as $\mathrm{ord}_-(F)$.

The *product* of two fls is defined in the usual way as a Cauchy product, for those cases where both have finite + order, finite - order or if one of them is a Laurent polynomial. This is all that we shall need. The *reciprocal* of a fls with finite + order (- order) is also defined in the usual way. It is again a fls with finite + order (- order). If it is a Laurent polynomial, it is not clear which reciprocal we want: the one with finite + order or the one with finite - order. To distinguish between both, we use the notation $L_+(1/F(z))$ to indicate the reciprocal with finite + order and $L_-(1/F(z))$ for the other one. Also if $F(z)$ and $G(z)$ are Laurent polynomials, then

$$L_+(F(z)/G(z)) \quad \text{means} \quad F(z)\, L_+(1/G(z)) \quad \text{etc...} \quad .$$

By a *ratio of two Laurent polynomials* we mean a complete equivalence class. That is, we do not distinguish between $P(z)/Q(z)$ and $[A(z)P(z)]/[A(z)Q(z)]$ for $A(z) \neq 0$ and $P_1(z)/Q_1(z) = P_2(z)/Q_2(z)$ means that they are in the same equivalence class, i.e. $P_1(z)Q_2(z) = P_2(z)Q_1(z)$.

We call the fls (1.1) normal if all the Toeplitz determinants $T_{m,n} = \det(f_{m+i-j})_{i,j=1}^{n}$ for $m = 0, \pm 1, \pm 2, \ldots;\ n = 1, 2, \ldots$ are nonzero. It will be convenient to define $T_{m0} = 1$.

2. Laurent-Padé Approximants and Laurent-Padé Forms

The notion of Padé approximant for a formal power series can be extended to what is called a Laurent-Padé approximant (LPA) for a fls. Let m and n be nonnegative integers and suppose $R(z)$ and $\hat{R}(z)$ are both ratios of two Laurent polynomials such that:

(2.1)

(1) $Z(z) - L_+(R(z)) = O_+(z^{m+n+1})$,

(2) $\hat{Z}(z) - L_-(\hat{R}(z)) = O_-(z^{-(m+n+1)})$,

(3) $R(z) + \hat{R}(z) = A_m(z)/B_n(z)$ where $A_m(z)$ and $B_n(z)$ are Laurent polynomials of degree at most m and n , respectively.

Then we call the pair $(R(z), \hat{R}(z))$ an (m,n) LPA for $F(z)$.

Remark 1: This is clearly an extension of the classical notion of Padé approximant since $L_+(R(z)) + L_-(\hat{R}(z))$ is a fls which matches the coefficients f_k for $|k| \leq m+n$ in the given fls while $R(z) + \hat{R}(z)$

is the ratio of two Laurent polynomials with numerator degree m and denominator degree n.

Remark 2: We preferred to define a LPA as a couple of ratios rather than as the sum of these because in our formal setting there is not an unambiguous way to return from $A_m(z)/B_n(z)$ to $R(z)$ and $\hat{R}(z)$, thus to the fls that we associate with it. A LPA may not exist. What always exists is a Laurent-Padé form (LPF) which we now define by relaxing the conditions (1) and (2).

We call $(R(z),\hat{R}(z))$ an (m,n) LPF if $R(z)$ and $\hat{R}(z)$ have representants $R(z) = P_{mn}(z)/Q_{mn}(z)$ and $\hat{R}(z) = \hat{P}_{-m,n}(z)/\hat{Q}_{-m,n}(z)$, such that

(1) $Z(z)\, Q_{mn}(z) - P_{mn}(z) = O_+(z^{m+n+1})$,

(2) $\hat{Z}(z)\, \hat{Q}_{-m,n}(z) - \hat{P}_{-m,n}(z) = O_-(z^{-(m+n+1)})$,

(3) is as above.

A constructive way to find the (m,n) LPF goes as follows:

<u>Theorem 2.1</u> Let $\mu = 0, \pm 1, \pm 2, \ldots$; $\nu = 0,1,2\ldots$.
Let $Q_{\mu\nu}(z)$ and $\hat{Q}_{\mu\nu}(z)$ be nontrivial polynomials of degree ν which are (not necessarily unique) solutions of

$$(2.2) \qquad \Pi_{\mu+1:\mu+\nu}(F(z)Q_{\mu\nu}(z)) = 0 \quad \text{and} \quad \Pi_{\mu:\mu+\nu-1}(F(z)\hat{Q}_{\mu\nu}(z)) = 0 \ .$$

Define Laurent polynomials $P_{\mu\nu}(z)$ and $\hat{P}_{\mu\nu}(z)$ by

$$(2.3a) \qquad P_{\mu\nu}(z) = \Pi_{k(\mu,\nu):\ell(\mu,\nu)}(Z(z)Q_{\mu\nu}(z))$$

and

$$(2.3b) \qquad \hat{P}_{\mu+1,\nu}(z) = \Pi_{k(\mu,\nu):\ell(\mu,\nu)}(\hat{Z}(z)\hat{Q}_{\mu+1,\nu}(z)) ,$$

where $k(\mu,\nu) = \min(0,\mu+\nu+1)$ and $\ell(\mu,\nu) = \max(\mu,\nu)$.

Then $K_{mn}(z) = (P_{mn}(z)/Q_{mn}(z)\ ,\ \hat{P}_{-m,n}(z)/\hat{Q}_{-m,n}(z))$ and $\hat{K}_{m+1,n}(z) = (\hat{P}_{m+1,n}(z)/\hat{Q}_{m+1,n}(z)\ ,\ P_{-1-m,n}(z)/Q_{-1-m,n}(z))$ are (m,n) LPF's for $m \geq 0$.

<u>Proof</u> We remark that equations (2.1) always have nontrivial solutions. We define auxiliary formal series by

$$(2.4) \qquad V_{\mu\nu}(z) = \Pi_{-\infty:\mu}(F(z)Q_{\mu\nu}(z)),\ \hat{V}_{\mu\nu}(z) = \Pi_{-\infty:\mu-1}(F(z)\hat{Q}_{\mu\nu}(z)) ,$$

$$(2.5)\qquad W_{\mu\nu}(z) = \Pi_{\mu+\nu+1:\infty}(F(z)Q_{\mu\nu}(z)),\ \hat{W}_{\mu\nu}(z) = \Pi_{\mu+\nu:\infty}(F(z)\hat{Q}_{\mu\nu}(z)) \quad .$$

Clearly $F(z)Q_{\mu\nu}(z) = V_{\mu\nu}(z) + W_{\mu\nu}(z)$ and $F(z)\hat{Q}_{\mu\nu}(z) = \hat{V}_{\mu\nu}(z) + \hat{W}_{\mu\nu}(z)$.

From this it follows that (remember $F(z) = Z(z) + \hat{Z}(z)$)

$$(2.6)\qquad Z(z)Q_{\mu\nu}(z) - W_{\mu\nu}(z) = -\hat{Z}(z)Q_{\mu\nu}(z) + V_{\mu\nu}(z) = P_{\mu\nu}(z) \quad ,$$

$$(2.7)\qquad Z(z)\hat{Q}_{\mu\nu}(z) - \hat{W}_{\mu\nu}(z) = -\hat{Z}(z)\hat{Q}_{\mu\nu}(z) + \hat{V}_{\mu\nu}(z) = -\hat{P}_{\mu\nu}(z) \quad .$$

These could be seen as alternatives for the definitions (2.3).

From this we find that

$$Z(z)Q_{mn}(z) - P_{mn}(z) = W_{mn}(z) = O_+(z^{m+n+1})$$

and

$$\hat{Z}(z)(z^{-n}\hat{Q}_{-m,n}(z)) - (z^{-n}\hat{P}_{-m,n}(z)) = z^{-n}\hat{V}_{-m,n}(z) = O_-(z^{-(m+n+1)}),$$

so that the order of approximation is as required. Consider now

$$\frac{P_{mn}(z)}{Q_{mn}(z)} + \frac{z^{-n}\hat{P}_{-m,n}(z)}{z^{-n}\hat{Q}_{-m,n}(z)} = \frac{z^{-n}(P_{mn}(z)\hat{Q}_{-m,n}(z) + \hat{P}_{-m,n}(z)\ Q_{mn}(z))}{z^{-n}Q_{mn}(z)\ \hat{Q}_{-m,n}(z)} \quad .$$

The degree of the Laurent polynomial in the denominator is as it should be. Let L_{mn} be the numerator. Then we only have to prove that it is a Laurent polynomial of degree at most m . We drop the argument in the notation for simplicity. For $m,n \geq 0$ we have

$$\begin{aligned} z^nL_{mn} &= [Z\ Q_{mn} - W_{mn}]\hat{Q}_{-m,n} + [\hat{Z}\ \hat{Q}_{-m,n} - \hat{V}_{-m,n}]Q_{mn} \\ &= F\ Q_{mn}\ \hat{Q}_{-m,n} - W_{mn}\ \hat{Q}_{-m,n} - \hat{V}_{-m,n}\ Q_{mn} \\ &= [V_{mn} + W_{mn}]\hat{Q}_{-m,n} - W_{mn}\ \hat{Q}_{-m,n} - \hat{V}_{-m,n}\ Q_{mn} \\ &= V_{mn}\ \hat{Q}_{-m,n} - \hat{V}_{-m,n}\ Q_{mn} = O_-(z^{m+n}) \\ &= [\hat{V}_{-m,n} + \hat{W}_{-m,n}]Q_{mn} - W_{mn}\ \hat{Q}_{-m,n} - \hat{V}_{-m,n}\ Q_{mn} \\ &= \hat{W}_{-m,n}\ Q_{mn} - W_{mn}\ \hat{Q}_{-m,n} = O_+(z^{n-m}) \quad . \end{aligned}$$

This completes the proof for the first part of the theorem. The second part has a similar proof. □

We shall call Laurent-Padé table, the table with entries $K_{\mu\nu}$, $\mu = 0, \pm 1, \pm 2, \ldots$; $\nu = 0, 1, 2, \ldots$. Before we investigate its structure, we consider:

3. The T-table

We define the T-table as the infinite matrix with (m,n)-th entry T_{mn} , $m = 0, \pm 1, \pm 2, \ldots$, $n = 0,1,2,\ldots$, where T_{mn} is the Toeplitz determinant defined in section 1. It is known that the corresponding table in the classical Padé case has a characteristic block structure. (see Theorem 3.2 of [1]). Let $\text{ord}_+ F(z) = m^+ < \infty$. Then, except for a shift over m^+ rows, the block structure of the T-table is as described and proved in Gragg's Theorem 3.2. Similarly with the transformation $z \to 1/z$, the case $\text{ord}_- F(z) = m^- > -\infty$ is covered by this theorem. In the regions $m > m^+$ and $m < m^-$, the T_{mn} are trivially zero. Therefore we concentrate on $T_{\mu\nu}$ with $m^+ \leq \mu \leq m^-$ and $\nu \geq 0$.

For such (μ,ν) we suppose that $Q_{\mu\nu}$, $P_{\mu\nu}$, $V_{\mu\nu}$, $W_{\mu\nu}$ are solutions of the equations (2.2-5). We shall associate with it a reduced solution as follows. Let $P_{\mu\nu}(z) = z^{k(\mu,\nu)} A_{\mu\nu}$ so that $A_{\mu\nu}$ is a genuine polynomial of degree at most $\ell(\mu,\nu) - k(\mu,\nu)$. Let A/Q be the unique reduced form of $A_{\mu\nu}/Q_{\mu\nu}$ such that $Q(0) = 1$. Set further

$$(3.1) \qquad P(z) = z^{k(\mu,\nu)} A(z), \; V(z) = \hat{Z}(z)Q(z) + P(z), \; W(z) = Z(z)Q(z) - P(z).$$

We call $S = (P,Q,V,W)$ the reduced solution for equations (2.2-5). Let the greatest common divisor of $A_{\mu\nu}$ and $Q_{\mu\nu}$ producing the reduced form be $z^\lambda D(z)$ with $D(z) = d_0 + d_1 z + \cdots + d_\kappa z^\kappa$, $d_0 d_\kappa \neq 0$. Then we clearly have for $S_{\mu\nu} = (P_{\mu\nu}, Q_{\mu\nu}, V_{\mu\nu}, W_{\mu\nu})$ that

$$(3.2) \qquad S_{\mu\nu}(z) = z^\lambda D(z) S(z) .$$

Now $\text{ord}_- V$ and $\text{ord}_+ W$ will be used to define the top row, respectively the bottom row of the blocks in the T-table. In a classical Padé situation $\text{ord}_- V$ will always be nonnegative, but now $\text{ord}_- V$ can be $-\infty$. In this case, it is impossible to define the bottom row as $\text{ord}_- V+k$ because k must then be ∞. Finally, we note that $\text{ord}_- V = -\infty$ and $\text{ord}_+ W = \infty$ cannot occur simultaneously if $F \neq 0$

because $FQ = V + W$.

Theorem 3.1 Let $S = (P, Q, V, W)$ be the reduced solution of equations (2.2-5). Let

(3.3) $\mathrm{ord}_- Q = n$, $\mathrm{ord}_- V = m$, $\mathrm{ord}_+ W = \hat{m} + n + 1$

and define $k = \hat{m} - m$. Then the following statements are true:

(a) $k \geq 0$.

(b) $S = (P, Q, V, W)$ is a reduced solution of equations (2.2-5) if and only if

(3.4) $m \leq \mu \leq \hat{m}$ and $n \leq \nu \leq n + k$.

For (μ, ν) satisfying (3.4):

(c) $S_{\mu\nu} = (P_{\mu\nu}, Q_{\mu\nu}, V_{\mu\nu}, W_{\mu\nu})$ is a solution of (2.2-5) if and only if

$$S_{\mu\nu} = z^{\lambda_{\mu\nu}} D(z)\ S(z)$$

with

(3.5) $$\lambda_{\mu\nu} = \max\{0, (\mu-m) + (\nu-n)-k\} \quad \text{if} \quad m > -\infty$$
$$= \max\{0, (\mu-\hat{m}) + (\nu-n)\} \quad \text{if} \quad \hat{m} < \infty$$

and D a nonzero polynomial of degree at most

(3.6) $$\kappa_{\mu\nu} = \nu - \mathrm{rank}\ (f_{\mu+i-j})_{i=1,\dots\nu;\, j=0,1,\dots,\nu} \ .$$

(d) $T_{\mu n} \neq 0$ $\quad m \leq \mu \leq \hat{m}$,

$T_{m\nu} \neq 0$ $\quad n \leq \nu \leq n + k$, if $m > -\infty$,

$T_{\hat{m}+1,\nu} \neq 0$ $\quad n \leq \nu \leq n + k$, if $\hat{m} < \infty$,

$T_{\mu,n+k+1} \neq 0$ $\quad m \leq \mu \leq \hat{m}$, if $k < \infty$,

$T_{\mu\nu} = 0$ $\quad m < \mu \leq \hat{m}$ and $n < \nu \leq n + k$.

Proof

We give the proof only for $m > -\infty$, in which case $\hat{m} = m + k$. If $m = -\infty$, $\hat{m}$ will be finite and we can rewrite the following proof with m replaced by $\hat{m} - k$. Like in [1,p14] we easily derive from (3.2) that we must have the relations

(3.7) $\kappa \geq 0$, $\lambda \geq \max\{0,(\mu-m)+(\nu-n)-k\}$, $\kappa+\lambda \leq \min\{\mu-m,\nu-n\}$.

Conversely, let $S = (P, Q, V, W)$ satisfy (2.2-5) with (μ,ν) replaced by (m,n) and suppose $Q(0) = 0$, $\text{ord}_- Q = n$, $\text{ord}_- V = m$ and $\text{ord}_+ W = m+n+k+1$. Then for λ and κ solutions of (3.7) and for an arbitrary polynomial $D(z) = d_0 + d_1 z + \cdots + d_\kappa z^\kappa$, $d_0 d_\kappa \neq 0$, $S_{\mu\nu}$ given by (3.2) will be a solution of (2.2-5). Since S is a reduced solution there must exist integers κ, λ, μ, ν satisfying (3.7). Thus

$$k \geq (\mu-m) + (\nu-n) - \lambda \geq 2\kappa + \lambda \geq 0 .$$

This proves (a).

Furthermore (3.7) can only have a solution if

(3.8) $\max\{0,(\mu-m)+(\nu-n)-k\} \leq \min\{\mu-m,\nu-n\}$,

which is equivalent with (3.4). Thus also (b) is proved.

To prove (c) we remark that the most general solution of (2.2-5) is obtained by choosing λ as small as possible and κ as large as possible. The minimal λ satisfying (3.7) is given by $\lambda_{\mu\nu}$ in (3.5) and the maximal κ is $\kappa_{\mu\nu}$ as in (3.6) because $\kappa_{\mu\nu}+1$ is then the number of degrees of freedom we have in the choice of a solution $Q_{\mu\nu}$ for (2.2).

We know that $T_{\mu\nu} = 0$ if and only if (2.2) has a nontrivial solution for $Q_{\mu\nu}$ with $Q_{\mu\nu}(0) = 0$, i.e. with $\lambda > 0$. Because of (3.7) this occurs if and only if $\min\{\mu-m,\nu-n\} > 0$ and (3.8) is satisfied. This is equivalent with statement (d). □

The numbers m, n, k and $\hat{m}$ are related to the unique reduced solution of (2.2-5) and depend for a given F only on μ and ν . Therefore we define with the notation of previous theorem:

$$N_{\mu\nu} = 0 \quad \text{if } \mu < m^+ \text{ or } \mu > m^-$$
$$= n \quad \text{otherwise,}$$
$$K_{\mu\nu} = \infty \quad \text{if } \mu < m^+ \text{ or } \mu > m^-$$
$$= k \quad \text{otherwise,}$$
$$M_{\mu\nu} = -\infty \quad \text{if } \mu < m^+$$
$$= m^- \quad \text{if } \mu > m^-$$
$$= m \quad \text{otherwise,}$$

$$\hat{M}_{\mu\nu} = m^+-1 \quad \text{if} \quad \mu < m^+$$
$$= \infty \quad \text{if} \quad \mu > m^-$$
$$= \hat{m} \quad \text{otherwise.}$$

With these we define square blocks of indices:

$$\mathcal{B}_{\mu\nu} = \{(m,n) \,|\, M_{\mu\nu} \leq m \leq \hat{M}_{\mu\nu},\; N_{\mu\nu} \leq n \leq N_{\mu\nu} + K_{\mu\nu}\}$$

and by deleting its first row and first column we obtain

$$\beta_{\mu\nu} = \{(m,n) \,|\, M_{\mu\nu} < m \leq \hat{M}_{\mu\nu},\; N_{\mu\nu} < n \leq N_{\mu\nu} + K_{\mu\nu}\} \;.$$

Part (d) of previous theorem says that $T_{mn} = 0$ for $(m,n) \in \beta_{\mu\nu}$ and that for all the finite couples (m,n) bordering $\beta_{\mu\nu}$, T_{mn} will be nonzero.

4. The Laurent-Padé table

We consider now the table with entries $K_{\mu\nu}$ as defined in Theorem 2.1. It is a simple consequence of Theorem 3.1 (c) that $P_{mn}/Q_{mn} = P_{\mu\nu}/Q_{\mu\nu}$ for all $(m,n) \in \mathcal{B}_{\mu\nu}$. Also it follows from defining equations (2.2-5) that $\hat{P}_{\mu+1,\nu}/\hat{Q}_{\mu+1,\nu} = P_{\mu\nu}/Q_{\mu\nu}$. Thus we must also have $\hat{P}_{mn}/\hat{Q}_{mn} = \hat{P}_{\mu\nu}/\hat{Q}_{\mu\nu}$ for all (m,n) in the shifted grid $(1,0) + \mathcal{B}_{\mu-1,\nu} = \{(p+1,q) \,|\, (p,q) \in \mathcal{B}_{\mu-1,\nu}\}$. The (rectangular) blocks of equal entries in the Laurent-Padé table are found as follows. Produce a plot of the square blocks $\mathcal{B}_{\mu\nu}$. Reflect this plot in the line $\mu = -1/2$ and take the intersections of the original and the reflected squares. These are rectangles with equal K_{mn} entries. More precisely we have

<u>Theorem 4.1.</u>

Define $\mathcal{D}_{\mu\nu} = \mathcal{B}_{\mu\nu} \cap \hat{\mathcal{B}}_{\mu\nu}$ with $\hat{\mathcal{B}}_{\mu\nu} = \{(m,n) \,|\, (-m,n) \in (1,0) + \mathcal{B}_{\mu-1,\nu}\}$.
Then $K_{mn} = K_{\mu\nu}$ for all $(m,n) \in \mathcal{D}_{\mu\nu}$.

<u>Proof</u>

This is very simple to prove. By the remark at the beginning of this section we know that $P_{mn}/Q_{mn} = P_{\mu\nu}/Q_{\mu\nu}$ for $(m,n) \in \mathcal{B}_{\mu\nu}$ and $\hat{P}_{-m,n}/\hat{Q}_{-m,n} = \hat{P}_{-\mu,\nu}/\hat{Q}_{-\mu,\nu}$ for $(m,n) \in \hat{\mathcal{B}}_{\mu\nu}$. This proves the theorem. □

We remark that not all of these LPF's are LPA's. As in a classical Padé table, only the elements with indices in the left upper triangular part of a singular block, diagonal included, are Padé approximants. Here we have a similar situation: Let $L_{\mu\nu}$ be this left upper part of $B_{\mu\nu}$ and $L_{\mu\nu}$ the left upper part of $\hat{B}_{\mu\nu}$. Then we have

Theorem 4.2

For $\mu \geq 0$

$$(\mu,\nu) \in L_{\mu\nu} \text{ iff } \operatorname{ord}_+ (Z - L_+ (P_{\mu\nu}/Q_{\mu\nu})) \geq \mu + \nu + 1 \quad ,$$

$$(\mu,\nu) \in \hat{L}_{\mu\nu} \text{ iff } \operatorname{ord}_- (\hat{Z} - \hat{L}_- (\hat{P}_{\mu\nu}/\hat{Q}_{\mu\nu})) \leq -(\mu + \nu + 1)$$

and

$$K_{\mu\nu} \text{ is a LPA iff } (\mu,\nu) \in L_{\mu\nu} \cap \hat{L}_{\mu\nu}$$

Proof

Let $S = (P, Q, V, W)$ be the reduced solution for (2.2-5) with $\operatorname{ord}_- V = m$; $\operatorname{ord}_- Q = n$ and $\operatorname{ord}_+ W = \hat{m} + n + 1$. Then $\operatorname{ord}_+ (Z - L_+ (P_{\mu\nu}/Q_{\mu\nu})) = \operatorname{ord}_+ (Z - L_+ (P/Q)) = \operatorname{ord}_+ (Z\,Q - P) = \operatorname{ord}_+ W = \hat{m} + n + 1$. Thus $\operatorname{ord}_+ (Z - L_+ (P_{\mu\nu}/Q_{\mu\nu})) \geq \mu + \nu + 1$ iff $\hat{m} + n \geq \mu + \nu$. That is, $(\mu,\nu) \in L_{\mu\nu}$. The second relation can be proved similarly. Both relations are true iff $K_{\mu\nu}$ is a LPA. □

5. The Chebyshev-Padé table

For the special case that $f_{-k} = f_k$ for all k, we may replace "Laurent-Padé" by "Chebyshev-Padé" in the previous text. It can directly be verified from their definitions that the symmetry of F implies the following properties:

Theorem 5.1

If $F(z) = F(1/z)$, then for all $\mu = 0, \pm 1, \pm 2, \ldots$, $\nu = 0, 1, \ldots$: $Q_{-\mu,\nu}(z) = z^{\nu} Q_{\mu\nu}(1/z)$; $V_{-\mu,\nu}(z) = z^{\nu}\hat{W}_{\mu\nu}(1/z)$; $W_{-\mu,\nu}(z) = z^{\nu}\hat{V}_{\mu\nu}(1/z)$ and $P_{-\mu,\nu}(z) = z^{\nu}\hat{P}_{\mu\nu}(1/z)$. Also $T_{-\mu,\nu} = T_{\mu\nu}$ and therefore $B_{\mu\nu} = \hat{B}_{\mu\nu}$, so that $D_{\mu\nu} = B_{\mu\nu}$.

This means that the Chebyshev-Padé table is symmetric with respect to $\mu = -1/2$. The square blocks $D_{\mu\nu}$ define equal entries and in $L_{\mu\nu}$ these entries are Chebyshev-Padé approximants in the strict sense.

6. Conclusion

We derived the block structure for a Laurent-Padé table. It

contains rectangular blocks but these become square again in the symmetric case of Chebyshev-Padé. We derived the results in a formal setting. If the fls is a Laurent series of a function convergent in an annular region of the complex plane one should be more careful in the definition of a LPA as was done in [3].

References

[1] W. B. Gragg: The Padé table and its relations to certain algorithms of numerical analysis. SIAM Review, 14, 1972, pp. 1-62.

[2] W. B. Gragg and G. D. Johnson: The Laurent-Padé table. In Proc. IFIP congress 1974, North-Holland, 1974, pp. 632-637.

[3] K. O. Geddes: Block structure in the Chebyshev-Padé table. SIAM J. Numer. Anal. 18, 1981, pp. 844-861.

SQUARE BLOCKS AND EQUIOSCILLATION IN THE PADÉ, WALSH, AND CF TABLES

Lloyd N. Trefethen*

Courant Institute of Mathematical Sciences

New York University

New York, NY 10012

Abstract. It is well known that degeneracies in the form of repeated entries always occupy square blocks in the Padé table, and likewise in the Walsh table of real rational Chebyshev approximants on an interval. The same is true in complex CF (Carathéodory-Fejér) approximation on a circle. We show that these block structure results have a common origin in the existence of equioscillation-type characterization theorems for each of these three approximation problems. Consideration of position within a block is then shown to be a fruitful guide to various questions whose answers are affected by degeneracy.

0. Introduction

Consider the following three problems in rational approximation. In each case m and n are nonnegative integers, except that m may be negative in the CF case.

CHEBYSHEV ("T"). Let f be real and continuous on $I = [-1,1]$, and let R^r_{mn} be the set of rational functions of type (m,n) with real coefficients. Problem: find $r^* \in R^r_{mn}$ such that

(1.T) $$\|f-r^*\|_I \le \|f-r\|_I \quad \forall r \in R^r_{mn} ,$$

where $\|\cdot\|_I$ is the supremum norm on I .

PADÉ ("P"). Let f be a complex formal power series in z , and let R_{mn} be the set of rational functions of type (m,n) with complex coefficients. Problem: find $r^P \in R_{mn}$ such that

(1.P) $$(f-r^P)(z) = O((f-r)(z)) \quad \text{as} \quad z \to 0 \quad \forall r \in R_{mn} .$$

---

*Supported by an NSF Postdoctoral Fellowship and by the U.S. Dept. of Energy under contract DE-AC02-76-ER03077-V.

CF ("K"). Let f be a continuous function on $S = \{z: |z|=1\}$ whose Fourier series converges absolutely. Let $\tilde{R}_{mn}$ be the set of "extended rational" functions representable in the form

$$(2) \qquad \tilde{r}(z) = \frac{\tilde{p}(z)}{q(z)} = \sum_{k=-\infty}^{m} a_k z^k \Big/ \sum_{k=0}^{n} b_k z^k ,$$

where q has all of its zeros in $|z| > 1$, and the series for $\tilde{p}$ converges there and is bounded except possibly near $z = \infty$. Problem: find $\tilde{r}^* \in \tilde{R}_{mn}$ such that

$$(1.K) \qquad \|f-\tilde{r}^*\|_S \le \|f-\tilde{r}\|_S \qquad \forall \tilde{r} \in \tilde{R}_{mn} ,$$

where $\|\cdot\|_S$ is the L^∞ norm on S.

See [12] for information on Chebyshev approximation, [1,6] for Padé, and [7,14,15] for CF. (The CF approximant defined above is actually the "extended CF approximant"; in practice it would be projected onto a function $r^{cf} \in R_{mn}$ to yield a near-best Chebyshev approximant on a disk.) All three problems have unique solutions, and these can be constructed numerically: r^p by solving a finite Hankel system of linear equations, r^* by a procedure such as the Remes algorithm, and $\tilde{r}^*$ via a singular value decomposition of an infinite Hankel matrix of Laurent series coefficients. We will not go into this.

The Padé table is the array obtained by arranging the various approximants r^p for a given f in sequence in the lower-right quadrant of the plane, with m as the column index and n as the row index. (Sometimes these indices are reversed [6].) The corresponding array of Chebyshev approximants r^* is called the Walsh table. The (extended) CF table is the analogous array of CF approximants $\tilde{r}^*$, except that since $m < 0$ is permitted, it fills the entire lower half plane instead of a quadrant.

The first purpose of this paper is to publicize the connection between equioscillation theorems and block structure in approximation tables (Secs. 1,2). We will show by examples that the equioscillation/square blocks point of view is of mnemonic and heuristic value in investigating problems whose answers are affected by degeneracy (Secs. 2,3).

The second purpose is to elucidate the close analogy between the Chebyshev, Padé, and CF problems. Many aspects of these problems depend only on the superficial structure imposed by an equioscillation theorem, not on the details of the type of approximation. Pursuing this analogy gives insight into how things are different in related problems where there is no equioscillation principle (Sec. 4). It also raises an interesting question of whether the Walsh, Padé, and CF tables all admit the same set of possible block configurations (Sec. 5).

1. Equioscillation characterizations

The word "equioscillation" comes from the Chebyshev problem, where it describes an error curve $(f-r^*)(I)$ that oscillates sufficiently many times between positive and negative extrema with equal magnitude. In CF approximation, the analogous object is an error curve $(f-\tilde{r}^*)(S)$ that is a perfect circle of sufficiently large winding number. In Padé approximation, it is an error function $f-r^p$ that is zero to sufficiently high order at the origin. One can think of this as a circular error curve condition too, for as $\varepsilon \to 0$, $(f-r^p)(\varepsilon S)$ approaches a circle with winding number equal to the degree of the first nonzero coefficient in $f-r^p$.

The question of how great an equioscillation number is "sufficient" depends on the <u>defect</u> δ. Given r ($\tilde{r}$), let it be expressed as a quotient p/q ($\tilde{p}/q$) in lowest terms, i.e. in which the numerator and denominator have no common zeros. Let $\mu \le m$ and $\nu \le n$ be the exact degrees of p ($\tilde{p}$) and q, with $\mu = -\infty$ if $p \equiv 0$ ($\tilde{p} \equiv 0$). Then δ is defined by

(3) $$\delta = \min\{m-\mu, n-\nu\}.$$

<u>THEOREM 1T</u>. *If $r \in R^r_{mn}$ has defect δ, then $r = r^*(f)$ if and only if the error curve $(f-r)(I)$ oscillates between $\pm \|f-r\|_I$ on some sequence of points $-1 \le x_0 < \ldots < x_N \le 1$ with $N \ge m+n+1-\delta$.*

<u>THEOREM 1P</u>. *If $r \in R_{mn}$ has defect δ, then $r = r^p(f)$ if and only if*

(4) $$(f-r)(z) = O(z^N) \quad \text{as} \quad z \to 0 \quad \text{with} \quad N \ge m+n+1-\delta .$$

<u>THEOREM 1K</u>. *If $\tilde{r} \in \tilde{R}_{mn}$ has defect δ, then $\tilde{r} = \tilde{r}^*(f)$ if and only if $\tilde{r}$ is continuous in $|z| \ge 1$ and the error curve $(f-\tilde{r})(S)$ is a circle of winding number $N \ge m+n+1-\delta$ in the positive sense.*

<u>Remark</u>. In each case we will assume N is chosen as large as possible (possibly ∞), and refer to it as "the equioscillation number".

<u>Proofs</u>. These assertions have two halves, namely "equioscillation implies best" and "best implies equioscillation". The result in the first direction is easily obtained by counting zeros. For example if $r \in R_{mn}$ satisfies (4), then by (1.P) one has also $f-r^p = O(z^N)$.

Therefore $r-r^p$ has at least $N \geq m+n+1-\delta$ zeros at the origin. But since $r-r^p \in R_{m+n-\delta,2n-\delta}$, this implies $r = r^p$. (This argument also shows that r^p is unique.) Analogous proofs work for Chebyshev and CF.

Showing "best implies equioscillation" is less trivial, but the threefold analogy can be maintained by arguing in the following way. Suppose r ($\tilde{r}$) does not equioscillate sufficiently many times. Then it can be perturbed slightly to a new function $r' = r+\Delta r$ ($\tilde{r}'$) which is a better approximant. The method of constructing this perturbation depends on which approximation problem is being considered. See [12] for the Chebyshev case (quite straightforward) and [8] for CF (trickier). For Padé approximation one could also write down a perturbation argument, but it is unnecessary since the problem of determining coefficients is actually linear. Therefore it is as well to obtain Thm. 1P as a corollary of the usual Padé table derivation via linear algebra. See [1] or [6].

2. Square blocks

The following arguments run the same way for Chebyshev, Padé, or CF; we consider the Padé case for definiteness. Given f, suppose a function r of exact type (μ,ν) with $\mu > -\infty$ happens to satisfy

$$(5) \qquad (f-r)(z) = O(z^N), \quad \neq O(z^{N+1}) \quad \text{as} \quad z \to 0 \quad \text{with} \quad N = \mu+\nu+1+\Delta$$

for some $\Delta \geq 0$ (possibly ∞, in which case the $\neq O(z^{N+1})$ clause is dropped). For which (m,n), if any, is r the Padé approximant r^P? The answer is, for precisely those (m,n) that lie in any of the positions of the following $(\Delta+1) \times (\Delta+1)$ block:

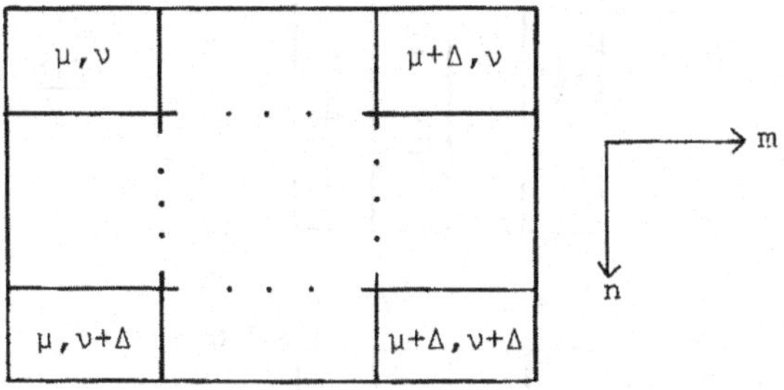

Figure 1

To verify this claim, combine (4) and (5) to obtain the following condition: $r = r^p$ if and only if $m \geq \mu$, $n \geq \nu$, and

$$\mu+\nu+\Delta \geq m+n-\delta ,$$

or by (3),

$$\mu+\nu+\Delta \;\geq\; m+n-\min\{m-\mu, n-\nu\} .$$

If $m-\mu \leq n-\nu$, this becomes $\nu+\Delta \geq n$, which gives the lower-left half of the square block of Fig. 1. The alternative $n-\nu \leq m-\mu$ leads to $\mu+\Delta \geq m$, which gives the upper-right half of the block.

This argument carries over directly to Chebyshev and CF approximation. We can summarize the situation for all three problems as follows: if $f-r$ equioscillates the "normal" number of times $N = \mu+\nu+1$, then r is the desired approximant in the (μ,ν) position but nowhere else. With each "extra" oscillation, the size of the block in which r is the approximant increases by 1.

In (5) we have excluded the possibility $\mu = -\infty$, i.e. $r \equiv 0$ ($\tilde{r} \equiv 0$). Here (3) gives $\delta = n$, and so 0 is the desired approximant if and only if f itself equioscillates with $N \geq m+1$. That is, the zero function fills all columns of the table with $m \leq N-1$, if any. This is the only situation in which non-square blocks occur.

Here is a general block structure statement.

THEOREM 2. *The Walsh, Padé, and CF tables all break down into precisely square blocks containing identical entries. (One of these may be infinite in extent, if f can be approximated exactly for large enough (m,n).) The only exception is that if an entry $r \equiv 0$ ($\tilde{r} \equiv 0$) appears in the table, then it fills all of the columns to the left of some fixed index $m = N$.*

To make these conclusions vivid, Fig. 2 shows how m, n, δ, and N are distributed within a square block.

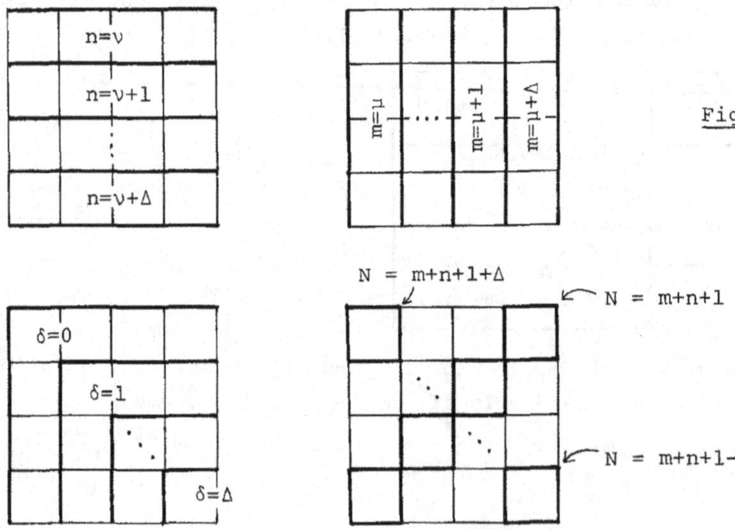

Figure 2

Here is a summary of the notation we have introduced:

f - function to be approximated

m,n - nonnegative integers (m may be negative in CF case)

$R^r_{mn}, R_{mn}, \tilde{R}_{mn}$ - spaces of real, complex, extended rational functions

$r^*, r^p, \tilde{r}^*$ - Chebyshev, Padé, CF approximants of type (m,n)

$\mu \le m, \nu \le n$ - exact degrees of approximant

$\delta = \min\{m-\mu, n-\nu\}$ - defect

$\Delta+1$ - dimension of square block

$N = \mu+\nu+1+\Delta$ - equioscillation number

3. Continuity of the Chebyshev, Padé, and CF operators

Complications often arise when one deals with degenerate approximants. For this reason we say that a Walsh, Padé, or CF table is normal if no entry appears twice, that is, if every block has size 1×1. The word "normal" has also been applied to individual entries in a table, but unfortunately its uses in the Chebyshev and Padé literature have been inconsistent. Perhaps the following problem-independent definitions make the most sense: an approximant is nondegenerate if $m = \mu$ or $n = \nu$ (i.e. if $\delta = 0$), and normal if $m = \mu$, $n = \nu$, and $\Delta = 0$. (This use of "normal" follows the Padé convention [6]; in Chebyshev approximation "normal" has meant what we call "nondegenerate" [12,17].) Figure 3 illustrates the two definitions.

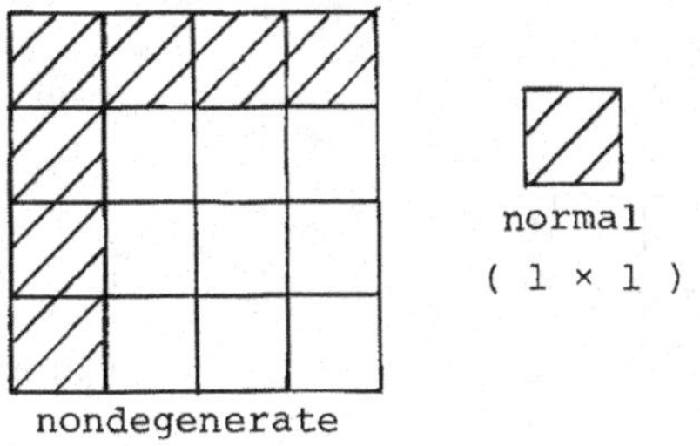

Figure 3

Various approximation results can be stated in terms of hypotheses on position within the square block. For example: (1) Walsh showed in 1974 that if r^p is nondegenerate, then $r^*_{\varepsilon I} \to r^p$ as $\varepsilon \to 0$, where $r^*_{\varepsilon I}$ is the Chebyshev approximant on $[-\varepsilon,\varepsilon]$; on the other hand if r^p is degenerate, this need not hold [16]. (2) The analogous result for CF approximation appears to be that $r^*_{\varepsilon S} \to r^p$ as $\varepsilon \to 0$ is guaranteed if r^p lies in the upper-right or lower-left corner of its block (Gutknecht and Trefethen, forthcoming). (3) A theorem of Ruttan states that if the Chebyshev approximant r^* of a real function f lies in the strict lower-right subtriangle of its square block, then f can be better approximated in R_{mn} than R^r_{mn} [13]. (4) In the corresponding strict lower-right subtriangle of a square block in the Padé table, the so-called Padé equations are inconsistent, and according to the "Baker definition" of the Padé approximant (different from ours), r^p does not exist [1].

We will now describe a particularly appealing application of block structure arguments. Let the Chebyshev, Padé, and CF approximation operators be defined by

$$T: f \mapsto r^* , \qquad P: f \mapsto r^p , \qquad K: f \mapsto \tilde{r}^* .$$

The question is, when are these operators continuous? We omit details concerning the precise definitions of continuity -- see [8,16].

For T and P the answers turn out to be the same, and were obtained by Werner [12,17] and by Werner and Wuytack [16,18], respectively:

<u>THEOREMS 3T, 3P</u>. <u>The operator</u> T (P) <u>is continuous at</u> f <u>if and only if</u> $T(f)$ ($P(f)$) <u>is nondegenerate</u>.

In contrast, the forthcoming paper [8] will establish the following result for K :

<u>THEOREM 3K</u>. <u>The operator</u> K <u>is continuous at</u> f <u>if and only if</u> $K(f)$ <u>is nondegenerate and in addition has equioscillation number exactly</u> $N = m+n+1$.

Figure 4 summarizes these results:

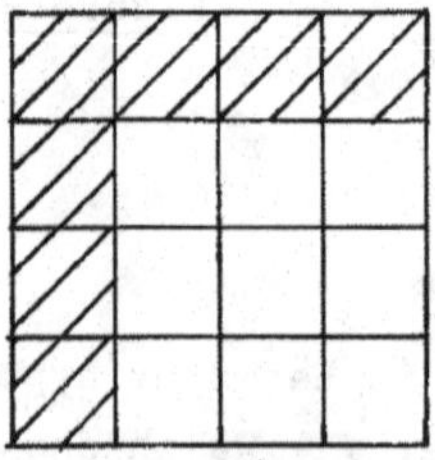

T,P continuous

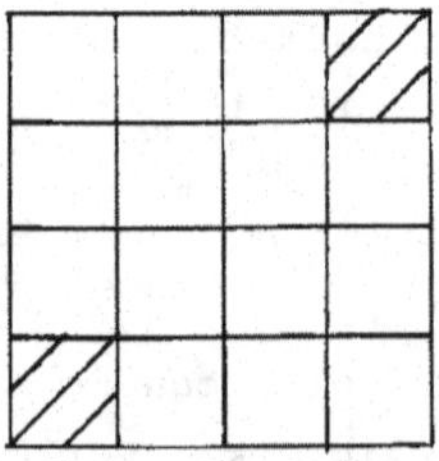

K continuous

<u>Figure 4</u>

The interesting thing is that despite the apparent discrepancy between Thms. 3T,3P and Thm. 3K, all of these results actually have a single explanation in terms of block structure. The underlying principle is this: small perturbations can break square blocks, but not make them (cf. [10]).

THEOREM 4. *Suppose* (m_1,n_1) *and* (m_2,n_2) *lie in distinct square blocks of the Walsh, Padé, or CF table for the function* f . *Then the same is true for all sufficiently small perturbations* $f' = f + \Delta f$.

Proof. Without loss of generality we can assume $m_1 \le m_2$ and $n_1 \le n_2$. (Otherwise, an easy argument based on block structure shows we can replace either (m_1,n_1) or (m_2,n_2) by $(\min\{m_1,m_2\},\min\{n_1, n_2\})$.) In the Chebyshev case, the hypotheses imply $\|f-r_2^*\|_I < \|f-r_1^*\|_I$, and by the definition of r^* , the inequality will persist under perturbations with $\|\Delta f\|_I < \frac{1}{2}(\|f-r_1^*\|_I - \|f-r_2^*\|_I)$. The same argument (with $\|\cdot\|_S$) works in the CF case. For Padé approximation an analogous proof can also be constructed, or one can appeal to known results about the linear algebra of the Padé table and use the fact that a small perturbation of a nonsingular matrix is nonsingular.

Theorem 4 now suggests the following idea for the proofs of Thms. 3T,3P,3K, though it is quite disguised in the original papers: assuming that perturbations of f can be constructed to fracture a square block in any desired way, what does this imply about discontinuity? As it happens, the following single fracture pattern is the only one needed, and it turns out that a perturbation can always be found that accomplishes it. Of course, constructing this perturbation requires consideration of problem-dependent details.

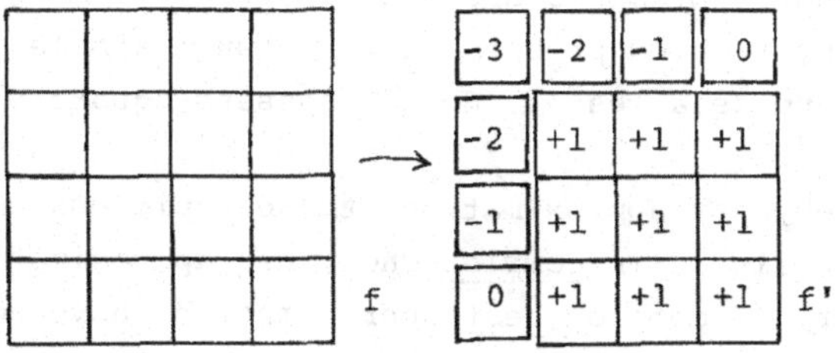

Figure 5

From Fig. 2 we know that in any square block, the equioscillation number N is determined by the position of the main cross-diagonal. Therefore under the perturbation of Fig. 5, the value of N in each position must change by exactly the quantity indicated. In particular, it increases in the lower-right subblock, stays constant in two corners, and decreases in the remaining positions on the upper-left border.

Now from Thms. 1, it can be seen that in all three cases T, P, K, an increase in N cannot be obtained through an arbitrarily small perturbation. For example, the winding number of a circular error curve with positive radius ρ cannot increase except by the addition of a function of norm at least ρ . Therefore the construction of Fig. 5 proves discontinuity of T , P , and K whenever the approximant to f is degenerate.

On the other hand, Thms. 1 also imply that N can decrease under a small perturbation in cases T and P , but not in case K . For example, a point x_j in the alternant set for the error curve $(f-r^*)(I)$ may cease to be extremal in response to arbitrarily small perturbations of f -- whereas a circular CF error curve cannot decrease smoothly to another circle of lower winding number. Therefore Fig. 5 proves discontinuity of K , but not of T or P , in the upper-left border positions away from the corners.

These arguments establish discontinuity in all of the unshaded positions of Fig. 4. A little experimentation quickly shows that no alternative fragmentations of the block produce discontinuity in any further positions; in fact the shaded positions of Fig. 4 can be characterized as those locations at which N cannot increase through an arbitrarily small perturbation of f (cases T , P), and those where it cannot change at all (case K). To complete the proofs of Thms. 3, all that remains is to verify continuity in these shaded positions. This is another problem-dependent argument, which we omit.

4. Related problems

The Chebyshev, Padé, and CF problems are unusual in obeying equioscillation principles. Many related problems have no such simple characterization theorems, and as a result, do not possess square block structure.

Within the realm of Chebyshev approximation, the conspicuous context where equioscillation fails is in complex Chebyshev approximation. On either a disk or an interval, complex best approximations have no simple characterization, need not be unique, and need not lie in square blocks [7,16]. A circular error curve of high enough winding number is here still sufficient for best approximation, but no longer necessary.

Both equioscillation and square blocks also vanish in general if one approximates in other norms, such as L^2 .

The equioscillation principle fails in Padé approximation as soon as one generalizes from interpolation just at the origin to any kind of multipoint (or "Newton-Padé") scheme. In fact in problems of this kind the interpolation table itself requires careful definition, since the approximation obtained depends not just on the number of interpolation conditions specified, but also on the order in which they are taken [3, 4].

The analogous multipoint version of CF approximation is the Pick-Nevanlinna problem, with its extension to the rational case due to Achieser. The block structure (not square) in the "PNA table" is currently being investigated by Gutknecht.

All of these remarks show that the ideas of this paper are far from universal in application. However, they are not exhausted, either. On one hand, there are probably other interesting problems besides the three we have mentioned in which one gets equioscillation and square blocks. (This appears to be at least nearly true in so-called Chebyshev-Padé approximation; see [5] and the paper by Bultheel in this volume.) If so, their analysis will be aided by recognition of the recurring patterns of reasoning illustrated here. On the other hand, the strength of the analogies between various problems with equioscillation theorems suggests that useful connections between more complicated problems are also worth looking for. The multipoint Padé/PNA analogy mentioned above is a step in this direction.

5. Which block patterns are possible?

We know that the Walsh, Padé, and CF tables break into square blocks, but what about the converse? Can an arbitrary tiling of a quadrant by squares of various sizes, say, be realized as the block pattern of the Padé table of some formal power series f ? It seems this question has not appeared in print before, but apparently it was asked and answered several years ago by A. Magnus [11] (and possibly others).

The answer is no, and a simple example proves it. Suppose that the top two rows of the Padé (or Walsh or CF) table of f are known to break precisely into a chain of 2×2 blocks. Then it is easily seen that f is even, and this implies that the rest of its table also divides into 2×2 blocks, or larger. Therefore any finer patterns of subdivision are impossible. Of course one can readily generalize this example in various directions.

Since arbitrary block patterns are not permitted, one is naturally led to the following

PROBLEM. *Characterize all patterns of square blocks that can occur in the Walsh, Padé, and CF tables.*

One reason this problem is of interest is that its solution would help one judge the extent of validity of the theme of this paper --

that block patterns are mainly a function of superficial structure, not of problem-dependent details. What is the most one could hope for in this direction? One possibility is the following, offered here not with conviction but as a stimulus to further thought.

CONJECTURE. *The sets of possible block patterns in the Walsh and Padé tables are identical.*

Acknowledgments

I am happy to thank Martin Gutknecht, William Gragg, William Jones, and Arne Magnus for many useful comments.

References

[1] G. Baker and P. Graves-Morris, *Padé Approximants* (2 vols.), Encyc. of Math. v. 13 and 14, Addison-Wesley, 1981.

[2] A. Bultheel, paper in this volume.

[3] G. Claessens, On the structure of the Newton-Padé table, J. Approx. Theory 22 (1978), 304-319.

[4] M. A. Gallucci and W. B. Jones, Rational approximations corresponding to Newton series, J. Approx. Theory 17 (1976), 366-392.

[5] K. O. Geddes, Block structure in the Chebyshev-Padé table, SIAM J. Numer. Anal. 18 (1981), 844-861.

[6] W. Gragg, The Padé table and its relation to certain algorithms of numerical analysis, SIAM Review 14 (1972), 1-62.

[7] M. H. Gutknecht, On complex rational approximation II, in *Computational Aspects of Complex Analysis*, H. Werner et al. (eds.), D. Reidel, Dordrecht/Boston/Lancaster, 1983.

[8] M. H. Gutknecht, E. Hayashi, and L. N. Trefethen, The CF table, in preparation.

[9] M. H. Gutknecht and L. N. Trefethen, Nonuniqueness of best rational Chebyshev approximations on the unit disk, J. Approx. Theory 39 (1983), 275-288.

[10] A. Magnus, The connection between P-fractions and associated fractions, Proc. Amer. Math. Soc. 25 (1970), 676-679.

[11] A. Magnus, private communication, 1983.

[12] G. Meinardus, *Approximation of Functions: Theory and Numerical Methods*, Springer, 1967.

[13] A. Ruttan, The length of the alternation set as a factor in determining when a best real rational approximation is also a best complex rational approximation, J. Approx. Theory 31 (1981), 230-243.

[14] L. N. Trefethen, Rational Chebyshev approximation on the unit disk, Numer. Math. 37 (1981), 297-320.

[15] L. N. Trefethen, Chebyshev approximation on the unit disk, in *Computational Aspects of Complex Analysis*, H. Werner et al. (eds.), D. Reidel, Dordrecht/Boston/Lancaster, 1983.

[16] L. N. Trefethen and M. H. Gutknecht, On convergence and degeneracy in rational Padé and Chebyshev approximation, SIAM J. Math. Anal., to appear.

[17] H. Werner, On the rational Tschebyscheff operator, Math. Zeit. 86 (1964), 317-326.

[18] H. Werner and L. Wuytack, On the continuity of the Padé operator, SIAM J. Numer. Anal. 20 (1983), 1273-1280.

PROPERTIES OF PADÉ APPROXIMANTS TO STIELTJES SERIES AND SYSTEMS THEORY

N.K. Bose*
Departments of Electrical Engg. and Mathematics
348 Benedum Hall
University of Pittsburgh
Pittsburgh,
Pennsylvania 15621
U.S.A.

Abstract. The fact that Padé approximants of certain orders to a matrix Stieltjes series are realizable as open-circuit impedance matrices of RC-ideal transformer multiports is used to derive properties of such approximants. These properties include coefficient inequalities, the univalent nature of the approximants in a specified region of the complex plane and the coprimeness properties of the polynomial matrices in the matrix fraction description of the rational matrix approximants to a matrix Stieltjes series. Some known results will be shown to follow through network theory interpretations and new results concerning the rational approximants considered are derived.

1. Introduction

Recently, the following result was proved [1] via two different approaches, namely the matrix continued-fraction expansion and the use of the matrix Cauchy index.

Theorem 1. The $[\frac{m-1}{m}]$ and $[\frac{m}{m}]$ order Padé approximants to a matrix Stieltjes series can be synthesized as multiports using resistors and capacitors (possibly including ideal transformers). In fact, such approximants are the open-circuit impedance matrices of RC-ideal transformer multiports.

The result stated above is important in the sense that the realizability theory of multiport (including, of course, single port) network [2] can be exploited to advantage to obtain physically motivated simple proofs of known results about the mathematical properties of Padé approximants to series of the Stieltjes types and, in addition, new results. With the objective of working towards the desired goal, we begin by proving known results about properties of Padé approximants to scalar Stieltjes series by solely using network-theoretic

*Research supported by National Science Foundation Grant 78-23141.

concepts. To facilitate reading, we first state as facts known results in the mathematical realizability theory of passive networks and infer or deduce from these, certain results arrived at in the mathematical literature [3] via totally different arguments, and more. The definition given for a matrix Stieltjes series in [1] will be adhered to and the definition for a scalar Stieltjes series will be an obvious specialization. Unless otherwise specified, the field of coefficients will be assummed to be real.

2. Network-Theoretical Interpretations of Known Properties

Fact 1 [4]. A rational function, $Z(s) = \frac{P(s)}{Q(s)}$, in the complex variable s is the driving-point impedance of a RC one-port if and only if the zeros of the polynomials P(s) and Q(s) are simple, alternate on the nonpositive real axis with the zero closest to the origin, belonging to Q(s). The residue at each pole of Z(s) is positive.

From the definition of a RC impedance function, it is clear that the formal power series expansion about s=0 of a RC impedance function need not be a Stieltjes series. For example, $Z(s) = \frac{1}{1+s}$ is a legitimate RC impedance function but its formal power series expansion,

$$\frac{1}{1+s} = 1 - s + s^2 - s^3 + s^4 - \dots$$

does not satisfy the definition of a Stieltjes series. Also, note that a pole at s=0 will be absent in the approximant to a Stieltjes series. From Fact 1 and Theorem 1 we directly infer the following proposition contained in [3, Theorem 15.1].

Proposition 1. The $[\frac{m-1}{m}]$ and $[\frac{m}{m}]$ order Padé approximants to a Stieltjes series each has simple interlacing poles and zeros on the negative real axis and the residue at each pole is positive.

Fact 2 [4, p. 73]. If Z(s), $s=\sigma+jw$ is a RC impedance function, then

$$\frac{dZ(\sigma)}{d\sigma} < 0$$

and in the absence of a pole at s=0

$$Z(0) > Z(\infty) \geq 0 .$$

We introduce, next, the notation $[\frac{n}{m}]_\sigma$ to denote the value of the $[\frac{n}{m}]$ order Padé approximant, evaluated at $s=\sigma$. Note that,

$$[\frac{m-1}{m}]_\infty = 0 < [\frac{m}{m}]_\infty$$

and

$[\frac{m-1}{m}]_0 = [\frac{m}{m}]_0$.

The above results in combination with Fact 2 ($Z(\sigma)$ vs σ should be plotted to facilitate comprehension, as in [5, p. 72]) and Theorem 1 lead to Proposition 2, contained in [3, Theorem 15.2].

Proposition 2. The $[\frac{m-1}{m}]$ and $[\frac{m}{m}]$ order Padé approximants to a Stieltjes series satisfy the following equality/inequalities, where F(s) stands for the limit, if it exists, as m goes to infinity of the approximants $[\frac{m}{m}]$ or $[\frac{m-1}{m}]$.

$$[\frac{m}{m}]_\sigma \geq F(\sigma) \geq [\frac{m-1}{m}]_\sigma, \quad \sigma \geq 0$$

$[\frac{m}{m}]^1_\sigma \geq F^1(\sigma) \geq [\frac{m-1}{m}]^1_\sigma$, $\sigma \geq 0$, where the superscript denotes derivative.

Propositions 1 and 2 have their matrix counterparts. These results follow from the simple fact that if $\underline{Z}(s)$ is the open-circuit impedance matrix of a RC-ideal transformer p-port the $\underline{x}^t \ \underline{Z}(s) \ \underline{x}$ is a RC impedance function [5, pp. 136-137 for $\underline{Z}(s)$ of order 2x2] for arbitrary values of the real vector $\underline{x} = [x_1, \ldots, x_p]^t$.

Fact 3. $\underline{Z}(s)$ is the open-circuit impedance matrix of a RC-ideal transformer p-port if and only if it has partial-fraction expansion,

$$\underline{Z}(s) = \underline{K}_\infty + \sum_{i=1}^{n} \frac{\underline{K}_i}{s+\sigma_i}, \quad \sigma_i \geq 0$$

where $\underline{K}_\infty$ and $\underline{K}_i$, $i = 1, \ldots, n$ are real nonnegative definite symmetric matrices.

The matrix counterpart of Proposition 1 is, then, obvious. The matrix counterpart of Proposition 2 is given next. By the derivative of a matrix we mean, here, the matrix obtained by differentiating each element. Also $\underline{A} \geq \underline{B}$ implies that $\underline{A} - \underline{B}$ is nonnegative definite.

Proposition 3. The $[\frac{m-1}{m}]$ and $[\frac{m}{m}]$ order matrix Padé approximants to a matrix Stieltjes series obey the inequalities in Proposition 2 under the notational interpretations for matrices stated in the preceding paragraph.

3. New Properties of Padé Approximants to Stieltjes Series

In [1], it was proved that the inverse of a matrix Stieltjes séries exists and is unique. From the duality theorem [3, p. 112] which relates Padé approximants to a series and its reciprocal, it easily follows that the reciprocal of a Stieltjes series cannot be a Stieltjes series. Otherwise, that would contradict the fact that the reciprocal of a RC admittance function is a RL impedance function. In fact RL impedance functions are characterized by the interlacing property of simple poles and zeros on the nonpositive real axis with a zero occuring closest to the origin.

Fact 4. If T(s) is a Stieltjes series, then the $[\frac{m+1}{m}]$ and $[\frac{m}{m}]$ order Padé approximants to $[T(s)]^{-1}$ are RL realizable. In the matrix case, ideal transformers may be needed.

The realizability of Padé approximants to a matrix Stieltjes series as stated in Theorem 1 yields other interesting results. The following result is well-known.

Fact 5 [6]. If $Z(s) = \frac{P(s)}{Q(s)}$ is a RC impedance function, then

$$Z_k(s) = \frac{\frac{d^kP(s)}{ds^k}}{\frac{d^kQ(s)}{ds^k}} \triangleq \frac{P^k(s)}{Q^k(s)}$$

is a RC impedance function, $k = 1,2, \ldots$.

Note that,

$$\frac{dZ(\sigma)}{d\sigma} = \frac{P(\sigma)}{Q(\sigma)}\left[\frac{P^1(\sigma)}{P(\sigma)} - \frac{Q^1(\sigma)}{Q(\sigma)}\right], \quad \sigma \geq 0.$$

Then, applying Fact 2,

$$\frac{P^1(\sigma)}{Q^1(\sigma)} < \frac{P(\sigma)}{Q(\sigma)}, \text{ since } \frac{P(\sigma)}{Q(\sigma)} > 0, \ \sigma \geq 0.$$

Repeatedly applying Fact 5 and the steps discussed, it follows that

$$\frac{P^k(\sigma)}{Q^k(\sigma)} < \frac{P^{k-1}(\sigma)}{Q^{k-1}(\sigma)} < \ldots < \frac{P^1(\sigma)}{Q^1(\sigma)} < \frac{P(\sigma)}{Q(\sigma)}.$$

Putting $\sigma=0$ in the preceding inequalities, yields the following result.

Proposition 4. Let

$$Z(s) = \frac{\sum_{k=0}^{m} a_k s^k}{\sum_{k=0}^{m} b_k s^k}, \qquad \hat{Z}(s) = \frac{\sum_{k=0}^{m-1} \hat{a}_k s^k}{\sum_{k=0}^{m} \hat{b}_k s^k}$$

be, respectively, the $[\frac{m}{n}]$ and $[\frac{m-1}{m}]$ order Padé approximants to a Stieltjes series. Then,

$$\frac{a_m}{b_m} < \frac{a_{m-1}}{b_{m-1}} < \ldots < \frac{a_1}{b_1} < \frac{a_o}{b_o}$$

and

$$0 < \frac{\hat{a}_{m-1}}{\hat{b}_{m-1}} < \ldots < \frac{\hat{a}_1}{\hat{b}_1} < \frac{\hat{a}_o}{\hat{b}_o} \quad .$$

It is simple to convert a RC-multiport to a one-port via use of ideal transformers and gyrators. In fact, if $[z_{ij}(s)] = [\frac{R_{ij}(s)}{K(s)}]$ is the open-circuit impedance matrix of a RC-ideal transformer n-port, where the elements $z_{ij}(s)$ have common poles, then

$$Z(s) = \sum_{i=1}^{n} \sum_{j=1}^{n} z_{ij}(s)\, x_i x_j$$

is the driving-point RC impedance function for all real values of x_i, x_j and vice versa. Then, Proposition 4 is generalized to the matrix case. In Proposition 5, below, only the $[\frac{m}{m}]$ order Padé approximant is considered for the sake of brevity.

Proposition 5. Let

$$[z_{ij}(s)] = [\frac{a_{ij}s^m + b_{ij}s^{m-1} + \ldots + n_{ij}}{as^m + bs^{m-1} + \ldots + n}]$$

be the $[\frac{m}{m}]$ matrix Padé approximant to a matrix Stieltjes series. Note that the denominator polynomial is common to each of the elements in the matrix approximant. Then,

$$[\frac{a_{ij}}{a}] < [\frac{b_{ij}}{b}] <\ldots< [\frac{n_{ij}}{n}] .$$

Another interesting property of RC impedance functions is their schlicht or univalent property in the right-half plane. Theorem 1 and this schlicht characterization immediately leads to Proposition 6.

Proposition 6. The $[\frac{m}{m}]$ and $[\frac{m-1}{m}]$ order Padé approximants to a Stieltjes series are univalent in the right-half plane; i.e.

$$[\tfrac{m}{m}]_{s_1} \neq [\tfrac{m}{m}]_{s_2}, \quad \mathrm{Re}\ s_i > 0,\ i = 1, 2,\ s_1 \neq s_2$$

$$[\tfrac{m-1}{m}]_{s_1} \neq [\tfrac{m-1}{m}]_{s_2}, \quad \mathrm{Re}\ s_i > 0,\ i = 1, 2,\ s_1 \neq s_2 \quad .$$

The preceding result has a natural extension to the matrix case. In fact, the schlicht property holds to the right of the line $s=\alpha_1$ where α_1 is the pole of least magnitude in the rational approximant.

In [1], the proof of Theorem [1] was also given by using the artifice of matrix Cauchy index. The matrix fraction descriptions of the approximant $\underline{Z}(s)$,

$$\underline{Z}(s) = [\underline{A}(s)]^{-1}\, \underline{B}(s) = [\underline{C}(s)][\underline{D}(s)]^{-1}$$

where $\underline{A}(s)$, $\underline{B}(s)$, $\underline{C}(s)$ and $\underline{D}(s)$ are polynomial matrices are useful in the realization theory of multi-input multi-out systems. In fact, the number of dynamic variables needed to realize a system is equal to the McMillan degree of $\underline{Z}(s)$, defined below.

Fact 6 [8]. The McMillan degree of $\underline{Z}(s)$ is the degree of polynomials, det $\underline{A}(s)$, (or degree of det $\underline{D}(s)$), provided $\underline{A}(s)$, $\underline{B}(s)$ are left coprime (or $\underline{C}(s)$, $\underline{D}(s)$ are right coprime).

If $\underline{Z}(s)$ is a square-matrix of order p, then the maximum degree for det $\underline{A}(s)$ (or det $\underline{D}(s)$) for this $[\frac{m-1}{m}]$ or $[\frac{m}{m}]$ order approximant is mp. The following result was proved in [1].

Fact 7 [1]. The $[\frac{m-1}{m}]$ and $[\frac{m}{m}]$ order Padé approximants to a matrix Stieltjes series has McMillan degree mp, where p is the order of the matrix approximants.

Fact 6 and Fact 7 lead immediately to the next proposition.

Proposition 7. $[\frac{m-1}{m}]$ or $[\frac{m}{m}]$ matrix Padé approximants occur in irreducible form.

The preceding proposition implies that the realizations of Padé approximants of orders $[\frac{m-1}{m}]$ and $[\frac{m}{m}]$ to a matrix Stieltjes series are controllable as well as observable [9].

4. Conclusions

Systems-theoretic interpretations of properties of Padé approximants to a Stieltjes series have been given. Known properties of such approximants are derivable from results in mathematical realizability theory, which also enables one to obtain new properties of the Padé approximants to series of the Stieltjes type. Multivariate Stieltjes series are known to have integral representations. It will be interesting to investigate any multivariate network-theoretic [8, Ch. 5] properties of such approximants. Also, the characterization of series whose approximants have interlacing pole-zero pattern on the unit circle is of some interest.

References

[1] Basu, S. and Bose, N.K., "Matrix Stieltjes series and network models," SIAM J. Math. Anal., 14, no. 2 (1983), 209-222.

[2] Newcomb, R.W., "Linear Multiport Synthesis," McGraw-Hill, New York, 1966.

[3] Baker, G.A., "Essentials of Padé Approximants," Academic Press, New York, 1975.

[4] Hazony, D., "Elements of Network Synthesis," Reinhold Publishing Corporation, New York, 1963.

[5] Temes, C.G. and Lapatra, W., "Introduction to Circuit Synthesis and Design," McGraw-Hill, New York, 1977.

[6] Talbot, A., "Some theorems on positive functions," IEEE Trans. Circuit Theory, 12, (1965), 607-608.

[7] Reza, F.M., "On the Schlicht behavior of certain impedance functions," IEEE Trans. Circuit Theory, 9, (1962), 231-232.

[8] Bose, N.K., "Applied Multidimensional Systems Theory," Van Nostrand Reinhold Co., New York, 1982.

[9] Kailath, T., "Linear Systems," Prentice-Hall, Englewood Cliffs, NJ, 1980.

DEGREE OF RATIONAL APPROXIMATION IN DIGITAL FILTER REALIZATION

Charles K. Chui[1]
Department of Mathematics
Texas A&M University
College Station
Texas 77843, U.S.A.

and

Xie-Chang Shen[2]
Department of Mathematics
Peking University
Beijing, China

Abstract. In recursive digital filter design, the amplitude characteristic of an ideal filter has to be translated into a rational function that is pole-free on the closed unit disk and preferably has real coefficients. This is possible through a causal transformation utilizing the tolerance allowance. In this paper we study the degree of uniform approximation by rational functions that fulfill the filter criteria and can be computed by interpolation or the method of least-squares inverses.

1. Introduction

In digital filter design, the (ideal) filter amplitude characteristic $|H_I(e^{it})|$ with certain tolerance $\varepsilon = (\varepsilon_s, \varepsilon_t)$, where the positive numbers ε_s and ε_t are maximum tolerance allowances in the stop-bands and transition regions respectively, is usually prescribed. Here, $|H_I(e^{it})|$ is a nontrivial piecewise linear polynomial and, in most applications, a characteristic function of a finite union of intervals. Since only positive frequency is discussed $|H_I(e^{it})|$ could also be assumed to be an even function in t. A

[1]The research of this author was supported by the U. S. Army Research Office under Contract No. DAAG 29-81-K-0133.

[2]This author is a Visiting Scholar at Texas A&M University during 1983-1984.

causal representation of $|H_I(e^{it})|$ is a function $H(z)$ analytic in the open unit disk $|z| < 1$ and continuous on the closed disk $|z| \leq 1$ with the exception of a finite number of points such that $|H(e^{it})| = |H_I(e^{it})|$ at all points t where $|H_I(e^{it})|$ is continuous. This is certainly impossible without utilizing the tolerance allowance ε_s on the stop-bands. In [1] a causal transform $\hat{H}_\varepsilon(z)$ of $|H_I(e^{it})|$ is given such that $\hat{H}_\varepsilon$ is continuous and zero-free on $|z| \leq 1$ and $|\hat{H}_\varepsilon(e^{it})| = |H_\varepsilon(e^{it})|$ for all t where $|H_\varepsilon(e^{it})|$ is an arbitrary zero-free continuous function that approximates $|H_I(e^{it})|$ within the tolerance allowance $\varepsilon = (\varepsilon_s, \varepsilon_t)$. Hence, it is even possible to make $|H_\varepsilon(e^{it})|$ a C^∞ function in t. For simplicity, we will only consider a low-pass filter with ideal amplitude characteristic

$$|H_I(e^{it})| = \begin{cases} 1 & \text{for} \quad |t| \leq \theta_t \\ 0 & \text{for} \quad \theta_t < |t| \leq \pi. \end{cases}$$

It will be clear that the results in this paper also hold for a much more general setting. Let $0 < \theta_2 < \theta_t < \theta_1 < \pi$, with $\theta_1 - \theta_2 \leq \varepsilon_t \leq 1$, and $0 < \delta \leq \varepsilon_s$. Then a C^∞ zero-free correction of $|H_I(e^{it})|$ may be given by $|H_\varepsilon(e^{it})|$ where

$$(1) \quad \ln|H_\varepsilon(e^{it})| = \begin{cases} 0 & \text{for} \quad |t| \leq \theta_2 \\ \ln\delta + e^{(|t|-\theta_1)^{-1}}\left[e^{(\theta_2-|t|)^{-1}} - (\ln\delta)e^{(\theta_1-\theta_2)^{-1}}\right] & \text{for} \quad \theta_2 < |t| < \theta_1 \\ \ln\delta & \text{for} \quad \theta_1 \leq |t| \leq \pi \end{cases}$$

and the corresponding transformed filter characteristic is defined in [1] by

$$(2) \qquad \hat{H}_\varepsilon(z) = \exp\left\{\frac{1}{2\pi}\int_{-\pi}^{\pi} \frac{e^{it}+z}{e^{it}-z}\ln|H_\varepsilon(e^{it})|\,dt\right\}.$$

To realize the causal representation $\hat{H}_\varepsilon$, it is necessary to "approximate" $\hat{H}_\varepsilon$ on $|z| = 1$ by rational functions

$$R_{mn}(z) = \frac{a_0 + \dots + a_m z^m}{b_0 + \dots + b_n z^n}$$

with real feed-forward and feed-back coefficients $a_0,\dots,a_m$ and $b_0,\dots,b_n$ respectively. In order to guarantee stability, we require that R_{mn} has no poles on $|z| \le 1$. This design criterion is automatically satisfied if $\{R_{mn}\}$ approximates $\hat{H}_\varepsilon$ uniformly on $|z| \le 1$.

The main purpose of this paper is to study the order of such approximation. A more general class of functions will be considered. These results will be contained in Theorems 1 and 2 to be found in Section 3. Problems of this type have been studied by Szabados [6,7] and Rusak [3,4]. However, since the functions such as $\hat{H}_\varepsilon$ to be approximated satisfy certain nice boundary conditions, we are able to give more precise estimates. The order of uniform approximation of two linear realization schemes will also be discussed: interpolation and least-squares inverse approximation. These procedures produce stable recursive digital filters which are more powerful than non-recursive filters resulted from polynomial approximation. The realization scheme of least-squares inverses was discussed in [1] but here we will give an order of its uniform approximation on $|z| \le 1$ and prove that this order is, in some sense, sharp. The next section will be devoted to the study of the analytic and boundary-value behaviour of the transformed filter characteristic.

2. Preliminary lemmas

In this section we will give upper bounds on the derivatives of the transformed filter characteristic $\hat{H}_\varepsilon$ and introduce two classes of functions $\mathcal{C}$ and $\mathcal{H}_m$. The order of uniform approximation by rational functions to these two classes will be delayed to Section 3. We need the following elementary result.

LEMMA 1. Let $\hat{H}_\varepsilon$ be the transformed filter characteristic as defined in (2). Then $\hat{H}_\varepsilon$ is analytic everywhere in the complex plane with the exception of the rays $\rho e^{i(\theta_1-\theta_2)/2}$, $\rho e^{i\theta_2}$, $\rho e^{i\theta_1}$, $\rho e^{i(\pi-(\theta_1-\theta_2)/2)}$, $\rho e^{i(\pi+(\theta_1-\theta_2)/2)}$, $\rho e^{-i\theta_1}$, $\rho e^{-i\theta_2}$, and $\rho e^{-i(\theta_1-\theta_2)/2}$, where $\rho \ge 1$. Furthermore, $\hat{H}_\varepsilon$ is infinitely continuously differentiable in $|z| \le 1$.

Proof. Write $\hat{H}_\varepsilon(z) = \exp(U(z) + iV(z))$. From (1) it is clear that U is infinitely continuously differentiable on $|z| \le 1$, so that its harmonic conjugate function $V(z)$ has the same property (cf. [8; p. 163]). Set $u(t) = U(e^{it})$ and $v(t) = V(e^{it})$. It is also well-known (cf. [8, p. 155]) that

$$(3) \qquad v(x) = -\frac{1}{\pi}\int_0^{\pi} \frac{u(x+t) - u(x-t)}{2\tan(t/2)}\,dt,$$

where $u(t) = \ln|H_\varepsilon(e^{it})|$ as defined in (2). For $0 \le x < (\theta_1-\theta_2)/2$, the integral in (3) is the sum of three integrals of real analytic functions on $(\theta_2-x, x+\theta_2)$, $(x+\theta_2, \theta_1-x)$, and (θ_1-x, θ_2+x); and for $(\theta_1-\theta_2)/2 < x \le \theta_2$, it is that of three integrals of real analytic functions on (θ_2-x, θ_1-x), (θ_1-x, θ_2+x), and $(\theta_2+x, x+\theta_1)$. In fact, similar results hold for x in $[\theta_2, \pi-(\theta_1-\theta_2)/2)$, $(\pi-(\theta_1-\theta_2)/2, \pi+(\theta_1-\theta_2)/2)$, $(\pi+(\theta_1-\theta_2)/2, 2\pi-(\theta_1-\theta_2)/2)$, and $(2\pi-(\theta_1-\theta_2)/2, 2\pi]$. From the definition of u and the relationships of v and u on these intervals, it is clear that $\hat{H}_\varepsilon = U + iV$ is analytic everywhere in the complex plane with the exception of the eight rays indicated in the statement of the lemma. That $\hat{H}_\varepsilon$ is in C^∞ at the initial points of these rays also follows from the definition of u and the integral representation of v in terms of u. This completes the proof of the lemma.

We also need the following estimate on the derivatives of $\hat{H}_\varepsilon$. Let $z = e^{it}$ and denote by $f_z^{(r)}$ and $f_t^{(r)}$ the corresponding r^{th} order derivativies of f with respect to z and t.

LEMMA 2. <u>The transformed filter characteristic</u> $\hat{H}_\varepsilon$ <u>satisfies the following</u>:

$$(4) \qquad \|(\hat{H}_\varepsilon)_z^{(r)}\|_\infty \le (B+r)!A^r r^{3r}$$

<u>and</u>

$$(5) \qquad \|(\hat{H}_\varepsilon)_t^{(r)}\|_\infty \le (B+r)!D^r r^{3r}$$

<u>for any non-negative integer</u> r, <u>where</u> $A > 24/e^3$, $D > 8/e^3$, <u>and</u> B <u>is some natural number depending on</u> A <u>or</u> D.

To facilitate the proof of Lemma 2, we first establish the following inequalities:

LEMMA 3. <u>Let</u> r <u>be a natural number</u>.

(a) <u>For</u> $c \ge 3$ <u>and</u> $q \ge \ln 3 / (\ln c + \ln 3)$,

$$\frac{1}{r+1} \sum_{j=0}^{r+1} 2^{r-j+1} \, j(cj)^{qj} \le [c(r+2)]^{q(r+2)} .$$

(b) <u>For</u> $c \ge 3$ <u>and</u> $q \ge \ln 6 / (\ln c + \ln 3)$,

$$\sum_{j=1}^{r} 2^{r-j+1}(r-j+1)(cj)^{qj} \le (r+1)[c(r+1)]^{q(r+1)} .$$

(c) <u>For any natural number</u> B,

$$B \sum_{j=0}^{r} \binom{r}{j}(B+j)! \le (B+r+1)! .$$

<u>Proof of Lemma 3</u>. We will use mathematical induction. It is clear that all these inequalities are valid for $r = 1$. We will prove the validity of these inequalities for $r = k+1$ assuming their validity for $1 \le r \le k$.

(a)
$$\frac{1}{k+2} \sum_{j=0}^{k+2} 2^{k-j+2} \, j(cj)^{qj}$$

$$= [c(k+2)]^{q(k+2)} + \frac{2}{k+2} \sum_{j=0}^{k+1} 2^{k-j+1} \, j(cj)^{qj}$$

$$\le [c(k+2)]^{q(k+2)} + \frac{2(k+1)}{k+2} [c(k+2)]^{q(k+2)}$$

$$\le 3[c(k+2)]^{q(k+2)} \le [c(k+3)]^{q(k+3)} ,$$

(b)
$$\sum_{j=0}^{k+1} 2^{k-j+2}(k-j+2)(cj)^{qj}$$

$$\le 2(k+1)[c(k+1)]^{q(k+1)} + 2 \sum_{j=0}^{k+1} 2^{k-j+1}(cj)^{qj}$$

$$\le 2(k+2)[c(k+1)]^{q(k+1)} + \frac{2}{k+1} \sum_{j=0}^{k+1} 2^{k-j+1} \, j(cj)^{qj}$$

$$\le 2(k+2)[c(k+1)]^{q(k+1)} + 2[c(k+2)]^{q(k+2)}$$

$$\le (k+2)[c(k+2)]^{q(k+2)} ,$$

since $q \ge \ln 6 / (\ln c + \ln 3)$, where (a) has been used.

(c) $$B\sum_{j=0}^{k+1}\binom{k+1}{j}(B+j)!$$

$$= B\cdot(B+k+1)! + (k+1)\,B\sum_{j=0}^{k}\frac{k!}{j(k-j+1)!}(B+j)!$$

$$\leq B\cdot(B+k+1)! + (k+1)\cdot(B+k+1)!$$

$$\leq (B+k+2)!\ .$$

We are ready to prove Lemma 2.

Proof of Lemma 2. For any $c \geq 3$ and $q \geq \ln 6 / \ln (3c)$, we will prove that for $|t| \leq 1$,

$$|(e^{t^{-1}})_z^{(k)}| \leq k!c^{qk}k^{qk}e^{t^{-1}}t^{-2k}, \quad k = 0,1,\ldots, \tag{6}$$

where $z = e^{it}$. First note that

$$(e^{t^{-1}})'_z = e^{t^{-1}}(t^{-1})'_z$$

so that

$$|(e^{t^{-1}})_z^{(r+1)}| = |\sum_{j=0}^{r}\binom{r}{j}(e^{t^{-1}})_z^{(j)}(t^{-1})_z^{(r-j+1)}|$$

$$\leq \sum_{j=0}^{r}\binom{r}{j}2^{r-j+1}(r-j+1)!\ t^{-(r-j+2)}|(e^{t^{-1}})_z^{(j)}|\ .$$

Hence (6) can be proved by mathematical induction. Indeed, by using the induction hypothesis, we have

$$|(e^{t^{-1}})_z^{(r+1)}| \leq \sum_{j=0}^{r}\binom{r}{j}2^{r-j+1}(r-j+1)!j!c^{qj}j^{qj}e^{t^{-1}}|t|^{-(r+j+2)}$$

$$\leq e^{t^{-1}}t^{-2(r+1)}\sum_{j=0}^{r}r!(r-j+1)2^{r-j+1}c^{qj}j^{qj}$$

$$\leq (r+1)!\ [c(r+1)]^{q(r+1)}\ e^{t^{-1}}\ t^{-2(r+1)}$$

from Lemma 3(b), establishing (6).

Next, consider $\theta_2 < t < \theta_1$, so that $|t - \theta_1|$, $|t - \theta_2| \leq \theta_1 - \theta_2 \leq 1$. We have, by using (6),

$$|(u(t))_z^{(r)}| = \left|\sum_{j=0}^{r} \binom{r}{j}\left(e^{1/(t-\theta_1)}\right)_z^{(j)}\left(e^{1/(\theta_2-t)} - c_1 e^{1/(\theta_1-\theta_2)}\right)_z^{(r-j)}\right|$$

$$\leq \sum_{j=0}^{r} \binom{r}{j}(cj)^{qj} j!\,(c(r-j))^{q(r-j)}(r-j)!\; \frac{e^{\frac{1}{t-\theta_1}}}{(t-\theta_1)^{2j}} \quad \frac{e^{\frac{1}{\theta_2-t}}}{(t-\theta_2)^{2(r-j)}} .$$

By majorizing first with respect to t and then with respect to j, it follows that

$$\frac{e^{1/(t-\theta_1)}}{(t-\theta_1)^{2j}} \cdot \frac{e^{1/(\theta_2-t)}}{(t-\theta_2)^{2(r-j)}} \leq \left(\frac{4}{e^2}\right)^r r^{2r} .$$

Hence, if we set $q = \ln 6 / \ln(3c)$, we have

$$|u(t)_z^{(r)}| \leq r!\,c^{qr}\left(\frac{4}{e^2}\right)^r r^{2r} \sum_{j=0}^{r} j^{qj}(r-j)^{qj}$$

$$\leq (r+1)!\; c^{qr} \left(\frac{4}{e^2}\right)^r r^{2r}\, r^{qr}$$

$$\leq (r+1)!\; r^{2r} \left(\frac{24}{e^2}\right)^r r^{qr} .$$

Let $c \to \infty$, so that $q \to 0$, yielding

(7) $$|u(t)_z^{(r)}| \leq (r+1)!\; r^{2r}\left(\frac{24}{e^2}\right)^r$$

$$\leq \sqrt{2\pi}\,(r+1)^{3/2} \left(\frac{24}{e^3}\right)^r r^{3r}$$

for $\theta_2 < t < \theta_1$. Since u is even, periodic, and is constant otherwise, this inequality holds for all t. The derivatives of the conjugate function v can be estimated by using the formulas (3) and (7), giving

$$|v(t)_z^{(r)}| \leq A_1(r+2)^{3/2}\left(\frac{24}{e^3}\right)^{r+1}(r+1)^{3(r+1)}$$

for some absolute constant A_1. Hence, for any constant $A > 24/e^3$, there is a natural number B such that

(8) $$|((u + iv)(t))_z^{(r)}| \leq BA^r\, r^{3r}, \quad r = 0,1,\dots .$$

We are now ready to establish (4). Since it holds for $r = 0$, we have, by induction hypothesis, for all $z = e^{it}$,

$$|\hat{H}_\varepsilon(z)_z^{(r+1)}| = |(e^{(u+iv)(t)})_z^{(r+1)}|$$

$$= |\sum_{j=0}^{r} \binom{r}{j}(e^{(u+iv)(t)})_z^{(j)}((u+iv)(t))_z^{(r-j+1)}|$$

$$\leq \sum_{j=0}^{r} \binom{r}{j}(B+j)!A^j j^{3j} BA^{r-j+1}(r-j+1)^{3(r-j+1)}$$

$$\leq A^{r+1}(r+1)^{3(r+1)}\ B \sum_{j=0}^{r} \binom{r}{j}(B+j)!$$

$$\leq (B+r+1)!\ A^{r+1}(r+1)^{3(r+1)},$$

where Lemma 3(c) has been used in the last inequality. Inequality (5) can be proved in the same manner.

As mentioned earlier, it is clear that we do not have to restrict ourselves to the low-pass filter. Indeed, any ideal filter amplitude characteristic $|H_I(e^{it})|$ has a transformed causal representation $\hat{H}_\varepsilon$ which satisfies the properties similar to those given in Lemmas 1 and 2. For this reason, we define the following classes of functions C_e and $H^R_{m,e}$ and will study their degree of approximation by rational functions satisfying the filter prescriptions.

Let $C = C(a,b,c)$ be the collection of functions f analytic in $|z| < 1$ and infinitely continuously differentiable in $|z| \leq 1$ satisfying the condition

(9) $\quad |f(e^{it})_t^{(r)}| \leq Mr^a b^r r^{cr}\ , \quad r = 0,1,\dots\ .$

Suppose that $z_j = e^{it_j}$, $0 < t_1 < \dots < t_m \leq 2\pi$, and $t_{m+1} = t_1 + 2\pi$. Let $H^R_m = H^R_m(a,b,c)$ be the collection of functions f analytic in $|z| < 1 + R$ for some $R > 0$ with the exception of the line segments ρe^{it_j}, $1 \leq \rho \leq 1+R$, $j = 1,\dots,m$, and infinitely continuously differentiable on $|z| \leq 1$, satisfying

(10) $\quad |f^{(r)}(z)| \leq Mr^a b^r r^{cr}, \quad r = 0,1,\dots\ ,$

for $|z| = 1$. In addition, we let $C_e = C_e(a,b,c)$ and

$H^R_{m,e} = H^R_{m,e}(a,b,c)$ be the sub-collections of C and H^R_m, respectively, of functions $f(re^{it})$ which are even in t. From Lemma 2, we have $\hat{H}_\varepsilon \in C_e\left(B + \frac{1}{2}, b_1, 4\right) \cap H^R_{m,e}\left(B + \frac{1}{2}, b_2, 4\right)$ for arbitrary numbers $b_1 > 8/e^4$ and $b_2 > 24/e^4$ and some natural number B depending on b_1 and b_2 respectively.

3. Order of uniform rational approximation

In this section we will study the degree of approximation of functions in H^R_m by rational functions whose poles lie outside the unit circle. As discussed earlier, we remark that problems of this type were also studied by Szabados [6,7] and Rusak [3,4]. In the proof of our first thoerem, we will use two basic lemmas of Rusak's and give more precise analysis to obtain better estimates for the function class H^R_m. It is worth mentioning that the techniques we use in this paper are quite different to those in the vast literature of rational approximation on a real interval. For simplicity, R_k will denote a rational function P_k/Q_k where P_k and Q_k are polynomials with complex coefficients and degree no greater than k.

THEOREM 1. Let $f \in H^R_m(a,b,c)$ where $a \geq 0$, $b > 0$, $c > 1$. Then there exist rational functions R_k (whose poles lie in $|z| > 1$) such that for any $\varepsilon > 0$,

$$|f(z) - R_k(z)| \leq M e^{-k^{1/2}/(\ln k)^{(3/2)+\varepsilon}},$$

and all integers $k \geq 2$ and $|z| \leq 1$, where the constant M depends on a, b, c, and ε.

To prove this result, we need the following preliminaries.

Let $\{w_k\}_{k=0}^n$, with $w_0 = 0$, $|w_k| < 1$, and

$$K(z) = \prod_{k=1}^{n} \frac{z - w_k}{1 - \bar{w}_k z}. \tag{11}$$

For any f analytic in $|z| < 1$ and continuous on $|z| \leq 1$, the rational function $R_n = R_n(\cdot,f)$ interpolating f at $\{w_k\}_0^n$ and having its poles at $\{1/\bar{w}\}_1^n$ has the representation

$$R_n(z,f) = \frac{1}{2\pi i}\int_{|\xi|=1} \frac{f(\xi)}{\xi - z}\left[\frac{K(\xi)}{K(z)} - \frac{zK(z)}{\xi K(\xi)}\right] d\xi, \quad |z| < 1,$$

and the corresponding error formula is

$$R_n(z,f) - f(z) = \frac{1}{2\pi i}\int_{|\xi|=1} [f(\xi) - f(z)]\left[\frac{K(\xi)}{K(z)} - \frac{zK(z)}{\xi K(\xi)}\right]d\xi, \quad |z| < 1 . \tag{12}$$

These formulas can be proved by using a result in Walsh [9] and Cauchy's formula as derived in [4]. The following lemmas can also be found in [4].

LEMMA A. *Let* r *and* N *be arbitrary natural numbers,* $n \geq 2N+r-1$, $0 \leq u_1 < u_2 < 2\pi$, *and* $\{w_k\}_1^n$ *be defined as follows:* $w_k = (1 - e^{-k/\sqrt{N}})e^{iu_1}$, *for* $k = 1,\ldots,N-1$; $w_k = (1 - e^{-(k-N+1)/\sqrt{N}})e^{iu_2}$ *for* $k = N, \ldots, 2N-2$; $|w_k| < 1$ *for* $k = 2N-1, \ldots, n-r$ *and* $w_{n-r} = \ldots = w_n = 0$. *Then for any function* ϕ *analytic in the sector* $u_1 < \arg \xi < u_2$, *and continuous on its closure, such that* $\phi(\xi) \leq M_\phi$ *for* $|\xi| \leq 1$, *and* $|\phi(\xi)\xi^{1-r}| \leq M_\phi$ *for* $|\xi| > 1$, ϕ *satisfies*

$$\left|\int_{\xi_1}^{\xi_2} \phi(\xi)\left[\frac{K(\xi)}{K(z)} - \frac{zK(z)}{\xi K(\xi)}\right] d\xi\right| \leq 8M_\phi e^{-\sqrt{N/2}} ,$$

for $|z| = 1$, *where* $\xi_1 = e^{iu_1}$, $\xi_2 = e^{iu_2}$, *and* K *as defined in* (11).

LEMMA B. *Let* ℓ, N, r *be natural numbers,* $n \geq \ell(N-1) + 1$, $n \geq r$ $0 \leq u_1 < \ldots < u_\ell < 2\pi$, *and* $\{w_k\}_1^n$, *be defined as follows:* $w_{j+(N-1)(p-1)} = (1 - e^{-j/\sqrt{N}})e^{iu_p}$, *for* $p = 1,\ldots,\ell$ *and* $j = 1,\ldots,N-1$, *and the remaining* $2n-\ell(N-1)$ w_k*'s which are not defined above equal to zero. Then the operator norm of the rational interpolator* R_n *on the Hardy space* H^∞ *satisfies*

$$\|R_{2n}\| \leq 2\ \ell n\left(2[n - 1 - \ell(N-1) + 2\ell\sqrt{N}\, e^{\sqrt{N}}]\right) . \tag{13}$$

We are now ready to prove our result.

Proof of Theorem 1. For any natural number r, write

$$f(\xi) = f(z) + f'(z)(\xi - z) + \ldots + \frac{f^{(r)}(z)}{r!}(\xi - z)^r$$

$$+ \frac{1}{r!}\int_z^\xi (\xi - \zeta)^r f^{(r+1)}(\zeta)d\zeta; \quad |\xi|, \ |z| \leq 1 .$$

Let $\varepsilon > 0$ be fixed. For any sufficiently large natural number k, let n be the integral part of $k/2$ and set r, $p(r)$, and N as the integral parts of $n^{1/2}/(\ln n)^{\frac{3}{2}+\frac{4}{5}\varepsilon}$, $(\ln r)^{1+\varepsilon/2}$, and $(n-1)/(2p(r)m)$ respectively. Let K be the product (11) where w_k's are defined in Lemma B with $\ell = 2p(r)m$. Using (12) and the fact that for any $j = 1,\ldots,r$,

$$\int_{|\xi|=1} (\xi-z)^j \left[\frac{K(\xi)}{K(z)} - \frac{zK(z)}{\xi K(\xi)}\right]\frac{d\xi}{\xi-z} = 0, \quad |z| \leq 1,$$

we obtain

$$(14)\quad R_{2n}(z,f)-f(z) = \frac{1}{2\pi i}\int_{|\xi|=1}\left[\int_z^\xi \frac{(\xi-\zeta)^r}{r!}f^{(r+1)}(\zeta)d\zeta\right]\left[\frac{K(\xi)}{K(z)} - \frac{zK(z)}{\xi K(\xi)}\right]\cdot\frac{d\xi}{\xi-z},$$

$|z| \leq 1$. Consider $|z| = 1$ and pick s such that $\arg z_s - \arg z < \pi$ and $\arg z_{s+1} - \arg z \geq \pi$, where $z_j = e^{it_j}$ and $z_{m+p} = z_p$, $j = 1,\ldots,m$. Divide the integral (14) on $|\xi| = 1$ into two parts I_1 and I_2, where I_1 is the integral from z to z_s and I_2 from z_s back to z. By interchanging the order of integration, we have

$$I_1 = \frac{1}{2\pi i}\int_z^{z_s}\left[\int_\zeta^{z_s} \frac{(\xi-\zeta)^r}{r!(\xi-z)}\left(\frac{K(\xi)}{K(z)} - \frac{zK(z)}{\xi K(\xi)}\right)d\xi\right] f^{(r+1)}(\zeta)d\zeta.$$

Let $0 < q < 1$. We divide each of the arcs $\{e^{it}: t_k \leq t \leq t_{k+1}\}$, $k = 1,\ldots,m$, into $2(p(r) + 1)$ subarcs with the points $z^k_{\pm j}$, $j = 0,\ldots,p(r)$, defined by

$$\arg z^k_{\pm j} = \frac{t_{k+1} + t_k}{2} \pm \frac{t_{k+1} - t_k}{2}(1 - q^j),$$

and $z^k_{-(p(r)+1)} = z_k$, $z^k_{p(r)+1} = z_{k+1}$. Assume, without loss of generality, that $\arg z^{k_0}_{j_0-1} < \arg z \leq \arg z^{k_0}_{j_0}$ for some k_0 and $j_0 \geq 1$, and write

$$(15)\qquad I_1 = \sum_{k=k_0}^{s-1}{}' \sum_{j=1}^{p(r)+1}{}' \int_{z_{j-1}^k}^{z_j^k} \Big[\int_\zeta^{z_j^k} \cdots \Big] d\zeta$$

$$+ \sum_{k=k_0+1}^{s-1} \sum_{j=1}^{p(r)+1} \int_{z_{-j}^k}^{z_{-(j-1)}^k} \Big[\int_\zeta^{z_{-(j-1)}^k} \Big] d\zeta$$

$$+ \sum_{k=k_0}^{s-1}{}' \sum_{j=1}^{p(r)+1}{}' \int_{z_{j-1}^k}^{z_j^k} \Big[\int_{z_j^k}^{z_s} \cdots \Big] d\zeta$$

$$+ \sum_{k=k_0+1}^{s-1} \sum_{j=1}^{p(r)+1} \int_{z_{-j}^k}^{z_{-(j-1)}^k} \Big[\int_{z_{-(j-1)}^k}^{z_s} \cdots \Big] d\zeta$$

$$= I_1' + I_1'' + I_1''' + I_1''''$$

where in I_1' and I_1''', the "prime" in the double summation signs means that for $k = k_0$ the outside integral is taken over first from z to $z_{j_0}^k$ and then from z_{j-1}^k to z_j^k, $j = j_0+1,\dots,p(r)+1$ and the inside integral is taken accordingly.

To estimate I_1', we first estimate the portion with $j = p(r) + 1$. We have by using (10) and (13),

$$(16)\qquad \Big| \int_{z_{p(r)}^k}^{z_{p(r)+1}^k} \Big[\int_\zeta^{z_{p(r)+1}^k} \cdots \Big] d\zeta \Big|$$

$$< \frac{C}{r!}\, r^{a+c}\, b^r r^{cr} \Big(\frac{t_{k+1}-t_k}{2}\, q^{p(r)}\Big)^{r+1} \|R_{2n}\|$$

$$\le \frac{C}{r!}\, r^{a+c}\, b^r\, r^{cr} (\alpha q^{p(r)})^{r+1} (n/(2p(r)m) + 1)^{1/2}$$

where

$$(17)\qquad \alpha = \max_k\, (t_{k+1} - t_k)/2\ .$$

Here and throughout the symbol C denotes various constants, independent of q and n. For $1 \le j \le p(r)$ in I_1', we have, using the inequality $\sin x \ge 2x/\pi$ for $0 \le x \le \pi/2$ and Cauchy's estimate,

$$\left|\int_{z_{j-1}^k}^{z_j^k} \left[\int_{\zeta}^{z_j^k} \cdots\right] d\zeta\right|$$

$$\leq C(r+1)\left(\frac{t_{k+1}-t_k}{2}\, q^{j-1}(1-q)\right)^{r+1} \Big/ \left(\min\left[R,\ \frac{t_{k+1}-t_k}{\pi}\, q^j\right]\right)^{r+1} \|R_{2n}\|$$

so that

$$(18) \qquad \sum_{j=1}^{p(r)} \left|\int_{z_{j-1}^k}^{z_j^k} \left[\int_{\zeta}^{z_j^k} \cdots\right] d\zeta\right| = \sum_{j\in S_k} + \sum_{j\notin S_k}$$

where $S_k = \{j:\ 1 \leq j \leq p(r),\ \ R \geq (t_{k+1} - t_k)q^j/\pi\}$ with

$$\sum_{j\in S_k} \leq C(r+1)\left(\frac{\pi}{2}\right)^r \sum_{j\in S_k} \left(\frac{1-q}{q}\right)^{r+1} \|R_{2n}\|$$

$$\leq Cr^2\left(\frac{\pi}{2}\right)^r\left(\frac{1-q}{q}\right)^{r+1}(n/(2p(r)m) + 1)^{1/2}$$

and

$$\sum_{j\notin S_k} \leq C(r+1) \sum_{j\notin S_k} \left[\frac{t_{k+1}-t_k}{2R}\, q^{j-1}(1-q)\right]^{r+1} \|R_{2n}\|$$

$$\leq \frac{C(r+1)}{R^{r+1}} \sum_{j=1}^{r} \left(\frac{1-q}{q}\right)^{r+1}\left(\frac{t_{k+1}-t_k}{2}\, q^j\right)^{r+1}(n/(2p(r)m) + 1)^{1/2}$$

$$\leq C\left(\frac{\alpha}{R}\right)^{r+1} r^2\left(\frac{1-q}{q}\right)^{r+1}(n/(2p(r)m) + 1)^{1/2} ,$$

α being defined in (17). Hence, combining (16) and (18), we have

$$(19) \qquad I_1' \leq (s-k_0)C(n/(2p(r)m) + 1)^{1/2}\Big[\frac{1}{r!}\, r^{a+c} b^r r^{cr} (\alpha q^{p(r)})^{r+1} + 2\left(\frac{\pi}{2}\right)^r\left(\frac{\alpha}{R}\right)^{r+1} r^2\left(\frac{1-q}{q}\right)^{r+1}\Big]$$

$$\leq C(n/(2p(r)m) + 1)^{1/2}\Big[\frac{1}{r!}\, r^{a+c}\, b^r\, r^{cr}\, (\alpha q^{p(r)})^{r+1} + r^2\left(\frac{\pi\alpha(1-q)}{2Rq}\right)^{r+1}\Big] ,$$

where, without loss of generality, we have assumed $\alpha \geq R$. The same estimates can be obtained in the same manner for I_1''. To estimate I_1''', we first estimate the portion with $j = p(r) + 1$ using (10) and Lemma A with at most $2p(r)m$ rays instead of two rays:

$$(20)\qquad \left|\int_{z^k_{p(r)}}^{z^k_{p(r)+1}} \left[\int_{z^k_{p(r)+1}}^{z_s} \quad \ldots \quad\right] d\zeta\right| = \left|\int_{z^k_r}^{z^k_{r+1}} \left[\int_{z_{k+1}}^{z_s} \quad \ldots \quad\right] d\zeta\right|$$

$$\leq \left(\frac{C}{r!}\, r^{a+c}\, b^r\, r^{cr}\, \alpha q^r\right)\left(8\cdot 2^{r-1} e^{-(n/(4p(r)m))^{1/2}}\right)$$

$$\leq \frac{C}{r!}\, r^{a+c} (2bq)^r\, r^{cr}\, e^{-(n/(4p(r)m))^{1/2}} .$$

For $1 \leq j \leq p(r)$ in I_1''', we have

$$(21)\qquad P_{kj} = \int_{z^k_{j-1}}^{z^k_j} \left[\int_{z^k_j}^{z_s} \quad \ldots \quad\right] d\zeta .$$

To estimate P_{kj}, we use Cauchy's estimate and Lemma A again with $2p(r)m$ rays and obtain

$$(22)\quad P_{kj} \leq Cq^{j-1}(1-q)(r+1)2^r e^{-(n/(4p(r)m))^{1/2}} / \left[\min\left(R, \frac{t_{k+1}-t_k}{\pi}\, q^j\right]^{r+1} ,\right.$$

so that

$$(23)\qquad \sum_{j=1}^{r} \left|\int_{z^k_{j-1}}^{z^k_j} \left[\int_{\zeta}^{z_s} \quad \ldots \quad\right] d\zeta\right| = \sum_{j\in S_k} P_{kj} + \sum_{j\notin S_k} P_{kj}$$

where, by using (23)

$$(24)\qquad \sum_{j\in S_k} P_{kj}$$

$$\leq C\,\frac{1-q}{q}\,(1+r)2^r e^{-(n/(4p(r)m)+1)^{1/2}} \sum_{j\in S_k} q^j / \left(\frac{t_{k+1}-t_k}{\pi}\, q^j\right)^{r+1}$$

$$\leq \frac{C}{q}\,(r+1)\left(\frac{\pi}{\beta}\right)^r q^{-p(r)r}\, e^{-(n/(4p(r)m))^{1/2}}$$

with

$$(25)\qquad \beta = \min_k\, (t_{k+1} - t_k)/2 .$$

As in the estimate in I', we can similarly obtain

$$(26) \qquad \sum_{j \notin S_k} P_{kj} \leq C\left(\frac{2\alpha}{\beta R}\right)^r \frac{r+1}{q} q^{-p(r)r} e^{-(n/(4p(r)m))^{1/2}} .$$

Hence, by using (20), (21), (23), (24), and (26), we have

$$(27) \qquad I_1''' \leq \sum_{k=k_0}^{s-1} \left[\frac{C}{r!} r^{a+c}(2bq)^r r^{cr} e^{-(n/(4p(r)m))^{1/2}}\right.$$

$$\left. + \sum_{k=1}^{r} (P_{kj} + Q_{kj})\right]$$

$$\leq (s-k_0)\left[\frac{C}{r!} r^{a+c}(2bq)^r r^{cr} e^{-(n/(4p(r)m))^{1/2}}\right.$$

$$\left. + C\left(\frac{\pi\alpha}{\beta R}\right)^r \frac{r+1}{q} q^{-p(r)r} e^{-(n/(4p(r)m))^{1/2}}\right] .$$

The estimate of I_1'''' is similar. By setting

$$\left(1 + \frac{2R}{\pi\alpha}\right)^{-1} < q < 1$$

in (19) and (27), and similar inequalities for I_1'' and I_1, we have, from (15),

$$I_1 \leq Ce^{n^{1/2}/(\ln n)^{\frac{3}{2}+\frac{7}{8}\varepsilon}} .$$

The same estimate holds for I_2. This completes the proof of the theorem.

Remark 1: In 1978, Rusak [3] stated three theorems on the order of uniform rational approximations on the closed unit disk. He gave proofs of the second and third results in 1979 [4], but said nothing about the first theorem, where the functions to be approximated are assumed to be "piecewise analytic" on the unit circle. In addition, the notion of piecewise analyticity was not clear, and to the best of our knowledge we still do not know the truth of this statement.

Remark 2: In Theorem 1, the poles of the rational approximants constructed lie on many rays; and since the functions to be approximated only have m "bad" points $z_j = e^{it_j}$, $j = 1,\dots,m$ on the unit circle, it would be interesting to require all the poles to lie on the rays ρe^{it_j}, $\rho > 1$, $j = 1,\ldots,m$. It should be emphasized that in digital filter realization these poles, which give rise to the feed-back coefficients, are very important. In fact, polynomial approximants which only give non-recursive digital filters are not as useful. We have the following

PROPOSITION 1. *Let* $f \in H_m^R(a,b,c)$ *where* $a \geq 0$ $b > 0$ *and* $c > 2$. *Then there exist rational functions* R_k, $k = 2m+1,\ldots,$ *which have at least* $\frac{k}{2} - m$ *poles lying on the rays* ρe^{it_j}, $1 < \rho < \infty$, $j = 1,\ldots,m$, *and the remaining poles at* ∞, *such that*

$$(28) \qquad |f(z) - R_k(z)| < M_R k^A e^{-Dk^{1/c}}$$

for $|z| \leq 1$ *where* M_R *is bounded as* $R \to \infty$ *and*

$$A = \max\left(\frac{2}{c}, \frac{2a+1}{2c}\right),$$

$$D = \frac{c}{2e}(\alpha b B_1)^{-1/c},$$

$$B_1 = \max\left(\frac{1}{R}, \frac{\pi}{2\beta}\right).$$

Proof. The proof of this proposition is similar to that of Theorem 1 with the exception of the following modifications:

Set $p(r) = r$, and the relation of r with n will be determined later. Let N be the integer part of n/m and choose w_k, $k = 1,\ldots,n$, as in Lemma B with $\ell = m$ and $u_\ell = t_\ell$ ($z_\ell = e^{it_\ell}$), and define $w_{n+1} = \ldots = w_{2n} = 0$. As in (15), write $I_1 = I_1' + \ldots + I_1''''$. To estimate I_1' we have, for $j = r+1$,

$$\left|\int_{z_r^k}^{z_{r+1}^k} \left[\int_\zeta^{z_{r+1}^k} \ldots\right] d\zeta\right| \leq C r^{a+c-1/2} (eb)^r r^{(c-1)r} (\alpha q^r)^{r+1} n^{1/2},$$

and for $1 \leq j \leq r$,

$$\sum_{j\in S_k} \left|\int_{z_{j-1}^k}^{z_j^k} \left[\int_\zeta^{z_j^k} \ldots\right] d\zeta\right| \leq C r^2 \left(\frac{\pi}{2}\right)^r \left(\frac{1-q}{q}\right)^{r+1} n^{1/2},$$

and

$$\sum_{j \notin S_k} \left| \int_{z_{j-1}^k}^{z_j^k} \left[\int_{\zeta}^{z_j^k} \cdots \right] d\zeta \right| \le C \left(\frac{\alpha}{R}\right)^{r+1} r^2 \left(\frac{1-q}{q}\right)^{r+1} n^{1/2} .$$

Hence,

$$(29) \qquad I_1' \le C\{r^{a+c-1/2}(eb)^r r^{(c-1)r}(\alpha q^r)^{r+1}$$

$$+ C_R[\max(\tfrac{\pi}{2}, \tfrac{\alpha}{R})]^r (\tfrac{1-q}{q})^{r+1}\} n^{1/2}$$

where $C_R = \max(1, 1/R)$. To estimate I''', we have, for $j = r+1$,

$$(30) \qquad \left| \int_{z_r^k}^{z_{r+1}^k} \left[\int_{z_{r+1}^k}^{z_s} \cdots \right] d\zeta \right| \le C r^{a+c-1/2} (2beq)^r r^{(c-1)r} e^{-\sqrt{n/2m}} ,$$

and for $1 \le j \le r$, the integral with respect to ζ in (21) can be divided into two parts V_{kj} and Q_{kj} where Q_{kj} has an estimate given above in the second half of (19), and V_{kj} can be estimated by using a result of Rusak (cf. [4; p.165])

$$\left| \int_{\zeta}^{z_s} \frac{(\xi-\zeta)^r}{(\xi-z)} \left[\frac{K(\xi)}{K(z)} - \frac{zK(z)}{\xi K(\xi)}\right] d\xi \right|$$

$$\le 2\left[\frac{(r-1)!}{(n+1)\ldots(n+r)} + 2^r e^{-\sqrt{n/2m}}\right] .$$

Here, the first integral was transformed to the integrals on the radii $[0, \zeta]$ and $[0, z_s]$, and the second integral to the integrals along the rays $[\zeta, \infty)$ and $[z_s, \infty)$. Hence,

$$(31) \qquad \sum_{j=1}^{r} |P_{kj}| \le CC_R r^2 B_1^2 q^{-r^2} \left[\left(\frac{r}{en}\right)^r + 2^r e^{-\sqrt{n/2m}}\right]$$

so that I_1''' can be estimated by using (29), (30), and (31). I_1'' and I_1'''' can be approximated in the same manner as I' and I_1''' respectively. Finally, by choosing q and r to satisfy

$$q^r = B_1 \frac{r}{n} e^{\frac{c}{2} - 1} \quad \text{and} \quad r = (\alpha b B_1)^{-1/c} e^{-1} n^{1/c}$$

and using the assumption $c > 2$, we obtain (28).

In [1], a very efficient method to give a stable all-pole filter was discussed. This method, a modification of that of Robinson [2], is called approximation by least-squares inverses. Let f be a function in the Hardy space H^2 with norm $\| \|_2$ such that $f(0) \neq 0$. Then the (unique) polynomial $Q_n = Q_n(f)$ of degree $\leq n$ such that $\|1 - fQ_n\|_2$ is the smallest among all polynomials of degree $\leq n$ is called a least-squares inverse of f. It was proved in [1] that if ϕ_k's are the orthonormal polynomials with respect to the inner product $\langle \, , \, \rangle_{d\mu}$, where $d\mu = |f^*(e^{i\theta})|^2 d\theta$, f^* being the a.e. radial limit of f, then $Q_n(z) = c_n z^n \overline{\phi_n(1/\bar{z})}$. Hence, Q_n is zero-free on $|z| \leq 1$, so that $1/Q_n$ can be considered as a rational approximant of the given function f. In the following, we give an order of this approximation.

THEOREM 2. _Let_ $f \in H^2$ _such that_ $\frac{1}{f} \in C(a,b,c)$ _when_ $a \geq 0$ _and_ $b, c > 0$, _and let_ $Q_n = Q_n(f)$ _be polynomials of degree_ n _which are least-squares inverses of_ f. _Then_

$$(32) \qquad |f(z) - \frac{1}{Q_n(z)}| \leq C n^{a/c} e^{-(ce^{-1}b^{-1/c})n^{1/c}}$$

for all $|z| \leq 1$ _where_ C _is independent of_ n. _Furthermore the exponent_ $\frac{1}{c}$ _of_ n _cannot be improved_.

Proof. From a result on orthogonal polynomials (cf. [5]), we have

$$(33) \qquad \phi_n(z) = g(z)\, z^n (1 + c_n(z)\, \ell n\, n / n), \quad |z| \geq 1,$$

where g is analytic in $|z| > 1$ and continuous on $|z| \geq 1$ such that

$$(34) \qquad |g(z)| = \frac{1}{|f(z)|}, \quad |z| = 1 ,$$

and $c_n(z)$ are uniformly bounded on $|z| \geq 1$. In addition, since Q_n best approximates $1/f$ in $\| \|_{d\mu}$ among all polynomials of degree $\leq n$, we have (cf. [5]),

$$(35) \qquad |\frac{1}{f(z)} - Q_n(z)| \leq C E_n(\tfrac{1}{f})\, \ell n\, n, \quad |z| \leq 1 ,$$

where C is an absolute constant and

$$E_n(\tfrac{1}{f}) = \inf_{T_n} \max_t |\frac{1}{f(e^{it})} - T_n(t)|,$$

the infimum being taken over all trigonometric polynomials T_n of degree $\leq$ n. It is well known (cf. [8]) that for any natural number r,

$$(36) \qquad E_n\left(\frac{1}{f}\right) \leq \frac{\pi}{2}\frac{1}{n^r} \max\left|\left(\frac{1}{f}\right)_t^{(r)}\right|$$

$$\leq \frac{\pi}{2} Mn^{-r} r^a b^r r^{cr}.$$

Now, let r be the smallest integer $\geq e^{-1} b^{-1/c} n^{1/c}$. Hence, from (35) and (36), we have

$$\left|\frac{1}{f(z)} - Q_n(z)\right| \leq Cn^{a/c} e^{-cn^{\frac{1}{c}}/eb^{\frac{1}{c}}}, \quad |z| \leq 1 .$$

Combining this inequality with (33) and using (34), we obtain

$$\left|f(z) - \frac{1}{Q_n(z)}\right| = \left|\frac{f(z)}{Q_n(z)}\right|\left|\frac{1}{f(z)} - Q_n(z)\right|$$

$$\leq C\, n^{a/c} e^{-cn^{\frac{1}{c}}/eb^{\frac{1}{c}}}$$

for all $|z| \leq 1$. This proves (32).

To establish the sharpness of this estimate we consider the following example. Let

$$F(z) = \frac{a_0}{2} + \sum_{k=1}^{\infty} a_k z^k$$

where, for $k \geq 3$,

$$a_k = e^{-k^{1/c}},$$

and a_0, a_1, a_2 are chosen such that $a_k \downarrow 0$ and the second difference $\Delta^2 a_k = a_{k+2} - 2a_{k+1} + a_k$ is positive for each k. Hence, using Abel's transformation twice, it is easy to prove that $F(z) \neq 0$ on $|z| = 1$. For $|z| < 1$, by considering $(1-z)F(z)$, it is again clear that $F(z) \neq 0$ for $|z| > 1$. In addition, $F \in C^\infty$ on $|z| = 1$ and, in fact, we have

$$|F(e^{it})_t^{(r)}| \leq \sum_{k=1}^{\infty} k^r a_k$$

$$\leq (cr)^{cr+1} + \int_{(cr)^c}^{\infty} t^{r - t^{1/c}/\ln t}\, dt$$

$$= (cr)^{cr+1} + c\int_{cr}^{\infty} x^{rc+c-1} e^{-x}\,dx$$

$$< (cr)^{cr+1} + c\Gamma(rc+c) \le \frac{\sqrt{2\pi}}{e} c^{c+1/2} r\, c^{cr} r^{cr}$$

so that $F \in C(1, c^c, c)$ and $f = \frac{1}{F} \in H^2$. Let $Q_n = Q_n(f)$ be the least-squares inverses of f. Then by using (33) and (34) it is clear that c_n is bounded so that

$$\max_{|z|\le 1} \left|f(x) - \frac{1}{Q_n(z)}\right| = \max_{|z|=1} \left|\frac{f(z)}{Q_n(z)}(F(z) - Q_n(z))\right|$$

$$\ge C \max_{|z|=1} |F(z) - Q_n(z)|$$

$$\ge C\|F - Q_n\|_2$$

for some positive constant C and all sufficiently large n. It is easy to see that

$$\|F - Q_n\|_2^2 = \sum_{k=n+1}^{\infty} e^{-2k^{1/c}}$$

$$> \int_{n+1}^{\infty} e^{-2t^{1/c}} dt \ge \frac{2}{c+1} n^{(c+1)/c} e^{-2(n+1)^{1/c}}$$

for all sufficiently large n. This completes the proof of the theorem.

4. Realization of recursive digital filters

Let $|H_I(e^{it})|$ be the amplitude characteristic of an ideal filter with tolerance allowance $\varepsilon = (\varepsilon_s, \varepsilon_t)$ and $\hat{H}_\varepsilon$ be a causal transformation of $|H_I(e^{it})|$ as described in the first section. Again, since only positive frequency is of interest, we also assume that $|H_\varepsilon(e^{it})|$ is even in t, so that $\hat{H}_\varepsilon$ is in $C_e(B + \frac{1}{2}, b_1, 4) \cap H^R_{m,e}(B + \frac{1}{2}, b_2, 4)$ for $b_1 > 8/e^4$ and $b_2 > 24/e^4$. For the low-pass filter, for example, we also have $m = 8$, and these eight rays are in complex conjugate pairs. From the construction of R_k in Theorem 1 and Proposition 1, the poles of R_k are therefore in conjugate pairs. Since the power series expansion of $\hat{H}_\varepsilon$ at 0 has real coefficients (cf. [1]), R_k also has real coefficients. Indeed, R_k was obtained from R_{2n}

which can also be viewed as the solution of

$$\|\hat{H}_\varepsilon(z) - R_{2n}(z)\|_2 = \min \|\hat{H}_\varepsilon(z) - \frac{P(z)}{Q(z)}\|_2 ,$$

(cf. [9]) or equivalently,

$$\|\overline{\hat{H}_\varepsilon(\bar{z})} - \overline{R_{2n}(\bar{z})}\|_2 = \min \|\overline{\hat{H}_\varepsilon(\bar{z})} - \frac{P(z)}{Q(z)}\|_2$$

where P and Q are polynomials of degree $\leq 2n$. Since $\overline{\hat{H}_\varepsilon(\bar{z})} = \hat{H}_\varepsilon(z)$ and the denominator of R_{2n} has the same property, so does the numerator of R_{2n}.

Furthermore, since the poles of R_k have been clearly pre-assigned, R_k can easily be obtained by "interpolation" of $\hat{H}(z)$ at the reflection of these poles across the unit circle.

Concerning the all-pole filter $1/Q_n$ where Q_n is a least-squares inverse of $\hat{H}_\varepsilon$, it has already been observed in [1] that it has real coefficients and can be obtained easily by inverting a Toeplitz matrix.

In this paper we have not only conidered two efficient methods of realization of stable recursive digital filters but have also studied their degrees of uniform approximation. It is not clear, however, how much the estimate in Theorem 1 can be improved. Perhaps $O(e^{-k^{1/2}})$ would be the correct order of approximation.

References

1. Chui, C. K. and Chan, A. K., Application of approximation theory methods to recursive digital filter design, IEEE Trans. on ASSP, 30 (1982), 18-24.
2. Robinson, E. A, Statistical Communication and Detection, Hafner, New York, 1967.
3. Rusak, V. N., Direct methods in rational approximation with free poles, Dokl. Akad. Nauk BSSR, 22 (1978), 18-20.
4. Rusak, V. N., Rational Functions as Approximation Apparatus, Beloruss. Gos. Univ., Minsk, 1979.
5. Suetin, P. K., Fundamental properties of polynomials orthogonal on a contour, Russian Math. Surveys, 21 (1966), 35-84.
6. Szabados, J., Rational approximation in complex domain, Studia Sci. Math. Hungarian, 4 (1969), 335-340.
7. Szabados, J., Rational approximation to analytic functions on an inner part of the domain of analyticity, in Approximation Theory, ed. by A. Talbot, Academic Press, New York, 1970, pp. 165-177.

APPLICATIONS OF SCHUR FRACTIONS TO DIGITAL FILTERING AND SIGNAL PROCESSING

William B. Jones*
Department of Mathematics
University of Colorado
Boulder, Colorado 80309
U.S.A.

Allan Steinhardt
School of Electrical and Computer Engineering
Oklahoma State University
Stillwater, Oklahoma 74078
U.S.A.

Abstract. Lattice digital filters are used as models in machine analysis and synthesis of signals such as speech. It is shown that rational functions expressed in the form of Schur type continued fractions have poles which contain the desired information in the input signals. Results are given to locate these poles in various regions (e.g., disks, annuli, or complements of disks) without having to compute the poles.

1. Introduction

A class of widely used digital filters are called lattice filters because of their implementation by lattice-shaped directed graphs. Lattice filters are used as models in the processing of data from such diverse areas as economics, medicine, radar detection, seismology and speech. The transfer function of the filter $G_n(z)$ is determined from the input data by using Levinson's method to compute the reflection coefficients γ_k. In the processing of speech data the zeros of $G_n(z)$ characterize individual vowel sounds. Thus the vowel sound of an input signal can be identified by finding the zeros of $G_n(z)$. Conversely, a vowel sound can be produced (electronically) by prescribing the appropriate zeros of $G_n(z)$.

In Section 2 we derive (Theorem 2.1) a Schur type continued fraction $H_n(z)$ (defined by the reflection coefficients γ_k) whose poles are the zeros of $G_n(z)$. This Schur type fraction is derived from a more general result (Theorem 2.1) for a family of lattice directed graphs. A quotient-difference type algorithm (related to the Schur fraction) for computing the poles of $H_n(z)$ was described in [4]. In Section 3, efficient methods are given (in terms of the reflection coefficients γ_k) to determine the number of poles of $H_n(z)$

---

*Research supported by the National Science Foundation under Grant MCS-8202230.

located in $|z| < 1$, $|z| > 1$, and in $R < |z| < 1$. Schur type continued fractions are also used (Corollary 3.3) to determine the number of zeros of a polynomial $Q_n(z)$ in $|z| < 1$ and in $|z| > 1$. This is the analogue for the unit disk of the theorem due to Wall and Frank for half planes (see, e.g., [12, Theorem 48.1] or [5, Theorem 7.35]). We give now some background material for readers not familiar with digital filters and signal processing (for further details, see [7], [8], [9], [11]).

Let ℓ denote the linear space over $\mathbb{R}$ consisting of $\ell := [\{x(m)\}_{m=0}^{\infty}: x(m) \in \mathbb{R}]$. A map $\mathcal{N} : \ell \to \ell$ of the form $\mathcal{N}(\{x(m)\}) = \{y(m)\}$ where

$$(1.1a) \qquad y(m) = \sum_{k=0}^{M} a_k x(m-k) - \sum_{k=1}^{N} b_k y(m-k) \ , \quad m = 0,1,2,\ldots,$$

$$(1.1b) \qquad a_k, b_k \in \mathbb{R} \ , \quad a_M \neq 0 \ , \quad b_N \neq 0 \ , \quad x(m) = y(m) = 0 \quad \text{if} \quad m < 0,$$

is called a <u>digital filter</u>. If $N = 0$, the second sum on the right in (1.1a) is zero and the filter is called <u>nonrecursive.</u> The filter is called <u>recursive</u> if $N > 0$. $\{x(m)\}$ and $\{y(m)\}$ are called, respectively, the <u>input</u> and <u>output</u> of $\mathcal{N}$. Every filter $\mathcal{N}$ is a linear transformation of ℓ into ℓ .

A convenient tool for digital filters is the Z-<u>transform</u>

$$(1.2) \qquad Z(\{x(m)\}) := \sum_{m=0}^{\infty} x(m) z^{-m} \ , \quad \{x(m)\} \in \ell \ .$$

It can be seen that Z is a linear, one-to-one map of ℓ onto the space L of formal Laurent series (1.2). Following Henrici [3, Section 10.10], we use the <u>Doetsch symbol</u> $Z(\{x(m)\} \bullet\!\!\overset{z}{\longrightarrow}\!\!\circ \{x(m)\}$ to indicate (1.2). Two operations on ℓ are of special interest, the <u>unit delay</u> $D\{x(m)\} := \{0, x(0), x(1), x(2), \ldots\}$ and the <u>convolution</u> $\{h(m)\} * \{x(m)\} := \{\sum_{k=0}^{m} h(k) x(m-k)\}$ of $\{h(m)\}$ and $\{x(m)\}$. If $X(z) \bullet\!\!\overset{z}{\longrightarrow}\!\!\circ \{x(m)\}$ and $H(z) \bullet\!\!\overset{z}{\longrightarrow}\!\!\circ \{h(m)\}$, then it can be seen that

$$(1.3) \qquad z^{-1} X(z) \bullet\!\!\overset{z}{\longrightarrow}\!\!\circ D(\{x(m)\})$$

and

$$(1.4) \qquad H(z) X(z) \bullet\!\!\overset{z}{\longrightarrow}\!\!\circ \{h(m)\} * \{x(m)\} \ .$$

Let $A(z) := \sum_{k=0}^{M} a_k z^{-k} \bullet\!\!\overset{z}{\longrightarrow}\!\!\circ \{a_k\}$ and $B(z) := \sum_{k=0}^{N} b_k z^{-k} \bullet\!\!\overset{z}{\longrightarrow}\!\!\circ \{b_k\}$, where $b_0 := 1$. Then (1.1a) can be expressed as

$$(1.5) \qquad \{b_k\} * \{y(m)\} = \{a_k\} * \{x(m)\} \ .$$

Taking the Z-transform of both sides and applying (1.4) yields $B(z)Y(z) = A(z)X(z)$, where $Y(z) \bullet\!\!\overset{z}{\longrightarrow}\!\!\circ \{y(m)\}$. Thus

$$(1.6) \qquad Y(z) = H(z)X(z) \text{ , where } H(z) := \frac{a_0 + a_1 z^{-1} + \ldots + a_M z^{-M}}{1 + b_1 z^{-1} + \ldots + b_N z^{-N}} .$$

The rational function $H(z)$ is called the <u>transfer function</u> of the filter $\mathcal{N}$ in (1.1). The sequence $\{h(m)\} \circ\!\!\overset{z}{\longrightarrow}\!\!\bullet H(z)$ is called the <u>shock response</u> of $\mathcal{N}$. Since $1 \bullet\!\!\overset{z}{\longrightarrow}\!\!\circ \delta := \{1,0,0,0,\ldots\}$, we see that the shock response $\{h(m)\}$ is the output of $\mathcal{N}$ resulting from the unit input δ. Moreover, (1.6) implies that

$$(1.7) \qquad \{y(m)\} := \mathcal{N}(\{x(m)\}) = \{h(m)\} * \{x(m)\} .$$

A filter $\mathcal{N}$ is called <u>stable</u> if the output $\{y(m)\}$ is a bounded sequence whenever the input $\{x(m)\}$ is bounded. Stability is characterized in terms of $H(z)$ and $\{h(m)\}$ in the following:

<u>Theorem</u> 1.1. <u>If $\mathcal{N}$ is a digital filter with transfer function</u> $H(z)$ <u>and shock response</u> $\{h(m)\}$, <u>then the following are equivalent:</u> (A) $\mathcal{N}$ <u>is stable</u>. (B) <u>All poles of</u> $H(z)$ <u>are in</u> $|z| < 1$. (C) $\sum_{m=0}^{\infty} |h(m)| < \infty$ [2].

The next theorem deals with input sequences of values of trigonometric series sampled at equally spaced instants of time. Such sequences frequently arise in signal processing.

<u>Theorem</u> 1.2 <u>Let</u>

$$(1.8) \qquad x(k) := \sum_{j=1}^{I} \mu_j \cos(\omega_j k + \phi_j) \text{ , } \quad k = 0,1,2,\ldots$$

<u>where</u> μ_j, ω_j <u>and</u> ϕ_j <u>are real numbers and the</u> ω_j <u>are distinct.</u> <u>Let</u> $\{\delta(k)\} = \{1,0,0,0,\ldots\}$ <u>and let $\mathcal{S}$ denote a digital filter with transfer function</u> $G(z)$ <u>such that</u>

$$(1.9) \qquad \mathcal{S}(\{x(k)\}) = \{\delta(k)\} .$$

<u>Then</u>

$$(1.10) \qquad \frac{1}{G(z)} = \sum_{j=1}^{I} \left[\frac{\alpha_j z}{z - e^{i\omega_j}} + \frac{\overline{\alpha}_j z}{z - e^{-i\omega_j}} \right] \text{ , } \quad \alpha_j := \frac{\mu_j e^{i\phi_j}}{2} .$$

<u>Proof outline</u>. Let $X(z) \bullet\!\!\overset{z}{\longrightarrow}\!\!\circ \{x(k)\}$ and note that $1 \bullet\!\!\overset{z}{\longrightarrow}\!\!\circ \{\delta(k)\}$. Then by (1.6),

$$(1.11) \qquad 1 = G(z)X(z) .$$

Let $H(z)$ denote the rational function on the right side of (1.10). Then using geometric series expansions of $\alpha_j/(1-e^{i\omega_j}z^{-1})$ and $\bar{\alpha}_j/(1-e^{-i\omega_j}z^{-1})$ (convergent for $|z| > 1$), interchanging order of summation and applying (1.11), we obtain $H(z) = \sum_{k=0}^{\infty} x(k)z^{-k} = X(z) = 1/G(z)$. ■

<u>Remarks</u>: From (1.9) we see that $\mathcal{S}$ filters out all information in the input sequence $\{x(k)\}$. As a consequence the transfer function $G(z)$ contains all of the information in $\{x(k)\}$, namely the amplitudes μ_j , frequencies ω_j and phase angles ϕ_j . In particular, the zeros $e^{\pm i\omega_j}$ of $G(z)$ give the frequencies ω_j . They always occur in conjugate pairs and lie on $|z| = 1$. Thus the filter is, in fact, a model of the input signal $\{x(k)\}$.

In practical situations the input has the form $\{x(k)\} = \{x_1(k)\} + \{x_2(k)\}$ where $\{x_1(k)\}$ is the true signal having the form (1.8) and $\{x_2(k)\}$ is white noise, small in amplitude relative to the μ_j and affecting all frequencies. In such a case we do not seek a filter $\mathcal{S}$ with the property (1.9), since then it would represent all of the noise as well as the true signal. Rather we seek a filter $\mathcal{S}_n$ with the property

$$(1.12) \qquad \mathcal{S}_n(\{x(k)\}) = \{\varepsilon_n(k)\}$$

where the $\varepsilon_n(k)$ are made to be small but not zero. Thus some of the noise is allowed to pass through the filter and $\mathcal{S}_n$ provides an approximate model of the signal. For this purpose we consider a transfer function $G_n(z)$ of $\mathcal{S}_n$ to be of the form

$$(1.13) \qquad G_n(z) = \sum_{j=0}^{n} g_j^{(n)} z^{-j} \ , \quad g_j^{(n)} \in \mathbb{R} \ , \quad g_0^{(n)} = 1 \ .$$

In this context it is useful to consider a sequence

$$\hat{x}(k) := -\sum_{j=1}^{n} g_j^{(n)} x(k-j) \ , \quad k = 1,2,3,\ldots \ ,$$

where we think of $\hat{x}(k)$ as being a <u>prediction</u> of $x(k)$ based on the sample values $x(k-1), x(k-2),\ldots,x(k-n)$. Thus the <u>residual</u> $\varepsilon(k) := x(k) - \hat{x}(k) = \sum_{j=0}^{n} g_j^{(n)} x(k-j)$, $k = 1,2,3,\ldots,$ satisfies (1.12). The criterion for determining the coefficients $g_j^{(n)}$ is to minimize the expected value $E[\varepsilon_n(k)^2]$. This gives rise to a linear least squares problem for which it is convenient to introduce the covariance terms $r_k := E[x(m)x(m+k)]$, $k = 0,\pm 1,\pm 2,\ldots$. Assuming $\{x(m)\}$ to be a stationary random process, we have

$$r_k = r_{-k}\;,\quad k = 1,2,3,\ldots \tag{1.14}$$

since $r_{-k} := E[x(m)x(m-k)] = E[x(j+k)x(j)] =: r_k$. In the following theorem, which shows how to obtain $G_n(z)$, we use the *Toeplitz determinants*

$$R_n^{(m)} := \begin{vmatrix} r_m & r_{m+1} & \cdots & r_{m+n-1} \\ r_{m-1} & r_m & \cdots & r_{m+n-2} \\ \vdots & \vdots & & \vdots \\ r_{m-n+1} & r_{m-n+2} & \cdots & r_m \end{vmatrix} . \tag{1.15}$$

Theorem 1.3. *Let* $\{x(k)\}$ *be a given input sequence with covariance terms* $r_k := E[x(m)x(m+k)]$. *Let* $\mathcal{F}_n$ *be a filter with transfer function* $G_n(z)$ *of the form* (1.13) *and let* $\{\varepsilon_n(k)\}$ *satisfy* (1.12). *Then:*

(A) $E[\varepsilon_n(k)^2]$ *takes its minimum value if and only if the* $g_j^{(n)}$ *satisfy the equations*

$$\sum_{j=1}^{n} r_{j-k} g_j^{(n)} = -r_k\;,\quad k = 1,2,\ldots,n\;. \tag{1.16}$$

(B) *Suppose the* $g_j^{(n)}$ *satisfy* (1.16) *for all* $n \geq 1$. *For* $n \geq 1$, *define*

$$P_0(z) := Q_0(z) := 1\;, \tag{1.17a}$$

$$Q_n(z) := z^n G_n(z) := \sum_{j=0}^{n} q_j^{(n)} z^n\;,\quad q_j^{(n)} := g_{n-j}^{(n)}\;,\quad q_n^{(n)} := 1, \tag{1.17b}$$

$$P_n(z) := z^n Q_n(1/z)\;. \tag{1.17c}$$

Then for $n = 1,2,3,\ldots,$

$$P_n(z) = P_{n-1}(z) + \gamma_n z Q_{n-1}(z),\quad Q_n(z) = \gamma_n P_{n-1}(z) + z Q_{n-1}(z), \tag{1.18a}$$

$$q_j^{(n)} = q_{j-1}^{(n-1)} + \gamma_n q_{n-1-j}^{(n-1)}\;,\quad j = 1,2,\ldots,n\;, \tag{1.18b}$$

where

$$\gamma_n := Q_n(0) = (-1)^n \frac{R_n^{(-1)}}{R_n^{(0)}} = -\frac{\sum_{j=0}^{n-1} q_j^{(n-1)} r_{j+1}}{E_{n-1}} \tag{1.18c}$$

$$(1.18d)\qquad E_n := \min E[\varepsilon_n(k)^2] = \sum_{j=0}^{n} q_j^{(n)} r_{n-j}$$

<u>and</u>

$$(1.19)\qquad Q_n(z) = \frac{1}{R_n^{(0)}} \begin{vmatrix} r_0 & r_1 & \cdots & r_{n-1} & 1 \\ r_{-1} & r_0 & \cdots & r_{n-2} & z \\ \vdots & \vdots & & \vdots & \vdots \\ r_{-n} & r_{-n+1} & \cdots & r_{-1} & z^n \end{vmatrix} .$$

Proofs of Theorem 1.3 can be found, for example, in [11, Section 7.6], [13, Appendix B, 131-139], and [6]; hence we omit a proof here. We note simply that (1.16) is obtained by setting equal to zero the partial derivatives of

$$(1.20)\qquad E[\varepsilon_n(k)^2] = E[(\sum_{j=0}^{n} g_j^{(n)} x(k-j))^2] = \sum_{j=0}^{n}\sum_{m=0}^{n} g_j^{(n)} g_m^{(n)} r_{j-m}$$

with respect to the different $g_j^{(n)}$. Then (1.18d) is obtained by applying (1.16) to (1.20). The formula (1.19) is an immediate consequence of (1.16) and (1.17b).

The function $G_n(z)$ of Theorem 1.3 could be obtained by solving the system of equations (1.16) directly. However, a more efficient procedure consists of applying the recurrence relations (1.18) to compute successively $\{q_j^{(n)}\}_{j=1}^{n}$, γ_n and E_n for $n = 1,2,3,\ldots$. This is referred to as Levinson's method (or the Levinson-Durbin algorithm). The numerical stability of this algorithm has been discussed by Cybenko [1]. The γ_n in (1.18c) are called the <u>reflection coefficients</u>. Unless E_n decreases appreciably when n increases, there is no benefit in increasing n. Thus by computing the successive values $E_1, E_2, E_3, \ldots$, one can determine a best value for n. In practice one has only a finite sequence $\{x(m)\}_{m=0}^{s}$ for input; hence one approximates the covariance terms by

$$(1.21)\qquad r_k \approx \frac{1}{s-k+1} \sum_{m=0}^{s-k} x(m)x(m+k) , \quad k = 0,1,\ldots,s .$$

2. Lattice Filters and Directed Graphs

Directed graphs are used to describe specific procedures for implementing a digital filter. In this section we describe a class of lattice directed graphs for which the corresponding filter is thus called a lattice filter. It is shown (Theorem 2.1) that the transfer function of such a filter can be expressed as a continued fraction whose elements are given directly in terms of the directed graph parameters. As a special case, we derive (Theorem 2.2) the lattice directed graph and continued fraction associated with the digital filter $\mathcal{L}_n$ of Theorem 1.3. For the reader who is not familiar with directed graphs, we begin with some introductory remarks.

A directed graph is a geometric configuration consisting of two types of elements: points (called nodes) and simple directed curves (called branches). Each branch connects two nodes, the direction of the branch being indicated by an arrow. The nodes in the lattice directed graph (Figure 1) are numbered from 1 to 4n . Every graph has a source node and at least one sink node; the source node has no branch originating from it. A graph may be described either in the sequence domain ℓ or in the Z-transform domain L . We describe the latter, but similar terminology applies to the former, if we replace the function $X_i(z) = \sum_{k=0}^{\infty} x_i(k)z^{-k}$ by the corresponding sequence $\{x_i(k)\}_{k=0}^{\infty}$, etc. In this context we call $X_i(z)$ a function even if the formal series is not convergent.

Each branch of the graph has associated with it a transmittance function $T_{jk}(z)$, usually of the form

$$(2.1)\qquad T_{jk}(z)=a_{jk} + b_{jk}z^{-1}\ , \quad a_{jk},b_{jk} \in \mathbb{R}\ ,$$

where either $a_{jk} = 0$ or $b_{jk} = 0$ but not both. The index j (k) indicates the node at which the branch originates (terminates). The effect of T_{jk} is a scalar multiplication by a_{jk} if $b_{jk} = 0$; it is a scalar multiplication by b_{jk} and a unit delay if $a_{jk} = 0$.

Each node has associated with it a node function; $X_j(z) = \sum_{k=0}^{\infty} x_j(k)z^{-k}$ denotes the node function of the jth node (Figure 1). The node functions are interrelated by means of the fundamental equations

$$(2.2)\qquad X_j(z) = \sum_{\substack{k=1\\k\neq j}}^{4n} T_{kj}(z)X_j(z)\ , \quad j = 1,2,\ldots,4n\ .$$

If there is no branch from the kth to the jth node, then it is understood that $T_{kj}(z) = 0$. In our discussion, node 1 is the source node and node 4n is a sink. Hence $\{x_1(k)\} \circ\!\xrightarrow{\;z\;}\!\bullet\, X_1(z)$ is the input and $\{x_{4n}(k)\} \circ\!\xrightarrow{\;z\;}\!\bullet\, X_{4n}(z)$ is the output. It follows that

(2.3) $X_{4n}(z) = H_n(z)X_1(z)$

where $H_n(z)$ denotes the transfer function of the filter. By setting $X_1(z) = 1$ in (2.2) and (2.3) and solving for $X_{4n}(z)$, one can obtain $H_n(z)$. It is well known that every digital filter can be described by equations of the form (2.2) and (2.3). Hence every digital filter can be represented by a directed graph; moreover, the specific procedures for implementing the filter are indicated by the graph.

It is sometimes useful to decompose a directed graph into subgraphs. A directed graph is called a subgraph of a (parent) graph if it satisfies the following conditions: (1) Exactly one branch of the parent graph enters the subgraph (at the subgraph source node). (2) Exactly one branch of the parent graph leaves the subgraph (at the subgraph sink node). The role of a subgraph in its parent graph is governed by the following result: Let $X_{SO}(z)$ (X_{SI}) denote the source (sink) node function of a subgraph of a given parent graph. Then there exists a transmittance function $T(z) = \sum_{k=0}^{\infty} t(k)z^{-k}$ with the property that if the original subgraph is replaced by a new subgraph consisting of a single branch, with transmittance function $T(z)$, going from the source to the sink node (of the original subgraph), then the node functions at all nodes of the parent graph exterior to the subgraph remain the same. In other words a subgraph can be replaced by a single branch and the remainder of the parent graph is unchanged. Moreover, $X_{SI}(z) = T(z)X_{SO}(z)$ so that $T(z)$ can be obtained by setting $X_{SO}(z) = 1$ and solving the fundamental equations.

In the lattice directed graph in Figure 1, the transmittance functions associated with the various branches are denoted by the Greek letters κ_j, λ_j, μ_j, ν_j, σ_j, and τ_j.

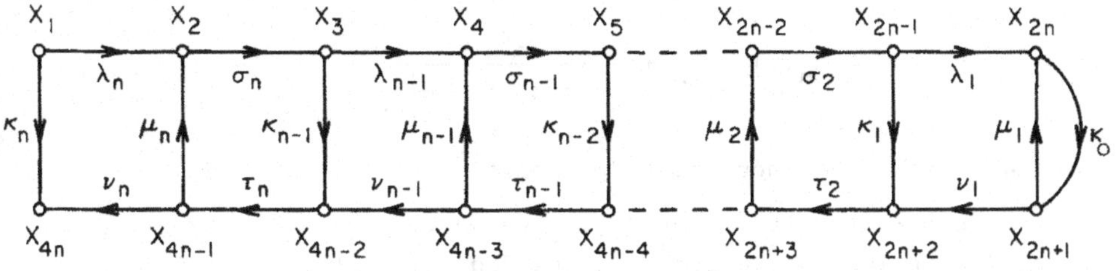

Figure 1. General lattice directed graph.

<u>Theorem</u> 2.1. <u>The transfer function $H_n(z)$ of the lattice digital filter represented by the directed flow graph in Figure 1 is given by the continued fraction</u>

$$(2.4)\qquad H_n(z) = \kappa_n + \frac{\lambda_n\nu_n\sigma_n\tau_n}{-\mu_n\sigma_n\tau_n} + \frac{1}{\kappa_{n-1}} + \frac{\lambda_{n-1}\nu_{n-1}\sigma_{n-1}\tau_{n-1}}{-\mu_{n-1}\sigma_{n-1}\tau_{n-1}} + \cdots$$

$$+ \frac{1}{\kappa_1} + \frac{\lambda_1\nu_1\sigma_1\tau_1}{-\mu_1\sigma_1\tau_1} + \frac{1}{\kappa_0}\ ,$$

where $\sigma_1 = \tau_1 = 1$.

<u>Proof by induction</u>. If $n = 1$, the fundamental equations (2.2) for the filter are: $X_2 = \lambda_1X_1 + \mu_1X_3$, $X_3 = \kappa_0X_2$ and $X_4 = \kappa_1X_1 + \nu_1X_3$. Then setting $X_1 = 1$ and $H_1 = X_4$ yields (2.4) with $n = 1$. Now suppose that (2.4) holds with n replaced by $n-1$ (induction hypothesis). Consider the graph in Figure 2, obtained from that in

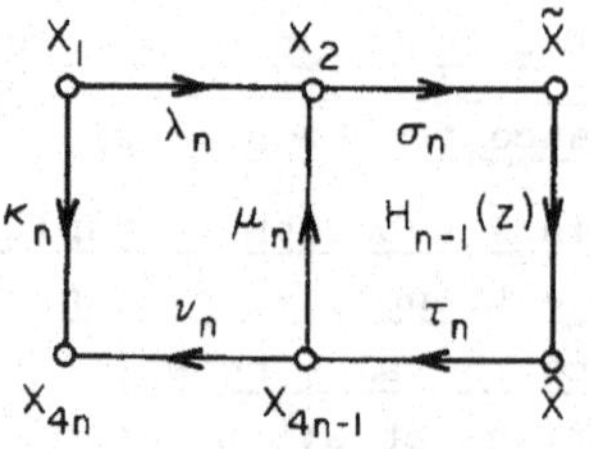

Figure 2. Reduced lattice directed graph.

Figure 1 by replacing a subgraph of the right side of Figure 1 by a single branch with transmittance function $H_{n-1}(z)$. By the induction hypothesis, Figures 1 and 2 represent the same filter. The fundamental equations (2.2) for the filter in Figure 2 are: $X_2 = \lambda_nX_1 + \mu_nX_{4n-1}$, $\tilde{X} = \sigma_nX_2$, $\hat{X} = H_{n-1}(z)\tilde{X}$, $X_{4n-1} = \tau_n\hat{X}$ and $X_{4n} = \kappa_nX_1 + \nu_nX_{4n-1}$. Setting $X_1 = 1$ and $H_n = X_{4n}$ and solving for H_n gives

$$H_n(z) = \kappa_n + \cfrac{\lambda_n\nu_n\sigma_n\tau_n}{-\mu_n\sigma_n\tau_n + \cfrac{1}{H_{n-1}(z)}}\ .$$

Applying to this the induction hypothesis gives (2.4). ■

In our next result (Theorem 2.2) we derive the Schur type continued fraction associated with the digital filter $\mathcal{S}_n$ of Theorem 1.3. The related lattice directed graph is shown in Figure 3.

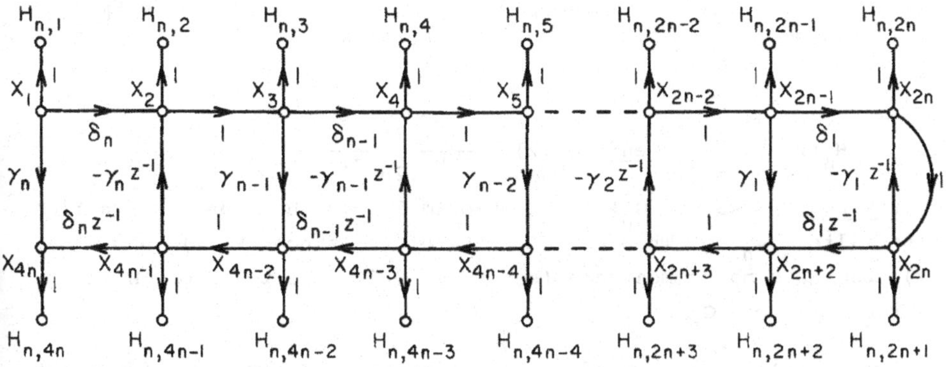

Figure 3. Lattice directed graph (in transform domain) for the analysis filter $\mathcal{J}_n$ of Theorem 1.3. The nodal functions are denoted by X_k and $H_{n,k}$. Next to each branch is the associated transmittance function, where the γ_k are the reflection coefficients (1.18c) and $\delta_k := \sqrt{1-\gamma_k^2}$. There is one source node, namely at X_1; there are $4n$ sink nodes at $H_{n,k}$, $k = 1,2,\ldots,4n$.

<u>Theorem</u> 2.2. <u>Let</u> γ_k <u>and</u> δ_k <u>be real numbers such that</u>

$$(2.5) \qquad |\gamma_k| \neq 1 \quad \underline{\text{and}} \quad \delta_k := \sqrt{1-\gamma_k^2}\,, \quad k = 1,2,\ldots,n\,.$$

<u>Let</u> $P_k(z)$ <u>and</u> $Q_k(z)$ <u>be polynomials defined by</u> $P_0(z) := Q_0(z) = 1$,

$$(2.6) \qquad P_k(z) := P_{k-1}(z)+\gamma_k z Q_{k-1}(z),\ Q_k(z) := \gamma_k P_{k-1}(z)+zQ_{k-1}(z)$$
$$k = 1,2,\ldots,n\,.$$

<u>For</u> $k = 1,2,\ldots,4n$, <u>let</u> $H_{n,k}(z)$ <u>denote the indicated transfer functions of the digital filters implemented by the lattice directed graph in Figure 3.</u> <u>Let</u> $H_n(z) := H_{n,4n}(z)$ <u>and</u> $G_n(z) := z^{-n}Q_n(z)$. <u>Then</u>: (A) $H_n(z)$ <u>has the Schur type continued fraction representation</u>

$$(2.7) \qquad H_n(z) = \gamma_n + \frac{(1-\gamma_n^2)z^{-1}}{\gamma_n z^{-1}} + \frac{1}{\gamma_{n-1}} + \frac{(1-\gamma_{n-1}^2)z^{-1}}{\gamma_{n-1}z^{-1}} + \cdots + \frac{1}{\gamma_1}$$
$$+ \frac{(1-\gamma_1^2)z^{-1}}{\gamma_1 z^{-1}} + \frac{1}{1}\,.$$

(B) <u>If</u> $\Delta_{n,k} := \delta_{k+1}\delta_{k+2}\cdots\delta_n$, $k = 0,1,\ldots,n-1$ <u>and</u> $\Delta_{n,n} := 1$, <u>then for</u> $k = 1,1,\ldots,n$,

$$(2.8)\quad H_{n,2(n-k)}(z) = \Delta_{n,k} z^{n-k}\frac{Q_k(z)}{Q_n(z)},\quad H_{n,2(n+k)}(z) = \Delta_{n,k} z^{n-k}\frac{P_k(z)}{Q_n(z)}.$$

(C)

$$(2.9)\qquad H_n(z) = \frac{P_n(z)}{Q_n(z)} \quad \underline{\text{and}} \quad G_n(z) = \frac{\delta_1\delta_2 \cdots \delta_n}{H_{n,2n}(z)}.$$

<u>Remarks</u>. Clearly $G_n(z)$ in Theorem 2.2 is the transfer function of the filter $\mathcal{S}_n$ of Theorem 1.3. Therefore, since $P_n(z)$ and $Q_n(z)$ can have no common zero (see Lemma 3.2), the poles of $H_n(z)$ give the zeros of $G_n(z)$.

<u>Proof</u>. (A) is an immediate consequence of Theorem 2.1 if we set $\kappa_k = \gamma_k$, $\lambda_k = \delta_k$, $\mu_k = -\gamma_k z^{-1}$, $\nu_k = \delta_k z^{-1}$, and $\sigma_k = \tau_k = \kappa_o = 1$, $k = 1,2,\ldots,n$. (B): Setting $X_o = 1$, we obtain from the fundamental equations (2.2)

$$(2.10a)\qquad X_{2k} = \delta_{n-k+1}X_{2k-2} - \gamma_{n-k+1}z^{-1}X_{4n-2k},\quad k = 1,2,\ldots,n$$

$$(2.10b)\qquad X_{2n+2k} = \gamma_k X_{2n-2k} + \delta_k z^{-1}X_{2n+2k-2},\quad k = 1,2,\ldots,n,$$

since $X_{2k+1} = X_{2k}$, $k = 1,2,\ldots,2n-1$. For $k = 0,1,\ldots,n$ let

$$(2.11)\qquad W_{n-k} := \Delta_{n,k}z^{n-k}\frac{Q_k(z)}{Q_n(z)} \quad \text{and} \quad W_{n+k} := \Delta_{n,k}z^{n-k}\frac{P_k(z)}{Q_n(z)}.$$

By induction on k (using (2.6)) it can be shown that the W's satisfy the recurrence relations

$$(2.12a)\qquad W_k = \delta_{n-k+1}W_{k-1} - \gamma_{n-k+1}z^{-1}W_{2n-k},\quad k = 1,2,\ldots,n,$$

$$(2.12b)\qquad W_{n+k} = \gamma_k W_{n-k} + \delta_k z^{-1}W_{n+k-1},\quad k = 1,2,\ldots,n.$$

By comparing (2.12) with (2.10) it can be seen that $W_k = X_{2k}$, $k = 0,1,\ldots,n$. If we now set $X_1 = 1$, then $H_{n,k}(z) = X_k(z)$, $k = 1,2,\ldots,4n$. Thus (2.8) holds. (C) follows immediately from (B). ■

3. Location of Zeros of $G_n(z)$

Connections between Schur type continued fractions, general T-fractions and two-point Padé approximants were discussed by Thron [10, 215-226]. The present authors showed [4, Section 4] that, using the general T-fraction representation of $H_n(z)$, the McCabe-Murphy version of the quotient difference algorithm can be employed in the computation of the poles of $H_n(z)$ (i.e., the zeros of $G_n(z)$).

Examples of applications of this algorithm to speech analysis and signal detection are given in the preceding reference. This section gives results on the location of the poles of $H_n(z)$, not requiring their actual computation. In view of the relations $G_n(z) := z^{-n}Q_n(z)$ and $H_n(z) = P_n(z)/Q_n(z)$, the poles of $H_n(z)$ (i.e., the zeros of $Q_n(z)$) give the zeros of $G_n(z)$. In this section the reflection coefficients γ_n are allowed to be complex.

Theorem 3.1. Let $H_n(z)$ be a Schur type continued fraction

$$(3.1)\qquad H_n(z) = \gamma_n + \frac{(1-|\gamma_n|^2)z^{-1}}{\bar{\gamma}_n z^{-1}} \; \frac{1}{+\;\gamma_{n-1}} + \frac{(1-|\gamma_{n-1}|^2)z^{-1}}{\bar{\gamma}_{n-1}z^{-1}} + \cdots$$

$$\frac{1}{+\;\gamma_1} + \frac{(1-|\gamma_1|^2)z^{-1}}{\bar{\gamma}_1 z^{-1}} \; \frac{1}{+\;1}$$

where

$$(3.2)\qquad |\gamma_k| \neq 1 \;,\quad k = 1,2,\ldots,n \;.$$

Let $L_1, L_2, \ldots, L_m$ denote the indices for which $|\gamma_{n-L_j+1}| > 1$, $j = 1,\ldots,m$, where $L_1 < L_2 < \ldots < L_m$. Let

$$(3.3)\qquad NP := \begin{cases} \sum_{j=1}^{m} (-1)^j L_j & \text{if } m \text{ is even} \\[2ex] n+1 + \sum_{j=1}^{m} (-1)^j L_j & \text{if } m \text{ is odd}\,. \end{cases}$$

Then: (A) $H_n(z)$ has NP poles in $|z| > 1$ and n-NP poles in $|z| < 1$. (B) $|H_n(z)| = 1$ for $|z| = 1$. (C) If $|\gamma_k| < 1$ for $k = 1,2,\ldots,n$, then $NP = 0$ and hence all poles of $H_n(z)$ lie in $|z| < 1$.

In our proof of Theorem 3.1 we use the following lemma which is of some interest in itself.

Lemma 3.2. Let $\{\gamma_k\}_{k=1}^n$ satisfy (3.2) and let P_k and Q_k be defined by $P_0(z) := Q_0(z) = 1$ and, for $k = 1,2,\ldots,n$,

$$(3.4)\qquad P_k(z) := P_{k-1}(z) + \gamma_k z Q_{k-1}(z),\quad Q_k(z) = \bar{\gamma}_k P_{k-1}(z) + zQ_{k-1}(z)\,.$$

Then for $k = 1,2,\ldots,n$, the polynomials P_k and Q_k satisfy the following:

(A) *degree $P_k(z) \leq k$ and degree $Q_k(z) = k$.*

(B) $P_k(0) = 1$ and $Q_k(0) = \overline{\gamma}_k$.

(C) $P_k^*(z) := z^k\overline{P_k(1/\overline{z})} = Q_k(z)$ and $Q_k^*(z) := z^k\overline{Q_k(1/\overline{z})} = P_k(z)$.

(D) If $Q_k(z) = \prod_{j=1}^{k}(z-z_j^{(k)})$ then $P_k(z) = \prod_{j=1}^{k}(1-\overline{z_j^{(k)}}z)$.

(E) $|P_k(z)/Q_k(z)| = 1$ for $|z| = 1$.

(F) $P_k(z)$ and $Q_k(z)$ have no common zero. Moreover, if $P_k(w) = 0$ then $Q_k(1/\overline{w}) = 0$.

(G)

$$\frac{P_k(z)}{Q_k(z)} = \gamma_k + \cfrac{(1-|\gamma_k|^2)z^{-1}}{\overline{\gamma}_k z^{-1} + \cfrac{1}{P_{k-1}(z)/Q_{k-1}(z)}} . \tag{3.5}$$

Proof. (A), (B) and (C) follow directly from (3.4). (D) is immediate from (C), and (E) follows easily from (D). To prove (F) we note from (3.4) that, for $k = 1,2,\ldots,n$,

$$P_{k-1}(z) = \frac{P_k(z)-\gamma_k Q_k(z)}{1-|\gamma_k|^2} , \quad Q_{k-1}(z) = \frac{Q_k(z)-\overline{\gamma}_k P_k(z)}{(1-|\gamma_k|^2)z} . \tag{3.6}$$

Clearly $P_0(z)$ and $Q_0(z)$ have no common zero. By induction (using (3.6)) one can thus show that $P_k(z)$ and $Q_k(z)$ have no common zero. We also make use of the fact that $P_k(0) = 1 \neq 0$. The second part of (F) follows from (C). (G) is a direct consequence of (3.4). ■

Proof of Theorem 3.1. For each polynomial $Q(z)$, let $N(Q)$ denote the number of zeros of $Q(z)$ inside $|z| < 1$. From (3.6) we have

$$P_{k-1}(z) = \frac{P_k(z)}{1-|\gamma_k|^2} - \frac{\gamma_k Q_k(z)}{1-|\gamma_k|^2} , \quad k = 1,2,\ldots,n .$$

If $|\gamma_k| < 1$, then by Lemma 3.2(E)

$$\frac{|P_k(z)|}{1-|\gamma_k|^2} > \frac{|\gamma_k Q_k(z)|}{1-|\gamma_k|^2} \quad \text{for} \quad |z| = 1 .$$

Thus Rouché's theorem implies that $N(P_{k-1}) = N(P_k)$. On the other hand, if $|\gamma_k| > 1$, then by Lemma 3.2(E),

$$\left|\frac{P_k(z)}{1-|\gamma_k|^2}\right| < \left|\frac{\gamma_k Q_k(z)}{1-|\gamma_k|^2}\right| \quad \text{for} \quad |z| = 1 .$$

Thus Rouché's theorem implies that $N(P_{k-1}) = N(Q_k)$. By Lemma 3.2(A), degree $Q_k(z) = k$ and by Lemma 3.2(C), $P_k(w) = 0$ and $|w| < 1$ imply $Q_k(1/\overline{w}) = 0$ where $|1/w| > 1$ and $w \neq 0$. It follows that $N(Q_k) = k - N(P_k)$. Thus we have shown that, for $k = 1,2,\ldots,n$,

$$(3.7) \qquad N(P_k) = \begin{cases} N(P_{k-1}) & \text{if } |\gamma_k| < 1 \\ k - N(P_{k-1}) & \text{if } |\gamma_k| > 1 . \end{cases}$$

This can be written as follows: for $j = 1,2,\ldots,n$

$$N(P_{n-j+1}) = \begin{cases} N(P_{n-j}) & \text{if } |\gamma_{n-j+1}| < 1 \\ n-j+1 - N(P_{n-j}) & \text{if } |\gamma_{n-j+1}| > 1 . \end{cases}$$

Applying these relations successively and using the fact that $N(P_0) = 0$ yields $N(P_n) = n - L_1 + 1 - N(P_{n-L_1}) = (n-L_1+1) - (n-L_2+1) + N(P_{n-L_2}) = \ldots = \sum_{j=1}^{m} (-1)^{j-1}(n-L_j+1)$. It follows that $N(P_n) = NP$ so that $N(Q_n) = n - N(P_n) = n - NP$, which proves (A). (B) follows from Lemma 3.2(E) and (G). (C): If $|\gamma_k| < 1$ for $k = 1,\ldots,n$, then (3.7) implies $N(P_n) = N(P_0) = 0$ and hence $N(Q_n) = n - N(P_n) = n$. ■

It was shown in [4, Theorem 3.1] that a monic polynomial $Q_n(z)$ has all of its zeros in $|z| < 1$ if and only if there exist complex constants γ_k satisfying $|\gamma_k| < 1$, $k = 1,\ldots,n$ such that the test function $Q_n^*(z)/Q_n(z)$ can be expressed as a Schur fraction (3.1). Here $Q_n^*(z) := z^n\overline{Q_n(1/\overline{z})}$ denotes the <u>reciprocal polynomial</u>. A similar result involving Schur type continued fractions is the following consequence of Theorem 3.1.

<u>Corollary</u> 3.3. *Let $Q_n(z)$ be a monic polynomial of degree $n > 1$. Suppose there exist complex constants γ_k satisfying (3.2) such that the test function $Q_n^*(z)/Q_n(z)$ can be expressed as a Schur type continued fraction (3.1). Let $L_1, L_2, \ldots, L_m$ and NP be defined as in Theorem 3.1. Then: (A) $Q_n(z)$ has NP zeros in $|z| < 1$ and $n - NP$ zeros in $|z| < 1$. (B) If $|\gamma_k| < 1$, $k = 1,\ldots,n$, then all zeros of $Q_n(z)$ lie in $|z| < 1$.*

The preceding corollary is the analogue for the unit disk of a theorem due to Wall and Frank for the half-plane Re $z < 0$ (see for example [12, Theorem 48.1] or [5, Theorem 7.35]). Corollary 3.3 gives rise to the question: For a given monic polynomial $Q_n(z)$ of degree n, when do there exist complex constants γ_k satisfying (3.2) such that $Q_n^*(z)/Q_n(z)$ can be expressed by the Schur type continued fraction (3.1)? The question is answered by the following:

Theorem 3.4. Let $Q_n(z)$ be a monic polynomial of degree n

(3.8) $$Q_n(z) = a_0^{(n)} + a_1^{(n)}z + \dots + a_{n-1}^{(n)}z^{n-1} + z^n \, , \quad a_j^{(n)} \in \mathbb{C} \, .$$

Then: (A) There exist complex constants γ_k satisfying (3.2) such that $Q_n^*(z)/Q_n(z)$ can be expressed as the Schur type continued fraction (3.1) if and only if the following condition (Q) holds: For each $k = n, n-1, \dots, 1$, the equation

(3.9) $$Q_{k-1}(z) = \frac{Q_k(z) - a_0^{(k)} Q_k^*(z)}{(1 - |a_0^{(k)}|^2) z}$$

defines a monic polynomial $Q_{k-1}(z)$ of degree $k-1$, $Q_{k-1}(z) = a_0^{(k-1)} + a_1^{(k-1)}z + \dots + a_{k-2}^{(k-1)}z^{k-2} + z^{k-1}$, where

(3.10) $$|a_0^{(k)}| \neq 1 \, , \quad k = 1,2,\dots,n \, .$$

(B) If condition (Q) holds, then for $k = n, n-1, \dots, 2$, we have

(3.11) $$\gamma_k := \overline{a_0^{(k)}} \quad \text{and} \quad a_j^{(k-1)} = \frac{a_{j+1}^{(k)} - a_o^{(k)} \overline{a_{k-j-1}^{(k)}}}{1 - |a_o^{(k)}|^2} \, , \quad j = 0,1,\dots,k-2,$$

and

(3.12) $$Q_{k-1}^*(z) = \frac{Q_k^*(z) - \overline{a_0^{(k)}} Q_k(z)}{1 - |a_0^{(k)}|^2} \, .$$

Proof. (A): Suppose complex constants γ_k exist satisfying (3.2) such that $Q_n^*(z)/Q_n(z)$ equals the continued fraction (3.1). For $k = 1,\dots,n$ let P_k and Q_k be defined by (3.4) and $P_0(z) := Q_0(z) := 1$. Then by Lemma 3.2, condition (Q) holds where $\gamma_k := \overline{a_0^{(k)}}$. Conversely, suppose condition (Q) holds. Define $\gamma_k := \overline{a_0^{(k)}}$, $k = 1,2,\dots,n$. Then the Q_k and $P_k := Q_k^*$ satisfy (3.4) and hence also (3.5). It follows that Q_n^*/Q_n can be expressed by the Schur type continued fraction (3.1). (B) is an immediate consequence of the given hypotheses. ■

We note in passing that (3.11) gives the Schur-Cohn algorithm for generating the coefficients γ_k. We describe now an interesting application of this algorithm and Corollary 3.3. Let $Q_n(z)$ be a given monic polynomial of degree n. Suppose (3.11) is applied and one finds that $|\gamma_k| < 1$, $k = 1,\dots,n$, so that all zeros of $Q_n(z)$

lie in $|\hat{z}| < 1$. Let $\hat{z} := z/R$, $\hat{Q}_n(\hat{z}) := Q_n(z)$, where $0 < R < 1$. Then the number of zeros of $Q_n(z)$ in the annulus $R < |z| < 1$ equals the number of zeros of $\hat{Q}_n(\hat{z})$ in $1 < |\hat{z}| < 1/R$ and hence in $|\hat{z}| > 1$. By applying (3.11) and Corollary 3.3 to $\hat{Q}_n(\hat{z})$ one can determine the number of zeros of $\hat{Q}_n(\hat{z})$ in $|\hat{z}| > 1$, which tells us the number of zeros of $Q_n(z)$ in $R < |z| < 1$. This information can be useful in signal processing since the zeros of $G_n(z) = z^{-n}Q_n(z)$ near the unit circle are the ones of most interest and we have here a method of determining how many such zeros there are.

We conclude with the following remark about Blaschke type products. From Lemma 3.3, every Schur type continued fraction (3.1) with γ_n satisfying (3.2) is the reciprocal of a Blaschke type product

$$(3.13) \qquad \prod_{j=1}^{n} \frac{1 - \bar{z}_j z}{z - z_j} \; , \quad |z_j| \neq 1 \; , \quad j = 1,2,\ldots,n \; .$$

That not every rational function of the form (3.13) can be expressed as a Schur type continued fraction (3.1) with γ_k satisfying (3.2) can be seen by the following simple example: Let $Q_2(z) := z^2 + iz + 1 = (z-a)(z-b)$ where $a := (-1+\sqrt{5})i/2$ and $b := (-1-\sqrt{5})i/2$. Since $Q_2^*(0) = 1$, the algorithm (3.11) breaks down and so there is no Schur type continued fraction representation of $Q_2^*(z)/Q_2(z)$. On the other hand, if in (3.13) we have $|z_j| < 1$, $j = 1,2,..,n$, then by the remark preceding Corollary 3.3, there exists a Schur fraction (3.1) with $|\gamma_k| < 1$, $k = 1,\ldots,n$ representing the rational function in (3.13).

References

1. Cybenko, George, The numerical stability of the Levinson-Durbin algorithm for Toeplitz systems of equations, SIAM J. Sci. Stat. Comput. 1, No. 3 (September 1980), 303-319.

2. Gutknecht, M., Ein Abstiegsverfahren für gleichmässige Approximation, mit Anwendungen. Diss. ETH No. 5006. aku-Fotodruck, Zürich.

3. Henrici, P., Applied and Computational Complex Analysis, Vol. 2, Special Functions, Integral Transforms, Asymptotics and Continued Fractions, John Wiley and Sons, New York (1977).

4. Jones, William B. and Allan Steinhardt, Digital filters and continued fractions, Analytic Theory of Continued Fractions, (W. B. Jones, W. J. Thron and H. Waadeland, eds.), Lecture Notes in Mathematics 932, Springer-Verlag, New York (1982), 129-151.

5. Jones, William B. and Thron, W. J., Continued Fractions: Analytic Theory and Applications, Encyclopedia of Mathematics and Its Applications, No. 11, Addison-Wesley Publishing Company, Reading, Mass. (1980).

6. Levinson, Norman, The Wiener RMS (root mean square) error criterion in filter design and prediction, J. of Math. and Physics 25, (1947), 261-278.

7. Oppenheim, A. V. and Schafer, R. W., Digital Signal Processing, Prentice Hall, New Jersey (1975).

8. Rabiner, L. R. and Schafer, R. W., Digital Processing of Speech Signals, Prentice-Hall, Inc., Englewood Cliffs, New Jersey (1978).

9. Stanley, W. D., Dougherty, G. R. and Dougherty, R., Digital Signal Processing, Reston Publishing Co., Inc., Reston, VA (1984).

10. Thron, W. J., Two-point Padé tables, T-fractions and sequences of Schur, Padé and Rational Approximation (ed. E.B. Saff and R.S. Varga), Academic Press, Inc., New York (1977), 215-226.

11. Tretter, Steven A., Introduction to Discrete-Time Signal Processing, John Wiley and Sons, New York (1976).

12. Wall, H. S., Analytic Theory of Continued Fractions, D. Van Nostrand Co., New York (1948).

13. Wiener, Norbert, Extrapolation, Interpolation and Smoothing of Stationary Time Series, John Wiley and Sons, Inc., New York (1949).

A de MONTESSUS THEOREM FOR VECTOR VALUED RATIONAL INTERPOLANTS

P. R. Graves-Morris*
Mathematical Institute
University of Kent
Canterbury
Kent
ENGLAND

and

E. B. Saff*
Center for Mathematical Services
University of South Florida
Tampa,
Florida 33620
U.S.A.

Abstract A convergence theorem for vector valued Padé approximants (simultaneous Padé approximants) is established. The theorem is a natural extension of the theorem of de Montessus de Ballore for a row sequence of (scalar) Padé approximants. The result is also generalised to the case of vector valued (N-point) rational interpolants.

1. Introduction

The theorem of R. de Montessus de Ballore [7] is a remarkably elegant theorem on the convergence of row sequences of Padé approximants to a meromorphic function. Here, in section 2, we present its extension to the case of simultaneous Padé approximation (see Theorem 1) and to vector valued Padé approximation (see Theorem 2). The generalisations of de Montessus' theorem to multipoint rational interpolation, as distinct from Padé approximation, derived by Saff [8] and Warner [9] are extended to the case of vector valued rational interpolation in Theorem 3.

Simultaneous Padé approximation involves approximation of several functions $\{f_i(z),\ i=1,2,\ldots,d\}$ by rationals of the form $\{P_{N,i}(z)/Q_N(z),\ i=1,2,\ldots,d\}$, where the denominator polynomial $Q_N(z)$ is common to each of the d rational approximants. A full specification of the problem of constructing such polynomials was given by Mahler [5] in 1968. He also considered the extension to the case of interpolating rationals,

---

*We are grateful to the SERC (UK) and to the NSF (US) for support from grants GR/C/41807 and MCS 80-03185.

and such problems were called "German Polynomial Approximation Problems", because a Gothic font was originally used for printing the polynomials. For the simultaneous Padé approximation problem, an explicit solution in terms of determinants was given by de Bruin [1], and the explicit solution for the corresponding vector valued rational interpolation problem was given by Graves-Morris [4].

The first extension of de Montessus' theorem to simultaneous Padé approximation was given by Mall [6] in 1934. His results are a special case of the theorems of this paper, as we point out in remark 3 of section 2. Gončar and Rahmanov [3] have recently presented a powerful convergence theorem for simultaneous Padé approximants of Stieltjes functions. It is an extension of the work of Chebyshev and Markov on the convergence of an [N-1/N] sequence of Padé approximants to a Stieltjes function in the cut plane $\mathbb{C}^{-} := \mathbb{C}-(-\infty,0]$. In the theorem of Gončar and Rahmanov, the Stieltjes functions are generated by measures supported on mutually disjoint intervals of the real axis. (It should be noted that there is a small but significant difference in the use of the parameters ρ_i in the equivalent definition of a Simultaneous Padé Approximant used by Gončar and Rahmanov and our own usage (see (2.6).)

The more elegant proof of de Montessus' theorem uses complex variable methods [8]. Nevertheless, the original proof using Hadamard determinants is also instructive. This is also true for our extension of de Montessus' theorem to vector valued rational interpolants. For conciseness, we present the proofs using complex variable methods only, knowing that results such as (2.8) below may be proved, and the detail of Definition 1 motivated by determinantal representations.

2. Extensions of de Montessus' Theorem

As stated in the introduction, the vector valued Padé approximation problem is concerned with simultaneous rational approximation of d functions, $f_1(z), f_2(z),\dots,f_d(z)$, which are analytic at the origin. The degrees of the polynomials involved in forming the approximants are specified by non-negative integers N and $\rho_1,\rho_2,\dots,\rho_d$. We use the symbol $\partial\{\pi(x)\}$ to denote the degree of a polynomial $\pi(x)$. By inspection of the determinants which occur in the construction of these approximants from the power series coefficients of $f_1,f_2,\dots,f_d$, (see Mall [6], de Bruin [1] or Graves-Morris [4]), we see that $f_1(z)$, $f_2(z)$, $\dots,f_d(z)$ must, in some sense, be quite different from each other for the set of rational approximants to be unique. In the context of de Montessus type theorems, the concept is made precise by the following.

Definition 1 Let each of the functions $f_1(z), f_2(z), \ldots, f_d(z)$ be meromorphic in the disc $\mathcal{D}_R := \{z : |z| < R\}$ and let non-negative integers $\rho_1, \rho_2, \ldots, \rho_d$ be given for which

$$(2.1) \qquad \sum_{i=1}^{d} \rho_i > 0 .$$

Then the functions $f_i(z)$ are said to be polewise independent, with respect to the numbers ρ_i, in $\mathcal{D}_R$ if there *do not exist* polynomials $\pi_1(z), \pi_2(z), \ldots, \pi_d(z)$, at least one of which is non-null, satisfying

$$(2.2a) \qquad \partial\{\pi_i(z)\} \leqslant \rho_i - 1, \quad \text{if } \rho_i \geqslant 1$$

$$(2.2b) \qquad \pi_i(z) \equiv 0 \quad , \quad \text{if } \rho_i = 0$$

and such that

$$(2.3) \qquad \Phi(z) := \sum_{i=1}^{d} \pi_i(z)\, f_i(z)$$

is *analytic* throughout $\mathcal{D}_R$.

Remark 1 Under the assumptions of Definition 1, each f_i must have poles of total multiplicity at least ρ_i in $\mathcal{D}_R$. On the other hand, a particular f_i may be analytic throughout $\mathcal{D}_R$, in which case, necessarily, $\rho_i = 0$. The power series coefficients of such an f_i do not appear in the standard determinantal representation of the denominator polynomial.

The theorem of de Montessus de Ballore [7] applies to the case where the degree of the denominator precisely matches the number of poles (counting multiplicity) of the given function in some disc $\mathcal{D}_R$. This is generalised to the case of simultaneous Padé approximation in the following main result.

Theorem 1 Suppose that each of the d functions $f_1(z), f_2(z), \ldots, f_d(z)$ is analytic in the disc $\mathcal{D}_R := \{z : |z| < R\}$, except for possible poles at the M (not necessarily distinct) points $z_1, z_2, \ldots, z_M$ in $\mathcal{D}_R$ which are different from the origin. (If z_k is repeated exactly p times, then each $f_i(z)$ is permitted to have a pole of order at most p at z_k.) Let $\rho_1, \rho_2, \ldots, \rho_d$ be non-negative integers such that

$$(2.4) \qquad M = \sum_{i=1}^{d} \rho_i$$

and such that the functions $f_i(z)$ are polewise independent in $\mathcal{D}_R$ with respect to the ρ_i's in the sense of Definition 1. Then, for each integer N sufficiently large, there exist polynomials $Q_N(z)$, $\{P_{N,i}(z)\}_{i=1}^{d}$ with

$$(2.5) \qquad \partial\{Q_N(z)\} = M ,$$

$$(2.6) \qquad \partial\{P_{N,i}(z)\} \leq N-\rho_i, \quad i=1,2,\ldots,d ,$$

such that

$$(2.7) \qquad f_i(z) - P_{N,i}(z)/Q_N(z) = O(z^{N+1}), \quad i=1,2,\ldots,d .$$

The denominator polynomials (suitably normalised) satisfy

$$(2.8) \qquad \lim_{N\to\infty} Q_N(z) = Q(z) := \prod_{j=1}^{M} (z-z_j), \quad \forall z \in \mathbb{C} .$$

Let $\mathcal{D}_R^- := \mathcal{D}_R - \bigcup_{j=1}^{M}\{z_j\}$. Then, for $i=1,2,\ldots,d$,

$$(2.9) \qquad \lim_{N\to\infty} P_{N,i}(z)/Q_N(z) = f_i(z), \quad \forall z \in \mathcal{D}_R^- ,$$

the convergence being uniform on compact subsets of $\mathcal{D}_R^-$. More precisely, if K is any compact subset of the plane,

$$(2.10) \qquad \limsup_{N\to\infty} \| Q_N-Q\|_K^{1/N} \leq \frac{1}{R} \max_{j=1}^{M} \{|z_j|\} < 1 ,$$

and if E is any compact subset of $\mathcal{D}_R^-$,

$$(2.11) \qquad \limsup_{N\to\infty} \| f_i-P_{N,i}/Q_N\|_E^{1/N} \leq \|z\|_E/R < 1$$

for $i=1,2,\ldots,d$.

In (2.10) and (2.11), the norm is taken to be the sup norm over the indicated set.

Remark 2 By the assumptions of Theorem 1, each $f_i(z)$ has poles in $\mathcal{D}_R$ of total multiplicity at most M. Furthermore, if z_k is repeated exactly p times, then at least one $f_i(z)$ has a pole of order p at z_k. The latter assertion is a consequence of the assumption of polewise independence, as is revealed in the following preliminary lemma.

Lemma 1 With the assumptions of Theorem 1, write the list $z_1, z_2, \ldots, z_M$ in the form $\{\zeta_k\}_{k=1}^{\nu}$, where the ζ_k's are distinct and each ζ_k is of multiplicity m_k, so that

$$(2.12)\qquad Q(z) = \prod_{j=1}^{M} (z-z_j) = \prod_{k=1}^{\nu} (z-\zeta_k)^{m_k}, \qquad \sum_{k=1}^{\nu} m_k = M.$$

Then for each $k=1,2,\ldots,\nu$ and each $s=1,2,\ldots,m_k$, there exists a function $F_{k,s}(z)$ of the form

$$(2.13)\qquad F_{k,s}(z) = \sum_{i=1}^{d} \pi_i(z)\, f_i(z) ,$$

where the π_i's satisfy (2.2), which is analytic in $\mathcal{D}_R$, except for a pole of order s at the point ζ_k.

Naturally, the polynomials $\pi_i(z)$ in (2.13) will, in general, depend on k and s.

Proof Consider the linear problem of finding d polynomials $\pi_i(z)$, satisfying (2.2), such that for each $j=1,2,\ldots,\nu$, $j \neq k$,

$$(2.14)\qquad \int_{|z-\zeta_j|=\varepsilon} (z-\zeta_j)^{\ell} \left\{ \sum_{i=1}^{d} \pi_i(z) f_i(z) \right\} dz = 0, \quad \ell=0,1,\ldots,m_j-1 ,$$

and

$$(2.15)\qquad \int_{|z-\zeta_k|=\varepsilon} (z-\zeta_k)^{\ell} \left\{ \sum_{i=1}^{d} \pi_i(z) f_i(z) \right\} dz = 0 , \quad \ell=0,1,\ldots,m_k-1, \ \ell \neq s-1,$$

where ε (>0) is sufficiently small. The system (2.14) and (2.15) has M unknowns (the coefficients of the π_i's) and consists of $M-1$ homogeneous equations. Hence it has a non-trivial solution. For such a solution, the function defined by

$$(2.16)\qquad F_{k,s}(z) := \sum_{i=1}^{d} \pi_i(z) f_i(z)$$

is either analytic throughout $\mathcal{D}_R$ or is analytic in $\mathcal{D}_R$ except for a pole of precise order s at the point ζ_k. The former possibility is excluded by the hypothesis of polewise independence. Thus $F_{k,s}(z)$ is the desired function. □

Having established the preliminary lemma, we now give the

Proof of Theorem 1 It is well known [6,5,1,4] that, for each integer N $(\geq M)$, polynomials $q_N(z)$ and $\{p_{N,i}(z)\}_{i=1}^{d}$ exist which satisfy $\partial\{q_N(z)\} \leq M$, $\partial\{p_{N,i}(z)\} \leq N-\rho_i$ for $i=1,2,\ldots,d$, and

(2.17) $$q_N(z)f_i(z) - p_{N,i}(z) = O(z^{N+1}), \quad i=1,2,\ldots,d,$$

with $q_N(z)\not\equiv 0$. We normalise $q_N(z)$ by setting

(2.18a) $$q_N(z) = \sum_{j=0}^{M} b_{N,j}z^j,$$

(2.18b) $$\sum_{j=0}^{M} |b_{N,j}| = 1, \qquad N=M,M+1,\ldots,$$

and then the $q_N(z)$ are uniformly bounded on each compact subset of the plane.

We first show that, for $k=1,2,\ldots,\nu$,

(2.19) $$\limsup_{N\to\infty} |q_N^{(j)}(\zeta_k)|^{1/N} \leq |\zeta_k|/R, \qquad j=0,1,\ldots,m_k-1,$$

where

(2.20) $$q_N^{(j)}(z) := \left(\frac{d}{dz}\right)^j q_N(z).$$

To establish (2.19), fix k and consider $F_{k,1}(z)$ of the lemma which is analytic in $\mathcal{D}_R$ except for a simple pole at ζ_k. Write

(2.21) $$F_{k,1}(z) = \frac{g_{k,1}(z)}{z-\zeta_k},$$

where $g_{k,1}(z)$ is analytic in $\mathcal{D}_R$ and $g_{k,1}(\zeta_k)\neq 0$. By using the polynomials $\pi_i(z)$ defined by (2.13) when s=1, we deduce from (2.17) that

(2.22) $$q_N(z)\, F_{k,1}(z) - \tilde{p}_{N,1}(z) = O(z^{N+1}),$$

(2.23) $$\tilde{p}_{N,1}(z) := \sum_{i=1}^{d} \pi_i(z)\, p_{N,i}(z).$$

From (2.2) and the fact that $\partial\{p_{N,i}(z)\}\leq N-\rho_i$, it follows that

(2.24) $$\partial\{\tilde{p}_{N,1}(z)\} \leq N-1,$$

and hence, from (2.22), $(z-\zeta_k)\tilde{p}_{N,1}(z)$ is the unique polynomial of degree at most N which interpolates $q_N(z)g_{k,1}(z)$ to order N (inclusively) at the origin. Thus, since $q_N(z)g_{k,1}(z)$ is analytic in $\mathcal{D}_R$, we use Hermite's formula to show that

$$(2.25)\qquad q_N(z)g_{k,1}(z) - (z-\zeta_k)\tilde{p}_{N,1}(z) = \frac{1}{2\pi i}\int_{C_{R'}} \frac{z^{N+1}}{t^{N+1}}\,\frac{q_N(t)g_{k,1}(t)}{t-z}\,dt ,$$

for all $|z|<R'$, where $|\zeta_k|<R'<R$ and $C_{R'}:\ |t| = R'$. On taking $z=\zeta_k$ in (2.25), we obtain by straightforward estimation of the integral and then by letting $R'\to R$,

$$\limsup_{N\to\infty} |q_N(\zeta_k)g_{k,1}(\zeta_k)|^{1/N} \leqslant |\zeta_k|/R .$$

As $g_{k,1}(\zeta_k)\neq 0$, this implies that

$$(2.26)\qquad \limsup_{N\to\infty} |q_N(\zeta_k)|^{1/N} \leqslant |\zeta_k|/R .$$

Proceeding by induction, we take $s \leqslant m_k$ and assume that

$$(2.27)\qquad \limsup_{N\to\infty} |q_N^{(j)}(\zeta_k)|^{1/N} \leqslant |\zeta_k|/R,\quad j=0,1,\ldots,s-2 ,$$

and we must show that (2.27) holds for $j = s-1$. Utilizing the function $F_{k,s}(z)$ of the lemma, we obtain as above,

$$(2.28)\qquad q_N(z)F_{k,s}(z) - \tilde{p}_{N,s}(z) = O(z^{N+1}) ,$$

where $\partial\{\tilde{p}_{N,s}(z)\} \leqslant N-1$. Express

$$(2.29)\qquad F_{k,s}(z) = \frac{g_{k,s}(z)}{(z-\zeta_k)^s} ,$$

where $g_{k,s}(z)$ is analytic in $\mathcal{D}_R$ and $g_{k,s}(\zeta_k)\neq 0$. By (2.28), the polynomial $(z-\zeta_k)\tilde{p}_{N,s}(z)$ interpolates $q_N(z)g_{k,s}(z)/(z-\zeta_k)^{s-1}$ to order N inclusively at the origin. For any given compact set K, where $K\subset\bar{\mathcal{D}}_R$, we may choose $R'(<R)$ and $\varepsilon(>0)$, so that, for all $z\in K$,

$$(2.30)\qquad q_N(z)\,\frac{g_{k,s}(z)}{(z-\zeta_k)^{s-1}} - (z-\zeta_k)\tilde{p}_{N,s}(z) = I_N(z) - J_N(z) ,$$

where

$$(2.31)\qquad I_N(z) := \frac{1}{2\pi i}\int_{C_{R'}} \frac{z^{N+1}}{t^{N+1}}\,\frac{q_N(t)g_{k,s}(t)}{(t-\zeta_k)^{s-1}(t-z)}\,dt ,$$

$$(2.32)\qquad J_N(z) := \frac{1}{2\pi i}\int_{|t-\zeta_k|=\varepsilon} \frac{z^{N+1}}{t^{N+1}}\,\frac{q_N(t)g_{k,s}(t)}{(t-\zeta_k)^{s-1}(t-z)}\,dt.$$

The result (2.30) is established using the Hermite formula and the residue theorem. To estimate $J_N(z)$ for $z \in K$, express

$$q_N(t) = \sum_{j=0}^{M} \frac{q_N^{(j)}(\zeta_k)}{j!} (t-\zeta_k)^j .$$

Then

$$(2.33) \qquad J_N(z) = \sum_{j=0}^{s-2} \frac{1}{2\pi i} \int_{|t-\zeta_k|=\varepsilon} \frac{z^{N+1}}{t^{N+1}} \frac{q_N^{(j)}(\zeta_k) g_{k,s}(t)}{j!(t-\zeta_k)^{s-1-j}(t-z)} dt.$$

By straightforward estimation of the integral in (2.33) and using the inductive hypothesis (2.27), we obtain

$$(2.34) \qquad \lim_{\varepsilon\to 0} \limsup_{N\to\infty} \|J_N(z)\|_K^{1/N} \leq \frac{\|z\|_K}{|\zeta_k|} \cdot \frac{|\zeta_k|}{R} = \frac{\|z\|_K}{R} .$$

Similarly,

$$(2.35) \qquad \lim_{R'\to R} \limsup_{N\to\infty} \|I_N(z)\|_K^{1/N} \leq \|z\|_K/R .$$

Hence, from (2.30), (2.34) and (2.35), we find that, for compact $K \subset \bar{\mathcal{D}}_R$,

$$(2.36) \qquad \limsup_{N\to\infty} \|q_N(z) g_{k,s}(z) - (z-\zeta_k)^s \tilde{p}_{N,s}(z)\|_K^{1/N} \leq \|z\|_K/R .$$

But since the function

$$\phi(z) := q_N(z) g_{k,s}(z) - (z-\zeta_k)^s \tilde{p}_{N,s}(z) ,$$

which appears in the left-hand side of (2.36) is analytic throughout $\mathcal{D}_R$, then (2.36) also holds for any $K \subset \mathcal{D}_R$. By using Cauchy's contour integral formula for the function $(d/dz)^{s-1}\phi(z)$, and (2.36), we find that

$$(2.37) \qquad \limsup_{N\to\infty} \left[\left(\frac{d}{ds}\right)^{s-1} \left[q_N(z) g_{k,s}(z)\right] \Big|_{z=\zeta_k} \right]^{1/N} \leq \frac{|\zeta_k|}{R} .$$

Using Leibniz's formula for differentiating the product (2.37), and the inductive hypothesis (2.27), we get

$$(2.38) \qquad \limsup_{N\to\infty} |g_{k,s}(\zeta_k)\, q_N^{(s-1)}(\zeta_k)|^{1/N} \leq |\zeta_k|/R .$$

As $g_{k,s}(\zeta_k)\neq 0$, it follows from (2.38) that (2.27) holds also for $j=s-1$, which completes the induction. This proves the claim (2.19).

Next, consider a basis of polynomials

$$\mathcal{B} = \{B_{k,s}(z),\ k=1,2,\ldots,\nu,\ s=0,1,\ldots,m_k-1\}$$

such that both

(2.39) $$\partial\{B_{k,s}(z)\} \leqslant M-1 \quad \text{for all } k,s$$

and the polynomials interpolate at the points ζ_i according to

(2.40) $$\left[\left(\frac{d}{dz}\right)^j B_{k,s}(z)\right]_{z=\zeta_i} = \delta_{ik}\cdot\delta_{js},\quad 1\leqslant i\leqslant\nu,\quad 0\leqslant j\leqslant m_i-1 .$$

Then we can write (see (2.12), (2.18a) and (2.39))

(2.41) $$q_N(z) = \sum_{k=1}^{\nu}\sum_{s=0}^{m_k-1} q_N^{(s)}(\zeta_k)\, B_{k,s}(z) + b_{N,M}Q(z) .$$

By (2.18b), we have $|b_{N,M}| \leqslant 1$. More importantly, however,

(2.42) $$\liminf_{N\to\infty} |b_{N,M}| > 0 ;$$

indeed, if this were not the case, (2.19) shows that some subsequence of indices $\{N_i\}$ exists for which

$$\lim_{i\to\infty} q_{N_i}(z) = 0, \qquad \text{for all } z\in\mathbb{C} ,$$

contradicting (2.18). Thus, for N sufficiently large, we define

(2.43) $$Q_N(z) := q_N(z)/b_{N,M}$$

(2.44) $$P_{N,i}(z) := p_{N,i}(z)/b_{N,M}, \qquad i=1,2,\ldots,d ,$$

and the assertions (2.5)-(2.8) all follow. Assertion (2.10) follows from (2.19) and (2.41).

Finally, to establish (2.11), (and hence (2.9)), let E be a compact subset of $\mathcal{D}_R^-$. Then, for $z\in E$ and $i=1,2,\ldots,d$,

(2.45) $$Q_N(z)f_i(z) - P_{N,i}(z) = I_{N,i}(z) - \sum_{k=1}^{\nu} J_{N,i,k}(z) ,$$

where

$$(2.46) \qquad I_{N,i}(z) := \frac{1}{2\pi i}\int_{C_{R'}} \frac{z^{N+1}}{t^{N+1}} \frac{Q_N(t)f_i(t)}{t-z}\,dt ,$$

$$(2.47) \qquad J_{N,i,k}(z) := \frac{1}{2\pi i}\int_{|t-\zeta_k|=\varepsilon} \frac{z^{N+1}}{t^{N+1}} \frac{Q_N(t)f_i(t)}{t-z}\,dt, \ k=1,2,\ldots,\nu.$$

Since the inequalities (2.19) also hold for the polynomials $Q_N(t)$, the integrals of (2.46) and (2.47) can be estimated in the same manner as used in the inductive portion of the proof. This gives

$$\limsup_{N\to\infty} \|Q_N(z)f_i(z) - P_{N,i}(z)\|_E^{1/N} \leqslant \|z\|_E/R, \quad i=1,2,\ldots,d .$$

The inequalities (2.11) now follow from (2.8). □

Corollary 1 (Uniqueness). Under the assumptions of Theorem 1, there exist, for each sufficiently large integer N, a *unique* set of rationals $\{R_{N,i}(z)\}_{i=1}^d$ of the form

$$R_{N,i}(z) = P_{N,i}(z)/Q_N(z) ,$$

such that

$$\partial\{Q_N(z)\} \leqslant M , \qquad \partial\{P_{N,i}(z)\} \leqslant N-\rho_i ,$$

and

$$(2.48) \qquad f_i(z) - R_{N,i}(z) = O(z^{N+1}), \quad i=1,2,\ldots,d .$$

Proof Assume, to the contrary, that, for some subsequence $\mathcal{N}$ of integers N, another set of rationals $\{\tilde{R}_{N,i}(z)\}_{i=1}^d$ exists of the form

$$\tilde{R}_{N,i}(z) = \tilde{P}_{N,i}(z)/\tilde{Q}_N(z) ,$$

such that

$$\partial\{\tilde{Q}_N(z)\} \leqslant M , \qquad \partial\{\tilde{P}_{N,i}(z)\} \leqslant N-\rho_i ,$$

$$(2.49) \qquad f_i(z) - \tilde{R}_{N,i}(z) = O(z^{N+1}) , \quad i=1,2,\ldots,d ,$$

but such that $\tilde{R}_{N,j}(z)\neq R_{N,j}(z)$ for some j. Then, necessarily, $Q_N(z)$ is not a constant multiple of $\tilde{Q}_N(z)$. The proof of Theorem 1 shows that, for $N\in\mathcal{N}$ and N sufficiently large, both $Q_N(z)$ and $\tilde{Q}_N(z)$ must be of precise degree M, and so, without loss of generality, we assume that both are monic and of degree M. From (2.48) and (2.49), we have

$$\{Q_N(z) - \tilde{Q}_N(z)\}f_i(z) - \{P_{N,i}(z) - \tilde{P}_{N,i}(z)\} = O(z^{N+1}) ,$$

for i=1,2,...,d. The proof of Theorem 1 now implies that

$$(2.50) \qquad \lim_{N\to\infty} c_N\{Q_N(z) - \tilde{Q}_N(z)\} = Q(z)$$

for $N\in\mathcal{N}$, $z\in\mathbb{C}$ and a suitable choice of normalizing constants c_N. But Q(z) is of degree M, whereas the polynomials $\{Q_N(z)-\tilde{Q}_N(z)\}$ are of degree M-1 at most, making (2.50) absurd. □

Remark 3 Consider the special case of Theorem 1, in which each $f_i(z)$ has poles of total multiplicity precisely equal to ρ_i in $\mathcal{D}_R$. Further, assume that the pole sets of each f_i (within $\mathcal{D}_R$) are mutually disjoint sets. Then it is clear that the f_i's are polewise independent with respect to the ρ_i's, and so the conclusions of Theorem 1 and its corollary are valid. In this special case, if we *further* assume that all the poles involved are simple, then Theorem 1 yields the result (Theorem VIII) in the dissertation of Mall [6].

Theorem 2 (for Directed Vector Padé Approximants) Let each of the d functions $f_1(z)$, $f_2(z)$,...,$f_d(z)$ be analytic in the disc $\mathcal{D}_R$ except for possible poles in the M (not necessarily distinct) points $z_1,z_2,\ldots,z_M$ in $\mathcal{D}_R$ which are different from the origin. Given ℓ constant d-dimensional vectors $\underline{w}^{(1)},\underline{w}^{(2)},\ldots,\underline{w}^{(\ell)}$, define a column vector $\underline{f}$ by

$$\underline{f} = (f_1(z),f_2(z),\ldots,f_d(z))^T$$

and set

$$(2.51) \qquad F_j(z) := \underline{f}^T.\underline{w}^{(j)}, \quad j=1,2,\ldots,\ell,$$

where T denotes transpose. Let $k_1,k_2\ldots,k_\ell$ be positive integers for which $\sum_{j=1}^{\ell} k_j = M$ and such that the functions $F_j(z)$ are polewise independent with respect to the k_j's in $\mathcal{D}_R$. Then, for each N large enough, there exist polynomials $Q_N(z)$, $\{P_{N,i}(z)\}_{i=1}^{d}$, for which

$$(2.52)\quad \partial\{Q_N(z)\} = M, \quad \partial\{P_{N,i}(z)\} \leq N, \qquad i=1,2,\ldots,d,$$

$$(2.53)\quad f_i(z) - P_{N,i}(z)/Q_N(z) = O(z^{N+1}), \quad i=1,2,\ldots,d,$$

$$(2.54)\quad \partial\{\underline{P}_N^T.\underline{w}^{(j)}\} \leq N-k_j, \qquad j=1,2,\ldots,\ell,$$

where $\underline{P}_N^T := (P_{N,1}(z),P_{N,2}(z),\ldots,P_{N,d}(z))$. Furthermore, the conclusions (2.8)-(2.11) of Theorem 1 hold.

Remark 4 The assumption of polewise independence implies that the vectors $\underline{w}^{(1)},\underline{w}^{(2)},\ldots,\underline{w}^{(\ell)}$ are linearly independent, and consequently $\ell \leq d$.

Proof In the case that $\ell<d$, the definitions need to be extended by taking

$$(2.55)\quad k_j := 0, \quad j=\ell+1,\ell+2,\ldots,d\ .$$

By suitably defining vectors $\underline{w}^{(\ell+1)},\underline{w}^{(\ell+2)},\ldots,\underline{w}^{(d)}$ such that $\underline{w}^{(1)},\underline{w}^{(2)},\ldots,\underline{w}^{(d)}$ are linearly independent, we define a non-singular matrix W whose columns are the vectors $\underline{w}^{(j)}$, i.e.

$$W_{ij} = w_i^{(j)}, \qquad i,j=1,2,\ldots,d\ .$$

In this way, we may extend (2.51) so that it becomes

$$(2.56)\quad F_j(z) = \sum_{i=1}^{d} w_i^{(j)} f_i(z) = (\underline{f}^T W)_j, \quad j=1,2,\ldots,d\ .$$

Furthermore, $\{F_j(z)\}_{j=1}^d$ remain polewise independent in $\mathcal{D}_R$ with respect to $\{k_j\}_{j=1}^d$, by (2.2b) and (2.55). We now infer from Theorem 1 that polynomials $Q_N(z)$ and $\{\tilde{P}_{N,i}(z)\}_{i=1}^d$ exist for which

$$(2.57)\quad \partial\{Q_N(z)\} = M$$

$$(2.58)\quad \partial\{\tilde{P}_{N,i}(z)\} \leq N-k_i \qquad i=1,2,\ldots,d\ ,$$

$$(2.59)\quad F_i(z) - \tilde{P}_{N,i}(z)/Q_N(z) = O(z^{N+1}) \qquad i=1,2,\ldots,d\ ,$$

and the corresponding equivalents of (2.8)-(2.11) hold too.

From (2.56), we have

$$f_i = (\underline{f}^T)_i = (\underline{F}^T W^{-1})_i = \sum_{j=1}^{d} F_j(z)(W^{-1})_{ji} \quad ,$$

and so we are led to define

$$P_{N,i}(z) := \sum_{j=1}^{d} (W^{-1})_{ji} \tilde{P}_{N,j}(z), \quad i=1,2,\ldots,d \ .$$

With this definition, (2.52) and (2.53) follow immediately; (2.54) follows from (2.58), and the other equivalent properties of Theorem 1 follow similarly by linearity. □

Theorems 1 and 2 can immediately be extended to include more general cases of Lagrange and Hermite rational interpolation. These extensions are analogues of the theorem of Saff [8] for equilibrium distributions of the interpolating points and its generalisation by Warner [9] for the case of regular interpolation schemes. Padé approximation problems are associated with osculatory interpolation at the origin, and we next consider the generalisation of Theorems 1 and 2 to Lagrange and Hermite interpolation on points which belong to a compact set S .

For each positive integer N, we consider an interpolating point set

$$S_N := \{\beta_{N,i}, \quad i=0,1,\ldots,N : \beta_{N,i} \in S\}$$

where the $\beta_{N,i}$'s need not necessarily be distinct, so that partially confluent cases of Hermite interpolation are included. Following Warner [9], we assume that the points $\{\{\beta_{N,i}\}_{i=0}^{N}\}_{N=1}^{\infty}$ have an associated sequence of elementary measures μ_N which is regular in the sense that $\mu_N \to \mu$. This ensures that the logarithmic potentials

$$u(z,\mu_N) := -\int \log|z-\zeta| d\mu_N(\zeta) \ , \qquad N=1,2,\ldots,$$

have the property that

$$\lim_{N\to\infty} u(z,\mu_N) = u(z,\mu), \ \forall \ z \in \mathbb{C}\backslash S \ ,$$

where

$$u(z,\mu) := -\int \log|z-\zeta| d\mu(\zeta) \ .$$

The convergence theorems are based on a nested set of regions $\mathbf{D}_\lambda$, defined for each $\lambda > 0$ by

$$\mathbf{D}_\lambda = \{z : e^{-u(z,\mu)} < \lambda\} .$$

With these preliminaries and the obvious extension of Definition 1 to the region $\mathbf{D}_\lambda$, we state

Theorem 3. Let each of the functions $f_1(z), f_2(z), \ldots, f_d(z)$ be analytic on S and also in the larger region $\mathbf{D}_R$, except for possible poles at the points $z_1, z_2, \ldots, z_M$ in $\mathbf{D}_R$. Define

$$\mathbf{D}_R^- := \mathbf{D}_R - \bigcup_{i=1}^{M} \{z_i\}$$

and let E by any compact subset of $\mathbf{D}_R^-$. Given d non-negative integers $\rho_1, \rho_2, \ldots, \rho_d$ satisfying $M = \sum_{i=1}^{d} \rho_i$, assume that the $f_i(z)$ are polewise independent with respect to the ρ_i's in $\mathbf{D}_R$. Then, for each N large enough, there exist polynomials $Q_N(z)$, $\{P_{N,i}(z)\}_{i=1}^{d}$, with

$$\partial\{Q_N(z)\} = M, \quad \partial\{P_{N,i}(z)\} \leq N-\rho_i, \qquad i=1,2,\ldots,d ,$$

such that $P_{N,i}(z)/Q_N(z)$ interpolates as

$$P_{N,i}(z)/Q_N(z) = f_i(z) \quad \forall \ z \in S_N , \qquad i=1,2,\ldots,d ,$$

in the Hermite sense. The denominator polynomials obey

$$\lim_{N\to\infty} Q_N(z) = Q(z) := \prod_{i=1}^{M} (z-z_i), \quad \forall \ z \in \mathbb{C}.$$

Furthermore,

$$\lim_{N\to\infty} P_{N,i}(z)/Q_N(z) = f_i(z), \qquad \forall \ z \in \mathbf{D}_R^- , \quad i=1,2,\ldots,d,$$

the convergence being uniform on E, which is an arbitrary compact subset of $\mathbf{D}_R^-$.

Define r to be the smallest number for which it is true that

$$\bigcup_{i=1}^{M} \{z_i\} \subset \mathbf{D}_{r'}, \qquad \forall \ r'>r ,$$

and define σ to be the smallest number for which it is true that

$$E \subset \mathbf{D}_{\sigma'} \qquad \forall\ \sigma' > \sigma \quad .$$

Then the rates of convergence of the interpolants are given by

$$\limsup_{N\to\infty} \|Q_N - Q\|_K^{1/N} \leqslant r/R \ ,$$

where K is any compact subset of $\mathbb{C}$, and

$$\limsup_{N\to\infty} \|f_i(z) - P_{N,i}(z)/Q_N(z)\|_E^{1/N} \leqslant \sigma/R, \quad i=1,2,\ldots,d \ .$$

A uniqueness assertion and the generalisation to directed vector valued rational interpolants also hold for Theorem 3.

Remark 5 According to the hypotheses of the theorem, each $f_i(z)$ is analytic in $\mathbf{D}_R^-$. A possible configuration is shown in Fig.1.

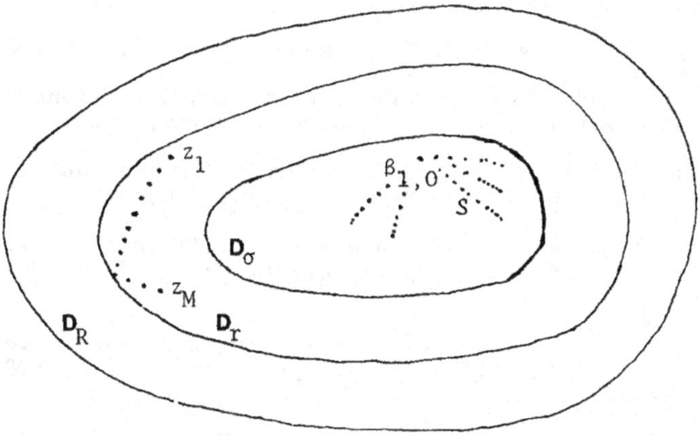

Fig. 1 The poles $\{z_i\}$, the point sequence $\{\beta_{N,i}\} \subset S$, and the boundaries of the domains $\mathbf{D}_\sigma$, $\mathbf{D}_r$ and $\mathbf{D}_R$ are shown schematically.

Postscript After this paper and the paper [W] by Hans Wallin had been presented at the Conference, we saw that we had adopted quite similar approaches to estimation of the rate of convergence of the denominator polynomials and related quantities.

References

1. de Bruin, M. G., Generalised Continued Fractions and a Multidimensional Padé Table, Thesis, Amsterdam University, 1974.

2. de Bruin, M. G., Some Convergence Results in Simultaneous Rational Approximation to the Set of Hypergeometric Functions $\{{}_1F_1(1;c_i,z)\}_{i=1}^{n}$, in Padé Seminar 1983, eds. H. Werner and H. J. Bünger, Bonn 1983, 95-117.

3. Gončar, A. A. and Rahmanov, E. A., On the Simultaneous Convergence of Padé Approximants for Systems of Functions of Markov type, Proc. Steklov Inst. Math. 157 (1981), 31-48.

4. Graves-Morris, P. R., Vector Valued Rational Approximants II, I.M.A.J. Numerical Analysis, 4, (1984), 209-224.

5. Mahler, K., Perfect Systems, Comp. Math. 19 (1968), 95-166.

6. Mall, J., Grundlagen für eine Theorie der mehrdimensionalen Padéschen Tafel, Thesis, Munich University, 1934.

7. de Montessus de Ballore, R., Sur les Fractions Continues Algébriques, Bull. Soc. Math. de France 30, (1902), 28-36.

8. Saff, E. B., An Extension of Montessus de Ballore's Theorem on the Convergence of Interpolating Rationals, J. Approx. Theory 6, (1972), 63-67.

9. Warner, D. D., An Extension of Saff's Theorem on the Convergence of Interpolating Rationals, J. Approx. Theory 18, (1976), 108-118.

W. Wallin, H., Convergence and Divergence of Multipoint Padé Approximants of Meromorphic Functions, These proceedings.

ON THE CONVERGENCE OF LIMIT PERIODIC CONTINUED FRACTIONS $K(a_n/1)$, WHERE $a_n \to -1/4$

Lisa Jacobsen
Matematisk Institutt
NLHT Trondheim
Norway

Arne Magnus
Department of Mathematics
Colorado State University
Ft. Collins
Colorado 80523
USA

Abstract. It is well known that the continued fraction $K(a_n/1)$, where $a_n \to -1/4$, converges, provided $|a_n + 1/4| \leq 1/16n(n+1)$ for all n. We show that the constant 1/16 is best possible in the sense that if $a_n = -1/4 - c/n(n+1)$, where $c > 1/16$ then $K(a_n/1)$ diverges by oscillation.

1. Basic Concepts and Notation

The continued fraction

(1.1) $$K(a_n/1) := \frac{a_1}{1} + \frac{a_2}{1} + \frac{a_3}{1} + \ldots = \cfrac{a_1}{1 + \cfrac{a_2}{1 + \cfrac{a_3}{1} + \ldots}}$$

where $a_n \neq 0$ for all n, may be generated as follows. Let $s_n(w) = a_n/(1+w)$, $n \geq 1$, and $S_1(w) := s_1(w)$, $S_n(w) := S_{n-1}(s_n(w))$, $n \geq 2$, then the nth approximant, f_n, of (1.1) is defined by

$$f_n := S_n(0) = \frac{a_1}{1} + \frac{a_2}{1} + \frac{a_3}{1} + \ldots + \frac{a_n}{1} \in \hat{\mathbb{C}},$$

where $\hat{\mathbb{C}}$ is the extended complex plane. The nth tail of (1.1) is defined by

(1.2) $$\underset{m=1}{\overset{\infty}{K}} (a_{n+m}/1) := \frac{a_{n+1}}{1} + \frac{a_{n+2}}{1} + \frac{a_{n+3}}{1} + \ldots.$$

$K(a_n/1)$ is said to converge, possibly to ∞, iff $\{f_n\}$ converges to some f in $\hat{\mathbb{C}}$. It is easily shown that this occurs iff each tail converges to some value $f^{(n)}$ in $\hat{\mathbb{C}}$, which then clearly satisfies the recurrence relation

$$f^{(n)} = a_{n+1}/(1 + f^{(n+1)})$$

with the convention

$$f^{(n)} = a_{n+1} \Longleftrightarrow f^{(n+1)} = 0 \Longleftrightarrow f^{(n+2)} = \infty \Longleftrightarrow f^{(n+3)} = -1.$$

If $K(a_n/1)$ converges, then $\{f^{(n)}\}$ is called the sequence of right tails of $K(a_n/1)$.

Any sequence $\{g^{(n)}\}$ in $\hat{\mathbb{C}}$ which satisfies

(1.3) $$g^{(n)} = a_{n+1}/(1 + g^{(n+1)}), \text{ that is, } a_{n+1} = g^{(n)}(1 + g^{(n+1)})$$

with the same convention, is called a sequence of wrong tails of $K(a_n/1)$ if $K(a_n/1)$ diverges or if it is not the sequence of right tails, $\{f^{(n)}\}$.

If $s_n(w)$ is independent of n,

$$s_n(w) = s(w) := a/(1 + w),$$

then $\{S_n(0)\}$ generates the periodic continued fraction

(1.4) $$K(a/1) = \frac{a}{1} + \frac{a}{1} + \frac{a}{1} + \cdots.$$

Here $s(w)$ is elliptic iff $\arg(a + 1/4) = \pi$, that is $a \in (-\infty, -1/4)$, in which case $S_n(w)$ rotates on an Appolonian circle surrounding one of the fixed points of $s(w)$. In particular, $\{S_n(0)\} = \{f_n\}$ diverges by oscillation since 0 is not a fixed point. If $s(w)$ is not elliptic, that is, $a \notin (-\infty, -1/4)$, then $K(a/1)$ converges.

If (1.1) is limit periodic with $a_n \to a \in \mathbb{C}$, then, for large n, (1.2) is "close to" the periodic continued fraction (1.4), and it is therefore not surprising that (1.1) converges when $|\arg(a + 1/4)| < \pi$, see [5, Satz 2.40].

2. Historical Remarks

In 1865 Worpitzky [13] proved that (1.1) converges if $|a_n| \leq 1/4$ for all n. That is, the closed disk $E := \{z: |z| \leq 1/4\}$ is a so-called simple convergence region for (1.1). No fixed disk $D := \{z: |z| \leq r\}$ with radius $r > 1/4$ can be such a simple convergence region, since the periodic continued fraction (1.4) with $-r < a < -1/4$ diverges.

In 1905 Pringsheim [6] proved that (1.1) converges if

$$|a_n| \leq (p_n - 1)/p_{n-1}p_n, \quad n = 1, 2, 3, \ldots,$$

where $p_0 = 1$ and $p_n > 1$ for $n \geq 1$. He chose $p_n = (2n + 1)/(n + 1)$ giving $|a_n| \leq n^2/(4n^2 - 1)$. That is, (1.1) converges if a_n lies in the variable element region

$$E_n := \{z: |z| \leq n^2/(4n^2 - 1)\}$$

which contracts towards the Worpitzky disk as $n \to \infty$. In particular, for a_n near -1/4 we find that (1.1) converges if

$$(2.1) \qquad |a_n + 1/4| \leq 1/4(4n^2 - 1),$$

as was pointed out by Thron and Waadeland [10] in 1980. It seems that this was not explicitly recognized earlier. Szasz who knew Pringsheim's work was not aware of this implication when he published in 1917 [8] that $|a_n + 1/4| < 2/9(n^2 - 1)n(n + 2)$ implies the convergence of (1.1), which is weaker than (2.1).

Generalizing earlier results of Scott and Wall [7], of Paydon and Wall [4] and of Leighton and Thron [3], Thron [9] proved in 1958 that (1.1) converges if $\{a_n\}$ is bounded and lies inside or on a parabola which has focus at 0, passes through -1/4 and has axis along $\arg z = -2\alpha$; that is, if $a_n \in E(\alpha)$ for all n, where

$$E(\alpha) := \{z: |z| - \mathrm{Re}(ze^{2i\alpha}) \leq \tfrac{1}{2}\cos^2\alpha\}, \quad |\alpha| < \pi/2.$$

Thus, if a_n approaches -1/4 in such a manner that a_n ultimately lies in one of the parabolic regions $E(\alpha)$, then (1.1) converges. If α is close to $\pi/2$ or to $-\pi/2$ then the parabola is nearly tangent to the real axis at -1/4, so (1.1) converges if a_n approaches -1/4 at an angle to the real axis or via the Worpitzky disk. See also Jones and Thron [1].

3. Main Results

We consider the case where $a_n \to -1/4$ and $a_n \leq -1/4$. We first note that if

$$-1/4 - 1/16(n^2 + pn + q) \leq a_n \leq -1/4$$

for all $n \geq n_0$ and some real p and q then there exists a positive integer m such that

$$-1/4 - 1/4(4(n-m)^2 - 1) \leq -1/4 - 1/16(n^2 + pn + q) \leq a_{m+(n-m)} \leq -1/4$$

for $n \geq n_0$, which implies, by Pringsheim's criterion, the convergence of the mth tail of (1.1) and therefore the convergence of (1.1) itself. We now show that the constant 1/16 is best possible.

<u>Theorem 3.1.</u> <u>The continued fraction</u> $K(a_n/1)$ <u>where</u> $a_n = -1/4 - c/n(n+1)$ <u>and</u> $c > 1/16$ <u>diverges by oscillation</u>.

For the proof we seek tails $g^{(n)}$ of the form

(3.1) $$g^{(n)} = L/(n+1) - 1/2$$

so that

$$-1/4 - c/n(n+1) = a_n = g^{(n-1)}(1 + g^{(n)}) = (L/n - 1/2)(L/(n+1) + 1/2)$$

which yields

$$L = -(1 \pm \sqrt{1 - 16c})/4.$$

Thus, there are solutions of (1.3) of the form (3.1) and $g^{(n)}$ is nonreal when $c > 1/16$, in which case they are wrong tails since the a_n's are real and therefore the right tails $f^{(n)}$ — if $K(a_n/1)$ converges — are real.

Waadeland [12] proved the following theorem.

The continued fraction $K(a_n/1)$ _converges iff_

(3.2) $$1 + \kappa_1 + \kappa_1\kappa_2 + \kappa_1\kappa_2\kappa_3 + \dots$$

converges, possibly to ∞, _where_ $\kappa_n = -(1 + g^{(n)})/g^{(n)}$ _and_ $\{g^{(n)}\}$ _is any sequence of right or wrong tails satisfying_

$$g^{(n)} \neq 0, -1, \infty \quad \text{for all } n.$$

The series (3.2) _converges to_ ∞ _iff_ $g^{(n)} = f^{(n)}$, _the right tails of_ (1.1).

To prove our theorem we will therefore show that (3.2) diverges by oscillation. We set

$$L = -(1 + i\beta)/4, \qquad \beta > 0$$

and find

$$g^{(n)} = -(2n + 3 + i\beta)/4(n+1),$$

$$\kappa_n = \frac{2n+1-i\beta}{2n+3+i\beta} = \frac{2n+1}{2n+3} \cdot \frac{r_n}{r_{n+1}} e^{-i(\theta_n + \theta_{n+1})},$$

where $r_n e^{i\theta_n} = 1 + i\beta/(2n+1)$. r_n is decreasing to 1 and $0 < \theta_n = \arctan \frac{\beta}{2n+1} < \frac{\beta}{2n+1}$ for all n. We then have

$$\left|\sum_{k=N}^{N+m} \prod_{n=1}^{k} \kappa_n\right| = \left|\prod_{n=1}^{N} \kappa_n\right| \left|\sum_{k=N}^{N+m} \prod_{n=N+1}^{k} \kappa_n\right|$$

$$= \frac{3}{2N+3}\frac{r_1}{r_{N+1}}\left|\sum_{k=N}^{N+m}\frac{2N+3}{2k+3}\frac{r_{N+1}}{r_{k+1}}\exp\left(-i\sum_{n=N+1}^{k}(\theta_n+\theta_{n+1})\right)\right|$$

$$\geqq \left|\sum_{k=N}^{N+m}\frac{3}{2k+3}\frac{r_1}{r_{k+1}}\cos\left(\sum_{n=N+1}^{k}(\theta_n+\theta_{n+1})\right)\right|.$$

Now choose ε so that $0 < \varepsilon < \pi/2$ and let $N > 2\beta/\varepsilon$. Then it is easy to show that there exists a nonnegative integer m such that

$$0 < \varepsilon/2 \leqq 2\beta \sum_{k=N+1}^{N+m+1} \frac{1}{2k+1} \leqq \varepsilon$$

which implies that, for $N \leqq k \leqq N + m$,

$$0 \leqq \sum_{N+1}^{k}(\theta_n + \theta_{n+1}) \leqq \sum_{N+1}^{k}\left(\frac{\beta}{2n+1} + \frac{\beta}{2n+3}\right) \leqq 2\beta\sum_{N+1}^{k}\frac{1}{2n+1} \leqq \varepsilon < \pi/2$$

so that

$$\cos\sum_{N+1}^{k}(\theta_n + \theta_{n+1}) > \cos\varepsilon > 0.$$

Thus,

$$\left|\sum_{k=N}^{N+m}\prod_{n=1}^{k}\kappa_n\right| > \sum_{N}^{N+m}\frac{3}{2k+3}\frac{r_1}{r_{k+1}}\cos\varepsilon \geqq 3\cos\varepsilon\sum_{N+1}^{N+m+1}\frac{1}{2k+1} \geqq 3\cos\varepsilon\frac{\varepsilon}{4\beta} > 0,$$

which shows that (3.2) is not convergent to a finite value. Nor can it converge to ∞ since the $g^{(n)}$'s are nonreal and therefore not equal to the $f^{(n)}$'s of a real continued fraction. The series (3.2) is therefore divergent by oscillation and Theorem 3.1 has been proven.

By a similar argument we may prove the following generalization of Theorem 3.1.

Theorem 3.2. The continued fraction $K(a_n/1)$, where $a_n = -1/4 - c/16(n + \theta)(n + \theta + 1)$ for $n = 1, 2, 3, \ldots$, $c \in \mathbb{C}$ and $\theta > -1$ diverges iff $c > 1$.

If $K(a_n/1)$ converges then it has the (right) tails

$$f^{(n)} = -\frac{1}{2} - \frac{1-\sqrt{1-c}}{4(n+\theta+1)} \quad \text{for } n = 0, 1, 2, \ldots,$$

where $\mathrm{Re}\sqrt{1-c} \geqq 0$ when $|c| \leqq 1$ and $\mathrm{Re}(\sqrt{1-c}\, e^{-i/2\arg(-c)}) > 0$ otherwise.

In particular, the value of $K(a_n/1)$, when convergent, is $f^{(0)} = -1/2 - (1 - \sqrt{1 - c})/4(\theta + 1)$. The proof is in part based on a paper by Thron and Waadeland [11].

References

1. Jones, W.B. and Thron, W.J., Convergence of continued fractions, Canad. J. Math 20 (1968), 1037-1055.

2. Jones, W.B. and Thron, W.J., Continued Fractions: Analytic Theory and Applications, Encyclopedia of Mathematics and Its Applications, V.11, Addison-Wesley, Reading, MA, 1980.

3. Leighton, W. and Thron, W.J., Continued fractions with complex elements, Duke Math. J. 9 (1942), 763-772.

4. Paydon, J.F. and Wall, H.S., The continued fraction as a sequence of linear transformations, Duke Math. J. 9 (1942), 360-372.

5. Perron, O., Die Lehre von den Kettenbrüchen, 3 Auflage Band II, Teubner, Stuttgart, 1957.

6. Pringsheim, A., Über die Konvergenzkriterien für Kettenbrüche mit komplexen Gliedern, Sb. Münch. 35 (1905).

7. Scott, W.T. and Wall, H.S., A convergence theorem for continued fractions, Trans. Amer. Math. Soc. 47 (1940), 155-172.

8. Szasz, O., Über die Erhaltung der Konvergenz unendlicher Kettenbrüche bei independenter Veränderlichkeit aller ihrer Elemente, J. f. Math. 147 (1917), 132-160.

9. Thron, W.J., On parabolic convergence regions for continued fractions, Math. Z. 69 (1958), 173-182.

10. Thron, W.J. and H.Waadeland, Accelerating convergence of limit periodic continued fractions $K(a_n/1)$, Numer. Math. 34 (1980), 155-170.

11. Thron, W.J. and H.Waadeland, On a certain transformation of continued fractions, Lecture Notes in Math. 932 (1982), 225-240, Springer-Verlag.

12. Waadeland, H., Tales about tails, to appear in Proc. Amer. Math. Soc.

13. Worpitzky, J., Untersuchungen über die Entwiklung der monodromen und monogenen Funktionen durch Kettenbrüche, Friedrichs-Gymnasium und Realschule Jahresbereicht (1865), 3-39, Berlin.

ON THE UNIFORM APPROXIMATION OF HOLOMORPHIC FUNCTIONS ON CONVEX SETS BY MEANS OF INTERPOLATION POLYNOMIALS

Thomas Kövari
Department of Mathematics
Imperial College
180 Queen's Gate
London
SW7 2BZ

Abstract:

The principal aim of this paper is to give an explicit construction for a sequence of polynomials that interpolates the function f on a given system of nodes and also converges to f uniformly. The functions we are able to approximate in this manner are continuous on the compact convex set K, and holomorphic in the interior of K.

1. INTRODUCTION.

For any closed Jordan domain K, we define $A(K)$ as the class of functions continuous on K, and holomorphic in the interior of K. We also define $E_n(f,K)$: the best uniform polynomial approximation of f on K by polynomials of degree at most n. In the special case when K is the closed unit disc, we simply write A and $E_n(f)$ respectively.

We state the following two results (the terms used will be defined later):

Theorem 1. Let Γ be a closed Jordan curve of bounded rotation without any zero (interior) angles, and let K be the closed interior of Γ. Let $T_n(f,z)$ denote the Faber-de la Vallée Poussin sums of the function f. Then, $\forall f \in A(K)$

$$||f - T_n(f)|| \le B\, E_n(f,K) ,$$

where $||g|| = \sup_{z \in K} |g(z)|$, and B is a constant that depends on K only.

Theorem 2. Let K be a compact convex set, and $\{z_k^{(n)}, 1 \le k \le n\}$ a regular system of nodes on $\partial K = \Gamma$. Then $\forall f \in A(K)$ and $\forall n \exists$ a polynomial $P_n(f,z)$ of degree $\le 2n - 1$ such that

$$P_n(f, z_j^{(n)}) = f(z_j^{(n)}) \qquad j = 1,2,\ldots,n,$$

and $||f - P_n(f)|| \le B\, E_n(f,K)$.

The constant B is independent of f and n.

Theorem 1 is a straightforward consequence of a recent result of Anderson and Clunie [1]. We include its proof for the sake of completeness. The Faber-de la Vallée Poussin sums were introduced in an earlier paper of the author [7].

The special case of Theorem 2 when Γ satisfies a smoothness condition, was already established by the author in an earlier paper [4]. It is an essential aspect of the proof of Theorem 2 that it is wholly constructive.

In both results the degrees of the approximating polynomials could be reduced to $n(1 + \varepsilon)$ without difficulty.

2. THE FABER-DE LA VALLÉE POUSSIN POLYNOMIALS, AND THE FABER OPERATOR.

Assuming only that K is a closed Jordan domain, every function $f \in A(K)$ is associated with a Faber series

$$\sum_{n=0}^{\infty} c_n F_n(z) \quad . \tag{1}$$

Here $F_n(z)$ is the n-th Faber polynomial of the set K, and

$$c_n = \frac{1}{2\pi} \int_0^{2\pi} f(\psi(e^{it}))e^{-int}dt ,$$

where $z = \psi(w) = \rho w + a_0 + \frac{a_1}{w} + \ldots$ maps $|w| > 1$ conformally onto the complement of K. $\psi(w)$ has a continuous extension to $|w| \ge 1$. If the partial sums of the series (1) are denoted by $S_n(f,z)$, one may define the Faber-de la Vallée Poussin sums of the function f by:

$$T_n(f,z) = \frac{1}{n} \sum_{k=0}^{n-1} S_{n+k}(f,z) \quad . \tag{2}$$

If Γ is of bounded rotation,

$$v(t,\theta) = \arg(\psi(e^{it}) - \psi(e^{i\theta}))$$

is a function of bounded variation $V = V(\Gamma)$. For every $g \in A$ and $z \in \Gamma$, define Tg by:

$$(Tg)(z) = (Tg)(\psi(e^{i\theta})) = \frac{1}{\pi}\int_0^{2\pi} g(e^{it})\,d_t V(t,\theta) - g(0) \quad .$$

T is a linear operator from A to $A(K)$ - the so-called Faber operator - with the following properties:

(i) $||T|| \leq \frac{V}{\pi}$

(ii) T is injective

(iii) $T(z^n) = F_n(z) \qquad (n = 0,1,2,\ldots)$.

3. PROOF OF THEOREM 1.

By a recent result of Anderson and Clunie [1] , if Γ is of bounded rotation and has no zero (interior) angles, then T is surjective (bijective) and (hence) T^{-1} is a bounded operator on $A(K)$. Then, if $T^{-1}f = \tilde{f}$,

$$(3) \qquad ||f - T_n(f)|| = ||T(\tilde{f} - \tau_n(\tilde{f}))|| \leq ||T||\;||\tilde{f} - \tau_n(\tilde{f})|| \;,$$

where τ_n is the classical de la Vallée Poussin sum of $\tilde{f}$. It is well-known that

$$(4) \qquad ||\tilde{f} - \tau_n(\tilde{f})|| \leq 4E_n(\tilde{f}) \quad .$$

On the other hand, if π_n is the n-th degree polynomial of best approximation to f on K , then

$$(5) \qquad E_n(\tilde{f}) \leq ||\tilde{f} - T^{-1}\pi_n|| = ||T^{-1}(f - \pi_n)|| \leq$$

$$\leq ||T^{-1}||\;||f - \pi_n|| = ||T^{-1}||\,E_n(f,K) \quad .$$

Combining (3), (4), and (5) we obtain

$$||f - T_n(f)|| \leq 4||T||\;||T^{-1}||\,E_n(f,K) \;,$$

which proves Theorem 1.

4. REGULAR NODES, AND AUXILIARY RESULTS.

Let

$$\ell_k^{(n)}(z) = \prod_{j\neq k} \frac{z - z_j}{z_k - z_j} \qquad (k = 1,2,\ldots n)$$

be the fundamental polynomials of the Lagrange interpolation associated with the nodes $\{z_k^{(n)}\}$. This system of nodes is said to be M-regular, if:

$$\sup_{k,n} \sup_{z\in K} |\ell_k^{(n)}(z)| \leq M \qquad (6)$$

There always exist regular nodes, since the system of Fekete points is trivially 1-regular.

Lemma 1 [5, Theorem 2]. If $z_k^{(n)} = \psi(e^{i\Theta_k(n)})$ is an M-regular system of nodes for the convex compact set K, then for $i\neq j$

$$n|\Theta_i^{(n)} - \Theta_j^{(n)}| > \frac{1}{4M} .$$

We shall assume from now on that Γ is a convex Jordan curve. This does not restrict the generality of our discussion, as the only convex compact set that is not the closed interior of such a curve, is a line segment, and for this special case the result is already known.

Lemma 2 [6, p.42-44] *. Let $|\Theta| \leq \pi$, $w_o = e^{i\phi}$, $w = e^{i(\phi+\Theta)}$, $z_o = \psi(w_o)$, $z = \psi(w)$. Write:

$$Q(z) = Q_m(z,z_o) = \frac{1}{m}\sum_{k=0}^{m-1} \frac{F_k(z)}{w_o^k} .$$

Q(z) has the following properties:

(i) Q is a polynomial of degree $m - 1$.

(ii) $\max_{z\in K}|Q(z)| \leq 2$

---

* There is an error in the proof presented in [6] which, however, can easily be corrected as follows: On page 43, line 4 reads: "Let $\alpha\pi$ denote the external angle of Γ at the point $\psi(e^{i\phi})$ (clearly $\alpha \geq 1$)". This should be replaced by: 'Let $\alpha\pi$ denote the internal angle of Γ at the point $\psi(e^{i\phi})$. Clearly $\alpha \geq \alpha_o > 0$ where α_o is the smallest angle of Γ (this always exists)'.

(iii) $|Q(z_o)| \geq \ell(K) > 0$

(iv) $|Q(z)| \leq a(K)(m|\theta|)^{-\frac{1}{2}}$.

Here $\ell(K)$ and $a(K)$ are constants depending on K only.

5. PROOF OF THEOREM 2.

Let $m = \left[\frac{n}{4}\right]$ and define for every $z_o \in \Gamma$:

$$p_n(z,z_o) = \left[\frac{Q_m(z,z_o)}{Q_m(z_o,z_o)}\right]^4 ,$$

where Q_m is the polynomial defined in Lemma 2. It follows at once from Lemma 2, that p_n has the following properties (the relationship between z and θ is given in Lemma 2):

(i) p_n is a polynomial of degree, at most, n .

(ii) $|p_n(z,z_o)| \leq \left(\frac{2}{\ell}\right)^4 = C_o$

(iii) $p_n(z_o,z_o) = 1$

(iv) $|p_n(z,z_o)| \leq \left(\frac{\sqrt{5}\,a}{\ell}\right)^4 (n\,\theta)^{-2} = C_1(n\,\theta)^{-2}$ for $n > 20$,

where C_o , C_1 depend on K only.

Lemma 3 (cf. [4, Lemma 2.3]). If $\{z_k^{(n)}\}$ is an M-regular system of nodes, then:

(7) $$\sup_{z\in K} \sum_{k=1}^{n} |p_n(z,z_k^{(n)})| < CM^2 ,$$

where C is a constant depending on K only.

Proof: Since the sum on the left hand side is subharmonic, it is sufficient to prove (6) for $z \in \Gamma$. Write:

$$z = \psi(w),\ z_k = z_k^{(n)} = \psi(w_k) \qquad (1 \leq k \leq n),\ w = e^{i\theta},\ w_k = e^{i\theta_k} .$$

We may assume without loss of generality that: $\theta_1 < \theta_2 < \ldots < \theta_n < \theta_1 + 2\pi$, and that $\theta_n - 2\pi < \theta \leq \theta_1$, and we may define ν so that: $\theta_\nu - \pi \leq \theta < \theta_{\nu+1} - \pi$. Using (ii), (iv), and Lemma 1, we obtain:

$$\sum_{k=1}^{n} |p_n(z,z_k)| = |p_n(z,z_1)| + \sum_{j=1}^{\nu-1} |p_n(z,z_{1+j})| +$$

$$+ \sum_{j=1}^{n-\nu-1} |p_n(z,z_{n-j})| + |p_n(z,z_n)| \le$$

$$\le C_o + C_1 \sum_{j=1}^{\nu-1} n^{-2}(\Theta_{1+j} - \theta)^{-2} + C_1 \sum_{j=1}^{n-\nu-1} n^{-2}(\theta + 2\pi - \Theta_{n-j})^{-2} + C_o \le$$

$$\le C_o + C_1 \sum_{j=1}^{\nu-1} n^{-2}(\Theta_{1+j} - \Theta_1)^{-2} + C_1 \sum_{j=1}^{n-\nu-1} (\Theta_n - \Theta_{n-j})^{-2} + C_o \le$$

$$\le 2C_o + C_1 16 M^2 \{ \sum_{j=1}^{\nu-1} \frac{1}{j^2} + \sum_{j=1}^{n-\nu-1} \frac{1}{j^2} \} <$$

$$< 2C_o + 32 C_1 M^2 \sum_{j=1}^{\infty} \frac{1}{j^2} < 2C_o + 52\, C_1 M^2 < 54\, C_1 M^2 \quad ,$$

since $C_o < C_1$ and $M \ge 1$. Thus the Lemma is established.

Consider finally the polynomial

$$P_n(f,z) = T_n(f,z) + \sum_{k=1}^{n} \{f(z_k^{(n)}) - T_n(f,z_k^{(n)})\} \ell_k^{(n)}(z) p_n(z,z_k^{(n)}) \; .$$

$P_n(f)$ is a polynomial of degree at most $2n - 1$. In view of (iii) and $\ell_k^{(n)}(z_j^{(n)}) = \delta_{kj}$, one finds at once that for $j=1,2,\dots n$

$$P_n(f,z_j^{(n)}) = f(z_j^{(n)}) \quad .$$

Further, by (6), (7), and Theorem 1, $\forall\, z \in K$:

$$|f(z) - P_n(f,z)| \le |f(z) - T_n(f,z)| +$$

$$+ \sum_{k=1}^{n} |f(z_k^{(n)}) - T_n(f,z_k^{(n)})| \, |\ell_k^{(n)}(z)| \, |p_n(z,z_k^{(n)})| \le$$

$$\le B(1 + C M^3)\, E_n(f,K)$$

which completes the proof of Theorem 2.

References

1. J.M. Anderson and J.G. Clunie: "Isomorphisms of the Disc Algebra and Inverse Faber sets" (to be published).

2. J.E. Andersson: Dissertation, Göteborg 1975.

3. T. Kövari and Ch. Pommerenke: "On Faber polynomials and Faber expansions", Math. Zeit., 99(1967), 193-206.

4. T. Kövari, "On the uniform approximation of analytic functions by means of interpolation polynomials", Commentarii Math. Helv. 43(1968), 212-216.

5. T. Kövari and Ch. Pommerenke, "On the distribution of Fekete points", Mathematika, 15(1968), 70-75.

6. T. Kövari, "On the distribution of Fekete points II", Mathematika, 18(1971), 40-49.

7. T. Kövari, "On the order of polynomial approximation for closed Jordan domains", Journ. of Approx. Theory 5(1972), 362-373.

ON EQUICONVERGENCE OF CERTAIN SEQUENCES OF RATIONAL INTERPOLANTS

E. B. Saff[1] and A. Sharma[2]
Center for Mathematical Services
University of South Florida
Tampa, Florida 33620

Mathematics Department
University of Alberta
Edmonton, Canada T6G 2G1

Abstract For a function $f(z)$ analytic on $|z| < \rho$, $\rho > 1$, we consider two schemes of rational interpolants which have poles equally spaced on the circle $|z| = \sigma$, $\sigma > 1$. The first scheme interpolates $f(z)$ in the roots of unity, while the second consists of best L^2-approximants to $f(z)$ on the unit circle. We obtain precise regions of equiconvergence for the two schemes of rational functions, thus extending a well-known result of J. L. Walsh.

1. Introduction

Let A_ρ denote the class of functions $f(z)$ which are analytic in the open disk $|z| < \rho$, but not on $|z| \leq \rho$. A fundamental result concerning the equiconvergence of certain sequences of polynomials is the following theorem of J. L. Walsh [5, p. 153]:

Theorem 1.1. Suppose $f \in A_\rho$ with $\rho > 1$. For each positive integer n, let $L_{n-1}(z)$ denote the Lagrange polynomial interpolant to f in the n th roots of unity, and denote by $s_{n-1}(z)$ the $(n-1)$th order Taylor polynomial of f about the origin. Then

$$(1.1)\qquad \lim_{n\to\infty} \{L_{n-1}(z) - s_{n-1}(z)\} = 0 \quad , \quad \forall\ |z| < \rho^2 \ ,$$

the convergence being uniform and geometric on any compact set in $|z| < \rho^2$. Moreover, the result is sharp in the sense that for any point z_0 on $|z| = \rho^2$, there is a function in A_ρ for which (1.1) does not hold at z_0.

---

[1]Research supported in part by the National Science Foundation.

[2]Research supported in part by the National Research Council of Canada.

For $f \in A_\rho$, $\rho > 1$, it is a simple consequence of the convergence properties of the two sequences $\{L_{n-1}(z)\}_1^\infty$ and $\{s_{n-1}(z)\}_1^\infty$ that (1.1) holds for $|z| < \rho$. The essential feature of Walsh's theorem is that equiconvergence holds in the larger disk $|z| < \rho^2$. A discussion of various extensions of Theorem 1.1 and related results can be found in [3] and [4].

The purpose of the present paper is to describe generalizations of Theorem 1.1 to the case of interpolating rational functions whose poles are equally spaced on a given circle $|z| = \sigma$, $\sigma > 1$. In place of the Lagrange polynomial $L_{n-1}(z)$, we will take the unique function $R_{n+m,n}(z)$ of the form

$$(1.2) \qquad R_{n+m,n}(z) = \frac{B_{n+m,n}(z)}{z^n - \sigma^n} , \quad B_{n+m,n}(z) \in \pi_{n+m} ,$$

which interpolates $f(z)$ in the $(n+m+1)$ th roots of unity; that is,

$$(1.3) \qquad B_{n+m,n}(z) = f(z)(z^n - \sigma^n) , \quad \text{if} \quad z^{n+m+1} - 1 = 0 .$$

(Here and below, π_k denotes the collection of all polynomials of degree at most k .) Since the $(n-1)$ th Taylor polynomial $s_{n-1}(z)$ is also the least squares approximation to $f(z)$ from π_{n-1} on the unit circle $|z| = 1$, we shall replace this polynomial by the unique rational function

$$(1.4) \qquad r_{n+m,n}(z) = \frac{P_{n+m,n}(z)}{z^n - \sigma^n} , \quad P_{n+m,n}(z) \in \pi_{n+m} ,$$

which minimizes the integral

$$(1.5) \qquad \int_{|z|=1} |f(z) - r(z)|^2 |dz|$$

over all rationals of the form $p(z)/(z^n - \sigma^n)$, $p(z) \in \pi_{n+m}$. From another elegant theorem of Walsh [5, §9.1], for each integer $m \geq -1$, the rational $r_{n+m,n}(z)$ must interpolate $f(z)$ in the $(n+m+1)$ roots of the equation $z^{m+1}(z^n - \sigma^{-n}) = 0$; that is, for $m \geq -1$,

$$(1.6) \qquad P_{n+m,n}(z) = f(z)(z^n - \sigma^n) , \quad \text{if} \quad z^{m+1}(z^n - \sigma^{-n}) = 0 .$$

In the spirit of Theorem 1.1, we shall examine the difference

$$R_{n+m,n}(z) - r_{n+m,n}(z) = \frac{B_{n+m,n}(z) - P_{n+m,n}(z)}{z^n - \sigma^n} ,$$

for each fixed integer m and show that the phenomenon of equiconvergence persists. In fact, if $\rho^2 > \sigma$, a new phenomenon arises which is described in Theorem 2.3 of §2.

Of special interest is the situation when $m < -1$ since, in this case, the interpolation property of (1.6) no longer holds. As we shall show in §4, the L^2-extremal rational function $r_{n+m,n}(z) = P_{n+m,n}(z)/(z^n - \sigma^n)$ for $m < -1$ has the following simple characterization. If we write

$$(1.7) \qquad P_{n-1,n}(z) = \sum_{k=0}^{n-1} b_{k,n} z^k ,$$

where (as in (1.6)) $P_{n-1,n}(z)$ interpolates $f(z)(z^n - \sigma^n)$ in the roots of $z^n - \sigma^{-n} = 0$, then for each $m = -2,-3,\ldots$ and $n \geq -m$, we have

$$(1.8) \qquad P_{n+m,n}(z) = \sum_{k=0}^{n+m} b_{k,n} z^k .$$

2. Equiconvergence of $\{R_{n+m,n}(z)\}$ and $\{r_{n+m,n}(z)\}$ for $m \geq -1$.

The first two theorems concern the separate convergence properties of the sequences $\{R_{n+m,n}(z)\}_{n=1}^{\infty}$ and $\{r_{n+m,n}(z)\}_{n=1}^{\infty}$ for fixed $m \geq -1$. We shall use the symbol $\|\cdot\|_A$ to denote the sup norm taken over the set A.

Theorem 2.1. Let $\rho > 1$, $\sigma > 1$ and an integer $m \geq -1$ be fixed. If $f \in A_\rho$ and if $R_{n+m,n}(z)$ is the rational function of the form (1.2) which interpolates $f(z)$ in the $(n+m+1)$ th roots of unity, then

$$(2.1) \qquad \lim_{n\to\infty} R_{n+m,n}(z) = f(z) , \quad \forall\ |z| < \min\{\sigma,\rho\} .$$

More precisely, if $\tau := \min\{\sigma,\rho\}$ and $K \subset \{z : |z| < \tau\}$ is compact, then

$$(2.2)\qquad \limsup_{n\to\infty} \|f(z) - R_{n+m,n}(z)\|_K^{1/n} \le \frac{1}{\tau}\max\{1, \|z\|_K\} < 1 .$$

Furthermore, if $\rho > \sigma$, then for all $|z| > \sigma$

$$(2.3)\qquad \lim_{n\to\infty} R_{n+m,n}(z) = \begin{cases} 0 , & \text{for } m = -1 \\ \sum_{k=0}^{m} a_k z^k , & \text{for } m = 0,1,\ldots , \end{cases}$$

where $f(z) = \sum_{k=0}^{\infty} a_k z^k$.

Theorem 2.2. Let $\rho > 1$, $\sigma > 1$ and an integer $m \ge -1$ be fixed. If $f \in A_\rho$ and $r_{n+m,n}(z)$ is the rational function of (1.4) of least squares approximation to f on the unit circle, then the conclusions (2.1), (2.2) and (2.3) of Theorem 2.1 remain valid if $R_{n+m,n}(z)$ is replaced by $r_{n+m,n}(z)$.

Remark 1. The proofs of Theorems 2.1 and 2.2 are immediate consequences of the following Hermite formula representations for $m \ge -1$:

$$(2.4)\qquad f(z) - R_{n+m,n}(z) = \frac{1}{2\pi i}\int_\Gamma \frac{(z^{n+m+1}-1)(t^n-\sigma^n)f(t)}{(z^n-\sigma^n)(t^{n+m+1}-1)(t-z)}\,dt ,$$

$$(2.5)\qquad f(z) - r_{n+m,n}(z) = \frac{1}{2\pi i}\int_\Gamma \frac{z^{m+1}(z^n-\sigma^{-n})(t^n-\sigma^n)f(t)}{(z^n-\sigma^n)(t^n-\sigma^{-n})t^{m+1}(t-z)}\,dt ,$$

where Γ is the circle $|t| = \hat{\rho}$, $1 < \hat{\rho} < \rho$, and $|z| < \hat{\rho}$. In writing (2.5) we have used the interpolation property of (1.6). From (2.4) and (2.5), one can obtain integral formulae for $R_{n+m,n}(z)$ and $r_{n+m,n}(z)$, valid for all $z \in \mathbb{C}$, which imply (2.3).

It follows from Theorems 2.1 and 2.2 that if $\rho \le \sigma$, then

$$(2.6)\qquad \lim_{n\to\infty}\{R_{n+m,n}(z) - r_{n+m,n}(z)\} = 0 , \quad \forall\ |z| < \rho ,$$

and, if $\rho > \sigma$, then (2.6) holds $\forall\ |z| \ne \sigma$. A better result is given by

Theorem 2.3. *Let* $\rho > 1$, $\sigma > 1$ *and an integer* $m \geq -1$ *be fixed. If* $f \in A_\rho$, *then the rational functions of* (1.2) *and* (1.4) *satisfy*

$$(2.7) \qquad \lim_{n\to\infty} \{R_{n+m,n}(z) - r_{n+m,n}(z)\} = 0 \quad \begin{cases} \forall\ |z| < \rho^2, & \text{if } \sigma \geq \rho^2 \\ \forall\ |z| \neq \sigma, & \text{if } \rho^2 > \sigma . \end{cases}$$

Moreover, the result is sharp.

Remark 2. The proof of Theorem 2.3 follows from the representations (2.4) and (2.5) which yield

$$(2.8) \qquad R_{n+m,n}(z) - r_{n+m,n}(z) =$$

$$= \frac{1}{2\pi i}\int_\Gamma \frac{f(t)}{t-z}\left(\frac{t^n-\sigma^n}{z^n-\sigma^n}\right)\left[\frac{t^{m+1}(t^n-\sigma^{-n})-z^{m+1}(z^n-\sigma^{-n})-t^{m+1}z^{m+1}(t^n-z^n)\sigma^{-n}}{t^{m+1}(t^n-\sigma^{-n})(t^{m+n+1}-1)}\right] dt.$$

One can also use (2.8) to obtain degree of convergence results. That (2.7) is sharp can be easily seen by taking $f(z) = (z - \rho e^{i\theta})^{-1}$.

Remark 3. Letting σ tend to infinity in Theorem 2.3 gives the classical result of Theorem 1.1.

Remark 4. For the case $\rho^2 > \sigma$, Theorem 2.3 asserts that equiconvergence holds at all points of the plane not on the circle $|z| = \sigma$. This is a new phenomenon which does not arise in Walsh's Theorem 1.1 where $\sigma = \infty$. (See also [2].)

3. Extension of Theorem 2.3.

Our next goal is to extend Theorem 2.3 in the spirit of Theorem 1 of [1]. The essence of the latter theorem is a representation of the Lagrange polynomial $L_{n-1}(z)$ interpolating $f(z)$ in the roots of $z^n - 1 = 0$. Namely, it is shown that, for each fixed n,

$$(3.1) \qquad L_{n-1}(z) = \sum_{\nu=0}^{\infty} s_{n-1}(z;\nu) ,$$

where $s_{n-1}(z;\nu) := \sum_{j=0}^{n-1} a_{j+\nu n} z^j$ is the shifted (n-1) th section of the Taylor expansion $\sum_{k=0}^{\infty} a_k z^k$ for $f(z)$. The representation (3.1) has two important properties. First, since $s_{n-1}(z;0) = s_{n-1}(z)$, equation (3.1) relates to an asymptotic formula for the difference $L_{n-1}(z) - s_{n-1}(z)$ occuring in Walsh's Theorem 1.1. Second, it yields a systematic way to construct the values of f in the nth roots of unity from the knowledge of the values of f and its derivatives at the origin; that is, if $\omega^n - 1 = 0$, then from (3.1) we have

$$f(\omega) = \sum_{\nu=0}^{\infty} \sum_{j=0}^{n-1} \frac{f^{(j+\nu n)}(0)}{(j+\nu n)!} \omega^j .$$

In a like manner, for the rational $R_{n+m,n}(z) = B_{n+m,n}(z)/(z^n - \sigma^n)$ which interpolates $f(z)$ in the roots of $z^{m+n+1} - 1 = 0$, we seek a representation

$$R_{n+m,n}(z) = \sum_{\nu=0}^{\infty} r_{n+m,n}(z;\nu) , \tag{3.2}$$

where $r_{n+m,n}(z;0) = r_{n+m,n}(z)$ and, for each $\nu = 0,1,\ldots,$ $r_{n+m,n}(z;\nu)$ is a rational function of the form

$$r_{n+m,n}(z;\nu) = \frac{P_{n+m,n}(z;\nu)}{z^n - \sigma^n} , \quad P_{n+m,n}(z;\nu) \in \pi_{n+m} \tag{3.3}$$

which is determined solely by the values of f and its derivatives in the roots of $z^{m+1}(z^n - \sigma^{-n}) = 0$.

For this purpose, it is convenient to have the following.

Lemma 3.1. For fixed integers $m \geq -1$, $n \geq 1$, set $N(\nu) := (\nu+1)(n+m+1)-1$, $\nu = 0,1,\ldots,$ and put

$$\alpha_{n,m}(z) := 1 - z^{m+1}\sigma^{-n} , \quad \beta_{n,m}(z) := z^{m+1}(z^n - \sigma^{-n}) , \quad \sigma > 1 . \tag{3.4}$$

Let $S_{N(\nu)}(z)$ denote the unique polynomial in $\pi_{N(\nu)}$ which interpolates the function $\{\alpha_{n,m}(z)\}^{\nu}(z^n - \sigma^n)f(z)$ in the Hermite sense in the $N(\nu)+1$ roots of $\{\beta_{n,m}(z)\}^{\nu+1} = 0$. If $f(z)$ is analytic in

$|z| \leq 1$, _then for each_ n _sufficiently large_,

$$(3.5) \qquad \lim_{\nu \to \infty} \frac{S_{N(\nu)}(z)}{\{\alpha_{n,m}(z)\}^{\nu}} = (z^n - \sigma^n) f(z) ,$$

uniformly on $|z| \leq 1$. _Furthermore_,

$$(3.6) \qquad S_{N(\nu)}(z) - \alpha_{n,m}(z) S_{N(\nu-1)}(z) = \{\beta_{n,m}(z)\}^{\nu} P_{n+m,n}(z;\nu) ,$$

where $P_{n+m,n}(z;\nu) \in \pi_{n+m}$, $\nu = 1,2,\ldots$.

Consequently, for $|z| \leq 1$,

$$(3.7) \qquad (z^n - \sigma^n) f(z) = \sum_{\nu=0}^{\infty} \left\{ \frac{\beta_{n,m}(z)}{\alpha_{n,m}(z)} \right\}^{\nu} P_{n+m,n}(z;\nu) ,$$

where $P_{n+m,n}(z;0) := S_{N(0)}(z)$.

Remark 5. Notice that since $S_{N(0)}(z)$ interpolates $(z^n - \sigma^n) f(z)$ in the zeros of $\beta_{n,m}(z)$, then $P_{n+m,n}(z;0) = P_{n+m,n}(z)$ which is the numerator polynomial in (1.4), i.e.,

$$(3.8) \qquad \frac{P_{n+m,n}(z;0)}{z^n - \sigma^n} \equiv r_{n+m,n}(z) .$$

Furthermore, since from (3.7), the polynomial $P_{n+m,n}(z;\nu) \in \pi_{n+m}$ interpolates the function

$$\left\{ \frac{\alpha_{n,m}(z)}{\beta_{n,m}(z)} \right\}^{\nu} \left[(z^n - \sigma^n) f(z) - \sum_{k=0}^{\nu-1} \left\{ \frac{\beta_{n,m}(z)}{\alpha_{n,m}(z)} \right\}^{k} P_{n+m,n}(z;k) \right]$$

in the zeros of $\beta_{n,m}(z)$, we see that $P_{n+m,n}(z;\nu)$ is determined only from the values of f and finitely many of its derivatives at these zeros.

Proof of Lemma 3.1. We first prove (3.6). Clearly, from the interpolation properties of the polynomials $S_{N(\nu)}(z)$, we see that $\{\beta_{n,m}(z)\}^{\nu}$ divides the polynomial $S_{N(\nu)}(z) - \alpha_{n,m}(z) S_{N(\nu-1)}(z)$. Hence

$$S_{N(\nu)}(z) - \alpha_{n,m}(z)S_{N(\nu-1)}(z) = \{\beta_{n,m}(z)\}^{\nu}P_{n+m,n}(z;\nu) \quad ,$$

where the degree of $P_{n+m,n}(z;\nu)$ is at most $N(\nu) - (n+m+1)\nu = n+m$.

In order to prove (3.5), we observe that for $|z| \leq 1$,

$$E_{\nu}(z) := (z^n - \sigma^n)f(z) - \frac{S_{N(\nu)}(z)}{\{\alpha_{n,m}(z)\}^{\nu}} =$$

$$= \frac{1}{2\pi i}\int_{|t|=\hat{\rho}} \left\{\frac{\beta_{n,m}(z)}{\beta_{n,m}(t)}\right\}^{\nu+1} \left\{\frac{\alpha_{n,m}(t)}{\alpha_{n,m}(z)}\right\}^{\nu} \frac{(t^n - \sigma^n)f(t)}{t-z}\, dt \quad ,$$

where $\hat{\rho} > 1$ is selected so that $f(t)$ is analytic on $|t| \leq \hat{\rho}$. Straightforward estimates then yield

$$\limsup_{\nu\to\infty} \|E_{\nu}(z)\|_{|z|\leq 1}^{1/\nu} \leq \frac{(1+\sigma^{-n})(1+\hat{\rho}^{m+1}\sigma^{-n})}{\hat{\rho}^{m+1}(\hat{\rho}^n - \sigma^{-n})(1-\sigma^{-n})} < 1$$

for $n > n_0(m,\hat{\rho},\sigma)$. This proves (3.5). Combining (3.5) and (3.6), we get (3.7). □

Corollary 3.2. *Let* $f \in A_{\rho}$, $\rho > 1$, *and* $B_{n+m,n} \in \pi_{n+m}$, $m \geq -1$, *interpolate* $(z^n - \sigma^n)f(z)$ *in the* $(n+m+1)$ *th roots of unity. Then, for each* n *large* $(n > n_0(m,\rho,\sigma))$, *we have*

$$B_{n+m,n}(z) = \sum_{\nu=0}^{\infty} P_{n+m,n}(z;\nu) \quad , \quad \forall\ z \in \mathbb{C} \quad , \tag{3.9}$$

where the polynomials $P_{n+m,n}(z;\nu) \in \pi_{n+m}$ *are defined in* (3.6).

Proof. If ω is an $(n+m+1)$th root of unity, then since $\beta_{n,m}(\omega) = \alpha_{n,m}(\omega)$, we deduce from (3.7) that

$$B_{n+m,n}(\omega) = (z^n - \sigma^n)f(z)\Big|_{z=\omega} = \sum_{\nu=0}^{\infty} P_{n+m,n}(\omega;\nu) \quad .$$

Thus, by the uniqueness of the interpolant, (3.9) follows. □

The next theorem gives a generalization of Theorem 2.3.

Theorem 3.3. *Let* $\rho > 1$, $\sigma > 1$ *and an integer* $m \geq -1$ *be fixed. If* $f \in A_\rho$ *and if* ℓ *is any given positive integer, then*

$$(3.10)\quad \lim_{n\to\infty}\left\{R_{n+m,n}(z) - \sum_{\nu=0}^{\ell-1} r_{n+m,n}(z;\nu)\right\} = 0 \begin{cases} \forall\ |z| < \rho^{\ell+1}, \text{ if } \sigma \geq \rho^{\ell+1} \\ \forall\ |z| \neq \sigma\ , \text{ if } \rho^{\ell+1} > \sigma\ , \end{cases}$$

where $R_{n+m,n}(z) = B_{n+m,n}(z)/(z^n - \sigma^n)$ *is defined in* (1.2) *and*

$$(3.11)\quad r_{n+m,n}(z;\nu) := P_{n+m,n}(z;\nu)/(z^n - \sigma^n)\quad ,\quad \nu = 0,1,\ldots\ .$$

The convergence in (3.10) *is uniform and geometric on compact subsets of the regions described. Moreover, the result is sharp.*

Notice from (3.8) that, in the case $\ell = 1$, Theorem 3.3 reduces to Theorem 2.3.

Proof of Theorem 3.3. For n sufficiently large, we have by Corollary 3.2,

$$(3.12)\quad B_{n+m,n}(z) - \sum_{\nu=0}^{\ell-1} P_{n+m,n}(z;\nu) = \sum_{\nu=\ell}^{\infty} P_{n+m,n}(z;\nu)\quad ,\quad \forall\ z \in \mathbb{C}\ .$$

Also, from the interpolating property of the polynomial $S_{N(\nu)}(z)$ defined in Lemma 3.1, we have

$$S_{N(\nu)}(z) = \frac{1}{2\pi i}\int_\Gamma \frac{f(t)(t^n - \sigma^n)\{\alpha_{n,m}(t)\}^\nu[\{\beta_{n,m}(t)\}^{\nu+1} - \{\beta_{n,m}(z)\}^{\nu+1}]}{(t-z)\{\beta_{n,m}(t)\}^{\nu+1}}\,dt,$$

where $\Gamma : |t| = \tau$, $1 < \tau < \rho$, and $\alpha_{n,m}$, $\beta_{n,m}$ are given in (3.4). Using this representation and equation (3.6), we obtain after some algebra the following integral representation for $P_{n+m,n}(z;\nu)$, $\nu \geq 1$:

(3.13) $P_{n+m,n}(z;\nu) =$

$$= \frac{1}{2\pi i} \int_{\Gamma} \frac{f(t)(t^n - \sigma^n)\{\alpha_{n,m}(z)\beta_{n,m}(t) - \alpha_{n,m}(t)\beta_{n,m}(z)\}}{(t-z)\alpha_{n,m}(t)\beta_{n,m}(t)} \left\{\frac{\alpha_{n,m}(t)}{\beta_{n,m}(t)}\right\}^{\nu} dt.$$

Thus, from (3.11), (3.12) and (3.13) we get, for n large and all $z \in \mathbb{C}$,

(3.14) $R_{n+m,n}(z) - \sum_{\nu=0}^{\ell-1} r_{n+m,n}(z;\nu) =$

$$= \frac{1}{2\pi i} \int_{\Gamma} \frac{f(t)(t^n - \sigma^n)\{\alpha_{n,m}(z)\beta_{n,m}(t) - \alpha_{n,m}(t)\beta_{n,m}(z)\}}{(t-z)(t^{m+n+1} - 1)\alpha_{n,m}(t)(z^n - \sigma^n)} \left\{\frac{\alpha_{n,m}(t)}{\beta_{n,m}(t)}\right\}^{\ell} dt.$$

A straightforward analysis of (3.14) then yields (3.10).

To prove the sharpness assertion of Theorem 3.3, we take $\hat{f}(z) := 1/(z - \rho)$. From (3.14), we obtain in this case

(3.15) $R_{n+m,n}(z) - \sum_{\nu=0}^{\ell-1} r_{n+m,n}(z;\nu) =$

$$= \frac{(\rho^n - \sigma^n)\{\alpha_{n,m}(z)\beta_{n,m}(\rho) - \alpha_{n,m}(\rho)\beta_{n,m}(z)\}}{(z-\rho)(\rho^{m+n+1} - 1)\alpha_{n,m}(\rho)(z^n - \sigma^n)} \left\{\frac{\alpha_{n,m}(\rho)}{\beta_{n,m}(\rho)}\right\}^{\ell},$$

from which it is easy to show that

$$\lim_{n\to\infty} \left\{R_{n+m,n}(\rho^{\ell+1}) - \sum_{\nu=0}^{\ell-1} r_{n+m,n}(\rho^{\ell+1};\nu)\right\} = \frac{1}{\rho - \rho^{\ell+1}}, \text{ if } \sigma > \rho^{\ell+1},$$

$$\lim_{n\to\infty} \left\{R_{n+m,n}(z) - \sum_{\nu=0}^{\ell-1} r_{n+m,n}(z;\nu)\right\} = \frac{z^{m+1}}{\sigma^{m+1}(z-\rho)}, \text{ if } \sigma = \rho^{\ell+1}, |z| > \sigma.$$

This completes the proof of Theorem 3.3. □

4. Equiconvergence of $\{R_{n-\mu,n}(z)\}$ and $\{r_{n-\mu,n}(z)\}$ for $\mu \geq 2$.

This case differs slightly from the case in §2 and §3. However, as we shall see, there is an essential continuity in the results which come out. We shall begin by proving a lemma.

Lemma 4.1. Let $\rho > 1$, $\sigma > 1$ and an integer μ , $2 \leq \mu \leq n$, be fixed. Let $P_{n-\mu,n}(z)$ denote the polynomial in $\pi_{n-\mu}$ for which

$$(4.1) \qquad \min_{Q \in \pi_{n-\mu}} \int_{|z|=1} \left| f(z) - \frac{Q(z)}{z^n - \sigma^n} \right|^2 |dz|$$

is attained, where $f(z) \in A_\rho$. Then $P_{n-\mu,n}(z)$ is given by the formula

$$(4.2) \qquad P_{n-\mu,n}(z) = \frac{1}{2\pi i} \int_\Gamma \frac{f(t)t^{\mu-1}(t^{n-\mu+1} - z^{n-\mu+1})(t^n - \sigma^n)}{(t-z)(t^n - \sigma^{-n})} dt$$

where $\Gamma : |t| = \tau$, $1 < \tau < \rho$.

Proof. The minimization problem (4.1) is equivalent to finding

$$(4.3) \qquad \min_{\{a_j\}} \int_{|z|=1} \left| f(z) - \sum_{j=0}^{n-\mu} a_j f_j(z) \right|^2 |dz| \quad ,$$

where $f_j(z) := z^j / (z^n - \sigma^n)$, $j = 0,1,\ldots,n-\mu$. It is easy to see that the minimum in (4.3) is attained if and only if

$$(4.4) \qquad \frac{1}{2\pi i} \int_{|z|=1} \{f(z) - \sum_{j=0}^{n-\mu} a_j f_j(z)\} \overline{f_\ell(z)} |dz| = 0 \ , \ (\ell = 0,1,\ldots,n-\mu).$$

Since

$$f_\ell(z) = \frac{z^\ell}{z^n - \sigma^n} = \frac{1}{n\sigma^{n-\ell-1}} \sum_{k=0}^{n-1} \frac{\omega^{k+k\ell}}{z - \sigma\omega^k} \ , \quad \omega := e^{2\pi i/n} \ ,$$

$$(\ell = 0,1,\ldots,n-\mu)$$

it follows that

$$\overline{f_\ell(z)} = -\frac{1}{n\sigma^{n-\ell}}\sum_{k=0}^{n-1}\frac{\omega^{-k\ell}z}{z-\frac{\omega^k}{\sigma}}\quad,\quad |z| = 1\quad .$$

From this observation, we see that

$$(4.5)\qquad \frac{1}{2\pi i}\int_{|z|=1} f(z)\overline{f_\ell(z)}\,|dz| = \frac{1}{2\pi i}\int_{|z|=1} f(z)\overline{f_\ell(z)}\,\frac{dz}{iz}$$

$$= \frac{i}{n\sigma^{n-\ell}}\sum_{k=0}^{n-1}\frac{\omega^{-k\ell}}{2\pi i}\int_{|z|=1}\frac{f(z)}{z-\omega^k/\sigma}\,dz$$

$$= \frac{i}{n\sigma^{n-\ell}}\sum_{k=0}^{n-1}\omega^{-k\ell}f\left(\frac{\omega^k}{\sigma}\right) .$$

Moreover, if we set $P_{n-\mu,n}(z) = \sum_{\nu=0}^{n-\mu} b_\nu z^\nu$, we get

$$\frac{1}{2\pi i}\int_{|z|=1}\frac{P_{n-\mu,n}(z)}{z^n-\sigma^n}\overline{f_\ell(z)}\,|dz| = \frac{i}{n\sigma^{n-\ell}(\sigma^{-n}-\sigma^n)}\sum_{k=0}^{n-1}\omega^{-k\ell}P_{n-\mu,n}\left(\frac{\omega^k}{\sigma}\right)$$

$$= \frac{i}{n\sigma^{n-\ell}(\sigma^{-n}-\sigma^n)}\sum_{\nu=0}^{n-\mu} b_\nu\sigma^{-\nu}\sum_{k=0}^{n-1}\omega^{k(\nu-\ell)} .$$

On using the properties of roots of unity, this yields

$$(4.6)\qquad \frac{1}{2\pi i}\int_{|z|=1}\frac{P_{n-\mu,n}(z)}{z^n-\sigma^n}\overline{f_\ell(z)}\,|dz| = \frac{ib_\ell}{\sigma^n(\sigma^{-n}-\sigma^n)}$$

$$(\ell = 0,1,\ldots,n-\mu) \quad .$$

From (4.4), (4.5) and (4.6), we see that

$$(4.7)\qquad b_j = \frac{1}{2\pi i}(\sigma^{-n}-\sigma^{n})\int_\Gamma \frac{f(t)t^{n-1-j}}{t^n-\sigma^{-n}}\,dt\ ,\ j=0,1,\ldots,n-\mu\ ,$$

since

$$\frac{z^{n-1-j}}{z^n-\sigma^{-n}} = \frac{\sigma^j}{n}\sum_{k=0}^{n-1}\frac{\omega^{-kj}}{z-\sigma^{-1}\omega^k}\ .$$

We now easily see from (4.7) that

$$(4.8)\qquad P_{n-\mu,n}(z) = \frac{\sigma^{-n}-\sigma^n}{2\pi i}\int_\Gamma \frac{f(t)t^{\mu-1}(t^{n-\mu+1}-z^{n-\mu+1})}{(t-z)(t^n-\sigma^{-n})}\,dt\ .$$

Since

$$\sigma^{-n}-\sigma^n = \sigma^{-n}-t^n+t^n-\sigma^n\ ,$$

the above integral splits up into two integrals, one of which is

$$-\frac{1}{2\pi i}\int_\Gamma \frac{f(t)t^{\mu-1}(t^{n-\mu+1}-z^{n-\mu+1})}{t-z}\,dt = 0\ .$$

This yields (4.2) and completes the proof. □

Remark 6. As stated in (1.6), the polynomial $P_{n-1,n}(z)$ interpolates $f(z)(z^n-\sigma^n)$ in the n roots of $z^n-\sigma^{-n}=0$, from which it follows that

$$(4.9)\qquad P_{n-1,n}(z) = \frac{1}{2\pi i}\int_\Gamma \frac{f(t)(t^n-z^n)(t^n-\sigma^n)}{(t-z)(t^n-\sigma^{-n})}\,dt\ .$$

Thus, equation (4.2) also holds for $\mu = 1$. Indeed, the derivation of (4.2) given above is valid in this case. Moreover, note that the formula (4.7) for the coefficients of $P_{n-\mu,n}(z)$ is independent of μ . Thus, if we write

$$(4.10)\qquad P_{n-1,n}(z) = \sum_{j=0}^{n-1} b_{j,n}z^j\ ,$$

it follows that $P_{n-\mu,n}(z)$ is just a partial sum of $P_{n-1,n}(z)$, i.e.,

$$(4.11) \qquad P_{n-\mu,n}(z) = \sum_{j=0}^{n-\mu} b_{j,n} z^j \quad , \quad \mu = 2,3,\ldots,n \quad ,$$

as claimed in (1.8). A similar situation arises in the case of discrete least squares approximation in the nth roots of unity. Namely, it is known (cf. [6,p.8]) that the polynomial $p_{n-\mu} \in \pi_{n-\mu}$, $\mu \geq 2$, for which the minimum

$$\min_{p \in \pi_{n-\mu}} \sum_{k=1}^{n} |F(\omega^k) - p(\omega^k)|^2 \quad , \quad \omega := e^{2\pi i/n} \quad ,$$

is attained is just a partial sum of the polynomial $p_{n-1} \in \pi_{n-1}$ which interpolates $F(z)$ in the nth roots of unity. This known characterization can be viewed as a limiting case of (4.11) where $\sigma \to 1$ and $f(z) = F(z)/(z^n - \sigma^n)$.

We can now prove that Theorem 2.3 holds for all negative integers m.

Theorem 4.2. Let $\rho > 1$, $\sigma > 1$ and an integer $\mu \geq 2$ be fixed. If $f \in A_\rho$, then the rational functions $R_{n-\mu,n}(z) = B_{n-\mu,n}(z)/(z^n - \sigma^n)$, $r_{n-\mu,n}(z) = P_{n-\mu,n}(z)/(z^n - \sigma^n)$ defined in (1.2) and (1.4) (with $m = -\mu$) satisfy

$$(4.12) \qquad \lim_{n\to\infty} \{R_{n-\mu,n}(z) - r_{n-\mu,n}(z)\} = 0 \quad \begin{cases} \forall\ |z| < \rho^2 \ , \text{ if } \sigma \geq \rho^2 \\ \forall\ |z| \neq \sigma \ , \text{ if } \rho^2 > \sigma \end{cases} .$$

Moreover, the result is sharp.

Proof. From formula (4.2) and the representation

$$(4.13) \qquad B_{n-\mu,n}(z) = \frac{1}{2\pi i} \int_\Gamma \frac{f(t)(t^n - \sigma^n)(t^{n-\mu+1} - z^{n-\mu+1})}{(t-z)(t^{n-\mu+1} - 1)} dt \quad ,$$

where $\Gamma : |t| = \hat{\rho}$, $1 < \hat{\rho} < \rho$, we find

$$(4.14)\quad B_{n-\mu,n}(z) - P_{n-\mu,n}(z) =$$

$$= \frac{1}{2\pi i} \int_{\Gamma} \frac{f(t)(t^n - \sigma^n)(t^{n-\mu+1} - z^{n-\mu+1})(t^{\mu-1} - \sigma^{-n})}{(t-z)(t^n - \sigma^{-n})(t^{n-\mu+1} - 1)}\, dt \quad .$$

Equation (4.12) then follows by estimating the integral in (4.14).

To prove the sharpness assertion, take $\hat{f}(z) = 1/(z-\rho)$. Then it is easy to verify from the interpolating properties that, for $n \geq 2(\mu-1)$, we have

$$(4.15)\quad B_{n-\mu,n}(z) = B_{n-\mu,n}(z;\hat{f}) = \frac{\sigma^n - z^{\mu-1}}{\rho - z} + \frac{z^{n-\mu+1}-1}{\rho - z}\left(\frac{\rho^{\mu-1} - \sigma^n}{\rho^{n-\mu+1} - 1}\right) \quad .$$

Moreover, from (4.8), we find for $\mu \geq 1$,

$$(4.16)\quad P_{n-\mu,n}(z) = P_{n-\mu,n}(z;\hat{f}) = \rho^{\mu-1}\left(\frac{\sigma^n - \sigma^{-n}}{\rho^n - \sigma^{-n}}\right)\left(\frac{\rho^{n-\mu+1} - z^{n-\mu+1}}{\rho - z}\right) \quad .$$

On subtracting (4.16) from (4.15), it can be shown that

$$(4.17)\quad \lim_{n\to\infty} \{R_{n-\mu,n}(\rho^2;\hat{f}) - r_{n-\mu,n}(\rho^2;\hat{f})\} = \frac{1}{\rho - \rho^2}\ , \text{ if } \sigma > \rho^2 \ ,$$

$$(4.18)\quad \lim_{n\to\infty} \{R_{n-\mu,n}(z;\hat{f}) - r_{n-\mu,n}(z;\hat{f})\} = \frac{z^{1-\mu}}{z-\rho}\ , \text{ if } \sigma = \rho^2\ ,\ |z| > \sigma\ ,$$

which proves that (4.12) is sharp. □

Theorem 4.2 can be extended in a manner similar to the generalization of Theorem 2.3, given in Theorem 3.3 by introducing the corresponding polynomials $P_{n-\mu,n}(z;\nu)$ defined by

$$(4.19)\quad P_{n-\mu,n}(z;\nu) := \frac{1}{2\pi i}\int_{|t|=1} \frac{f(t)t^{\mu-1}(t^n - \sigma^n)(t^{n-\mu+1} - z^{n-\mu+1})(t^{\mu-1} - \sigma^{-n})^{\nu}}{(t-z)(t^n - \sigma^{-n})^{\nu+1}}\, dt,$$

$$\nu = 0,1,2,\ldots \quad .$$

The details are left for the reader.

References

1. A. S. Cavaretta Jr., A. Sharma and R. S. Varga. Interpolation in the roots of unity: An extension of a theorem of J. L. Walsh. Resultate der Math. 1981, vol. 3, 155-191.

2. G. López Lagomasino and René Piedra de la Torre, Sobre un teorema de sobreconvergencia de J. L. Walsh. Revista Ciencias Matemáticas. 1983, vol. IV, No. 3, 67-78.

3. E. B. Saff, A. Sharma and R. S. Varga, An extension to rational functions of a theorem of J. L. Walsh on differences of interpolating polynomials, R.A.I.R.O. Analyse numérique, 1981, vol. 15, 371-390.

4. R. S. Varga. Topics in Polynomial and Rational Interpolation and Approximation. Séminaire de Math. Supérieures, Les Presses de L'Univ. de Montréal, Montréal, 1982, 69-93.

5. J. L. Walsh. Interpolation and Approximation by Rational Functions in the Complex Domain. A.M.S. Colloq. Publ. Vol XX, Providence, R.I., 5th ed. 1969.

6. A. Zygmund, Trigonometric Series, Vol. II, University Press, Cambridge, 2nd ed. 1959.

CONVERGENCE AND DIVERGENCE OF MULTIPOINT PADÉ APPROXIMANTS OF MEROMORPHIC FUNCTIONS

Hans Wallin
Department of Mathematics
University of Umeå
S-901 87 UMEÅ
SWEDEN

Abstract. The fact that the Taylor series expansion of an analytic function converges inside the largest disk of analyticity of the function and diverges outside the disk is generalized to interpolation with rational functions where the points of interpolation are chosen in a very general way.

0. Introduction

Let n and ν be non-negative integers, let $\beta_{jn\nu}$, $1 \leq j \leq n+\nu+1$, be given points in the complex plane $\mathbb{C}$, not necessarily distinct, and let f be a function which is analytic at least at the interpolation points. Then there exist polynomials $P_{n\nu}$ and $Q_{n\nu}$, $Q_{n\nu} \not\equiv 0$, of degree at most n and ν, respectively, such that $fQ_{n\nu}-P_{n\nu}$ is zero at $\beta_{jn\nu}$ for all j; if two interpolation points coincide, we require the corresponding zero to have multiplicity two, etc. It is also easy to prove that $R_{n\nu} = P_{n\nu}/Q_{n\nu}$ is uniquely determined by f and $\beta_{jn\nu}$, $1 \leq j \leq n+\nu+1$; $R_{n\nu}$ is the *multipoint Padé approximant of type* (n,ν) *of* f *and* $\{\beta_{jn\nu}\}$, and when $\beta_{jn\nu} \equiv 0$, we get the classic Padé approximant of type (n,ν) of f. When $\nu=0$, $R_{n\nu}$ is a general interpolation polynomial of f which, for $\beta_{jn\nu} \equiv 0$, becomes the Taylor polynomial of f.

We have the basic fact that the Taylor polynomial of f of degree n around zero converges, as $n\to\infty$, inside the largest disk around zero where f is analytic, that f has a singularity on the boundary of that disk and that the Taylor series diverges outside the disk. For the Padé approximants we have an analogous result by de Montessus de Ballore saying that if f is analytic at zero and meromorphic with

ν poles in the open disk $E(\rho)$ around zero with radius ρ, then the Padé approximant $R_{n\nu}$ of type (n,ν) of f converges, as n tends to infinity, to f in $E(\rho)$ except at the poles of f. Furthermore, if $E(\rho')$ is the largest such disk, then f has a singularity on the boundary of $E(\rho')$ and $R_{n\nu}$ diverges in $|z| > \rho'$. We refer to [4], §2 or [1], or to §§3-4 in this paper for the proof of these results; a related divergence result is in [6]. For the multipoint Padé approximant $R_{n\nu}$ we have an analogous convergence result (Theorem 1 in §1 and [10], Theorem 1) essentially going back to Walsh [11] in the polynomial case ($\nu=0$), and to Saff [2] and Warner [13] in the case $\nu > 0$. The purpose of this paper is to prove a corresponding divergence theorem (Theorem 3 in §1) for the multipoint Padé approximant in the case when the interpolation points $\beta_{jn\nu}$ are independent of n; a related but different divergence theorem was proved in [9]. In order to prove this divergence result we need a further fact on the convergence behaviour (Theorem 2 in §1). The proofs of Theorem 2 and 3 given below are generalizations of the corresponding proofs in the Padé case ($\beta_{jn\nu} \equiv 0$) given in [4]. In §1 the results are formulated and some further references are given. After some preparation in §2 the theorems are proved in §3 and §4.

1. Definitions and results

The definition of the multipoint Padé approximant $R_{n\nu} = P_{n\nu}/Q_{n\nu}$ of type (n,ν) of f and $\beta_{jn\nu}$, $1 \leq j \leq n+\nu+1$, may also be stated in the following way by using the auxiliary polynomial

$$\omega_{n\nu}(z) = \prod_{j=1}^{n+\nu+1} (z - \beta_{jn\nu}). \tag{1.1}$$

Determine $P_{n\nu}$ and $Q_{n\nu}$, $Q_{n\nu} \not\equiv 0$, as polynomials of degree at most n and ν, respectively, so that

$$(fQ_{n\nu} - P_{n\nu})/\omega_{n\nu} \text{ is analytic at } \beta_{jn\nu},\ 1 \leq j \leq n+\nu+1. \tag{1.2}$$

We assume throughout that the interpolation points $\beta_{jn\nu}$ all belong to a fixed compact subset E of $\mathbb{C}$ and that f is analytic at the interpolation points. We let $\nu \geq 0$ be fixed and define the *associated measure* $\mu_n=\mu_{n\nu}$ to $\beta_{jn\nu}$, $1 \leq j \leq n+\nu+1$, as the probability measure on E which distributes the point mass $1/(n+\nu+1)$ at each of the points $\beta_{jn\nu}$, $1 \leq j \leq n+\nu+1$. We shall assume that μ_n, $n=1,2,\ldots$, converges in a certain sense to a probability measure μ on E. This convergence is defined by means of the logarithmic potential of μ,

$$u(z;\mu) = \int \log \frac{1}{|z-t|} d\mu(t),$$

and the analogous logarithmic potential of μ_n, $u(z;\mu_n)$, which has the following fundamental relation to $\omega_{n\nu}$

$$u(z;\mu_n) = -\frac{1}{n+\nu+1} \log|\omega_{n\nu}(z)|. \tag{1.3}$$

The convergence we have in mind is stated in the following definition which gives a condition on the asymptotic distribution of the interpolation points.

DEFINITION. $\{\mu_n\}$ is (μ,E)-<u>regular</u> if

$$\liminf_{n\to\infty} u(z;\mu_n) \geq u(z;\mu) \quad \text{for } z \in \mathbb{C} \tag{1.4}$$

and

$$\lim_{n\to\infty} u(z;\mu_n) = u(z;\mu) \qquad \text{for } z \in CE. \tag{1.5}$$

In (1.5), $CE = \mathbb{C} \setminus E$. We note that by a compactness argument, (1.5) implies that

$$u(z;\mu_n) \to u(z;\mu) \text{ uniformly on compact subsets of } CE. \tag{1.5'}$$

We shall need (1.4) in the following equivalent form (for a proof of the equivalence, see [8], Proposition 3):
For every real α and compact $K \subset \mathbb{C}$, there exists a constant $n(\alpha,K)$ so that

$$u(z;\mu) > \alpha \text{ on } K \Rightarrow u(z;\mu_n) > \alpha \text{ on } K \text{ for } n > n(\alpha,K). \tag{1.4'}$$

The concept of (μ,E)-regularity is discussed in §2 and its relevance for the convergence of $R_{n\nu}$ to f is analyzed in [10].

Before stating the theorems we sum up our assumptions: Let $\nu \geq 0$ be fixed and $\mu_n = \mu_{n\nu}$, $n=1,2,\ldots$, the associated measure to $\beta_{jn\nu} \in E$, $1 \leq j \leq n+\nu+1$, where $E \subset \mathbb{C}$ is compact. Assume that μ is a fixed probability measure on E so that $\{\mu_n\}$ is (μ,E)-regular and introduce the open set

$$E(\rho) = \{z\in\mathbb{C}: u(z;\mu) > \log 1/\rho\} \quad \text{for} \quad \rho>0$$

and the set

$$F(\rho) = \{z\in\mathbb{C}: u(z;\mu) \geq \log 1/\rho\} \quad \text{for} \quad \rho>0.$$

Assume that f is a function which for some $\rho>0$ is analytic in an open set containing $\overline{E(\rho)\cup E}$ except at ν poles $z_1,\ldots,z_\nu$ $(\neq\beta_{jn\nu})$, counted with their multiplicities so that $z_j\in E(\rho)$ for all j. Let ρ', $0 < \rho' \leq \infty$, be the supremum of such numbers ρ. Let $R_{n\nu} = P_{n\nu}/Q_{n\nu}$ be the multipoint Padé approximant of type (n,ν) of f and $\{\beta_{jn\nu}\}$ and let $\|g\|_K$ stand for the supremum norm of g on K.

<u>Example</u>. In the Padé case we may take $E = \{0\}$. Then μ_n and μ are the unit mass at 0, $u(z;\mu_n) = u(z;\mu) = \log 1/|z|$, $E(\rho) = \{|z| < \rho\}$ and $F(\rho) = \{|z| \leq \rho\}$.

THEOREM 1. <u>With the assumptions just summed up we have, if</u> K <u>is a compact subset of</u> $F(\rho'')\setminus\{z_j\}_1^\nu$ <u>and</u> $\rho'' < \rho'$,

$$\limsup_{n\to\infty} \| f-R_{n\nu}\|_K^{1/n} \leq \rho''/\rho'. \tag{1.6}$$

<u>Furthermore,</u> $R_{n\nu}$ <u>has exactly</u> ν <u>poles in</u> $\mathbb{C}$ <u>if</u> n <u>is large and these converge to the poles</u> z_j, $1 \leq j \leq \nu$, <u>of</u> f, <u>as</u> $n\to\infty$.

We note that $E(\rho)$ grows with ρ and that $E(\rho')=\cup E(\rho)$, $\rho < \rho'$. $E(\infty)$ is interpreted as $\mathbb{C}$. From the fact that a logarithmic potential is lower semicontinuous and hence assumes a minimum on any compact set, we conclude that any compact subset of $E(\rho')$ is also a subset of $F(\rho'')$ for some $\rho'' < \rho'$. Hence, (1.6) means that $R_{n\nu} \to f$, as $n\to\infty$, uniformly on compact subsets of $E(\rho')\setminus\{z_j\}_1^\nu$.

THEOREM 2. <u>We keep the notation and assumptions from Theorem 1 and assume furthermore that the multipoint Padé approximant</u> $R_{n\nu}=P_{n\nu}/Q_{n\nu}$, <u>is normalized so that</u> $Q_{n\nu}$ <u>has leading coefficient</u> 1. <u>We introduce</u>

$$h_\nu(z) = \prod_{j=1}^{\nu} (z-z_j)$$

<u>where</u> z_j <u>are the poles of</u> f <u>and the non-negative numbers</u> ρ_j, $1 \leq j \leq \nu$, <u>defined by</u> $u(z_j;\mu) = \log 1/\rho_j$, <u>and we assume that</u> ρ_1 <u>is the largest of the numbers</u> ρ_j, $1 \leq j \leq \nu$. <u>Then</u>

$$\limsup_{n\to\infty} \|Q_{n\nu}-h_\nu\|_K^{1/n} \le \rho_1/\rho' \qquad (1.7)$$

<u>for all compact</u> $K\subset\mathbb{C}$.

We note that $\rho_j < \rho'$ for all j since every $z_j \in E(\rho)$ for some $\rho < \rho'$. Hence, the right-hand member of (1.7) is less than 1. We put $\rho_j=0$ if $u(z_j;\mu)=\infty$. For (1.7) in the Padé case, see [4]. If $\rho'=\infty$, Theorem 1 gives convergence of $R_{n\nu}$ to f everywhere in $\mathbb{C}$ except at the poles of f, but if $\rho' < \infty$ we have the following divergence theorem.

THEOREM 3. <u>We keep the notation and assumptions from Theorem 2. Furthermore, we assume that</u> $\rho' < \infty$ <u>and that</u> $\beta_{jn\nu}=\beta_j$ <u>are independent of</u> n. <u>Then</u> $R_{n\nu}=P_{n\nu}/Q_{n\nu}$, $n=1,2,\ldots$, <u>diverges on the set</u>

$$\{z\in\mathbb{C}: u(z;\mu) < \log 1/\rho'\}\setminus E$$

<u>and</u> $h_\nu f$ <u>has at least one singular point on</u> $F(\rho')\setminus E$.

On the set $\{z\in\mathbb{C}: u(z;\mu) < \log 1/\rho'\}\cap E$ we may have either convergence or divergence; see the remarks in §4.3. It is an open question what happens in Theorem 3 if $\beta_{jn\nu}$ are not independent of n, even, as far as the author knows, for the case $\nu=0$.

2. On (μ,E)-regularity

Assuming that $\{\mu_n\}$ is (μ,E)-regular we shall derive two estimates which are needed later. After that we give as a general background two examples of (μ,E)-regularity. These are, however, not needed later on.

Let K be a compact subset of $E(\rho)$, $0 < \rho < \infty$. Then $u(z;\mu)$, being lower semicontinuous, assumes a minimum larger than $\log 1/\rho$ on K and hence, for some $r < \rho$, we get by (1.4') that $u(z;\mu_n) > \log 1/r$ on K for $n > n(r,K)$. From (1.3) we now get the desired estimate: *If* $K\subset E(\rho)$ *is compact then, for some* $r < \rho$,

$$|\omega_{n\nu}(z)| < r^{n+\nu+1} \quad \textit{on } K \textit{ for } n > n(r,K). \qquad (2.1)$$

As a variant we also get that (2.1) *is true for every* $r > \rho$ *if* $K\subset F(\rho)$, K *compact*.

To get an estimate of $\omega_{n\nu}$ in the other direction we assume that

$\Gamma \subset C(E(\rho) \cup E)$, Γ compact. Then, for every $r < \rho$, $u(z;\mu) < \log 1/r$ on Γ and hence, by (1.5'), $u(z;\mu_n) < \log 1/r$ on Γ for $n > n(r,\Gamma)$. Again (1.3) gives the desired estimate: *If* $\Gamma \subset C(E(\rho) \cup E)$ *is compact, then, for every* $r < \rho$,

$$|\omega_{n\nu}(z)| > r^{n+\nu+1} \quad \text{on} \quad \Gamma \quad \text{for} \quad n > n(r,\Gamma). \tag{2.2}$$

As a variant we get that (2.2) *is true for some* $r > \rho$ *if* $\Gamma \subset C(F(\rho) \cup E)$.

Example 1. $\{\mu_n\}$ is (μ,E)-regular if $\mu_n \to \mu$ in the sense that $\int g d\mu_n \to \int g d\mu$ for all continuous functions g on E (see [8], Proposition 1, for a proof). This is the convergence in the naive sense of measures and it is straightforward to prove that $\mu_n \to \mu$ is equivalent to requiring that $\mu_n(B) \to \mu(B)$ for all closed disks B in $\mathbb{C}$ such that $\mu(\partial B) = 0$ where ∂B is the boundary of B.

Example 2. For the general polynomial interpolation $(\nu=0)$ the usual condition on the interpolation points is as follows. Let E have positive logarithmic capacity $\mathrm{cap}E$ and let CE be connected and regular in the sense that it has an ordinary Green's function $G(z)$ with pole at infinity. The condition is that (see [11], §7.2)

$$\lim_{n \to \infty} |\omega_{n0}(z)|^{1/(n+1)} = \mathrm{cap}E \exp G(z) \tag{2.3}$$

uniformly on compact subsets of CE. However, if τ is the equilibrium distribution and $u(z;\tau)$ the equilibrium potential of E and $V(E) = -\log(\mathrm{cap}E)$, then ([5], Theorem III.37) $u(z;\tau) = V(E) - G(z)$. This and (1.3) give that (2.3) holds if and only if $u(z;\mu_n) \to u(z;\tau)$ uniformly on compact subsets of CE. By using special properties of the logarithmic potential it may be proved ([8], Proposition 2), that the last condition implies (1.4) with $\mu = \tau$. Hence, (2.3) is equivalent to the assumption that $\{\mu_n\}$ is (τ,E)-regular. It should be observed, however, that (2.3) cannot in general be described by means of the type of convergence $\mu_n \to \mu$ described in Example 1. Hence, (2.3) and the condition in Example 1 are not comparable.

3. Proof of the convergence results

Theorem 1 was stated and proved in [10], Theorem 1, and in a somewhat weaker form in [7], Theorem 1. Because of that we give in §3.1 only those parts of the proof which are needed in the proof of

Theorem 2 given in §3.2 and §3.3.

3.1. The proof of Theorem 1 proceeds as follows. Take ρ, $0 < \rho < \rho'$, and consider h_ν introduced in Theorem 2. By (1.2) and the assumption it follows that $h_\nu(fQ_{n\nu}-P_{n\nu})/\omega_{n\nu}$ is analytic in an open set D containing $\overline{E(\rho)} \cup E$. By Cauchy's formula

$$\frac{h_\nu(fQ_{n\nu}-P_{n\nu})}{\omega_{n\nu}}(z) = \frac{1}{2\pi i}\int_\Gamma \frac{h_\nu(t)(fQ_{n\nu}-P_{n\nu})(t)}{\omega_{n\nu}(t)(t-z)}\,dt \tag{3.1}$$

if $z\in\overline{E(\rho)}\cup E$ and Γ is a cycle in $D\setminus(\overline{E(\rho)}\cup E)$ with index $\mathrm{ind}_\Gamma(a)=0$ for $a\notin D$ and $=1$ for $a\in\overline{E(\rho)}\cup E$. By deforming Γ into infinity we find that the term in the right-hand member of (3.1) containing $P_{n\nu}$ is zero and then (3.1) gives if $K\subset F(\rho'')$, K compact, $\rho''<\rho<\rho'$, with a constant $c(K,\Gamma,f) < \infty$,

$$\|h_\nu(fQ_{n\nu}-P_{n\nu})\|_K \le c(K,\Gamma,f)\|Q_{n\nu}\|_\Gamma\|\omega_{n\nu}\|_K\|1/\omega_{n\nu}\|_\Gamma \,. \tag{3.2}$$

By a suitable normalization of $R_{n\nu}=P_{n\nu}/Q_{n\nu}$ so that $\|Q_{n\nu}\|_\Gamma = \|h_\nu\|_\Gamma$ we conclude by means of (3.2), (2.1), and (2.2) by letting $\rho\to\rho'$

$$\limsup_{n\to\infty}\|h_\nu(fQ_{n\nu}-P_{n\nu})\|_K^{1/n} \le \rho''/\rho'. \tag{3.3}$$

The proof of Theorem 1 is now essentially completed along the following lines; we refer to [7], Theorem 1, for the details. By using (3.3) and going to a subsequence of $Q_{n\nu}$, $n=0,1,\dots$, one can conclude that $Q_{n\nu}\to h_\nu$ uniformly on compact subsets of $\mathbb{C}$. This fact and (3.3) then finally give (1.6).

3.2. As a preparation for the proof of Theorem 2 we prove the following lemma.

LEMMA. _We use the same notation and make the same assumptions as in Theorem 2. If_ z_j _is a pole of multiplicity_ m_j _of_ f, _then_

$$\limsup_{n\to\infty}|Q_{n\nu}^{(k)}(z_j)|^{1/n} \le \rho_j/\rho' \tag{3.4}$$

for $k=0,1,\dots,m_j-1$, _where_ $Q_{n\nu}^{(k)}$ _denotes the derivative of_ $Q_{n\nu}$ _of order_ k.

Proof. Remembering that $u(z_j;\mu)=\log 1/\rho_j$, $h_\nu(z_j)=0$ and $(h_\nu f)(z_j)\neq 0$, we get (3.4) for $k=0$ by using (3.3) with $\rho''=\rho_j$ and K consisting of the single point z_j. If $m_j \geq 2$, we prove (3.4) for $k=1$ as follows. From (3.1) we get for any $\rho < \rho'$ by differentiation, remembering that the term containing P_n vanishes,

$$\frac{d}{dz}(h_\nu fQ_{n\nu}-h_\nu P_{n\nu})(z) = \frac{1}{2\pi i}\int_\Gamma \frac{\omega_{n\nu}(z)}{\omega_{n\nu}(t)}\,\frac{(h_\nu fQ_{n\nu})(t)}{(t-z)^2}dt +$$

$$+\frac{1}{2\pi i}\int_\Gamma \frac{\omega'_{n\nu}(z)}{\omega_{n\nu}(t)}\,\frac{(h_\nu fQ_{n\nu})(t)}{t-z}\,dt = I + II.$$

The first term $I=I_n(z)$ is estimated as in §3.1 when we started from (3.1) to get the estimate (3.3). Exactly as in the proof of (3.4) for $k=0$ this then gives

$$\limsup_{n\to\infty}|I_n(z_j)|^{1/n} \leq \rho_j/\rho'. \tag{3.5}$$

Essentially $II=II_n(z)$ is handled in the same way with the difference that we now have to estimate $\omega'_{n\nu}(z)$ instead of $\omega_{n\nu}(z)$. To do that we put $z=z_j$; remember that $u(z_j;\mu)=\log 1/\rho_j$ where $\rho_j < \rho'$, and choose ρ and $\varepsilon > 0$ so that $\rho_j+\varepsilon < \rho < \rho'$. Then z_j belongs to the open set $E(\rho_j+\varepsilon)$ which means that we can choose $\delta > 0$, independent of n, so small that $B(z_j,\delta)\subset E(\rho_j+\varepsilon)$ where $B(z_j,\delta) = \{z:|z-z_j| \leq \delta\}$. Then, by (2.1), for $n > n(\varepsilon)$,

$$|\omega'_{n\nu}(z_j)| = \left|\frac{1}{2\pi i}\int_{B(z_j,\delta)} \frac{\omega_{n\nu}(t)}{(t-z_j)^2}dt\right| \leq \frac{(\rho_j+\varepsilon)^{n+\nu+1}}{\delta}.$$

Using this we can now proceed as we did when we proved (3.5) to get the analogous estimate for $II_n(z_j)$. Together these estimates give

$$\limsup_{n\to\infty}\left|\frac{d}{dz}(h_\nu fQ_{n\nu}-h_\nu P_{n\nu})(z_j)\right|^{1/n} \leq \rho_j/\rho'.$$

From this we now get (3.4) for $k=1$ by observing that $(h_\nu f)(z_j)\neq 0$ and $h_\nu(z_j)=h'_\nu(z_j)=0$ and by using that we have already proven (3.4) for $k=0$. If $m_j \geq 3$ we may then prove (3.4) for $k=2$ by differentiating once more and using that (3.4) is now proved for $k=0$ and 1, and so on. This proves the lemma.

3.3. Theorem 2 follows easily from the lemma in §3.2. We put $\Pi_{n\nu}= Q_{n\nu}-h_\nu$. By the assumption $Q_{n\nu}$ has leading coefficient 1 and, by Theorem 1, $Q_{n\nu}$ has degree ν for large n. Hence, for large n,

$\Pi_{n\nu}$ is a polynomial of degree at most $\nu-1$ and by the lemma we have, if z_j is a zero of multiplicity m_j of h_ν,

$$\limsup_{n\to\infty} |\Pi_{n\nu}^{(k)}(z_j)|^{1/n} \le \rho_j/\rho' \tag{3.6}$$

for $k=0,1,\dots,m_j-1$. Since $\Pi_{n\nu}$ has degree at most $\nu-1$ for large n, $\Pi_{n\nu}$ is determined by the ν values $\Pi_{n\nu}^{(k)}(z_j)$, $0 \le k < m_j$, $1 \le j \le \nu$ by means of the Lagrange-Hermite interpolation formula (see for instance [3], §5.2) as a sum of ν fundamental polynomials which are independent of n, times the coefficients $\Pi_{n\nu}^{(k)}(z_j)$. Since these are estimated by (3.6) and ρ_1/ρ' is the largest of the numbers ρ_j/ρ', $1 \le j \le \nu$, we get (1.7) and Theorem 2 is proved.

4. Proof of the divergence result

We use the notation and assumption from Theorem 3. Since $\beta_{jn\nu}=\beta_j$ we now prefer the notation $\omega_{n+\nu+1}$ for $\omega_{n\nu}$ defined by (1.1),

$$\omega_k(z) = \prod_{j=1}^{k} (z-\beta_j) \quad \text{for} \quad k \ge 1, \text{ and } \omega_0(z)=1. \tag{4.1}$$

We note that μ_n is the associated measure to β_j, $1 \le j \le n+\nu+1$. Theorem 3 is proved in §4.2 and as a preparation we collect in §4.1 some facts on Newton series.

4.1. The following proposition and its straightforward proof is given in [12], p. 150 in the case when $\mu_n \to \mu$ as in Example 1, §2.

PROPOSITION. Suppose that $\{\mu\}$ is (μ,E)-regular and that a_j are complex numbers satisfying

$$\limsup_{j\to\infty} |a_j|^{1/j} = 1/\rho < \infty. \tag{4.2}$$

Then $\sum_{j=0}^{\infty} a_j\omega_j(z)$ converges uniformly on compact subsets of $E(\rho)$ and diverges on $\{z\in\mathbb{C} : u(z;\mu) < \log 1/\rho\} \setminus E$.

Proof. By using (4.2) and (2.1) and comparing with a geometric series we get that $\Sigma a_j\omega_j$ converges uniformly on any compact subset of $E(\rho)$. Similarly, if $z\in C(F(\rho)\cup E)$, (4.2) and (2.2) mean that $\limsup|a_j\omega_j(z)| = \infty$ proving the proposition.

Now, let g be a function which is analytic in an open set $D \supset \overline{(E(\rho)\cup E)}$ for some ρ, $0 < \rho < \infty$, and let Γ as in §3.1 be a cycle in $D\setminus(\overline{E(\rho)}\cup E)$ with index 0 in CD and 1 in $\overline{E(\rho)}\cup E$. Then (see for instance [11], pp. 52-54 or [12], Chapter II) the unique polynomial of degree at most n interpolating to g at β_j, $1 \leq j \leq n+1$, has the form $\sum_{j=0}^{n} a_j\omega_j(z)$ where

$$a_j = \frac{1}{2\pi i}\int_{\Gamma} \frac{g(t)}{\omega_{j+1}(t)}dt. \tag{4.3}$$

Furthermore, by Theorem 1, $\sum_{0}^{n} a_j\omega_j \to g$ in $E(\rho)$; $\sum_{0}^{\infty} a_j\omega_j$ is the Newton series of g with respect to $\{\beta_j\}$.

4.2. Let $\Sigma a_j\omega_j$ be the Newton series of $h_\nu^2 f$ and $\Sigma a_{jn}\omega_j$ the Newton series of $h_\nu Q_{n\nu} f$ with respect to $\{\beta_j\}$. Since $h_\nu P_{n\nu}$ has degree at most $n+\nu$ and interpolates to $h_\nu Q_{n\nu} f$ at β_j, $1 \leq j \leq n+\nu+1$, by the definition of the multipoint Padé approximant we conclude (see §4.1) that

$$h_\nu P_{n\nu} = \sum_{j=0}^{n+\nu} a_{jn}\omega_j. \tag{4.4}$$

The idea of the proof of Theorem 3 is as follows: We shall see that $a_{jn} \to a_j$ fast since, by Theorem 2, $Q_{n\nu} \to h_\nu$ fast. It turns out that this means, since we can determine the convergence and divergence behaviour of $\Sigma a_j\omega_j$, that we can also determine the divergence behaviour of (4.4), as $n \to \infty$, and hence of $R_{n\nu}$. We now turn to the details and start by proving that

$$\limsup_{j\to\infty} |a_j|^{1/j} = 1/\rho'. \tag{4.5}$$

To do that we note that by (4.3)

$$a_j = \frac{1}{2\pi i}\int_{\Gamma} \frac{(h_\nu^2 f)(t)}{\omega_{j+1}(t)} dt,$$

and hence by (2.2) we find that, for $\rho < \rho'$, $\limsup |a_j|^{1/j} \leq 1/\rho$, and consequently the left hand member of (4.5) is at most equal to the right hand member of (4.5). Suppose that we have strict inequality, i.e. that (4.5) is not valid. Then, by the proposition in §4.1, $\Sigma a_j\omega_j$ converges in $E(r)$ for some $r > \rho'$ uniformly on compact subsets, and since, by Theorem 1, $\Sigma a_j\omega_j = h_\nu^2 f$ in $E(\rho')$, this gives an analytic continuation of $h_\nu^2 f$ to $E(r)$; we denote also this continuation by $h_\nu^2 f$. We claim that $h_\nu^2 f$ is analytic on $\overline{E(\rho)}\cup E$ for all $\rho < r$. In fact, for $\rho < r$, $h_\nu^2 f$ is analytic in E, in $\overline{E(\rho)}\cap E \subset E$, and in

$\overline{E(\rho)}\cap(CE)\subset E(r)$ since $u(z;\mu)$ is continuous in $\overline{E(\rho)}\cap(CE)$. Hence, $h_\nu^2 f$ is analytic in $\overline{E(\rho)}\cup E$ for all $\rho < r$; in particular, also for $\rho' < \rho < r$. This gives a contradiction to the definition of ρ' which proves (4.5).

Next we fix a point z in the set $\{u(z;\mu) < \log 1/\rho'\}\setminus E$ and we introduce r, $r > \rho'$, by $u(z;\mu) = \log 1/r$. We shall prove that $R_{n\nu}(z)$, $n=0,1,\ldots$, diverges. Since $Q_{n\nu}(z)\to h_\nu(z)$ it is enough to prove that $\{P_{n\nu}(z)\}$ diverges. By (4.4)

$$h_\nu P_{n\nu} = \sum_{j=0}^{n+\nu} a_{jn}\omega_j = \sum_0^{n+\nu} a_j\omega_j + \sum_0^{n+\nu} (a_{jn}-a_j)\omega_j$$

which gives

$$h_\nu(P_{n\nu}-P_{n-1,\nu})(z) = a_{n+\nu}\omega_{n+\nu}(z) + \sum_{j=0}^{n+\nu} (a_{jn}-a_j)\omega_j(z) - \\ - \sum_{j=0}^{n+\nu-1} (a_{j,n-1}-a_j)\omega_j(z) = I+II+III. \tag{4.6}$$

In the estimate of I, II and III below we let ε denote different, arbitrarily small positive constants and c different positive constants, independent of n. Then, by (4.5) and (2.2) we get for certain arbitrarily large n,

$$|I| = |a_{n+\nu}\omega_{n+\nu}(z)| > c\left(\frac{r(1-\varepsilon)}{\rho'}\right)^n.$$

In order to estimate II we first note that by (4.3) used with ρ close to ρ', Theorem 2 and (2.2)

$$|a_{jn}-a_j| = \left|\frac{1}{2\pi i}\int_\Gamma \frac{h_\nu f(Q_{n\nu}-h_\nu)(t)}{\omega_{j+1}(t)}\,dt\right| \le c\,\|Q_{n\nu}-h_\nu\|_\Gamma\;\|1/\omega_{j+1}\|_\Gamma \le$$

$$\le c\left(\frac{\rho_1(1+\varepsilon)}{\rho'}\right)^n\left(\frac{1+\varepsilon}{\rho'}\right)^j.$$

This and (2.1) give

$$|II| \le c\left(\frac{\rho_1(1+\varepsilon)}{\rho'}\right)^n \sum_{j=0}^{n+\nu}\left(\frac{r(1+\varepsilon)}{\rho'}\right)^j \le c\left(\frac{\rho_1 r(1+\varepsilon)}{\rho'\rho'}\right)^n$$

and we get the same estimate for III. From these estimates and (4.6) we get for certain arbitrarily large n

$$|h_\nu(P_{n\nu}-P_{n-1,\nu})(z)| \geq c\left(\frac{r(1-\varepsilon)}{\rho'}\right)^n \left|1 + O\left(\frac{\rho_1(1+\varepsilon)}{\rho'(1-\varepsilon)}\right)^n\right|$$

and since $r > \rho' > \rho_1$ this tends to infinity with n if ε is small. This proves that $\{P_{n\nu}(z)\}$ and hence $\{R_{n\nu}(z)\}$ diverges as desired.

It remains to prove that $h_\nu f$ has a singularity on $F(\rho')\setminus E$. We assume that this is false, i.e. that $h_\nu f$ is analytic on an open set $D\supset(F(\rho')\cup E)$. But on the closed set CD the function $u(z;\mu)$ is continuous and has a largest value which is less than $\log 1/\rho'$. This means that on CD we have $u(z;\mu) < \log 1/r$ for some $r > \rho'$. For this r, $E(r)\subset D$, i.e. $h_\nu f$ is analytic on $E(r)$. For $z\in\partial(E(r))\setminus E$, $u(z;\mu)=\log 1/r$, i.e. $z\in D$ which means that $h_\nu f$ is analytic at z. Also, $h_\nu f$ is analytic on $\partial(E(r))\cap E\subset E$. Taken together, this means that $h_\nu f$ is analytic on $\overline{E(r)}\cup E$ for some $r > \rho'$ contradicting the definition of ρ'. The proof of Theorem 3 is complete.

4.3. On the set $\{u(z;\mu) < \log 1/\rho'\}\cap E$ our theorems do not give any information. If a point β from that set belongs to the set of interpolation points $\beta_{jn\nu}$, $1 \leq j \leq n+\nu+1$, for all n, then $R_{n\nu}(\beta)\to f(\beta)$ with the assumptions in Theorem 2; in fact since $(fQ_{n\nu})(\beta)=P_{n\nu}(\beta)$ by definition and $Q_{n\nu}(\beta)\to h_\nu(\beta)\neq 0$ we have $R_{n\nu}(\beta)=f(\beta)$ for large n. On the other hand, with the assumptions in Theorem 3 let us consider the case $\nu=0$. By arguing as in the proof of the proposition in §4.1 we see that the Newton series of f diverges at those points $z\in\{u(z;\mu) < \log 1/\rho'\}\cap E$ where $\limsup u(z;\mu_n) \leq u(z;\mu)$; this last condition is equivalent to $\limsup u(z;\mu_n)=u(z;\mu)$. It is straightforward to prove that this condition holds if $\mu_n\to\mu$ in the sense of Example 1 in §2 and

$$\limsup_{n\to\infty} \int_0^\delta \mu_n(B(z;r))\frac{dr}{r} \to 0 \quad \text{as } \delta \to 0 \tag{4.7}$$

where $B(z;r)$ is the closed disk with center z and radius r.

Example. Chose $E=[-1,1]$ and β_j, $j=1,2,\ldots$, so that β_j, $1 \leq j \leq$ $\leq n+1$, are equidistributed on E for $n=2^m$, $m=0,1,\ldots$. Then a straightforward estimate shows that (4.7) holds on E except on a subset having 1-dimensional Lebesgue measure zero; in fact (4.7) holds on E except on a set having Hausdorff dimension zero.

Postscript. The paper by P.R. Graves-Morris and E.B. Saff presented at the Conference and in these Proceedings overlaps with this paper as concerns content and methods.

References

1. Gončar, A.A., Poles of rows of the Padé table and meromorphic continuation of functions, Mat. Sb. 115 (157), 1981; English transl. in Math. USSR Sb. 43, 1982, 527-46.

2. Saff, E.B., An extension of Montessus de Ballore's theorem on the convergence of interpolating rational functions, J. Approx. Th. 6, 1972, 63-67.

3. Schönhage, A., Approximationstheorie, Walter de Gruyter & Co., Berlin, 1971.

4. Stahl, H., Beiträge zum Problem der Konvergenz von Padéapproximierenden, Thesis, Technische Universität Berlin, 1976.

5. Tsuji, M., Potential theory in modern function theory, Maruzen, Tokyo, 1959.

6. Vavilov, V.V., On the convergence of the Padé approximants of meromorphic functions, Mat. Sb. 101 (143), 1976; English transl. in Math. USSR Sb. 30, 1976, 39-49.

7. Wallin, H., Potential theory and approximation of analytic functions by rational interpolation, in Proc. Coll. Complex Anal., Joensuu, Finland, Lecture Notes in Math., 747, Springer-Verlag, Berlin, 1979, 434-50.

8. Wallin, H., Rational interpolation to meromorphic functions, in Proc. Conf. Padé approximation and its applications, Amsterdam, Lecture Notes in Math. 888, Springer-Verlag, Berlin, 1981, 371-83.

9. Wallin, H., Divergence of multipoint Padé approximation, in Proc. Conf. Complex Analysis - Fifth Romanian - Finnish Seminar, Part 2, Bucharest. Lecture Notes in Math. 1014, Springer-Verlag, Berlin, 1983, 246-55.

10. Wallin, H., Multipoint Padé approximation of meromorphic functions for general interpolation schemes, Department of Math., Univ. of Umeå, No. 1, 1984.

11. Walsh, J.L., Interpolation and approximation by rational functions in the complex domain, Colloq. Publ. XX, Amer. Math. Soc., Providence, 1969.

12. Warner, D.D., Hermite interpolation with rational functions, Thesis, Univ. of California, San Diego, 1974.

13. Warner, D.D., An extension of Saff's theorem on the convergence of interpolating rational functions, J. Approx. Th. 18, 1976, 108-18.

APPROXIMATE ANALYTIC CONTINUATION BEYOND THE FIRST RIEMANN SHEET*

George A. Baker, Jr.
Theoretical Division
Los Alamos National Laboratory
University of California
Los Alamos, NM 87545

Abstract. A brief review of the classical theory of multivalued functions of a complex variable is used to introduce the classification of monogenic analytic functions by their monodromic dimension. Riemann's monodromy theorem is used to set the stage for a class of Hermite-Padé multiform approximants. The approximation problem for functions meromorphic on a Riemann surface of m-sheets with a finite number of singular points is completely solved. A general uniqueness of convergence theorem and another convergence theorem are given.

The subject of this paper is a possible method for the approximation of multiform functions of a complex variable. The motivation for this study lies in the applications, where one frequently wants to evaluate a function along a path between two closely spaced branch points, or on a branch cut, or even in a limit as a branch point is approached. Since there are a variety of notations currently in use for this problem, I will begin by introducing the notation [1] that I use. First I start by defining a functional element

$$f_{z_0}(z) = \sum_{n=0}^{\infty} f_n(z_0)(z-z_0)^n \quad . \tag{1}$$

This Taylor series expansion about point z_0 is intended to converge in a disk of nonzero radius of convergence and defines a unique, single-valued, regular function of the variable z in this disk. We know from

* Work performed under auspices of the US DOE.

the standard theory of analytic continuation that at any point in this disk, say z_1, we may re-expand that functional element and create a new functional element. It may be that the disk of convergence of the new functional element includes points which were not in the disk of convergence of the original functional element. If we repeat this proccess in all possible ways we will get a function defined over some Riemann surface. In fact only a denumerable set of steps is required for this process as has been proven [1]. Therefore there may only be at most a denumerable number of functional elements which are required to complete the process of analytic continuation. Therefore at every point in the z plane which is regular after continuation along any path from the initial functional element, there are only a finite or at most denumerably infinite number of different coverings. Let these coverings be

$$f_1(z;z_0),\ f_2(z;z_0) \quad , \ldots \quad . \tag{2}$$

The various f's are regular in the neighborhood of z_0. All the pairs

$$(z_0,w_0^{(1)}),\ (z_0,w_0^{(2)}) \quad , \ldots \quad , \tag{3}$$

constitute the *monogenic analytic function* generated by the original functional element. We define in equation (3), $w_0^{(n)} = f_n(z_0;z_0)$. Fundamental to this discussion of the theory of analytic continuation is the following classical theorem.

Monodromy theorem: If f(z) is regular in a simply connected region, G, then f(z) is uniform (single valued) there.

If we consider a region which is not simply connected, then the situation is entirely different. As can be seen by the following example:

$$w = z^{1/m}, \quad f_2(z) = e^{2\pi i/m} f_1(z) \quad , \tag{4}$$

where m is a positive integer and the region is the punctured complex plane. Here if one encircles the branch point at $z = 0$, one obtains f_2 as given in equation (4) from f_1. Continuing in this manner we find precisely m coverings, but only one linearly independent covering of the punctured complex plane. As a second example we consider

$$w = \ln z, \quad f_n(z) = f_1(z) + 2\pi i(n-1) \quad , \tag{5}$$

where n is any integer. In this case there are an infinite number of coverings of the complex plane, but only two linearly independent ones. From this point of view, it makes sense to classify all the monogenic analytic functions by the number of linearly independent coverings. We call this number the *monodromic dimension* of the monogenic analytic function.

Following further this line of reasoning I now remind you of Riemann's monodromy theorem [2]. First let us consider a monogenic analytic function of monodromic dimension m with exactly n singular points. In the complete, that is to say including the point at infinity, complex plane, we can, at some regular point z_0, from each of the m coverings define m functional elements, and hence m monogenic analytic functions, $y_1, \ldots, y_m$ with the property that if we encircle, for example, the *ith* of the n different branch points, we find that y_j changes into

$$\sum_{k=1}^{m} M_{jk}^{(i)} y_k \quad . \tag{6}$$

One can prove by the application of Cauchy's theorem that

$$M^{(1)} \cdots M^{(n)} = I \tag{7}$$

holds. It follows from equation (7) that for each of the n matrices there exists an inverse matrix. The matrices $M^{(i)}$ generate the Monodromy Group M of the $\vec{y}$ system. Using this concept let us now define a class

$$Q\begin{pmatrix} a_1, & \ldots & a_n \\ M^{(1)}, & \ldots, & M^{(n)} \end{pmatrix}; z \Bigg) \tag{8}$$

of all the M-systems $\vec{y}$ with these properties plus the additional property that there are no singularities of infinite order. That is to say, there exist an A and r, both finite, such that, as $z \to a_i$

$$\left|\frac{y'_k}{y_k}\right| \leq \frac{A}{|z-a_i|^r} \tag{9}$$

holds for all i and k.

<u>Theorem (Riemann)</u> For any m + 1 systems of functions $\vec{y}_j$, j=1, ..., m + 1 belonging to the same class Q, there exists a linear homogeneous relation *with polynomial coefficients* in z, the independent variable, such that

$$\sum_{j=1}^{m+1} A_j(z)\,\vec{y}_j(z) = 0 \quad . \tag{10}$$

Corollary. Note that if $\vec{y}$ belongs to Q then so also do $\vec{y}'$, $\vec{y}''$, ..., $\vec{y}^{(m)}$. This result can be shown simply by differentiating the Monodromy Group equations (6). Therefore

$$\sum_{j=0}^{m} A_j(z)y_k^{(j)}(z) = 0, \quad k = 1, 2, \ldots, m \quad . \tag{11}$$

Remark: Nuttall [3] has studied Hermite-Padé approximation to functions meromorphic on a Riemann surface of m sheets with a finite number of singularities. Clearly the relevant monogenic analytic function has at most m linearly independent coverings and so this problem admits an exact finite order solution by the above corollary.

Since the convergence question is avoided in the above class of functions, we will try to look at a wider class of functions. From the view point of applications there are two main and one minor restrictions to the above results

1. only a finite number of singular points are considered
2. only finite monodromic dimension is considered
3. only finite order singularities

Here will we attempt to think about the first restriction which is important from the point of view of applications. The second restriction is probably also important but will not be treated here. And so far as I know the third restriction is of only minor importance in applications. An example of the problem of the second restriction is

$$\frac{1}{1 + \log z} \quad . \tag{12}$$

In this case not only are there an infinite number of Riemann sheets but there are also an infinite number of linearly independent coverings even though this is just a simple function. The coverings are linearly independent and are given as

$$f_m = \frac{1}{1 + \log z + 2\pi i(m-1)} \quad , \tag{13}$$

where m is any integer.

Let us consider the function classes defined by: f(z) is contained in class m if and only if

$$\sum_{j=0}^{m} E_j(z)\, f^{(m-j)}(z) + E_{m+1}(z) = 0 \qquad (14)$$

where the E_j are entire functions, and $E_0 \neq 0$. Class 0 consists of the class of meromorphic functions for, dividing equation (14), we find $f = \frac{E_1}{E_0}$ which is of course meromorphic. Further by the Weierstrass theorem [1] on constructibility of an entire function with prescribed zeros, any meromorphic function (we allow no limit point of poles in the finite plane in our definition) can be so written. Class $m > 0$ consists of functions with m linearly independent coverings of the complex plane plus an analytic background function. If E_{m+1} is identically 0 then there is no analytic background.

Now in many ways an entire function is like a high order polynomial. We wish to construct the particular case of Hermite-Padé approximants which has been called integral approximants, and are defined by the following procedure: First define polynomials by

$$P(z) \frac{d^m y}{dz^m} + Q(z) \frac{d^{m-1} y}{dz^{m-1}} + \ldots + S(z)y + T(z) = O(z^{p+\ldots+t+m}) \qquad (15)$$

where P, Q ... are polynomials of degrees p, q, ... with the normalization $P(0) = 1.0$. Then define the intergal approximants as the solution, or integral curve $w(x)$, of the equation

$$P(z) \frac{d^m w}{dz^m} + Q(z) \frac{d^{m-1} w}{dz^{m-1}} + \ldots + S(z)w + T(z) = 0 \qquad (16)$$

subject to the initial conditions $w(0) = y(0), \ldots, w^{(m-1)}(0) = y^{(m-1)}(0)$. We denote these approximants by $[t/s; \ldots; q; p]$.

From the theory of differential equations, this solution will be the monogenic analytic function defined by an initial functional element and have at most m linearly independent coverings of the complex plane. One gets the other coverings simply by integrating around the singular points of the differential equation. Clearly since the equation is of mth order there can be at most m independent solutions of the differential equation, and hence at most m linearly independent coverings of the complex plane. One can see that approximations of the above form are capable of approximating functions of class m with arbitrary accuracy. This can be seen, for example, by using for the polynomials, instead of those determined by equation (15), simply the Taylor series truncations of the entire functions E_j which define class m.

<u>Invariance property</u>. If the degrees of the polynomials are as given by the superscripts in

$$P^{(N+2m)}(z)\frac{d^m y}{dz^m} + Q^{(N+2m-1)}(z)\frac{d^{m-1}y}{dz^{m-1}} + \ldots$$

$$\ldots + S^{(N+m+1)}(z)\frac{dy}{dz} + T^{(N)}(z) = O(z^{(m+2)(N+1)+m(m+1)}) \quad , \quad (17)$$

where in addition Q is subject to 1 linear restriction, ..., and S has m-1 linear restrictions, e.g. the highest order coefficients of Q and P in (17) are related by

$$q_{N+2m-1} = m(m-1)p_{N+2m} \cdots \quad , \quad (18)$$

then the above approximation procedure is invariant under the linear fractional transformation

$$u = \frac{az}{1+bz} \quad , \quad (19)$$

as has been previously discussed by Baker [4]. A number of the algebraic and other properties of these approximants have been studied by Hermite [5], Padé [6], Della Dora and Di-Crescenzo [7], Burley et al. [8].

I will now give a couple of theorems about the convergence properties of these approximants.

<u>Theorem (Uniqueness of Convergence)</u>. If there exists a sequence

$$Z_k = [t_k/s_k; \ldots; q_k; p_k] \quad , \quad (20)$$

where $t_k + s_k + \ldots + q_k + p_k \to \infty$ with k such that, in a region $\mathcal{R}$, possibly multiply connected and containing the origin as an interior point, and

$$\left|\frac{Q}{P}\right|, \ldots, \left|\frac{S}{P}\right|, \left|\frac{T}{P}\right| \quad (21)$$

are uniformly bounded in $\mathcal{R}$ and k, then that sequence converges to the monogenic analytic function defined by the original Taylor series in the interior of $\mathcal{R}$.

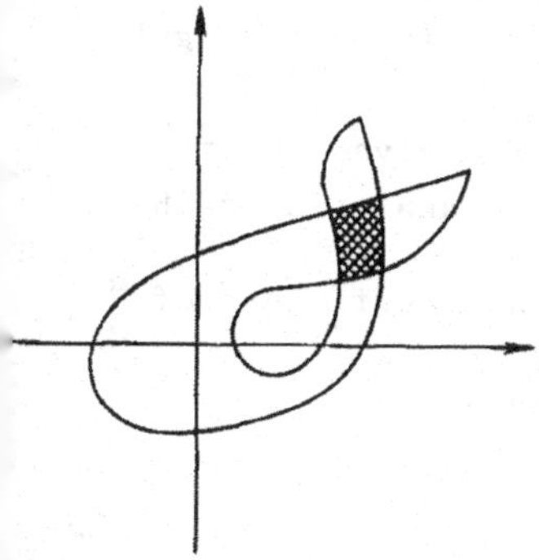

Fig. 1 Illustration of an allowed region R for the Theorem on Uniqueness of Convergence

Sketch of Proof: By standard estimation theorems for the theory of differential equations [9], the boundedness conditions of the theorem imply that the approximant is bounded throughout R. Since Q, P, ..., S, T are polynomials, the approximant is locally analytic in R and therefore analytic (but not necessarily uniform) throughout R. (See Fig. 1 where the shaded region could represent two distinct coverings of the complex plane.)

Since the origin is an interior point of R, there exists a disk about the origin in which all Z_k are regular. In a disk ½ that diameter about the origin, is easy to show, using the uniform boundedness and the fact that as there exists k_0, such that the Taylor series of each Z_k, $k > k_0$ agrees with that of the original Taylor series to any required degree of accuracy, the $Z_k \to y(z)$ in that disk. Thus we conclude $\lim_{k\to\infty} Z_k = y(z)$ in the disk and so by Vitali's (or even Stieltjes') convergence continuation theorem we can conclude

$$\lim_{k\to\infty} Z_k = y(z) \tag{22}$$

in the interior of R.

Theorem (pointwise convergence). Let us be given f(z) which satisfies

$$P_m(z)\frac{d^m}{dz^m} f + P_{m-1}(z)\frac{d^{m-1}}{dz^{m-1}} f + \ldots + P_1(z)\frac{df}{dz} + P_0(z) f + \phi(z) = 0 \quad , \tag{23}$$

where the P_j are polynomials of degree ν_j and $\phi(z)$ is an entire function and let the equation be irreducible. Then,

$$\lim_{L\to\infty} [L/\nu_0; \nu_1; \ldots; \nu_{m-1}; \nu_m] = f(z) \tag{24}$$

in any compact subset of the complex plane not containing a singular point of $f(z)$.

Sketch of Proof: The proof depends on a detailed analysis of the exact structure of the coefficients and a detailed analysis of the explicit solution for the approximant. If Q_j is the coefficient of $\frac{d^j f}{dz^j}$ in the defining equation for $[L/\nu_0; \ldots ; \nu_m]$, then

$$s = \sum_{i=0}^{m} (\nu_i+1) - 1 + L \quad , \tag{25}$$

$$Q_j(z) = \det \left|\begin{array}{ccc|c|ccc|c} f_{L+1} & f_L \;\ldots & f_{L+1-\nu_0} & & f^{(j)}_{L+1} & \ldots & f^{(j)}_{L+1-\nu_j} & \\ & & & \ldots & & & & \ldots \\ f_s & f_{s-1} \ldots & f_{s-\nu_0} & & f^{(j)}_{s} & \ldots & f^{(j)}_{s-\nu_j} & \\ 0 & 0 & 0 & 0\ldots 0 & 1, & z, \;\ldots, & z^{\nu_j} & 0\ldots 0 \end{array}\right| \tag{26}$$

If we use elementary column operations and the equation for $f(z)$ we can reduce Q_j to a form where the x^{ν_j} column is replaced by

$$p_{\nu_j}^{-1} \begin{pmatrix} \phi_{L+1-\nu_j} \\ \vdots \\ \phi_{s-\nu_j} \\ P_j(z) \end{pmatrix} \tag{27}$$

Here p_{ν_j} is the coefficient of z^{ν_j} in $P_j(z)$

Since, from the equation for $f(z)$, we know that there are only ν_m singular points which are located at the zeros, z_j, of $P_m(z)$ we can write

$$f_n = \sum_{j=1}\sum_{k} \binom{-\gamma_j+k}{n} A_{jk}\, z_j^{-n} + g_n \quad , \tag{28}$$

where the g_n are a contribution from an entire function and γ_j are the exponents at the various singularities. Now as the equation for f is irreducible, the sub-determinants can't cancel out completely the singular terms and therefore the worst divisor we have to deal with is of the form

$$\frac{1}{n^{\text{fixed power}}}\frac{1}{R^n} \tag{29}$$

where R is the largest absolute value of the zeros of $P_m(z)$. Since ϕ is entire, ϕ_n is arbitrarily small ($n \to \infty$) compared to equation (29) so the error term goes to 0. (The details are like the proof of Wilson's Theorem as given in Baker [10].)

Since the zeros of the polynomial appear on all Riemann sheets, in cases where the monodromic dimension of the function exceeds that of the integral approximant, there will surely be cuts that will complicate the procedure. The complete solution to the location problem is not yet known, but Nuttall [3,11] has solved it in the case of meromorphic functions on a finite-sheeted, Riemann surface with a finite number of branch points for equations one degree less than would give an exact solution.

References

1. K. Knopp, *Theory of Functions, parts I and II*, translated by F. Bagemihl, Dover Publications, New York, 1945.

2. B. Riemann, (1857) *Collected Works of Bernhard Riemann*, H. Weber, ed., pp. 379-390, Dover Pub. Inc., New York, 1953, and in English, G. V. Chudnovsky, in *Bifurcation Phenomena in Mathematical Physics and Related Topics*, C. Bardos and D. Bessis, eds. pp. 449-510 D. Reidel Publishing Co. Boston, 1980.

3. J. Nuttall, *Hermite-Padé Approximants to Functions Meromorphic on a Riemann Surface*, J. Approx. Theory 32 (1981) 233-240.

4. G. A. Baker, Jr., *Invariance Properties in Hermite-Padé Approximation Theory*, to be published, J. Comp. Appl. Maths.

5. C. Hermite, *Sur la généralisation des fractions continues algebriques*, Ann. Math. Sér. 2, 21 (1893) 289-308.

6. H. Padé, *Sur la généralisation des fractions continues algebriques*, J. Math. Ser. 4, 10 (1894) 291-329.

7. J. Della Dora and C. Di-Crescenzo, *Approximation de Padé-Hermite* in *Padé Approximation and its Applications*, L. Wuytack, ed., *Lecture Notes in Mathematics 765*, A. Dold and B. Eckmann, eds., pp. 88-115, Springer-Verlag, New York, 1979.

8. S. K. Burley, S. O. John and J. Nuttall, *Vector Orthogonal Polynomials*, SIAM J. Numer. Anal. 18 (1981), 919-924.

9. See, for example, E. Kamke, *Differentialgleichungen Lösungsmethoden und Lösungen*, Vol. 1, Akad. Verlagsgesellschaft, Leipzig (1951).

10. G. A. Baker, Jr., *Essentials of Padé Approximants*, Academic Press, New York, 1975.

11. J. Nuttall, *Asymptotics of Diagonal Hermite-Padé Polynomials*, Univ. of Western Ontario Preprint (1983).

CRITICAL EXPONENTS FOR THE GENERAL SPIN ISING MODEL USING THE RATIONAL APPROXIMATION METHOD

J. L. Gammel
Saint Louis University
St. Louis, Missouri 63103, U.S.A.

J. Nuttall*
University of Western Ontario
London, Ontario, Canada N6A 3K7

D. C. Power
McDonnell Aircraft Company
McDonnell Douglas Corporation
St. Louis, Missouri 63166, U.S.A.

Abstract. The rational approximation method is used to investigate the problem of calculating the critical indices of the Ising model. The method is applied to several test cases thought to be similar to the Ising model and then is applied to the susceptibility of the general spin Ising model on the body centered cubic lattice. It is found that for one particular spin (S = 0.73) convergence is rapid and smooth, allowing the estimate $\gamma = 1.2411 \pm 0.0002$ in agreement with the renormalization group and with recent analyses of the bcc series. If the existence or value of this 'critical' spin is not correct, or is not accepted, it is still possible to estimate $\gamma = 1.241 \pm 0.001$ if the universality of the subdominant index θ is assumed. Similar analysis of the series for the second moment correlation function M_2, leads to the estimate $\nu = 0.6335 \pm 0.0003$ (for S = 0.73). The implications of these results on the scaling and hyperscaling hypotheses are discussed.

* Supported in part by Natural Science and Engineering Research Council Canada.

1. Introduction

In a recent publication [2], we used a rational approximation method to analyze the low temperature Ising model series for the spontaneous magnetization. In this paper, we apply the method to the high temperature series for the susceptibility and other quantities.

If a function f(x) behaves like $(x_c-x)^q$ near a critical point x_c and this behaviour is manifested in the Maclaurin series of f(x), then it is familiar that the critical point and index may be estimated by Baker's "D-log" Padé method [3]. We form

$$(1) \qquad \ell(x) := \frac{f'(x)}{f(x)} = \ell_0 + \ell_1 x + \ell_2 x^2 + \dots$$

and construct Padé approximants to the series in (1). The values of the critical point and index are found from the appropriate pole of the Padé approximant and its residue:

$$\ell(x) \approx \frac{q}{x-x_c} \quad \text{for} \quad x \approx x_c \; .$$

To begin our analysis, we assume that the thermodynamic function $\ell(x)$ in (1) has analytic properties which permit an integral representation based on an application of Cauchy's theorem in the $t=x^{-1}$ plane, of the form

$$G(t) := x\ell(x) = \frac{qt_c}{t-t_c} + \frac{1}{2\pi i}\int_\Gamma \frac{\Delta G(t')}{t'-t}\,dt'$$

where Γ denotes the position of the cuts of G(t) and $\Delta G(t)$ equals the discontinuity of G(t) across its cuts. The antiferromagnetic Curie point has an associated cut $[-t_c,0]$ which is thought to give a significant contribution to the integrand, and there may be other contributions from [0,a] where $0<a<t_c$. We take account of such contributions by using an orthogonal polynomial $p_N(t)$ based on the interval $[-t_c,a]$. If f(x) possesses confluent singularities at x_c (see equn.(15)), then the cut on $[a,t_c]$ must also be included. A principal aim of this paper is an evaluation of the subdominant singularity at $x=x_c$, [14]. The method we use is a Padé-type approximation method [1,7]. We use the polynomial $P_N(x) := x^N p_N(^1/x)$ and define

$$\ell^{(N)}(x) := P_N(x)\,\frac{f'(x)}{f(x)} = t^{1-N}\,p_N(t)\;G(t).$$

We analyze the analytic behaviour of $\ell^{(N)}(x)$ from its representation

$$(2) \quad \ell^{(N)}(x) = \frac{q t_c^{2-N} p_N(t_c)}{t_c - t} + \frac{1}{2\pi i}\int_{-t_c}^{a} \frac{p_N(t')\ \Delta G(t')}{(t'-t)\ t'^{N-1}}\,dt'$$

$$+ \frac{1}{2\pi i}\int_{a}^{t_c} \frac{p_n(t')\ \Delta G(t')}{(t'-t)\ t'^{N-1}}\,dt' .$$

We choose $p_N(t)$ to be a Chebychev polynomial shifted to the interval $[-t_c, a]$, so that its characteristic mini-max property ensures that the contribution of the first integral on the right-hand side of (2) is very small for large N. We may regard our method as one in which the poles of the approximants of $\ell(x)$ are placed on the anti-ferromagnetic cut. The multipole at the origin ($t'=0$) contributes a polynomial $\pi_{N-1}(t^{-1})$ of degree N-1 at most in t^{-1} to (2). For large N, the dominant contribution to the integrand of the second integral comes from near $t'=t_c$. If f(x) has confluent singularities of the form

$$f(x) \simeq (x_c - x)^q + B(x_c - x)^{q+\theta} \quad \text{for} \quad x \simeq x_c ,$$

for some $\theta>0$, then for $t \simeq t_c$ and $t \le t_c$, $\Delta G(t) \simeq A(t_c - t)^{\theta}$. For large N and $t \simeq t_c$, we have the estimate

$$p_N(t) \simeq p_N(t_c)\ e^{N\phi'(t_c)(t-t_c)}$$

where

$$(3) \quad \phi'(t_c)\ t_c = \frac{1}{\sqrt{2}\sqrt{1-a/t_c}} .$$

From (2), we may now deduce that

$$(4) \quad \ell^{(N)}(x) \simeq \frac{q t_c^{2-N} p_N(t_c)}{t_c - t} + A p_N(t_c)\int_{a}^{t_c} \frac{e^{N\phi'(t_c)(t'-t_c)}(t_c - t')^{\theta}}{(t'-t)\ t'^{N-1}}\,dt'$$

$$+ \pi_{N-1}(t^{-1}).$$

Let $\ell^{(N)}(x)$ have the Maclaurin expansion

$$\ell^{(N)}(x) = \ell_0^{(N)} + \ell_1^{(N)} x + \ell_2^{(N)} x^2 + \dots .$$

Then, for $k \ge N$, we deduce from (4) that the Maclaurin coefficients of $\ell^{(N)}(x)$ are given by

$$\ell_k^{(N)} \simeq -q p_N(t_c) t_c^{k-N+1} - A p_N(t_c)\int_{a}^{t_c} e^{N\phi'(t_c)(t'-t_c)} t'^{k-N} (t_c - t')^{\theta} dt'.$$

Using a standard asymptotic analysis, [6], one finds that

$$(5) \quad \ell_k^{(N)} \simeq p_N(t_c)\ t_c^{k-N+1}\left[-q + \frac{C}{(k-N+N\phi'(t_c)t_c)^{\theta}}\right] ,$$

which is valid for large N, $k \geq N$, and where C is a constant independent of N.

Approximations $x_{ck}^{(N)}$ to x_c and $q_k^{(N)}$ to q are defined as follows:

$$(6)\quad x_{ck}^{(N)} = \ell_k^{(N)} / \ell_{k+1}^{(N)},$$

$$(7)\quad q_k^{(N)} = -\ell_k^{(N)} x_c^{k+1} / p_N(x_c) \ .$$

The "diagonal" approximants, that is with k-N fixed, provide approximations which approach x_c and q respectively, with correction terms which may be shown from (5) to be

$$(8)\quad x_{ck}^{(N)} \simeq x_c + \frac{D}{(N\phi'(t_c)t_c)^{\theta+1}} ,$$

$$(9)\quad q_k^{(N)} \simeq q + \frac{E}{(N\phi'(t_c)t_c)^{\theta}} ,$$

where D and E are constants independent of a. When θ is small, the $q_k^{(N)}$ approach q slowly.

The exponent θ can be determined in the following way. From (9) we expect that

$$(10)\quad q_k^{(N)} \simeq q + \frac{E'}{N^\theta} ,$$

where E' is a constant and θ is called the "subdominant critical exponent" which in general is not an integer. The sequence of approximations to $q_k^{(N)}$ is generally smooth enough that we may calculate approximations to θ as follows (for ease of notation, rename $q_k^{(N)}$ as q_N).

Since

$$(11)\quad \frac{d^2 q_N}{dN^2} \bigg/ \frac{dq_N}{dN} \simeq -\frac{(\theta + 1)}{N}$$

we take, as an approximation θ_N to θ,

$$(12)\quad \frac{q_{N+1} + q_{N-1} - 2q_N}{(q_{N+1} - q_{N-1})/2} = -\frac{\theta_N + 1}{N} .$$

It is important to note that the x_c which appears on the right hand side of (7) may be $x_{ck}^{(N)}$. If a better value of x_c than $x_{ck}^{(N)}$ is known, it may be used. In [2], we used the best estimate of x_c derived from the appropriate high temperature series since we considered it more accurate than the $x_{ck}^{(N)}$ calculated from low temperature series. In this work, we use the $x_{ck}^{(N)}$ because no better value of x_c is known and because this choice frees the method of a priori knowledge of x_c. It appears that the lower limit of the interval of orthogonality requires a knowledge of $t_c = 1/x_c$. However, the scheme is very insensitive to the value of this lower limit, and the scheme could be made self consistent in this regard by an iterative procedure. Because of the insensitivity of the approximants to the lower limit, previous [9] estimates of x_c suffice.

Equation (3) appears to state that the correction terms on the right hand sides of (8) and (9) could be made small by choosing a close to t_c. However, the asymptotics which lead to (5) fail as a approaches t_c because $p_N(t_c)$ becomes small. There is an optimum a which has to be determined experimentally.

When the asymptotics required by (5) are satisfied, then according to (8) and (9) the errors in the approximations depend only on the single variable $N\phi'(t_c)t_c$. That is, the accuracy of approximations obtained by varying N and a separately with k-N fixed depends only on the smallness of one variable $1/N\phi'(t_c)t_c$, which is trivially related to

$$(13)\quad X = \frac{15}{N} \sqrt{1 - a/t_c}$$

by a change of scale adjusted so that X = 1 for N = 15, a = 0. A tedious but straightforward study of the asymptotics shows that more precisely the error in the approximations depends only on the single variable

$$(14)\quad Z = X/(1 + \frac{1}{20\sqrt{2}} \frac{3 - a/t_c}{1 - a/t_c} X) ,$$

which is never far from X provided a is not too close to t_c. The only use we make of this fact is in the preparation of Fig. 6 which is self-justifying because the results for different a's join smoothly.

2. Application of the Method to Test Functions

To examine the effectiveness of the method described in Section 1, two test functions, believed to be similar to the Ising model, are investigated in this section. The first test function is the susceptibility series for the two dimensional Ising model on the simple quadratic lattice. The susceptibility is thought to behave as

$$(15)\quad \chi(y) = A_o y^{-\gamma}[1+A_1 y+a_2 y^2+\ldots+By^{\tau}]+c+Dy\ln y$$

where $y = v_c - v$ $\qquad \gamma = 1.75$

$v = \tanh(J/kT)$ $\qquad \tau > 1.00$

$v_c = \sqrt{2} - 1$

To investigate this function rational approximants were formed to the logarithmic derivative series using Chebyshev polynomials orthogonal on the interval $-1/v_c = -t_c < t < 0$. The results, shown in Figure 1, are seen to be converging in a smooth, rapid fashion to the known exact value $\gamma=1.75$. The boxed area in Figure 1 is shown with magnified scale in Figure 2.

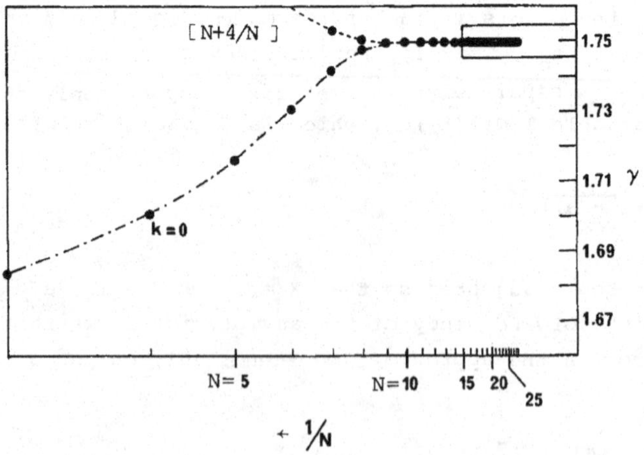

Figure 1. Estimates of the critical exponent γ for the simple quadratic lattice from successive [N/N] and [N+4/N] approximants.

The exact value of v_c was used when evaluating the approximations for γ (see Eq. (9) and the discussion following). Therefore these are termed 'biased' rational approximants.

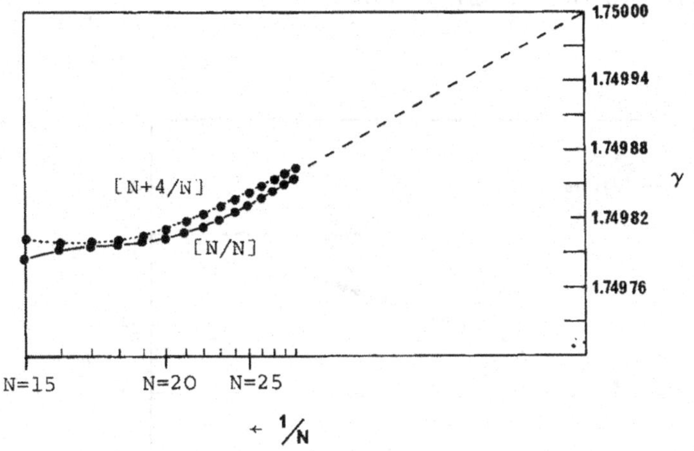

Figure 2. Estimates of the critical exponent γ for the simple quadratic lattice from successive [N/N] and [N+4/N] approximants.

The extremely fast convergence evident in Figure 1 suggests that the logarithmic derivative of χ has no significant confluent singularity. This is not a difficult case and all previously used methods, such as the ratio and Padé methods, also treat it successfully.

For this reason, a less favorable test-function was constructed of the form

$$(16)\quad \frac{d}{dx}\ln f(x) = \frac{\gamma}{1-x}\,[1-A(1-x)^{\theta}]^{-1}$$

$$\sim\frac{\gamma}{1-x}[1+A(1-x)^{\theta}+A^2(1-x)^{2\theta}+A^3(1-x)^{3\theta}+\ldots]$$

with $A = \frac{1}{2}$, $\theta = \frac{1}{2}$, $\gamma = \frac{1}{2}(1 - \frac{1}{\sqrt{2}}) = 0.1464466\ldots$

Chebyshev polynomials were placed on the interval $-1 < t < 0$ and diagonal [N/N] rational approximations for γ_N and θ_N were calculated using (9) and (12). In evaluating (9), the approximate value of x_c obtained from (8) was used so these approximants are "unbiased". The results for the θ_N are plotted in Figure 3 vs 1/N.

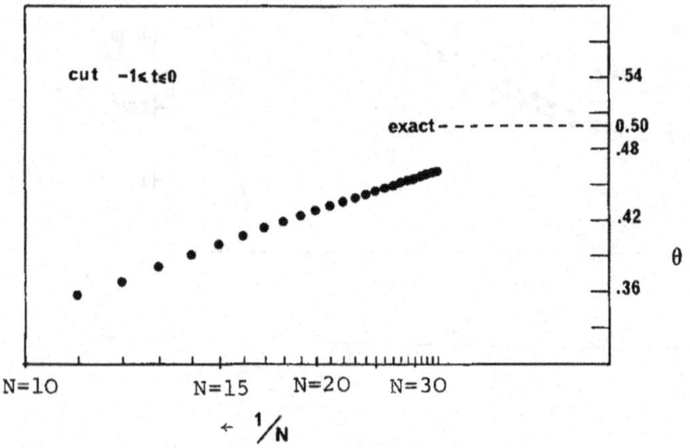

Figure 3. Successive approximations for the subdominant critical index θ for the test function plotted vs N^{-1}.

Examination of the figure shows that the approximations appear to have converged and a simple linear extrapolation of the sequence results in an estimate $\theta \sim 0.51 \pm 0.01$ in agreement with the known exact result.

It is seen from (9) that plotting the γ_N versus $N^{-\frac{1}{2}}$ should result in an asymptotically linear curve. This plot is given in Figure 4. Convergence to the exact answer is clear. A linear extrapolation of $N^{-\frac{1}{2}}$ to 0 using only the last two available γ_N yields a value of $\gamma=0.1466$, an error of 2 in fourth decimal place, or an error of approximately 0.1%.

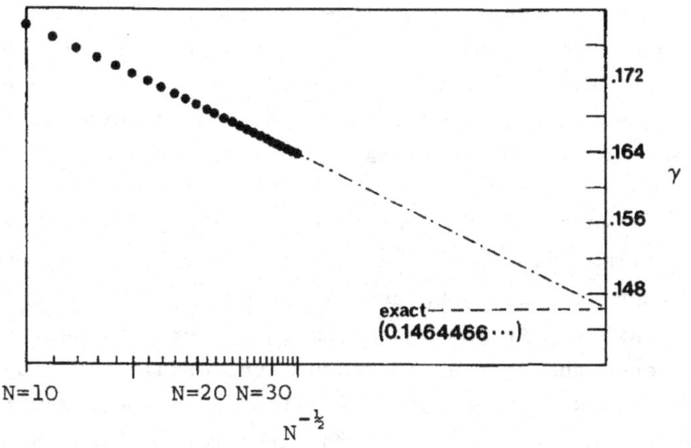

Figure 4. Successive approximations for the dominant critical index γ for the test function plotted vs $N^{-\frac{1}{2}}$.

From the results of the analysis of the two test functions it is clear that the rational approximation method is capable of providing critical indices with great precision, even in the presence of confluent singularities with significant amplitude.

3. Application to the High Temperature Expansion of the Susceptibility in the General Spin Ising Model, Body Centered Cubic Lattice

We use Nickel's [12] twenty one term series for the susceptibility and form the series for the logarithmic derivative. The nearest singularities are the Curie point x_c and the antiferromagnetic singularity $-x_c$.

We use Chebyshev polynomials orthogonal on $-1/x_c = -t_c < t < 0$. The operation $P_N(x) = x^N p_N(1/x)$ amounts to reversing the order of the coefficients in P_N, so we do this and multiply the logarithmic derivative by $P_N(x)$ to obtain the $\ell_k^{(N)}$. Then (8), (9) and (12) give the various approximants.

The standard notation is followed, the dominant critical exponent is denoted by γ and the subdominant index by θ. The sequences are smoothly varying because choosing the $p_N(t)$ is equivalent to choosing the denominators of the Padé approximants to $\ell(x)$. There are no spurious poles to slow convergence nor any need to solve even in principle a large set of linear equations.

The θ_N calculated from (12) are plotted versus N^{-1} in Figure 5 for various values of the spin S. If the value of θ were known, even approximately, then (12) could be used, to integrate the $\gamma_N \equiv q_N$ forward to $N=\infty$ and so to obtain γ. It is clear from the figure however, that there is no evidence that the θ_N approach a universal spin-independent value in the limit. Nevertheless, the following hopeful result emerges. It is clear that the θ_N versus 1/N curve for S = 1/2 has a pole at about N = 12.5. As the spin increases toward S=1, this pole moves toward 1/N = 0 as the following numerical data shows.

N = 13	S = 0.52953
14	0.56624
15	0.59228
16	0.61220
17	0.62797
18	0.64091

For S = 1, it is likely that there is no pole. Thus there will be some critical spin S_c between S = 1/2 and S = 1 such that the pole occurs at $N = \infty$. For this critical spin, the integration of Eq. (12) forward from N = 18 to $N = \infty$ cannot result in a long extrapolation of the γ_N sequence because large θ_N result in a slowly varying γ_N sequence. For the critical spin, the θ_N are all large and positive between N = 18 and $N = \infty$ no matter what θ may be. The extrapolation is short and independent of θ.

We find the critical spin by plotting the spins given in the numerical data immediately above versus (1/N) raised to some power. We find that when the power is three the resulting graph is a straight line. This straight line extrapolates to $S_c = 0.73 \pm 0.01$ as $N \to \infty$. Other evidence for this value of the critical spin arises in subsequent calculations described below (Section 4).

We have extrapolated the γ_N sequences for S = 0.72 and S = 0.74 to $N = \infty$ by integrating Eq. (12) using θ_N vs 1/N curves which are in rea-

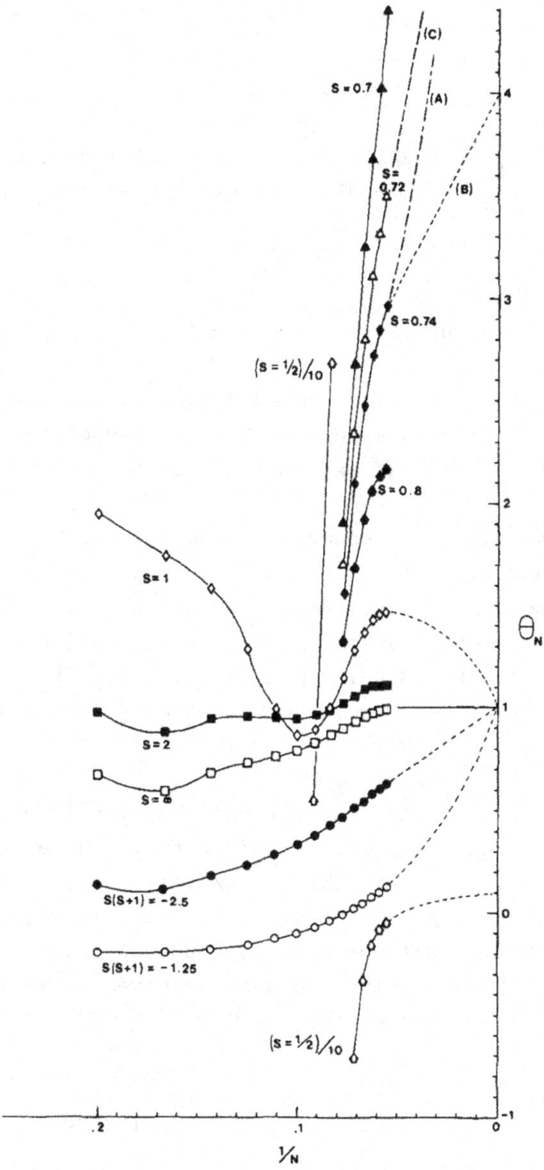

Figure 5. Successive approximants to the subdominant critical exponent θ for the magnetic susceptibility on the general spin Ising model on the body centered cubic lattice. Values of θ_N which are negative or unduly large occur because the corresponding N are too small for an indication of the true asymptotic behaviour, except by arguments of numerical continuity.

sonable agreement with Fig. 5. Two θ_N versus 1/N curves were conjectured for S = 0.74 (marked (A) and (B) in Fig 5.) and one for S = 0.72 (marked (C) in Fig. 5). The results are

S = 0.72	γ_{19} = 1.2410178	extrapolates to 1.2412201	(C)
S = 0.74	γ_{19} = 1.2405911	extrapolates to 1.2409039	(A)
		1.2409464	(B)

Thus we conclude that

$$\gamma(S_c) = 1.2411 \pm 0.0002 \quad , \quad S_c = 0.73 \pm 0.01 \quad .$$

The extrapolations are complete to 3 parts in 10^6 by N=30 because the θ_N are large and rising steeply. The details of the extrapolation cannot affect the outcome, as extrapolations (A) and (B) show.

So far, we have made no assumptions about the universality of γ. However, if universality is valid, then we conclude that $\gamma = 1.2411 \pm 0.0002$ for all models, spins, and lattices. We claim much more accuracy than previous calculations ($\gamma = 1.241 \pm 0.003$ [11], $\gamma = 1.2402 \pm 0.0009$ [5], $\gamma = 1.2385 \pm 0.0015$ [8], $\gamma = 1.241^{+0.003}_{-0.005}$ [10],). Our claim is valid <u>only</u> if our conclusions that S_c exists and has the value $S_c = 0.73 \pm 0.01$ are valid.

We now investigate whether or not universality is a necessary consequence of the series analyses. In order to get a value of γ for the spins other than S_c, in particular S = 1/2, 1, 2, and ∞, we require θ_N versus 1/N curves in order to integrate Eq. (12) forward to N = ∞. In these cases, the result will depend very much on the details of the θ_N versus 1/N curve. Conjecturing the dashed curves shown in Fig. 5, all of which have $\theta = 1$, we calculate

S	1/S(S+1)	γ
---	-0.4	1.2300
∞	0	1.2351
0.73	0.792	1.2411
0.5	1.333	1.2433

These results are in obvious disagreement with the universality of γ. That is, not any θ_N versus 1/N curves will lead to the universality of γ. Since the dashed extrapolations are reasonable, we must conclude that universality is not a necessary consequence of any analyses of these series.

We now investigate what assumptions are consistent with a universal γ and what such a value of γ may be. We do not assume the existence of a critical spin in this investigation. In particular, we investigate whether or not a universal θ will lead to a universal γ. Thus we suppose that for two different spins

$$(17)\quad D_N(S,S') = \gamma_N(S) - \gamma_N(S') = b/N^{\theta}$$

that is, that the two γ_N sequences approach each other as $N \to \infty$. We may calculate approximants to θ as follows:

$$(18)\qquad \frac{\ell n(D_N(S,S')/D_{N'}(S,S'))}{\ell n(N/N')} = -\theta(S,S',N,N').$$

Equation (18) depends only on first differences (in N) of the γ_N sequences rather than the second differences which appear in (12). For this reason, we are able to get reliable numerical results when the right hand edge of the interval on which the p_N are orthogonal is 1/2 t_c. Results of calculations for several pairs of spins and N's are given in Table I. From these results, which are decreasing with increasing N,N', we estimate that

$$\theta = 0.55 \pm 0.05,$$

is consistent with the universality of γ.

In order to obtain the universal γ which results from such a θ, we plot γ_N versus $1/N^{0.55}$. The result should be a straight line for each spin, each line extrapolating to a universal γ. Such a plot is shown in Fig. 6 except that in order to include results for several a's (the right hand edge of the interval on which the p_N are orthogonal), we have plotted versus $Z^{0.55}$ (see Eq.(14) for Z).

Table I. Approximants to the Sub-Dominant Exponent θ.

$$a = \frac{1}{2} t_c$$

N	N'	S	S'	$\theta_N(S,S',N,N')$
19	15	1	2	0.594
19	15	1	∞	0.599
19	18	2	∞	0.602
19	18	1	2	0.598
16	15	2	∞	0.606
16	15	1	2	0.589
19	15	0.73	∞	0.589
19	18	0.73	∞	0.595
19	10	0.73	∞	0.620
15	10	0.73	∞	0.638

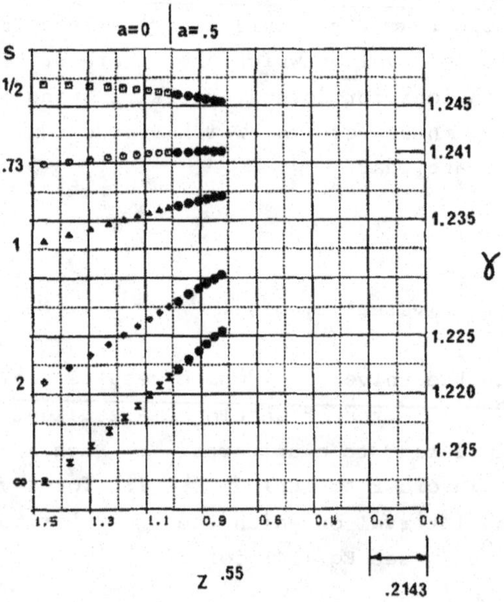

Figure 6. Successive approximants for the dominant critical exponent γ for the magnetic susceptibility on the general spin Ising model on the body centered cubic lattice. The independent variable is $z^{0.55}$ (see (14)) and the darkened points correspond to $a = t_c/2$ and the open points to $a = 0$.

The result is quite satisfying. Approximately straight lines do result and the results for different a's join smoothly. Even more satisfying is that the straight lines extrapolate to $\gamma = 1.241 \pm 0.001$ (the quoted error also allows for the results of plots versus $Z^{0.6}$ or $Z^{0.5}$ instead of $Z^{0.55}$) in agreement with our previous result $\gamma = 1.2411 \pm 0.0002$. Even if one does not believe our conclusions about the critical spin and so does not believe $\gamma = 1.2411 \pm 0.0002$, still one may believe (assuming the universality of γ and θ) $\gamma = 1.241 \pm 0.001$. This latter sort of accuracy is in accord with the sort of accuracies claimed in references [5,8,11,13].

4. Analysis of the Series for the Second Moment Correlation Function

The critical exponent for the second moment correlation function M_2, is $\gamma+2\nu$ where ν is the exponent of divergence of the correlation length, ξ.

We have analyzed the series in a variety of ways all of which lead to the same conclusion. These ways are i) analysis of M_2 directly as in Section 3, ii) dividing the series for M_2 by the series for the susceptibility to obtain a series for a quantity with critical exponent 2ν, and iii) dividing each term in the series for M_2 by the corresponding term in the series for γ, obtaining a series which has critical point equal to 1 and a critical exponent $1-\gamma+(\gamma+2\nu)=1+2\nu$ (this method is described in [4].

In each case, the critical spin $S_c = 0.73$ emerges from the work in exactly the same way that it arises in the analysis of the series for the susceptibility.

The detailed analysis indicates that

$$\nu(S_c) = 0.6335 \pm 0.0003 \,, \quad S_c = 0.73 \pm 0.01$$

5. Scaling and Hyperscaling

The predictions of scaling and hyperscaling are

$$(19) \quad \alpha + 2\beta + \gamma = 2 \,, \qquad 3\nu = 2 - \alpha \,,$$

These are summarized in Fig. 7 in terms of α, β, γ, and

(20) $x = \frac{\gamma + 2\nu}{\gamma} - 2$.

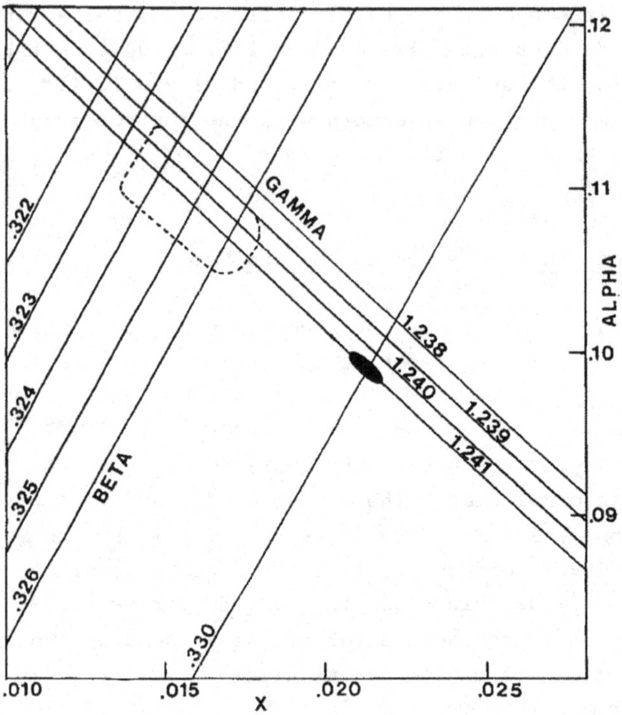

Figure 7. Predictions of the scaling and hyperscaling hypotheses. Each point on the graph represents a choice of α, β, γ, and x which satisfies these hypotheses.

According to our results,

$$x = \frac{1.2411 + 2(0.6335)}{1.2411} - 2 = 0.0209 \pm 0.0006 ,$$

$$\gamma = 1.2411 \pm 0.0002 ,$$

which is the smaller of the two regions shown in Fig. 7. If scaling and hyperscaling hold, then β must be near $\beta = 0.330$, whereas we previously reported $\beta = 0.323 \pm 0.003$ in [1] . However, a θ_N analysis of our BCC β results leads to Fig. 8, from which it is clear that the subdominant index cannot be determined. However, if one accepts a sub-dominant index $\theta \sim 0.5$ and makes the imaginative extrapolation of θ_N shown in Fig. 8, one easily calculates by integration of (12) that $\beta \sim 0.330$.

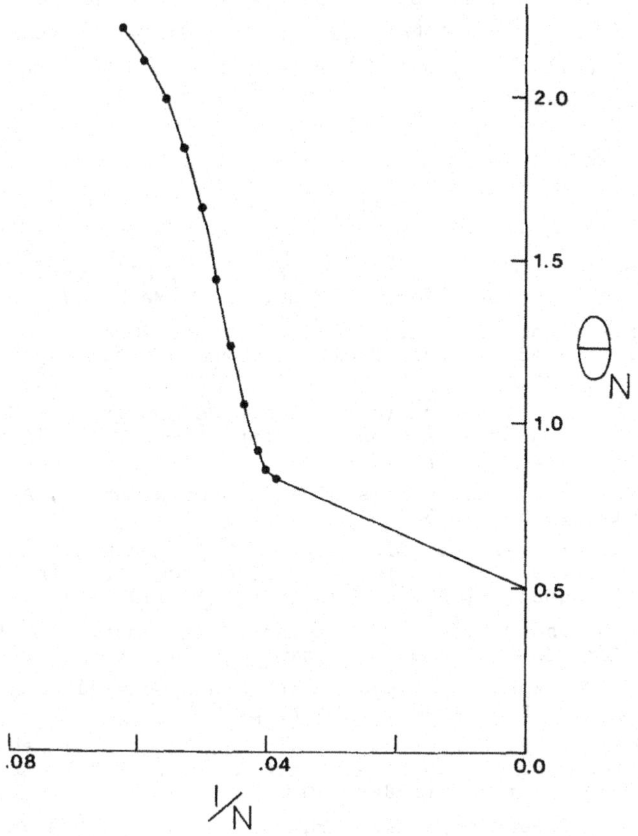

Figure 8. Successive approximations for the sub-dominant critical index, θ_N obtained from the β_N given in reference [2] for the body centered cubic lattice.

It is unfortunate that the high temperature specific heat series do not lead to a reliable result.

The larger region in Fig. 7 is based on results given in [10]. The fact that the two regions do not overlap means nothing about scaling and hyperscaling, but only that other analyses of the series are not in agreement with our results.

5. Acknowledgments

We thank Bernie G. Nickel and Marty Ferer for supplying the series expansions for the general spin S model. We thank Bernie G. Nickel, Michael Fisher and George Baker for correspondence and comments. For expediency of publication, an editor (P.R.G.-M.) was entrusted with revision of this paper.

References

1. Baumel, R.T., Gammel, J.L. and Nuttall, J., Placement of Cuts in Padé-like Approximation, J. Comp. Appl. Math., 7 (1981) 135.

2. Baumel, R.T., Gammel, J.L., Nuttall, J. and Power, D.C., The Calculation of Critical Indices by the Rational Approximation Method, J. Phys. A: Math. Gen. 15 (1982) 3233.

3. Baker, G.A., Jr., Application of the Padé Approximant Method to the Investigation of some Magnetic Properties of the Ising Model, Phys. Rev. 124 (1961) 768.

4. Baker, G.A., Jr., "Essentials of Padé Approximants", Academic Press, New York (1975) p.275.

5. Baker, G.A., Jr., Nickel, G.B., Green, M.S. and Meiron, D.I., Ising Model Critical Indices in Three Dimensions from the Callan-Symanzik Equation, Phys. Rev. Lett, 23, (1976) 1351.

6. Bender, C.M. and Orszag, S.A., Advanced Mathematical Methods for Scientists and Engineers, McGraw-Hill, New York (1978), p.265.

7. Brezinski, C., Padé-type Approximation and General Orthogonal Polynomials, Birkhäuser, (1980), p.12.

8. Chen Jing-Huei, Fisher, M.E. and Nickel, B.G., Unbiased Estimation of Corrections to Scaling by Partial Differential Approximants, Phys. Rev. Lett. 48 (1982) 630.

9. Domb, C. and Green, M. (eds.), Phase Transitions and Critical Phenomena, Academic Press N.Y., (1974) Vol. 3.

10. Ferer, M. and Velgakis, M., Hyperscaling in the Three Dimensional Ising Model, Phys. Rev. B, 27 (1983) 2839.

11. LeGuillou, J.C. and Zinn-Justin, J., Critical Exponents from Field Theory, Phys. Rev. B, 21 (1980) 3976.

12. Nickel, B.G., Phase Transitions, Cargèse 1980, M. Levy, J.-C. le Guillou and J. Zinn-Justin (eds.), Plenum Press N.Y., (1982) 291.

13. Wegner, F.J., Corrections to Scaling Laws, Phys. Rev. B, 5 (1972) 4529.

14. Zinn-Justin, J., Analysis of high temperature series of the spin S Ising model on the body-centred cubic lattice, J. Phys. (Paris) 42, (1981) 783.

PARTIAL DIFFERENTIAL APPROXIMANTS
AND THE ELUCIDATION OF MULTISINGULARITIES

Daniel F. Styer
Hill Center, Busch Campus
Rutgers University
New Brunswick, New Jersey 08903 USA

and

Michael E. Fisher
Baker Laboratory
Cornell University
Ithaca, New York 14853 USA

Abstract. A partial differential approximant, or PDA, $F(x,y)$, can accurately approximate a two-variable function, $f(x,y)$, on the basis of its power series expansion even near a *multisingular point* where the function is intrinsically nonanalytic in both variables. This brief review argues that multisingularities occur frequently in two-variable functions arising in practical situations. Partial differential approximants are defined and it is shown why they can approximate multisingularities. The invariance of PDAs under a change of variables is discussed and new results are presented concerning functions exactly representable by PDAs. Finally, several applications of PDAs are mentioned.

1. Introduction

Any effective approximant to any function must be capable of exhibiting the important behavior displayed by the function being approximated. This known functional behavior, which must be imitated by the approximant, typically consists of two parts: first, there is the behavior near the origin, which we suppose given by a finite number of low order terms in the formal power series expansion

$$f(x,y) = \sum_{i,i'=0}^{\infty} f_{i,i'} x^i y^{i'} . \qquad (1)$$

Second, in any practical calculation there is always some idea of how the function will behave away from the origin. For example, one might know the behavior of $f(x,0)$ as $x \to \infty$, or one might suspect that $f(x,y)$ has some particular singularity structure. In the following Section we first briefly review how this situation may be handled effectively in the single-variable case. We then argue heuristically that a particular but rather general singularity structure, the so-called scaling form, will frequently be found in two-variable functions arising in various different practical contexts. Lastly, a partial differential approximant is defined as the solution of an appropriate first order two-variable boundary value problem constructed from the coefficients $f_{i,i'}$ (and possibly embodying other known information).

The third section concerns the theory of partial differential approximants. After establishing that a partial differential approximant can, in fact, fulfill the two requirements for an effective approximant mentioned above, the invariance properties of partial differential approximants under a change of variable, such as an Euler transformation, are examined. This section also includes a short discussion of some practical techniques which permit significant information to be gleaned from an approximant with relatively little numerical labor.

Finally, Section 4 surveys some of the problems to which partial differential approximants have been applied. This section is short but the interested reader may consult other recent reviews which emphasize applications [12,26].

2. Background Considerations

Many functions which arise in physics, engineering, combinatorics, and related fields exhibit branch point singularities of the form

$$f(x) \approx A(x)(x-x_c)^{-\gamma} + B(x), \quad \text{as} \quad x \to x_c, \tag{2}$$

where $A(x)$ and $B(x)$ are functions analytic at $x=x_c$. It has long been known that when $\gamma > 0$ and the "background function", $B(x)$, is small, such functions can be approximated effectively by Baker's Dlog Padé technique [1]. More recently [11,19], non-negligible background functions have been treated successfully using the method of "inhomogeneous differential approximants" (or "integral curve approximants") in which $f(x)$ is approximated by the solution, $F(x)$, of the ordinary differential equation

$$U_J(x) + P_L(x)F(x) = Q_M(x)\frac{dF}{dx} , \qquad (3)$$

the so-called *defining equation*, subject to the initial condition

$$F(0) = f(0) . \qquad (4)$$

The polynomials $U_J(x)$, $P_L(x)$, and $Q_M(x)$ (of order J, L, and M, respectively) in the defining equation are selected to respect the *generating relation*

$$U_J(x) + P_L(x)f(x) = Q_M(x)\frac{df}{dx} + O(x^{J+L+M+2}) . \qquad (5)$$

An approximant constructed in this manner is denoted

$$F(x) = [J/L;M]_{f(x)} , \qquad (6)$$

and it can be found only if the series for f(x) is known to order J+L+M+2 or higher. In addition, one can easily show [3,15] that if both f(x) and F(x) are expanded about the origin, the two series will agree through order J+L+M+2 : in other words every expansion coefficient of f(x) used in the computation of F(x) is correctly reproduced by F(x), whence the approximant is termed *faithful* or *accurate through order.* While an effective approximant must be faithful, it must also be able to mimic the behavior of the function being approximated away from the origin. Inhomogeneous differential approximants can effectively approximate functions with branch point singularities because the initial value problem specified by (3) and (4) must take on precisely the form (2) near any zero of $Q_M(x)$ [11,19]. Conversely, the appropriate zero of $Q_M(x)$ provides an estimate for the anticipated singular point x_c and from the behavior of U_J, P_L, and Q_M near the zero, the exponent, γ, and the background, B, can also be estimated simply [11,19].

To find effective approximants for singular functions of *two* variables, we must first inquire into the expected form of the singularity, that is, we must find the two-variable analog of equation (2). One simple and obvious generalization of (2) is

$$f(x,y) \approx A(x,y)[x-x_c(y)]^{-\gamma} + B(x,y) , \quad \text{as} \quad x \to x_c(y) , \qquad (7)$$

where $x_c(y)$, A(x,y) and B(x,y) are analytic functions. While such behavior certainly does arise in practice, it is, in a real sense, hardly two-dimensional at all: thus a change in y merely changes the parameters associated with the singularity, but does not change its basic structure. While one would like to be able to approximate such functions well, they do not provide a sufficiently stringent challenge relative to what may be encountered in practice.

To see this, let us, instead, construct a bolder generalization of (2) which reflects what is known to occur in a variety of practical situations [9,10]. Thus, suppose $f(x,y)$ behaves like (7) but with two distinct singular loci, $x_{1,c}(y)$ and $x_{2,c}(y)$, each with its own exponents, γ_1 and γ_2, respectively. If the two singular loci meet, say at a point (x_c,y_c), the function $f(x,y)$ is likely to have more complex nonanalytic behavior *at* (x_c,y_c) since, when viewed from the origin side of the combined singular locus, $f(x,y)$ undergoes a discontinuous change in analytic behavior at (x_c,y_c); see Fig. 1. Such a point, marking an abrupt change in singular behavior and involving intrinsically, both variables, may be called a *multisingular point* [9,10,12, 14]. Because of the complex nature of a typical multisingular point, it turns out to be unreasonable, in general, to expect that the two singular loci remain analytic there. Instead, one should anticipate that the compound locus is continuous but singular at (x_c,y_c); for example, it may exhibit a sharp *cusp* as indicated in Fig. 1 [9,12].

For a more precise specification of "typical" multisingular behavior we introduce the difference variables

$$\Delta x = x - x_c \quad \text{and} \quad \Delta y = y - y_c , \tag{8}$$

and their special linear combinations, the so-called *linear scaling variables* [8,9,12]

$$\tilde{x} = \Delta x - \Delta y/e_2 \quad \text{and} \quad \tilde{y} = \Delta y - e_1\Delta x . \tag{9}$$

(These are sometimes denoted $\Delta\tilde{x}$ and $\Delta\tilde{y}$, respectively.) The two coefficients, e_1 and e_2, represent the slopes in the (x,y) plane of the two *scaling axes* through (x_c,y_c) which embody a characteristic feature of the multisingularity in question: see Fig. 1. Thus if the singular loci meet in a cusp, e_1 would normally specify the common tangent of the loci at (x_c,y_c). The variable $\tilde{y}$ then measures deviations away from the common tangent, which is the first scaling axis. In certain circumstances e_1 and e_2 may be known *a priori* as the result of some symmetry in the problem; but in other cases one will be concerned to estimate their values, as also the location, (x_c,y_c), of the multisingularity itself.

The singularity structure described above, while at first glance rather complex and seemingly arbitrary, in fact follows naturally from

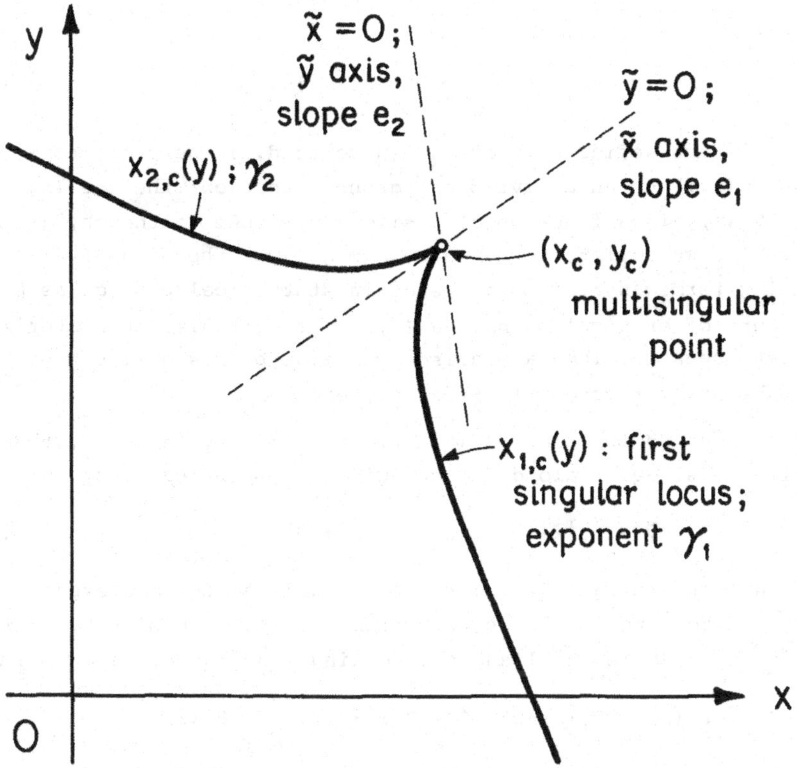

Figure 1. Illustration of a multisingular point, (x_c, y_c) in the two-variable plane (x, y). The solid curves represent loci of branch point singularities while the dashed lines depict the two scaling axes.

the asymptotic "scaling form"

$$f(x,y) \approx \frac{1}{|\tilde{x}|^{\gamma}} Z_{\pm}\left(\frac{\tilde{y}}{|\tilde{x}|^{\phi}}\right) + B(x,y) , \tag{10}$$

where γ and ϕ are definite exponents (unrelated, in general, to γ_1 and γ_2), while $B(x,y)$ is an analytic background function; the *scaling function*, $Z_{\pm}(z)$, is a function of a single variable (although, as indicated by the subscript, it may also depend upon the sign of $\tilde{x}$). This scaling form was first discovered in statistical physics as a basic feature of critical phenomena [8,9]: but it also characterizes the "catastrophes" of Thom and arises in Feigenbaum's analysis of period-doubling in functional iteration, etc..

If the scaling function is itself singular, say, at $z=z_1$, then $f(x,y)$ will be singular along a locus given asymptotically by

$$\tilde{y} \approx z_1|\tilde{x}|^{\phi} \qquad \text{as} \qquad \tilde{x} \to 0 . \tag{11}$$

Thus the cusp in the example in Fig. 1, as well as the two exponents γ_1 and γ_2 on the singular locus away from the multisingularity, will follow if, for example, $\phi > 1$ and the scaling function takes on the form

$$Z_{-}(z) = P_1(z_1-z)^{-\gamma_1} + P_2(z-z_2)^{-\gamma_2} + P_0(z) , \tag{12}$$

for $z_1 > z > z_2$, where P_1 and P_2 are constants and $P_0(z)$ is a less singular background function. Other scaling functions and multisingular exponents will, of course, lead to other sorts of multisingular behavior: for example, the singular locus is not cusped if $\phi < 1$.

We have asserted that multisingularities of the scaling form are reasonable and natural, and should thus be anticipated in a variety of situations. Readers unconvinced by our condensed and heuristic reasoning may wish to peruse the renormalization group arguments [8] which show why scaling multisingularities must be anticipated generally in the study of critical phenomena. Alternatively, they may simply accept the fact that such multisingularities have, indeed, arisen in practical problems, and that the desire to elucidate their structure and parameters led to the invention of partial differential approximants.

How can functions of the scaling form be approximated effectively? Just as inhomogeneous differential approximants are effective because solutions of the defining equation (3) can display the anticipated branch point form (2), so a defining equation for effective approximants for multisingular functions must admit solutions of the scaling form (10). Such a defining equation is not hard to find: in fact, it proves to be a simple generalization of (3). Thus a *partial differential approximant* (or PDA), $F(x,y)$, for a function $f(x,y)$ is defined [10,11] as a solution of the *defining equation*

$$U_{\underset{\sim}{J}}(x,y) + P_{\underset{\sim}{L}}(x,y)F(x,y) = Q_{\underset{\sim}{M}}(x,y)\frac{\partial F}{\partial x} + R_{\underset{\sim}{N}}(x,y)\frac{\partial F}{\partial y} . \tag{13}$$

The coefficient functions $U_{\underset{\sim}{J}}(x,y)$, $P_{\underset{\sim}{L}}(x,y)$, $Q_{\underset{\sim}{M}}(x,y)$ and $R_{\underset{\sim}{N}}(x,y)$ in (13) are the *defining polynomials*. A little thought reveals that the "order" of such two-variable polynomials cannot be specified by a single integer, as sufficed in the one-variable case. Rather, a polynomial $U_{\underset{\sim}{J}}(x,y)$ of form $\sum u_{j,j'}x^j y^{j'}$ is described by the set, $\underset{\sim}{J}$, of integer pairs $\{(j,j')\}$ which specify those powers of x and y which may appear in the polynomial. Because $\underset{\sim}{J}$ labels the monomials $x^j y^{j'}$ permitted in $U_{\underset{\sim}{J}}(x,y)$, we term it a *label set*. Label sets are conveniently represented by graphical arrays: a practical convention for such arrays is shown in the example below, which illustrates a polynomial and the array associated with its label set:

$$\begin{aligned} U_{\underset{\sim}{J}} = 1 + 2x &- 3x^2 + 7x^3 \\ + 4xy &\qquad\;\; + 3x^3y \\ &- 2x^2y^2 \end{aligned} \qquad \underset{\sim}{J} = \left|\begin{array}{cccc} \times & \times & \times & \times \\ \cdot & \times & \cdot & \times \\ \cdot & \cdot & \times & \cdot \end{array}\right. . \tag{14}$$

While the most natural label sets are either "triangular" or "rectangular", such as the two arrays

$$\left|\begin{array}{cccc} \times & \times & \times & \times \\ \times & \times & \times & \\ \times & \times & & \\ \times & & & \end{array}\right. \quad \text{and} \quad \left|\begin{array}{cccc} \times & \times & \times & \times \\ \times & \times & \times & \times \\ \times & \times & \times & \times \end{array}\right. , \tag{15}$$

other sorts of label sets frequently arise in practice and they must also be treated by the theory. For example "upper triangular" label sets, such as

$$\left|\begin{array}{ccccc} \times & \times & \times & \times & \times \\ \cdot & \times & \times & \times & \times \\ \cdot & \cdot & \times & \times & \times \\ \cdot & \cdot & \cdot & \times & \times \\ \cdot & \cdot & \cdot & \cdot & \times \end{array}\right. , \tag{16}$$

have proven useful in approximating functions whose series expansions are themselves upper triangular in structure [5].

The coefficients, $u_{j,j'}$, $p_{\ell,\ell'}$, $q_{m,m'}$ and $r_{n,n'}$ of the defining polynomials for partial differential approximants are selected, like those for inhomogeneous differential approximants, by imposing the *generating relation*

$$U_{\underset{\sim}{J}}(x,y) + P_{\underset{\sim}{L}}(x,y)f(x,y) = Q_{\underset{\sim}{M}}(x,y)\frac{\partial f}{\partial x} + R_{\underset{\sim}{N}}(x,y)\frac{\partial f}{\partial y} + \mathcal{E}_{\underset{\sim}{K}}(x,y). \tag{17}$$

The error symbol $\mathcal{E}_{\underset{\sim}{K}}(x,y)$ here generalizes the standard order symbol $O(x^{J+L+M+2})$ in (4), and stands for a (possibly infinite) sum of terms like $e_{k,k'}x^k y^{k'}$, where (k,k') falls *outside* the *matching set* $\underset{\sim}{K}$. The coefficients of the defining polynomials will generally be uniquely determined by (17) when the number of elements in $\underset{\sim}{K}$ is one less than the total number of elements in the sets $\underset{\sim}{J}$, $\underset{\sim}{L}$, $\underset{\sim}{M}$ and $\underset{\sim}{N}$. While the number of elements in $\underset{\sim}{K}$ is thus restricted, its "shape" is not : different matching sets, $\underset{\sim}{K}$, will produce different defining polynomials and hence different approximants. This dependence is recognized in the notation

$$F(x,y) = [\underset{\sim}{J}/\underset{\sim}{L};\underset{\sim}{M},\underset{\sim}{N}|\underset{\sim}{K}]_{f(x,y)} . \tag{18}$$

In addition, the approximant $F(x,y)$ will depend upon which boundary conditions are used in the integration of the defining equation (13). The selection of an appropriate boundary and of the boundary values on it is an important and sometimes tricky issue in a practical calculation. However, there is comparatively little that can be said in general about this selection [15,29], so we will simply assume here that appropriate boundary conditions can be found naturally from the practical context.

The definition of a partial differential approximant as given above allows a wide latitude in the choice of polynomial shapes. One consequence of this is that many other types of approximants are just special cases of PDAs. The simplest such reduction is achieved when the label set $\underset{\sim}{J}$ is taken to be the empty set $\emptyset$, so that the defining polynomial $U_{\underset{\sim}{J}}(x,y)$ vanishes. The resulting approximants are called homogeneous partial differential approximants and they have proven useful in studying multisingular points when the multisingular exponents γ and ϕ are positive and the background function, B, is small [14]. On the other hand, the choice $\underset{\sim}{M} = \underset{\sim}{N} = \emptyset$ yields a two-variable rational approximant, like those studied by the Canterbury group [6,7] and others [4,20,22]. An inhomogeneous differential approximant is simply a PDA with $\underset{\sim}{N} = \emptyset$ and with $\underset{\sim}{J}$, $\underset{\sim}{L}$, $\underset{\sim}{M}$ and $\underset{\sim}{K}$ chosen as single, gapless rows. The Dlog Padé and ordinary Padé approximants are, in turn, special cases of the inhomogeneous differential approximant [19]. This feature is of more than academic interest since it means that theorems proved for partial differential approximants automatically apply to all these other approximants as well. Thus the result on Euler invariance presented below encompasses all the results for Euler invariance in all five types of "reduced" approximants mentioned above.

3. Theory of Partial Differential Approximants

As we have stressed previously, an effective approximant must be faithful, so that it accurately mimics the function near the origin, and it must also be able to exhibit the singularity structure possessed by the function. In this section we ask whether partial differential approximants can fill the dual role required and we show that they can. We examine also the invariances of PDAs under an Euler transformation or other change of variables, and end by posing some open theoretical questions.

Faithfulness. An approximant $F(x,y)$ is *faithful* if it reproduces those series coefficients, $f_{i,i'}$, which were used in calculating the defining polynomials. If the labels of the coefficients used constitute the set $\underset{\sim}{H}=\{(i,i')\}$, then faithfulness simply means that

$$F(x,y) = f(x,y) + \mathcal{E}_{\underset{\sim}{H}}(x,y) \ . \tag{19}$$

Note that $\underset{\sim}{K}$ is generally a proper subset of $\underset{\sim}{H}$. Because the approximant is affected by the boundary conditions used in integrating the defining equation, its faithfulness will also depend upon those boundary conditions. For example, if boundary conditions are imposed on the x-axis, say as

$$F(x,0) = g(x) \tag{20}$$

for some given $g(x)$, then $F(x,y)$ cannot possibly be faithful unless the expansion $g(x) = \Sigma g_i x^i$ is itself "faithful", that is, if $g_i = f_{i,0}$ for those labels $(i,0)$ within $\underset{\sim}{H}$. In fact, it transpires [15] that if such faithful boundary values are used, then faithfulness is guaranteed by choosing a label set $\underset{\sim}{K}$ which is *echelon-on-x*. A technical definition of an echelon label set is given in [15]; for our purposes it suffices to say that an echelon-on-x array is one constructed by removing arbitrarily chosen rows from a triangular set and then pushing the remaining elements straight up to fill the vacated spaces. For example, the following array is echelon on x:

$$\begin{array}{|lllll} \hline \times & \times & \times & \times & \times \\ \times & \times & \times & \times & \\ \times & \times & & & \end{array} \ . \tag{21}$$

Because any number of rows may be removed, both triangular and empty arrays are echelon on x.

A question closely related to that of faithfulness concerns the approximation of upper triangular functions [see (16)]. In this case, a good approximant will itself be upper triangular. Conditions which ensure this are presented in Ref. 15.

Scaling form of approximants. Is a partial differential approximant capable of mimicking the scaling behavior (10)? The answer to this question is "Yes". In fact, if $(x_c,y_c) = (x_0,y_0)$ is a common zero of the defining polynomials $Q_{\underset{\sim}{M}}$ and $R_{\underset{\sim}{N}}$ for a PDA, so that $Q_{\underset{\sim}{M}}(x_0,y_0) = R_{\underset{\sim}{N}}(x_0,y_0) = 0$, then, in general, any of the corresponding PDAs will be of scaling form in the vicinity of (x_c,y_c). The aptness of PDAs for representing scaling multisingularities has been demonstrated in various different ways [10; 12, Sec. 4; 27, Appendix A]. We answer the question here in yet another way, through the following proposition which generalizes a result of Au-Yang and Fisher [11;12, Sec. 5] and exhibits "nonlinear scaling fields".

Proposition *Exact Representation of a Scaling Form* Suppose $\tilde{A}(x,y)$, $\tilde{C}(x,y)$, $\tilde{D}(x,y)$ and $\tilde{G}(x,y)$ are functions of the form

$$\tilde{A}(x,y) = \exp[A_0(x,y)/A_{-1}(x,y)] \prod_{k=1}^{K} [A_k(x,y)]^{\Gamma_k} , \tag{22}$$

in which A_{-1}, A_0, A_1, ..., A_K are polynomials and where $[A_k(x,y)]$ may represent either A_k or, keeping x and y real, $|A_k|$ and suppose that f(x,y) is of the scaling form

$$f(x,y) = \frac{\tilde{A}(x,y)}{|\tilde{G}(x,y)\tilde{x}|^{\gamma}} Z\left(\frac{\tilde{C}(x,y)\tilde{y}}{|\tilde{D}(x,y)\tilde{x}|^{\phi}}\right) + \frac{B_1(x,y)}{B_2(x,y)} , \tag{23}$$

where Z(z) is an arbitrary differentiable function, $B_1(x,y)$ and $B_2(x,y)$ are polynomials and $\tilde{x}$ and $\tilde{y}$ are the linear scaling fields given by (9). Then f(x,y) is exactly representable by some partial differential approximant.

Proof The result follows if we can produce polynomials U(x,y), P(x,y), Q(x,y) and R(x,y) which satisfy

$$U(x,y) + P(x,y)f(x,y) = Q(x,y)f_x(x,y) + R(x,y)f_y(x,y), \tag{24}$$

since appropriate boundary conditions can clearly be imposed away from the multisingular point at $\tilde{x} = \tilde{y} = 0$. (Here and in the following, partial differentiation is represented by a subscript.) Note, first, that the functions $\tilde{A}(x,y)|\tilde{G}(x,y)\tilde{x}|^{-\gamma}$ and $\tilde{C}(x,y)\tilde{y}|\tilde{D}(x,y)\tilde{x}|^{-\phi}$ are, in fact, functions of the form (22), so it suffices to prove the exact representability of

$$f(x,y) = \tilde{A}(x,y)\ Z[\tilde{C}(x,y)] + B_1(x,y)/B_2(x,y). \tag{25}$$

To form the derivatives of (25), notice that

$$\tilde{A}_x(x,y) = \tilde{A}(x,y)\left[\frac{A_{0,x}}{A_{-1}} - \frac{A_0 A_{-1,x}}{A_{-1}^2} + \sum_{k=1}^{K} \frac{\Gamma_k A_{k,x}}{A_k}\right]$$

$$= a_1(x,y)\tilde{A}(x,y)/a_0(x,y)\ , \tag{26}$$

where $a_0(x,y)$ is the polynomial $A_{-1}^2 \prod A_k$ and $a_1(x,y)$ is also a polynomial. (This holds whether $[A_k]^{\Gamma_k}$ represents $A_k^{\Gamma_k}$ or $|A_k|^{\Gamma_k}$.) By symmetry, there exists a polynomial $a_2(x,y)$ such that

$$\tilde{A}_y(x,y) = a_2(x,y)\tilde{A}(x,y)/a_0(x,y)\ , \tag{27}$$

and similarly for $\tilde{C}(x,y)$. It follows from (25) that

$$f_x = (a_1/a_0)\tilde{A}\ Z(\tilde{C}) + \tilde{A}(c_1/c_0)\tilde{C}\ Z'(\tilde{C}) + [B_{1,x}B_2 - B_1 B_{2,x}]/B_2^2\ , \tag{28}$$

and similarly for f_y. Thus (24) is satisfied for arbitrary Z if and only if one has the polynomial relations

$$U + PB_1/B_2 = Q(B_{1,x}B_2 - B_1B_{2,x})/B_2^2 + R(B_{1,y}B_2 - B_1B_{2,y})/B_2^2\ , \tag{29}$$

$$P\tilde{A} = Q\tilde{A}a_1/a_0 + R\tilde{A}a_2/a_0\ , \tag{30}$$

and

$$Q\tilde{A}\tilde{C}c_1/c_0 + R\tilde{A}\tilde{C}c_2/c_0 = 0\ . \tag{31}$$

The last equation is satisfied if

$$Q(x,y) = c_2(x,y)W(x,y) \quad \text{and} \quad R(x,y) = -c_1(x,y)W(x,y), \tag{32}$$

where $W(x,y)$ is any polynomial. Substituting these expressions into (30) gives

$$P = (a_1c_2W - a_2c_1W)/a_0\ , \tag{33}$$

so that (30) will be satisfied if $W(x,y) = a_0(x,y)V(x,y)$ with $V(x,y)$ a polynomial. Finally, (29) is satisfied if $V(x,y) = B_2^2(x,y)$. Thus the polynomials

$$\begin{aligned} U(x,y) = &-B_1B_2[a_1c_2 - a_2c_1] + a_0c_2[B_{1,x}B_2 - B_1B_{2,x}] \\ &- a_0c_1[B_{1,y}B_2 - B_1B_{2,y}]\ , \end{aligned} \tag{34}$$

$$P(x,y) = B_2^2(x,y)[a_1(x,y)c_2(x,y) - a_2(x,y)c_1(x,y)]\ , \tag{35}$$

$$Q(x,y) = a_0(x,y)B_2^2(x,y)c_2(x,y), \qquad R(x,y) = -a_0(x,y)B_2^2(x,y)c_1(x,y), \tag{36}$$

do satisfy (24), and the proposition is established.∎

While on the topic of scaling solutions, let us stress that, although the integrated approximant $F(x,y)$ and the scaling function, $Z(z)$, depend upon the boundary conditions imposed on the defining equation, the other asymptotic multisingular properties of $F(x,y)$ follow, in general, from the defining polynomials alone. This situation parallels that of the inhomogenous differential approximant: the theory of ordinary differential approximants tells us that any solution of (3) is analytic except at the zeros of $Q_M(x)$, where the solution is of form (2)[19]. As mentioned, the zeros of $Q_M(x)$ thus constitute estimates for the singular point x_c of the function $f(x)$, and these estimates are available without integrating the initial value problem (3), (4). Similarly, in general, any solution of (13) takes on the form (10) near a common zero of $Q_{\underset{\sim}{M}}$ and $R_{\underset{\sim}{N}}$ and so these zeros provide estimates for the true multisingular location. Explicit formulae for the singular parameters γ, ϕ, e_1 and e_2 of an approximant can likewise be given directly in terms of the values and derivatives of the defining polynomials at the common zero [12], so the integration of the defining equation is for many purposes unnecessary. On the other hand, estimates for the scaling function $Z(z)$ can be obtained only through an integration which requires appropriate boundary conditions.

Experience has shown that when such an integration is needed, it can be effectively implemented by using the method of characteristics [see e.g. 16]. The characteristics are the solutions of the coupled ODEs

$$\frac{dy}{d\tau} = Q_{\underset{\sim}{M}}[x(\tau),y(\tau)], \qquad \frac{dy}{d\tau} = R_{\underset{\sim}{N}}[x(\tau),y(\tau)], \tag{37}$$

by which the defining equation reduces to

$$\frac{dF}{d\tau} = P_{\underset{\sim}{L}}[x(\tau),y(\tau)]\; F(\tau) + U_{\underset{\sim}{J}}[x(\tau),y(\tau)] \quad . \tag{38}$$

The defining equation on a characteristic (or "trajectory") has thus been reduced to three coupled ODEs, which can be integrated using standard numerical techniques (with due care taken to avoid loss of accuracy near the multisingular point, where the approximant may well diverge).

<u>Invariance properties</u>. Extensive experience with Padé approximation has revealed [2,3,18] that the most reliable approximants are frequently those which are invariant under the *Euler transformation*

$$x \Rightarrow \bar{x} = Ax/(1 + Bx) \quad . \qquad (39)$$

Thus it is natural to ask whether two-variable approximants can be invariant under this Euler transformation of x, together with $y \Rightarrow \bar{y} = y$, or under the corresponding Euler transformation of y, or under both together. PDAs can, indeed, possess such invariance properties. Just as the Euler invariance of a Padé approximant [L/M] is ensured by the assignment L=M (regardless of the function being approximated), so Euler invariance of a PDA can be guaranteed by selecting appropriate label sets.

The label set restrictions which guarantee Euler invariance are easily stated. (Their proof is also not hard: see [29].) We first define the *degrees* $\hat{j}$ and $\hat{j}'$ of a label set $\underset{\sim}{J}$ as $\hat{j}=\max j$, $\hat{j}'=\max j'$, $(j,j') \in \underset{\sim}{J}$. The restrictions for Euler invariance in x are then:

(i) The polynomial label sets $\underset{\sim}{J}$, $\underset{\sim}{L}$, $\underset{\sim}{M}$ and $\underset{\sim}{N}$ possess a rectangular outline with $\hat{j} = \hat{\ell} = \hat{n}$ but with $\hat{m} = \hat{n} + 2$. (See figure 2.)

(ii) Elements may be removed from the rectangular outline, but if one element is removed, then all the elements to its left must also be removed. Such arrays are termed *flush-right*. Any number of elements may be removed, so the empty set $\emptyset$ is considered to be a flush-right set of any degree.

(iii) The matching set $\underset{\sim}{K}$ must be *flush-left*: that is, if some element falls outside $\underset{\sim}{K}$, then all the elements to its right must also fall outside $\underset{\sim}{K}$.

$$\underset{\sim}{J} = \begin{array}{|ccc} \times & \times & \times \\ \times & \times & \times \end{array} \qquad \underset{\sim}{L} = \begin{array}{|ccc} \cdot & \times & \times \\ \times & \times & \times \end{array} \qquad \underset{\sim}{M} = \begin{array}{|ccccc} \times & \times & \times & \times & \times \\ \cdot & \cdot & \cdot & \times & \times \end{array} \qquad \underset{\sim}{N} = \begin{array}{|ccc} \times & \times & \times \\ \cdot & \cdot & \times \\ \times & \times & \times \end{array}$$

$$\underset{\sim}{K} = \begin{array}{|cccccccc} \times & \times & \times & \times & \times & \times & \times & \cdot \\ \times & \times & \times & \times & \times & \cdot & \cdot & \cdot \\ \times & \times & \times & \times & \times & \times & \times & \times \\ \times & \times & \times & & & & & \\ \times & & & & & & & \end{array}$$

Figure 2. An example of label sets which serve to ensure that the PDA produced is Euler invariant in x. Note that $\hat{j} = \hat{\ell} = \hat{n} = 2$ while $\hat{m} = 4$.

The requirements for Euler invariance in y follow by symmetry from the above result. Because the requirements for Euler invariance in x and y are compatible, approximants can also be invariant under the combined change of variable

$$x \Rightarrow \bar{x} = Ax/(1 + Bx), \qquad y \Rightarrow \bar{y} = Cy/(1 + Dy). \tag{40}$$

In these approximants, the elements removed from the polynomial label sets will cluster toward the upper left: such arrays may be called *flush-high*.

Two variables may be subjected to many transformations which have no analog in the single-variable case. Thus it is reasonable to look for PDA invariances beyond Eulerian. A natural possibility is the *generalized rotation*

$$x \Rightarrow \bar{x} = a_{11}x + a_{12}y , \qquad y \Rightarrow \bar{y} = a_{21}x + a_{22}y , \tag{41}$$

and indeed PDAs are invariant under such transformations provided that $\underset{\sim}{M} = \underset{\sim}{N}$ and that all the sets $\underset{\sim}{J}$, $\underset{\sim}{L}$, $\underset{\sim}{M}$, $\underset{\sim}{N}$ and $\underset{\sim}{K}$ are chosen as *gapped triangles* [29]. A gapped triangle is just a triangular array [see (15)] with some (or all) of its counterdiagonals completely removed, as in the example:

$$\begin{array}{|llll} \hline \times & \times & \cdot & \times \\ \times & \cdot & \times & \\ \cdot & \times & & \\ \times & & & \end{array} \quad . \tag{42}$$

Another possible transformation is the skew Euler transformation

$$x \Rightarrow \bar{x} = Cy/(1 + Dy), \qquad y \Rightarrow \bar{y} = Ax/(1 + Bx) \ . \tag{43}$$

Although this transformation again admits PDA invariances [29], the necessary restrictions on the label sets are somewhat involved and so they will not be reproduced here.

It has been shown [28] that, under certain mild conditions, the only variable transformations which can admit PDA invariances are the three presented here [(40), (41) and (43)] and two others which have not proven important in practical applications and so will not be exposed here.

Open questions. The theory of partial differential approximants is far from complete: many open questions remain. To tempt the reader, we mention three.

An obviously important problem concerns the question of convergence of a PDA to the function being approximated. We know of no general results.

Can anything be gained by taking the functions $U(x,y)$, $P(x,y)$, $Q(x,y)$ and $R(x,y)$ analytic but not necessarily polynomials? Corresponding generalizations of ordinary Padé approximants have been useful in certain contexts.

Several techniques are known for rapid computation of the defining polynomial coefficients of Padé approximants (see the review [17]) and of Canterbury approximants (the "prong structure", see [7]). Can similar algorithms be developed for PDAs? They could be useful in cases of high order where the generating equations may be poorly conditioned.

4. Applications of Partial Differential Approximants

The theoretical questions discussed in the previous section are of interest in their own right, but their answers may be of narrow significance unless partial differential approximants prove to be a practical tool; such a proof can come only from real applications.

Traditionally, the first application of a new approximation technique is to study exactly known test functions. This can, of course, be done for partial differential approximants, but because many of the test functions that most naturally come to mind can be represented exactly by PDAs [9] the results may not prove very informative! Some work in this direction has been performed by Stilck and Salinas [25], but it still remains to test the technique against a wide variety of functions, both exhibiting asymptotic scaling forms, and with more complex nonscaling multisingularities involving, for example, logarithmic behavior.

The first significant application of PDAs was to the problem of "bicriticality" in the theory of phase transitions. In this case the function under study was believed [23] to possess precisely the cusped singularity structure described in the Section 2, and the problem was to find reliable, accurate estimates for the exponents γ and ϕ, and for the scaling function $Z(z)$. An initial study [14] focused on the exponents, and produced estimates in good agreement with the experimental values observed in antiferromagnetic crystals by King and Rohrer [21]. After this success, approximants were integrated [13] to evaluate the scaling functions which in this case are of form (12). An important descriptor of such a scaling function is the ratio $Q = -z_2/z_1$ of the two singular points of $Z(z)$. This value, which is expected to be a universal number, was estimated at $Q = 2.34 \pm 0.08$ in good agreement with values based on quite different theoretical considerations. Unfortunately, naive statistical fits to the experiments [21] suggested $Q = 1.56 \pm 0.35$. While this discrepancy has not yet been fully resolved, a study [13] of the data fitting techniques used by King and Rohrer has shown that even very small amounts of uncertainty in the experimental data can bias the fitted values of Q significantly below the true value!

Partial differential approximants have also been applied successfully to the problem of elucidating the "corrections to scaling" at the critical point of the three-dimensional Ising model [5]. In this problem the quantity of interest, namely the susceptibility $\chi(T)$ as a function of the temperature T, is expected to behave as

$$\chi(T) = At^{-\gamma}[1 + a_\theta t^\theta + a_1 t + \cdots] , \qquad (44)$$

when $t = (T - T_c)/T_c \to 0$. The issue is the appropriateness of this form and, in particular, the value of the (positive) correction exponent, θ, and the influence that it has on estimates for the leading exponent, γ. As posed, this represents a very badly conditioned problem in double exponential fitting, the more so since T_c must also be estimated. Indeed, single-variable analyses using series in powers of $x \propto 1/T$ for this (and related problems) have produced reasonable, consistent results only when reinforced by techniques which significantly bias the results towards some form of assumed behavior. On the other hand, it has proved possible to introduce a natural second variable, y, and thereby embed the Ising model in two families of models, the so-called double-Gaussian and Klauder models [5]. Within this context a single multisingular point is expected in the "physical range" $0 < y \lesssim 2$ if the prejudice (44), with A, a_θ, and a_1 now functions of y, is correct. However, the multisingular "crossover" exponent ϕ should now be *negative* and equal to $-\theta$. It follows that the scaling function, Z(z), should have no accessible singularities and that the singular locus passing through (x_c, y_c) should be smooth rather than cusped, as in the bicritical problem. As a consequence, the multisingular structure is more subtle and difficult to study than at a bicritical point. However, by analyzing two-variable upper triangular series to order x^{21}, the anticipated multisingular structure and, thereby, the prejudices embodied in (44) have been confirmed. By the same token, estimates for θ ($= 0.54 \pm 0.02$) and for γ could be obtained free of any special biassing assumptions [5]. A similar analysis for the two-dimensional Ising model (to be published) suggests $\theta \cong 1.3$ in accord with interesting theoretical speculations which yield $\theta = 4/3$. A study of the scaling functions for these problems is under way and should throw further light on the difficulties encountered in some of the single-variable work.

Finally, PDAs have recently been used to study the phenomenon of "dimensional crossover" observed in various statistical mechanical models in which change of system dimensionality and, hence, of singular behavior occurs when some coupling parameter (e.g., between lattice layers) goes to zero. This question had previously been analyzed in the context of anisotropic bond percolation [27] in two dimensions by single-variable techniques [24]. In this case much exact information is available: for example, it is known that the principal function of interest is symmetric, $f(x,y) = f(y,x)$, and reduces to a simple, explicit form when $x = 0$. By restricting the defining polynomials appropriately all this information can be incorporated precisely into the approximants. Study of the resulting PDAs confirmed the results of the single-variable work but went on to reveal features of the crossover in a detail inaccessible to the simpler techniques.

Acknowledgements

We are indebted to the U.S. National Science Foundation for its support both through grant DMR81-14726-01 to Rutgers University, and through grants from the Applied Mathematics Program and the Materials Research Laboratory Program at Cornell University.

References

1. G.A. Baker, Jr., Phys. Rev. 124, 768-774 (1961).

2. G.A. Baker, Jr., *Essentials of Padé Approximants*, New York: Academic Press (1975).

3. G.A. Baker, Jr. and P.R. Graves-Morris, *Padé Approximants*, parts I & II, Reading, Massachusetts: Addison-Wesley (1981).

4. N.K. Bose and S. Basu, IEEE Transactions on Automatic Control AC-25, 509-514 (1980).

5. J.-H. Chen, M.E. Fisher and B.G. Nickel, Phys. Rev. Lett. 48, 630-634 (1982).

6. J.S.R. Chisholm, Math. Comp. 27, 841-848 (1973).

7. J.S.R. Chisholm, In *Padé and Rational Approximation* (eds. E.B. Saff and R.S. Varga), pp. 23-42. New York: Academic Press (1977).

8. M.E. Fisher, Rev. Mod. Phys. 46, 597-616 (1974).

9. M.E. Fisher, Amer. Inst. Phys. Conf. Proc. No. 24, *Magnetism and Magnetic Materials*, 1974, pp. 273-280, A.I.P., New York (1975).

10. M.E. Fisher, Physica B 86-88, 590-592 (1977); also in *Statistical Mechanics and Statistical Methods in Theory and Application* (ed. U. Landman), pp. 3-31, New York: Plenum Press (1977).

11. M.E. Fisher and H. Au-Yang, J. Phys. A 12, 1677-1692; 13, 1517 (1979).

12. M.E. Fisher and J.-H. Chen, In Proceedings 1980 Cargèse Summer Institute on *Phase Transitions* (ed. M. Lévy, J.-C. Le Guillou and J. Zinn-Justin), pp. 169-216. New York: Plenum Press (1982).

13. M.E. Fisher, J.-H. Chen and H. Au-Yang, J. Phys. C 13, L459-464 (1980).

14. M.E. Fisher and R.M. Kerr, Phys. Rev. Lett. 32, 667-670 (1977).

15. M.E. Fisher and D.F. Styer, Proc. Roy. Soc. A 384, 259-287 (1982).

16. P.R. Garabedian, *Partial Differential Equations*. New York: John Wiley & Sons (1964).

17. P.R. Graves-Morris, In *Padé Approximation and its Applications* (ed. L. Wuytack), pp. 231-245. Berlin: Springer-Verlag (1979).

18. D.L. Hunter and G.A. Baker, Jr., Phys. Rev. B 7, 3346-3376, 3377-3392 (1973).

19. D.L. Hunter and G.A. Baker, Jr., Phys. Rev. B 19, 3808-3821 (1979).

20. J. Karlsson and H. Wallin, In *Padé and Rational Approximation* (ed. E.B. Saff and R.S. Varga), pp. 83-100. New York: Academic Press (1977).

21. A.R. King and H. Rohrer, Phys. Rev. B 19, 5864-5876 (1979).

22. C.H. Lutterodt, J. Phys. A 7, 1027-1037 (1974).

23. P. Pfeuty, D. Jasnow and M.E. Fisher, Phys. Rev. B 10, 2088-2112 (1974).

24. S. Redner and H.E. Stanley, J. Phys. A 12, 1267-1283 (1979).

25. J.F. Stilck and S.R. Salinas, J. Phys. A 14, 2027-2046 (1981).

26. D.F. Styer, In Proceedings 1983 Geilo Study Institute on *Multicritical Phenomena*. (To be published by Plenum Press.)

27. D.F. Styer, *Partial Differential Approximants and Applications to Statistical Mechanics*. Ph.D. thesis, Cornell University (1984).

28. D.F. Styer, Proc. Roy. Soc. A 390, 321-339 (1983).

29. D.F. Styer and M.E. Fisher, Proc. Roy. Soc. A 388, 75-102 (1983).

ZEROS OF POLYNOMIALS GENERATED BY 4-TERM RECURRENCE RELATIONS

Marcel G. de Bruin
Department of Mathematics
University of Amsterdam
Roetersstraat 15, 1018 WB Amsterdam
The Netherlands

Abstract. In Padé approximation the study of the location of the poles of the approximants, i.e. the zeros of the denominators, is of some importance. Loosely speaking, one might say that a (compact) set in the complex plane which is zero-free for a sequence of Padé approximants is the set where this sequence converges to the function it is derived from (under some suitable restrictions of course). As the most widely used sequences (e.g. steplines, diagonals) lead to certain "well-balanced" recurrence relations containing three terms, the study of the behaviour of the zeros of polynomials satisfying 3-term recurrence relations *without thinking about a connection with any Padé table whatsoever* has received some attention (cf. the famous Parabola theorem due to Saff & Varga [9] and Henrici [5], Leopold [6], Runckel [8]). The aim of this paper is to study the location of the zeros of polynomials generated by certain recurrence relations arising from sequences of approximants in a simultaneous Padé table.

1. Introduction

For sake of simplicity the case of simultaneous approximation to two functions will be treated here. Consider two formal power series with complex coefficients

$$(1)\quad f^{(i)}(z) = \sum_{n=0}^{\infty} c_n^{(i)} z^n \ , \ c_0^{(i)} \neq 0 \qquad (i=1,2)$$

and define the Padé-2-table (cf. Mall [7], De Bruin [1]) in the following manner from any triple k_0,k_1,k_2 of nonnegative integers. Put $s = k_0 + k_1 + k_2$, then the problem is to find three polynomials $P^{(i)}(z) = P^{(i)}(k_0,k_1,k_2;z)$ with coefficients satisfying

$$(2)\quad \begin{cases} \text{(a)}\ \deg P^{(i)} \leqslant s - k_i & (i=0,1,2) \\ \text{(b)}\ P^{(0)} f^{(i)} - P^{(i)} = O(z^{s+1}) & (i=1,2) \end{cases}$$

Without going into details (cf. [1], [7]) connected with existence and uniqueness (under mild restrictions) of a solution, the Padé-2-table is then defined to be a table with three entries (k_0,k_1,k_2) where at each point its unique solution of (2), normalized by $P^{(0)}(0) = 1$, is located. Assume for the time being tnat the table is normal (equality in (2a) and in at least one of (2b), solution unique up to a multiplicative constant).

Consider now a generalized stepline: a sequence of points, starting at $(k_0,0,0)$, where at each step the next point is found by increasing - in a cyclic manner - the coordinates by unity; the first increase is in the first coordinate i.e.

$$(3)\quad \begin{cases} (k_0,0,0), & (k_0+1,0,0), & (k_0+1,1,0), & (k_0+1,1,1), & (k_0+2,1,1), & (k_0+2,2,1), \ldots \\ \text{nr. } 0 & 1 & 2 & 3 & 4 & 5 \quad \ldots \end{cases}$$

The three sequences of polynomials $\{\{P_n^{(i)}(z)\}_{n=0}^{\infty}\}_{i=0}^{2}$, where the index refers to the number of the point the polynomials belong to, all satisfy the same recurrence relation given below for $P_n^{(0)}$:

$$(4a)\qquad P_n^{(0)}(z) = P_{n-1}^{(0)}(z) + a_n z\, P_{n-2}^{(0)}(z) + b_n z^2\, P_{n-3}^{(0)}(z) \quad (b_n \neq 0;\ n=2,3,\ldots)$$

with initial values for $P^{(0)}$:

$$(4b)\qquad P_{-1}^{(0)}(z) = 0,\ P_0^{(0)}(z) = 1,\ P_1^{(0)}(z) = 1.$$

For an extensive treatment of this type of recurrence relation, sometimes called generalized-C-fraction, cf. [2].

However, given the normality of the table, the degrees of the polynomials $P_n^{(0)}$ do not increase by unity at each step along the stepline; they are 0,0,1,2,2,3,4,4,... Although this does not look promising for the sequel, the behaviour of the zeros still turns out to be nice in many cases from the viewpoint of finding zero-free regions; for an example the reader is referred to section 3. But even worse things might happen if the recurrence relation (4) is studied away from the context of the Padé-2-table! For instance if $a_n=1$ $(n=2,3,4,5)$, $b_2=1, b_3=2, b_4=-3$ and $b_5=-2$ one finds easily

$$P_2(z)=1+z,\ P_3(z)=1+2z+2z^2,\ P_4(z)=1+3z,\ P_5(z)=1+4z,$$

showing a rather erratic behaviour of the degrees (even more while any choice of $b_6 \neq 0$ leads to deg $P_6=4$).

To avoid the aforementioned difficulties this type of recurrence relation will not be studied here; other steplines in the Padé-2-table will be the principal topic of this paper, as in Theorem 1.

There is a sequence of points in the table for which the degrees of the denominators build up as in the Parabola theorem: deg $P_n = n$. From (2) it is obvious that in this case at each step on the path in the table either the second or the third coordinate has to increase by unity (possibly accompanied by an - arbitrary - increase in the first coordinate). Actually there are two possibilities (each leading to two cases in its turn: a basic path and a path that arises from interchanging the roles of the second and third coordinate) of which only that case will be given that leads to a so-called regular algorithm (cf. [3]).

Consider a "restricted stepline" where the first coordinate is fixed and where the second and third are increased by unity in a cyclic manner, starting with the second coordinate i.e.

$$(5)\quad \begin{cases} (k_0,0,0),(k_0,1,0),(k_0,1,1),(k_0,2,1),(k_0,2,2),(k_0,3,2),\dots \\ \text{nr. } 0 \qquad\qquad 1 \qquad\qquad 2 \qquad\qquad 3 \qquad\qquad 4 \qquad\qquad 5 \end{cases}$$

The recurrence relation, as in (4) for the three sequences of polynomials the same, is

$$(6)\quad \begin{cases} P_n^{(0)}(z) = (1+a_nz)\,P_{n-1}^{(0)}(z) + b_nz\,P_{n-2}^{(0)}(z) + c_nz^2\,P_{n-3}^{(0)}(z) \\ \qquad\qquad\qquad\qquad\qquad\qquad\qquad (a_n, c_n \neq 0;\ n=2,3,4,\dots) \\ P_{-1}^{(0)}(z) = 0,\ P_0^{(0)}(z) = 1,\ P_1^{(0)}(z) = 1+a_1z \quad (a_1 \neq 0) \end{cases}$$

This time a treatment of (6) away from the context of Padé approximation is possible also; it is simple to verify that, given the restrictions on the coefficients as stated, the degree of P_n is n.

In the next section some theoretical results will be given concerning (6) and a variant of (6) which leads to rather interesting pictures.

2. Main results

The results of this section, the proofs of which will be deferred to section 4, have been obtained using elementary properties of bilinear fractional transformations only. The matter of sharpness will not be touched upon, although there can be defined a "sense" in which the results are sharp: but that would be nonsense. For the reader who is well-versed in the methods of the study of the location of zeros, it is obvious that much sharper results (in an absolute sense) can be obtained by exploiting differential equations techniques. This will be the subject of subsequent research. Moreover, knowing the coefficients explicitly, a considerable improvement of the estimates can be obtained in some cases.

For sake of simplicity we will restrict ourselves to the case of *real coefficients* only; as will become clear from the method of proof in section 4, complex coefficients are allowed too after minor modifications.

Theorem 1 Consider the sequence of polynomials $\{P_n(z)\}_{n=0}^{\infty}$ generated by the following recurrence relation

$$P_n(z) = (1 + a_n z)\, P_{n-1}(z) + b_n z\, P_{n-2}(z) + c_n z^2\, P_{n-3}(z) \tag{7}$$

$$(a_n, c_n \in \mathbb{R}\setminus\{0\}, b_n \in \mathbb{R};\ n \geqslant 3)$$

with initial values

$$P_0(z) = 1,\ P_1(z) = 1 + a_1 z,\ P_2(z) = (1 + a_1 z)\,(1 + a_2 z) + b_2 z \tag{8}$$

$$(a_1, a_2 \in \mathbb{R}\setminus\{0\}, b_2 \in \mathbb{R}) .$$

Let $\{A_n^{(k)}\}_{n=1}^{\infty}$ (k=1,2,...,5) be seqences of positive real numbers and define the sets V_k of complex numbers z as follows

$$V_1 = \begin{cases} 1 + a_1 \mathrm{Re} z \geqslant A_1^{(1)};\ 1 + a_2 \mathrm{Re} z + b_2 \dfrac{\mathrm{Re} z + a_1 |z|^2}{|1 + a_1 z|^2} \geqslant A_2^{(1)} \\[2ex] 1 + a_n \mathrm{Re} z + \dfrac{2 b_n A_{n-2}^{(1)} \mathrm{Re} z + c_n \mathrm{Re}(z^2)}{4 A_{n-2}^{(1)} A_{n-1}^{(1)}} - \dfrac{2|b_n z| A_{n-2}^{(1)} + 3|c_n z^2|}{4 A_{n-2}^{(1)} A_{n-1}^{(1)}} \geqslant A_n^{(1)} \end{cases}$$

$$(n \geqslant 3)$$

$$V_2 = \begin{cases} 1 + a_1 \mathrm{Re} z \leqslant - A_1^{(2)} ; \; 1 + a_2 \mathrm{Re} z + b_2 \dfrac{\mathrm{Re} z + a_1 |z|^2}{|1 + a_1 z|^2} \leqslant -A_2^{(2)} \\ 1 + a_n \mathrm{Re} z - \dfrac{2b_n A_{n-2}^{(2)} \mathrm{Re} z - c_n \mathrm{Re}(z^2)}{4A_{n-2}^{(2)} A_{n-1}^{(2)}} + \dfrac{2|b_n z| A_{n-2}^{(2)} + 3|c_n z^2|}{4A_{n-2}^{(2)} A_{n-1}^{(2)}} \leqslant \\ \qquad \leqslant - A_n^{(2)} \quad (n \geqslant 3) \end{cases}$$

$$V_3 = \begin{cases} \mathrm{Re} z + a_1 |z|^2 \geqslant A_1^{(3)} |z|^2 ; \; \mathrm{Re} z + a_2 |z|^2 + b_2 \dfrac{1 + a_1 \mathrm{Re} z}{|1 + a_1 z|^2} |z|^2 \geqslant \\ \qquad \geqslant A_2^{(3)} |z|^2 \\ \mathrm{Re} z + a_n |z|^2 + \dfrac{2b_n A_{n-1}^{(3)} + c_n}{4A_{n-2}^{(3)} A_{n-1}^{(3)}} \mathrm{Re} z - \dfrac{|2b_n A_{n-2}^{(3)} + c_n| + 2|c_n|}{4A_{n-2}^{(3)} A_{n-1}^{(3)}} |z| \geqslant \\ \qquad \geqslant A_n^{(3)} |z|^2 \quad (n \geqslant 3) \end{cases}$$

$$V_4 = \begin{cases} \mathrm{Re} z + a_1 |z|^2 \leqslant -A_1^{(4)} |z|^2 ; \; \mathrm{Re} z + a_2 |z|^2 + b_2 \dfrac{1 + a_1 \mathrm{Re} z}{|1 + a_1 z|^2} |z|^2 \leqslant \\ \qquad \leqslant -A_2^{(4)} |z|^2 \\ \mathrm{Re} z + a_n |z|^2 - \dfrac{2b_n A_{n-2}^{(4)} - c_n}{4A_{n-2}^{(4)} A_{n-1}^{(4)}} \mathrm{Re} z + \dfrac{|2b_n A_{n-2}^{(4)} - c_n| + 2|c_n|}{4A_{n-2}^{(4)} A_{n-1}^{(4)}} |z| \leqslant \\ \qquad \leqslant -A_n^{(4)} |z|^2 \quad (n \geqslant 3) \end{cases}$$

$$V_5 = \begin{cases} |1 + a_1 z| \geqslant A_1^{(5)} ; \; |1 + a_2 z| \geqslant A_2^{(5)} + \dfrac{|b_2 z|}{A_1^{(5)}} \\ |1 + a_n z| \geqslant A_n^{(5)} + \dfrac{|b_n z|}{A_{n-1}^{(5)}} + \dfrac{|c_n z^2|}{A_{n-2}^{(5)} A_{n-1}^{(5)}} \quad (n \geqslant 3) \end{cases} .$$

Then the set $\mathbb{C} \setminus \bigcup_{k=1}^{5} V_k$ contains all zeros of all of the polynomials $\{P_n(z)\}_{n=1}^{\infty}$.

Another type of recurrence relation (not arising from a regular algorithm in a normal Padé-2-table) led to very interesting results.

Theorem 2 Let the sequence $\{Q_n(z)\}_{n=0}^{\infty}$ be generated by

$$(9) \qquad Q_n(z) = (1 + a_n z)\, Q_{n-1}(z) + b_n z^2\, Q_{n-2}(z) + c_n z^3\, Q_{n-3}(z) \quad (n \geqslant 3)$$

with

$$(10) \qquad Q_0(z) = 1,\; Q_1(z) = 1 + a_1 z,\; Q_2(z) = (1 + a_1 z)(1 + a_2 z) + b_2 z^2.$$

Furthermore assume

$$(11) \qquad a_n > 0\ (n \geqslant 1),\; b_n > 0\ (n \geqslant 2),\; c_n > 0\ (n \geqslant 3).$$

Define the sets W_k of complex numbers z as follows

$$W_1 = \begin{cases} \mathrm{Re}z + a_1|z|^2 \geqslant A_1^{(1)}|z|^2 \\[2ex] \mathrm{Re}z + a_2|z|^2 + b_2 \dfrac{\mathrm{Re}z + a_1|z|^2}{|1 + a_1 z|^2}|z|^2 \geqslant A_2^{(1)}|z|^2 \\[2ex] \mathrm{Re}z + \left(a_n - \dfrac{c_n^2}{8A_{n-1}^{(1)} A_{n-2}^{(1)} (2A_{n-2}^{(1)} b_n + c_n)}\right)|z|^2 \geqslant A_n^{(1)}|z|^2 \quad (n \geqslant 3) \end{cases}$$

$$W_2 = \begin{cases} \mathrm{Re}z + a_1|z|^2 \leqslant -A_1^{(2)}|z|^2 \\[2ex] \mathrm{Re}z + a_2|z|^2 + b_2 \dfrac{\mathrm{Re}z + a_1|z|^2}{|1 + a_1 z|^2}|z|^2 \leqslant -A_2^{(2)}|z|^2 \\[2ex] \mathrm{Re}z + \left(a_n + \dfrac{P_n}{4A_{n-1}^{(2)} A_{n-2}^{(2)}}\right)|z|^2 \leqslant -A_n^{(2)}|z|^2 \quad (n \geqslant 3);\ \text{where} \end{cases}$$

$$P_n = 4(c_n - A_{n-2}^{(2)} b_n) \text{ for } A_{n-2}^{(2)} b_n \leqslant 3c_n/4;\; P_n = \frac{c_n^2}{2(2A_{n-2}^{(2)} b_n - c_n)}$$

for $A_{n-2}^{(2)} b_n \geq 3c_n/4$

$$W_3 = \begin{cases} |1 + a_1 z| \geq A_1^{(3)}|z|^2; |1 + a_2 z| \geq (A_2^{(3)} + \frac{b_1}{A_1^{(3)}})|z|^2 \\ |1 + a_n z| \geq (A_n^{(3)} + \frac{b_n}{A_{n-1}^{(3)}} + \frac{c_n}{A_{n-1}^{(3)} A_{n-2}^{(3)}})|z|^2 \quad (n \geq 3) \end{cases}$$

$$W_4 = \begin{cases} |\operatorname{Im} z| \geq A_1^{(4)}|z|^2; |\operatorname{Im} z(1 - \frac{b_2|z|^2}{|1 + a_1 z|^2})| \geq A_2^{(4)}|z|^2 \\ |\operatorname{Im} z| \geq (A_n^{(4)} + \frac{2(b_n A_{n-2} + c_n) + \sqrt{(2b_n A_{n-2}^{(4)})^2 + c_n^2}}{4A_{n-1}^{(4)} A_{n-2}^{(4)}})|z|^2 \quad (n \geq 3). \end{cases}$$

Again the sequences $\{A_n^{(k)}\}_{n=1}^{\infty}$ $(k = 1,2,3,4)$ are sequences of positive real numbers. Then the set $\mathbb{C} \setminus \bigcup_{k=1}^{4} W_k$ contains all zeros of all of the polynomials $\{Q_n(z)\}_{n=1}^{\infty}$.

It is obvious that the explicit form of the set containing all zeros depends heavily upon the values of the coefficients; therefore the theorems are given in their rather uninviting form; and simplification of the formulae will only be given by treating some examples.

3. Examples

In Figs. 1-4, we show the zeros of the denominators associated with selected points on the *main* stepline (defined by taking k=0 in (3)) of the Padé-2-table for $f^{(1)}(z) = \exp(-z)$ and $f^{(2)}(z) = {}_1F_1(1;c;-z)$, with $c=\pm 4.5$. The denominator polynomials are generated by the recurrence relations (4a,4b). Although consideration of the main stepline is not the principal theme of this paper, it is interesting to see the way in which the poles of the associated simultaneous Padé approximants cluster in these figures. Furthermore, these figures illustrate some of the results of the detailed analysis of this problem given in [4].

As an application of Theorem 1, certain restricted steplines in the same tables are treated in Figs. 5-8. The zeros of the polynomials generated by the recurrence relation (6) are shown; space does not allow us to write down all the coefficients explicitly. For comparison, the zeros of the polynomials generated by (7), (8) are treated in Figs. (9), (10) for $a_n=1.0$, $b_n=-0.5$, $c_n=1.0$ and in Figs. (11), (12) for $a_n=1.0$, $b_n=-0.5$, $c_n=1/n$.

Finally an example of theorem 2 is given in two cases in figs. 13-16. To compare the actual location of the zeros with that following from theorem 2, figs. 13, 14 - where all a_n, b_n and c_n are 1 - will be treated in some detail.

(a) With $A_n^{(1)} = .534 \ (n \geqslant 1)$: $W_1 \supset \{z : |z + 1.97| \geqslant 1.97\}$.

(b) With $A_n^{(2)} = .886 \ (n \geqslant 1)$: $W_2 \supset \{z : |z + .239| \leqslant .239\}$.

(c) With $A_n^{(4)} = 1.355 \ (n \geqslant 1)$: $W_4 \supset \{z : |z + .201i| \leqslant .210$ or $|z - .210i| \leqslant .210\}$.

It is easy to check that

(d) $W_3 \subset W_1 \cup W_2 \cup W_4$.

Of course it must be possible to improve upon these sets (especially W_1 as can be seen in figs. 13, 14); for instance by choosing A's *depending upon* z.

The example treated in figs. 15 and 16 leads to zero-free regions

$W_1 \supset \{z : |z + 0.76| \geqslant 0.76\}$ (via $A_1 = A_2 = 0.338$, A_3 etc. optimal)

$W_2 \supset \{z : |z + 0.354| \leqslant 0.354\}$ (via $A_1 = 0.414, A_2 = 0.25$, then optimal)

$W_4 \supset \{z : |z + 0.244i| \leqslant 0.244$ or $|z - 0.244i| \leqslant 0.244\}$ ($A_1 = 2.05$, $A_2 = 1.562$ etc.)

With "optimality" the following is meant: choose A_1 (or A_1 and A_2 if the conditions are the "same" cf. W_1 for $|1 + a_1 z|$ large) and choose subsequent A's in such a manner that the bounds on z turn out to be the same for all n.

It will be clear from the examples given, that the zero free regions depend heavily upon the value of the coefficients (cf. figs. 13, 14 versus figs. 15, 16). Moreover it is obvious that other methods than used in this paper probably will lead to explicit expressions for the limiting curve of the zeros (cf. the work by Saff & Varga [10]).

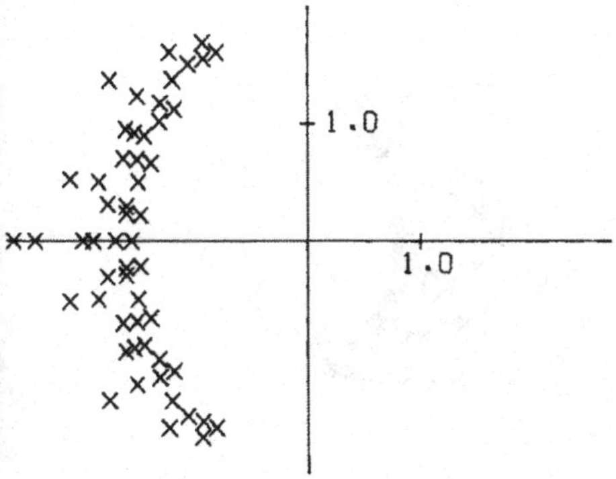

C = -4.5 ;n = 2,3,....,13

FIG. 1

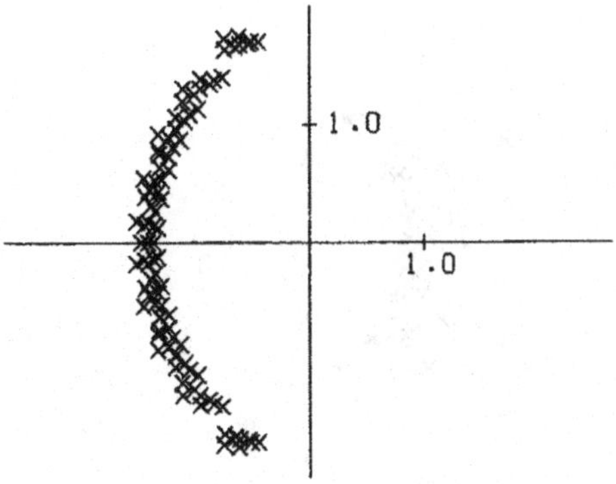

C = -4.5 ;n = 14,15,....,21

FIG. 2

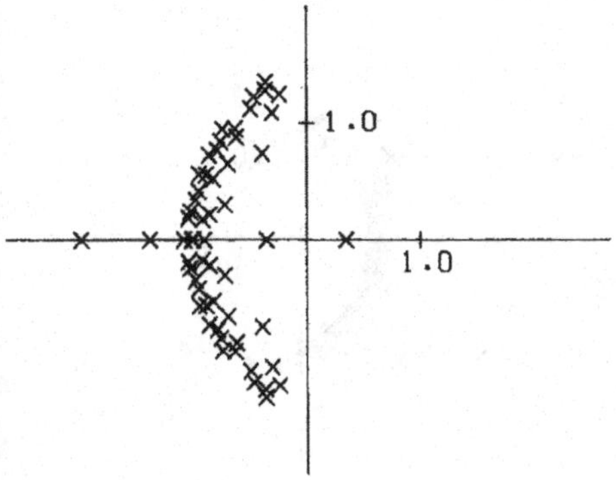

C = 4.5 ; n = 2,3,....,13

FIG. 3

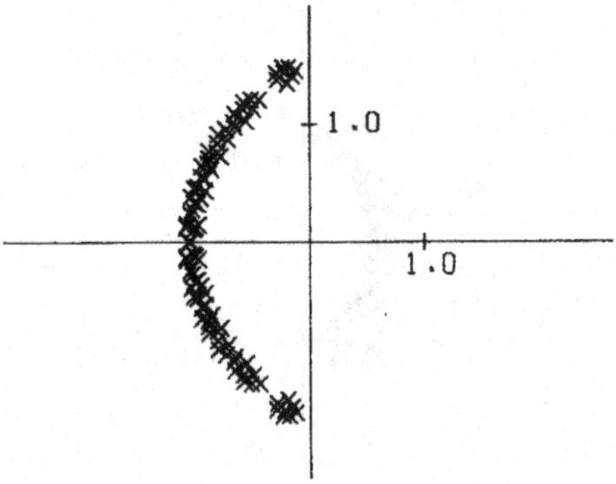

C = 4.5 ; n = 14,15,....,21

FIG. 4

Zeros of $P_n^{(0)}(z)$ versus z/n for the main stepline of the Padé-2-table for exp(-z) and ${}_1F_1(1;c;-z)$

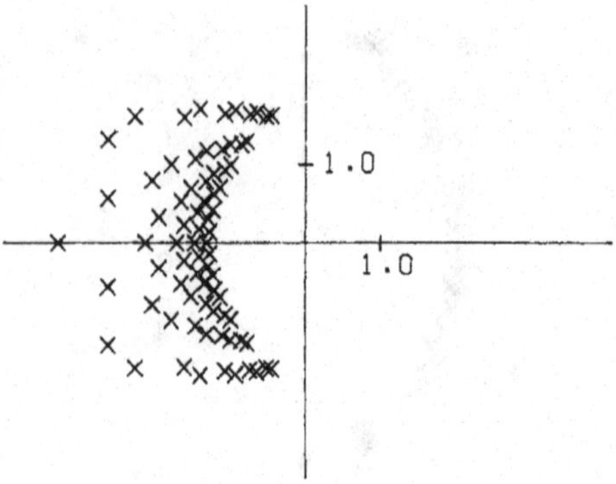

C=-4.5;POINTS (5,1,0) TO (5,6,6)

FIG. 5

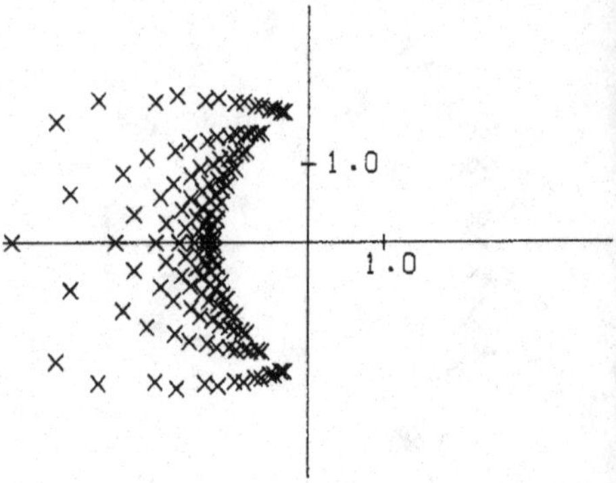

C=-4.5;POINTS (7,1,0) TO (7,8,8)

FIG. 6

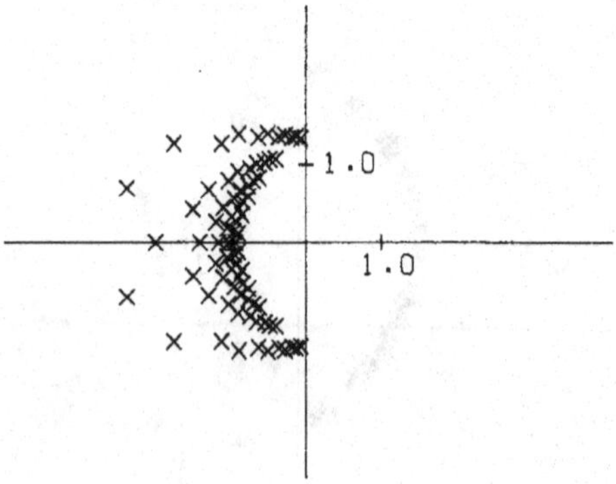

C= 4.5;POINTS (5,1,0) TO (5,6,6)

FIG. 7

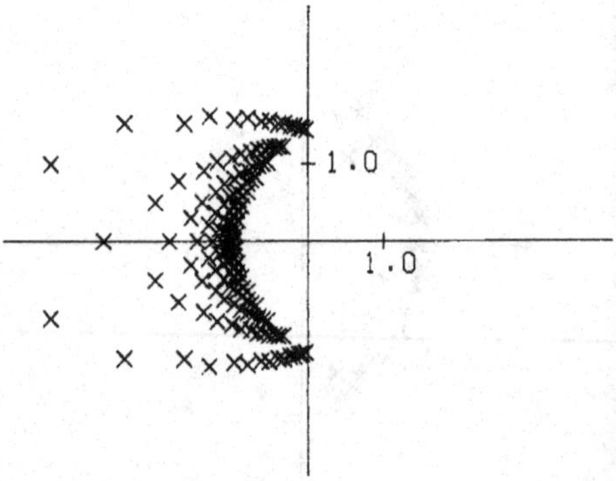

C= 4.5;POINTS (7,1,0) TO (7,8,8)

FIG. 8

Zeros of $P_n(z)$ versus z/n for the restricted stepline (5) of the Padé-2-table for exp(-z) and ${}_1F_1(1;c;-z)$

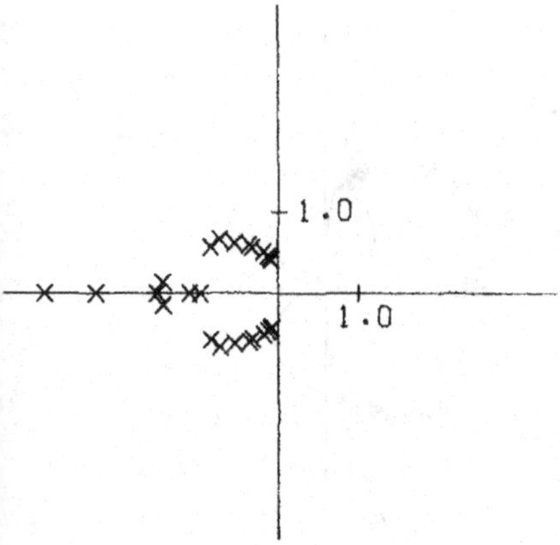

DEGREES RUNNING FROM 1 TO 7

FIG. 9

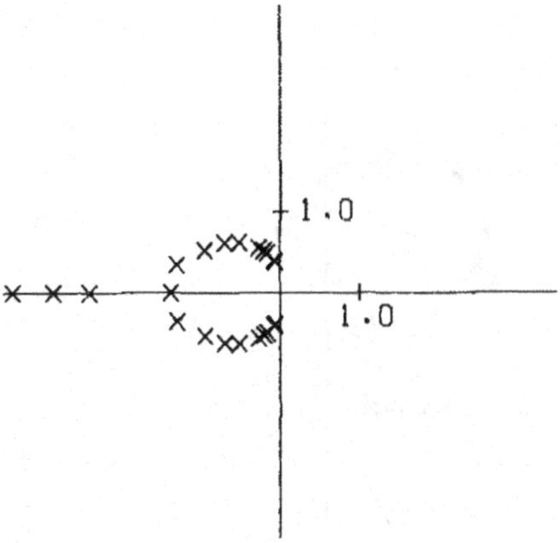

DEGREES RUNNING FROM 8 TO 10

FIG. 10

COEFFICIENTS: $a_n = 1.0$, $b_n = -0.5$, $c_n = 1.0$

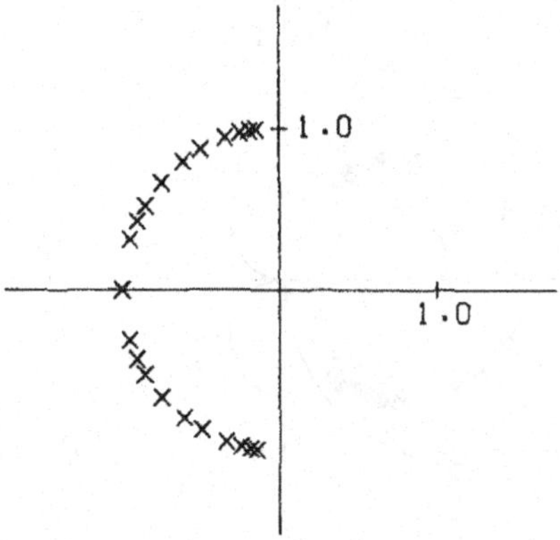

DEGREES RUNNING FROM 1 TO 7

FIG. 11

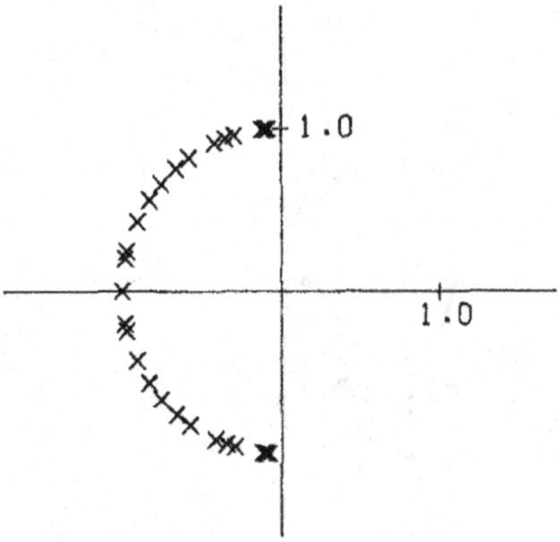

DEGREES RUNNING FROM 8 TO 10

FIG. 12

COEFFICIENTS: $a_n = 1.0$, $b_n = -0.5$, $c_n = 1/n$

Zeros of polynomials generated by a 4-term recurrence relation with coefficients $1 + a_n z$, $b_n z$, $c_n z^2$

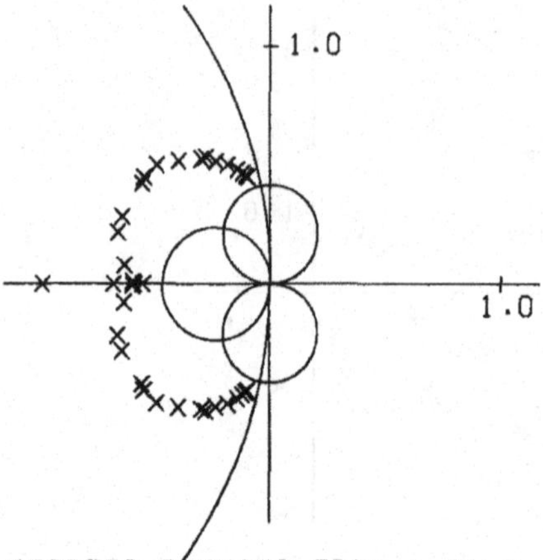

DEGREES RUNNING FROM 1 TO 8

FIG. 13

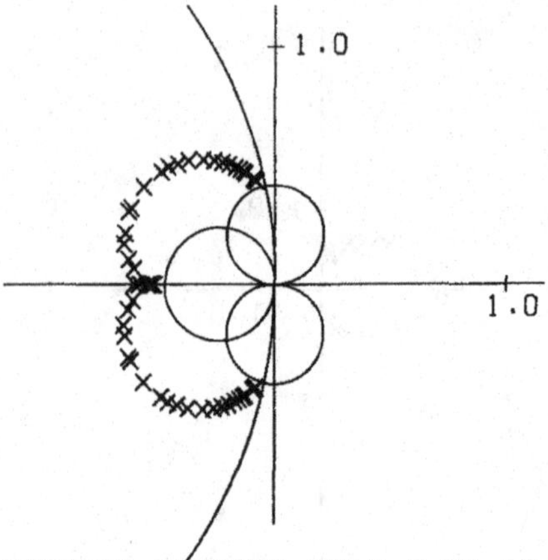

DEGREES RUNNING FROM 9 TO 13

FIG. 14

COEFFICIENTS: $a_n = 1.0$, $b_n = 1.0$, $c_n = 1.0$

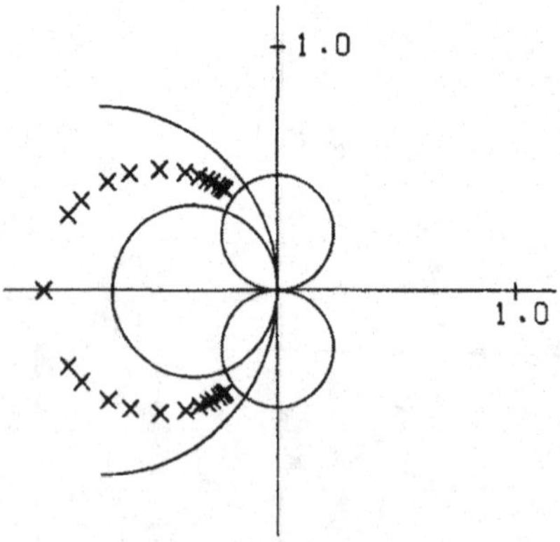

DEGREES RUNNING FROM 1 TO 8

FIG. 15

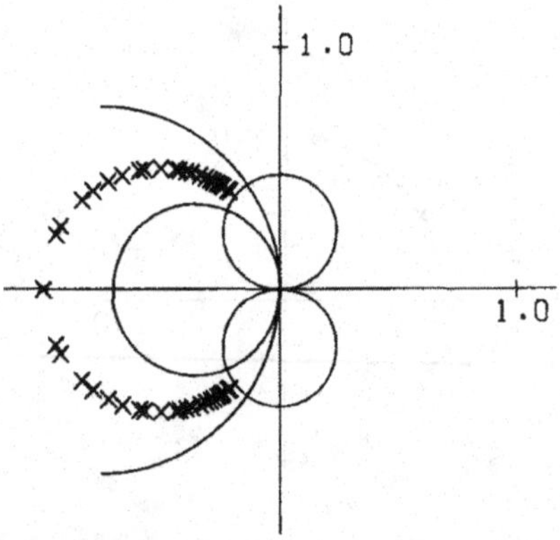

DEGREES RUNNING FROM 9 TO 13

FIG. 16

COEFFICIENTS: $a_n = 1.0$, $b_n = 1.0$, $c_n = 1/n$

Zeros of polynomials generated by a 4-term recurrence relation with coefficients $1 + a_n z$, $b_n z^2$, $c_n z^3$

4. Proofs

As the derivation of the sets V_k and W_k run along the same lines each time, only one will be treated explicitly while for the others the central quantity will be given only. The reader will be able to fill in quite easily the details that are omitted.

The set V_1

Define $\mu_n = \dfrac{P_n(z)}{P_{n-1}(z)}$ $(n \geq 1)$, then we have from (7) and (8) :

(a) $$\mu_1 = 1 + a_1 z,\ \mu_2 = 1 + a_2 z + \frac{b_2 z}{1 + a_1 z}.$$

(b) $$\mu_n = 1 + a_n z + \frac{b_n z}{\mu_{n-1}} + \frac{c_n z^2}{\mu_{n-1}\mu_{n-2}} \quad (n \geq 3).$$

Now restrict z in such a way that we have $\mathrm{Re}\mu_n \geq A_n^{(1)}$ $(n \geq 1)$ for a sequence of positive real numbers $\{A_n^{(1)}\}_{n=1}^{\infty}$.

From (a) we find

(c) $$1 + a_1 \mathrm{Re} z \geq A_1^{(1)},\quad 1 + a_2 \mathrm{Re} z + b_2 \frac{\mathrm{Re} z + a_1|z|^2}{|1 + a_1 z|^2} \geq A_2^{(1)}.$$

For the next step we use that the image of $\mathrm{Re}\, w_1 \geq A$ under the mapping $w_2 = \frac{1}{w_1}$ is the closed disc $|w_2 - \frac{1}{2A}| \leq \frac{1}{2A}$.

Inserting $\dfrac{1}{\mu_{n-1}} = \dfrac{1 + e^{i\psi}}{2A_{n-1}^{(1)}}$, $\dfrac{1}{\mu_{n-2}} = \dfrac{1 + e^{i\theta}}{2A_{n-2}^{(1)}}$, $z = re^{i\phi}$ and minimizing the resulting form for μ_n over $(\psi,\theta) \in [0,2\pi] \times [0,2\pi]$ we find the condition

(d) $$1 + a_n \mathrm{Re} z + \frac{2b_n A_{n-2}^{(1)} r \cos\phi + c_n r^2 \cos 2\phi - 2r|b_n|A_{n-2}^{(1)} - 3|c_n|r^2}{4A_{n-1}^{(1)} A_{n-2}^{(1)}} \geq$$
$$\geq A_n^{(1)}.$$

Now (c), (d) are the defining inequalities for V_1 and the proof follows deriving a contradiction from $z_0 \in V_1$ and $P_n(z_0) = 0$ in the manner given below:

- given $z_0 \in V_1$, let $P_n(z)$ be the first polynomial with $P_n(z_0) = 0$ (i.e. $P_k(z_0) \neq 0 \quad (k = 0,1,\ldots, n-1)$, $P_n(z_0) = 0$): thus $\mu_n(z_0) = 0$.
- $z_0 \in V_1$ implies $\quad \mathrm{Re}\,\mu_n(z_0) \geq A_n^{(1)} > 0$.

The other sets in the theorem 1 and 2 are derived in a similar manner using

- for V_2 : $\mathrm{Re}\,\dfrac{P_n(z)}{P_{n-1}(z)} \leq -A_n^{(2)}$
- for V_3 : $\mathrm{Re}\,\dfrac{P_n^*(z)}{P_{n-1}^*(z)} \geq A_n^{(3)}$
- for V_4 : $\mathrm{Re}\,\dfrac{P_n^*(z)}{P_{n-1}^*(z)} \leq -A_n^{(4)}$
- for V_5 : $\left|\dfrac{P_n(z)}{P_{n-1}(z)}\right| \geq A_n^{(5)}$
- for W_1 : $\mathrm{Re}\,\dfrac{Q_n^*(z)}{Q_{n-1}^*(z)} \geq A_n^{(1)}$
- for W_2 : $\mathrm{Re}\,\dfrac{Q_n^*(z)}{Q_{n-1}^*(z)} \leq -A_n^{(2)}$
- for W_3 : $\left|\dfrac{Q_n^*(z)}{Q_{n-1}^*(z)}\right| \geq A_n^{(3)}$
- for W_4 : $\mathrm{Im}\,\dfrac{Q_n^*(z)}{Q_{n-1}^*(z)} \geq A_n^{(4)}$ (and <u>real</u> coefficients imply the symmetry with respect to the real axis).

Here the standard notation

$$F_n^*(z) := z^n F_n\left(\frac{1}{z}\right)$$

for the inverted polynomial associated with a polynomial of degree n is used.

References

1. Bruin, M.G. de, Generalized C-fractions and a multidimensional Padé table, Thesis, Amsterdam, 1974.
2. Bruin, M.G. de, Convergence of generalized C-fractions, J. Approx. Theory <u>24</u> (1978), 177-207.
3. Bruin, M.G. de, Generalized Padé tables and some algorithms therein,

Proc. 1st French-Polish meeting on Padé approximation and convergence acceleration techniques, Warsaw 1981 (ed. J. Gilewicz), Centre de Physique Théorique C.N.R.S., Marseille, CPT-'81/PE. 1354, May 1982.

4. Bruin, M.G. de, Some convergence results in simultaneous rational approximation to the set of hypergeometric functions $\{{}_1F_1(1;c_i;z)\}_{i=1}^{n}$, in "Padé-Seminar 1983" (Vorlesungsreihe SFB 72; H. Werner, H.J. Bünger eds. Bonn, 1983), 95-117 [to appear in LNM].

5. Henrici, P., Note on a theorem of Saff and Varga, Padé and Rational Approximation, Theory and Applications, Academic Press, New York, 1977, 157-161.

6. Leopold, E., Approximants de Padé pour les fonctions de classe S et localisation des zéros de certains polynomes, Thèse de 3ième cycle, Univ. de Provence, January 1982.

7. Mall, J., Grundlagen für eine Theorie der mehrdimensionalen Padéschen Tafel, Inaugural Dissertation, München, 1934.

8. Runckel, H.-J., Zero-free parabolic regions for polynomials with complex coefficients, Proc.Am.Math.Soc. 88 (1983), 299-304.

9. Saff, E.B. and R.S. Varga, Zero-free parabolic regions for sequences of polynomials, SIAM J. Math. Anal. 7 (1976), 344-357.

10. Saff, E.B. and R.S. Varga, On the zeros and poles of Padé Approximants III, Numer. Math. 30 (1978), 241-266.

A LOWER BOUND FOR THE NUMBER OF ZEROS OF A FUNCTION ANALYTIC IN A DISK

Albert Edrei
Department of Mathematics
Syracuse University
Syracuse
New York 13210
U.S.A.

Abstract. Let $\phi(z)$ be a nonconstant analytic function regular for $|z| \leq 1$. The author shows that, with little additional information regarding $\phi(z)$, it is possible to obtain a lower bound for the number of zeros of $\phi(z)$ in the disk $|z| \leq t$ $(0 < t < 1)$.

1. Introduction. Let $\phi(z)$ be a nonconstant analytic function regular for $|z| \leq 1$.

It is almost trivial to derive, from Jensen's formula, upper bounds for the number $n(t)$ of zeros of $\phi(z)$ in the disk $|z| \leq t$ $(0 < t < 1)$.

Since lower bounds for $n(t)$ are not as obvious, I propose to establish in this note a lower bound for $n(t)$ which requires but scant information regarding $n(t)$. I prove

Lemma A. *Let* $\phi(z)$ *be a nonconstant function, regular for* $|z| \leq 1$. *Let* z_0 *be a point such that*

$$(1.1) \qquad |z_0| < 1, \qquad |\phi(z_0)| \geq 1,$$

and let ρ, R *and* $\varkappa$ *be real quantities such that*

$$(1.2) \qquad |z_0| < |z_0| + \rho < R < 1, \qquad 0 < \varkappa < 1.$$

Research supported in part by a grant from the National Science Foundation, Grant MCS-8301380.

Put

$$(1.3)\qquad M_1 = \max_{|z|=1} \log|\phi(z)|\,, \qquad M_0 = \max_{\theta} \log|\phi(z_0+\rho e^{i\theta})|$$

and let

$$(1.4)\qquad \max_{|z|=R} \log|\phi(z)| \geq \varkappa M_1\,.$$

Denote by n the number of zeros of $\phi(z)$ in the disk $|z|\leq 1$ and let

$$(1.5)\qquad \xi = \left(\frac{\varkappa\,\rho\,(1-R)^2}{50}\right)^{1/c}, \qquad c = \frac{\log\left(1+\frac{(1-R)^2}{4}\right)}{\log(4/\rho)}\,.$$

Then, if

$$(1.6)\qquad M_0 \leq \xi M_1\,,$$

we have

$$(1.7)\qquad n \geq \frac{\xi M_1}{\log(4/\rho)}\,.$$

The usefulness of a result such as Lemma A depends on the possibility of successfully applying it to the study of questions which present themselves naturally. As an illustration of the possible applications of Lemma A, I mention a property of the expansion

$$\frac{1}{\Gamma(z)} = \sum_{j=0}^{\infty} b_j z^j\,.$$

More precisely, I consider the partial sums

$$s_m(z) = \sum_{j=0}^{m} b_j z^j\,,$$

of the above series and study, for large values of m, the distribution of zeros of $s_m(z)$. I prove the following

Proposition. Given φ real, $|\varphi|<\pi$, there exist

(i) an infinite sequence $\mathfrak{M}$ of positive increasing integers;

(ii) an associated positive sequence $\{t_m\}_m$ $(m\in\mathfrak{M})$;

and

(iii) two positive constants γ_1, γ_2,

such that, under the restrictions

$$m \in \mathfrak{M}, \quad m > m_0, \quad R_m = \frac{m}{\log m},$$

there are, in the disk

$$|z - R_m t_m e^{i\varphi}| \leq \gamma_1 R_m t_m \frac{\log m}{m},$$

at least $\gamma_2 \log m$ zeros of $s_m(z)$.

The sequences $\mathfrak{M}$ and $\{t_m\}$ depend on φ and we always have

$$t_m = t_m(\varphi) \to \tau(\varphi) \qquad (m \to \infty,\ m \in \mathfrak{M}),$$

where $\tau = \tau(\varphi)$ is the unique positive solution, $\tau \in (0,1)$, of the equation

$$-\tau \cos\varphi - 1 - \log \tau = 0 .$$

The proofs of the above Proposition, and of other considerably more general theorems, make essential use of Lemma A. These results, to appear in [1], are suggested by (but do not follow from) some recent work of Edrei, Saff and Varga [2].

An inspection of the statement of Lemma A shows that it is of some importance to obtain, for the constant ξ in (1.6), as large a value as possible. This raises the question (probably difficult) of finding a sharp version of the Lemma. The value of the positive constant ξ becomes irrelevant if (as is the case in our proof of the Proposition) we apply Lemma A to each member of a sequence $\{\phi_k(z)\}_k$ such that

$$\frac{M_{0k}}{M_{1k}} \to 0 \qquad (k \to +\infty),$$

where M_{1k} and M_{0k} are given by (1.3) with ϕ replaced by ϕ_k.

2. Proof of Lemma A. It is convenient to introduce the bilinear transformation

$$z = \frac{z_0 + w}{1 + w\bar{z}_0}, \tag{2.1}$$

which maps the disk $|w| \leq 1$ onto the disk $|z| \leq 1$.

The auxiliary function

$$F(w) = \phi\left(\frac{z_0 + w}{1 + w\bar{z}_0}\right), \tag{2.2}$$

of the complex variable w has the following properties.

I. It is regular for $|w| \leq 1$,

II. $\max\limits_{|w|=1} \log|F(w)| = M_1$,

III. $\log|F(o)| = \log|\phi(z_0)| \geq 0$,

IV. The number of zeros of $F(w)$ is $|w| \leq 1$ is exactly n.

The above properties are immediate. We also need two elementary estimates; for the convenience of the reader we sketch brief proofs.

V. $\max\limits_{|w|=(\rho/2)} \log|F(w)| \leq M_0$.

To prove V it suffices to show that, if z is defined by (2.1) and

$$|w| \leq \frac{\rho}{2}, \tag{2.3}$$

then

$$|z-z_0| = \frac{|w|(1-|z_0|^2)}{|1+w\bar{z}_0|} < |w| \frac{1-|z_0|^2}{1-|z_0|} = |w|(1+|z_0|) < 2|w| \leq \rho . \tag{2.4}$$

It is now obvious that V follows from (1.3), (2.2) and (2.4).

VI. The quantity

$$R_1 = \frac{R+|z_0|}{1+R|z_0|} \tag{2.5}$$

satisfies the inequalities

$$|z_0| + \frac{\rho}{2} < R_1 < 1 - \frac{(1-R)^2}{2}, \tag{2.6}$$

$$\max\limits_{|w|=R_1} \log|F(w)| \geq \kappa M_1 . \tag{2.7}$$

Proof of VI. From (2.5) and one of the inequalities (1.2)

$$R_1 > \frac{\rho + 2|z_0|}{2},$$

which is equivalent to the first of the inequalities (2.6).

Similarly, since $|z_0| < R$, (2.5) yields

$$1 - R_1 = \frac{(1-R)(1-|z_0|)}{1+R|z_0|} > \frac{(1-R)^2}{2},$$

and the second inequality in (2.6) is proved.

To prove (2.7) we verify that the image in the z-plane, of $|w| \leq R_1$, contains the disk $|z| \leq R$. Starting from the elementary relation

$$\min_{\varphi} \left| \frac{z_0 + R_1 e^{i\varphi}}{1+\bar{z}_0 R_1 e^{i\varphi}} \right| = \min_{\varphi} \left\{ \frac{R_1^2 + |z_0|^2 + 2|z_0|R_1 \cos\varphi}{1+|z_0|^2 R_1^2 + 2|z_0|R_1 \cos\varphi} \right\}^{1/2},$$

we see that the above minimum is exactly

$$\frac{R_1 - |z_0|}{1-|z_0|R_1} = R .$$

Hence (2.7) follows from (1.4) and (2.2). The proof of assertion VI is now complete.

From this point on we study $F(w)$ and need never return to $\phi(z)$.

Let

$$(2.8) \qquad w_1, w_2, \ldots, w_n, \qquad (n \geq 0)$$

be all the zeros of $F(w)$ in $|w| \leq 1$, and let the first q of them lie in the disk $|w| \leq (\rho/2)$.

Define

$$(2.9) \qquad P(w) = \prod_{j=1}^{n} \left(1 - \frac{w}{w_j}\right).$$

Jensen's formula, III and V yield

$$(2.10) \qquad \sum_{j=1}^{q} \log \left| \frac{\rho}{2w_j} \right| \leq \frac{1}{2\pi} \int_0^{2\pi} \log |F(\tfrac{\rho}{2} e^{i\varphi})| \, d\varphi \leq M_0 .$$

From (2.9) and (2.10)

$$(2.11) \qquad \max_{|w| \leq 1} \log |P(w)| \leq n \log 2 - \sum_{j=1}^{q} \log |w_j| + (n-q) \log (\tfrac{2}{\rho}) =$$

$$= \sum_{j=1}^{q} \log \left| \frac{\rho}{2w_j} \right| + n \log (\tfrac{4}{\rho}) \leq M_0 + n \log (\tfrac{4}{\rho}) .$$

Since $F(w)/P(w)$ has no zeros in the disk $|w| \leq 1$, any branch of the auxiliary function $\log\{F(w)/P(w)\}$ is regular in the closed unit disk. Select some specific branch and represent it by the Poisson integral in complex form.

Under the assumptions

$$(2.12) \qquad |w| < t \leq 1,$$

we find

$$(2.13) \qquad \log\left(\frac{F(w)}{P(w)}\right) = \frac{1}{2\pi}\int_0^{2\pi} \log\left|\frac{F(te^{i\varphi})}{P(te^{i\varphi})}\right| \frac{te^{i\varphi}+w}{te^{i\varphi}-w}\,d\varphi + i\,C(t)$$

where $C(t)$ is a real constant. Taking $w=0$ we find

$$C(t) = \arg(F(0)/P(0)) = \gamma$$

and now introduce the new auxiliary function

$$(2.14) \qquad g(w) = \log\left(\frac{F(w)}{P(w)}\, e^{-i\gamma}\right), \qquad (g(0) = \log|F(0)| \geq 0).$$

To estimate $g(w)$ we use in (2.13) the well known notations of Nevanlinna's theory and some elementary properties of the proximity function $m(t,F)$. By (2.13) and (2.14)

$$(2.15) \qquad |g(w)| \leq \frac{1}{\pi(t-|w|)}\int_0^{2\pi} \left\{\left|\log|F(te^{i\varphi})|\right| + \left|\log|P(te^{i\varphi})|\right|\right\} d\varphi =$$

$$= \frac{2}{t-|w|}\,\{m(t,F) + m(t,\tfrac{1}{F}) + m(t,P) + m(t,\tfrac{1}{P})\}.$$

From Jensen's theorem we deduce

$$m(t,\tfrac{1}{P}) \leq m(t,P), \qquad 0 \leq \log|F(0)| \leq m(t,F) - m(t,\tfrac{1}{F}),$$

which used in (2.15) yield

$$(2.16) \qquad |g(w)| \leq \frac{4}{t-|w|}\,(m(t,F) + m(t,P)), \qquad (0 < |w| < t).$$

We now apply Hadamard's three circle theorem to the function $g(w)$ on the circles $|w| = \sigma_j$ $(j = 0,1,2)$ defined by

$$(2.17) \qquad \sigma_0 = \frac{\rho}{4}, \qquad \sigma_1 = R_1, \qquad \sigma_2 = \frac{1+R_1}{2}.$$

Putting

$$(2.18)\qquad \alpha = \frac{\log(\sigma_2/\sigma_1)}{\log(\sigma_2/\sigma_0)}, \qquad \beta = \frac{\log(\sigma_1/\sigma_0)}{\log(\sigma_2/\sigma_0)},$$

and

$$\mu_j = \max_{|w|=\sigma_j} |g(w)|, \qquad (j = 0,1,2),$$

we find

$$(2.19)\qquad 1 \le \left(\frac{\mu_0}{\mu_1}\right)^{\alpha} \left(\frac{\mu_2}{\mu_1}\right)^{\beta}, \qquad (\alpha > 0,\ \beta > 0,\ \alpha + \beta = 1).$$

To estimate μ_0 and μ_2 we use (2.16). With $t = (\rho/2)$ and $|w| = \sigma_0$ we obtain

$$\mu_0 \le \frac{4}{\frac{\rho}{2} - \frac{\rho}{4}} \left(m(\tfrac{\rho}{2}, F) + m(\tfrac{\rho}{2}, P)\right),$$

and hence, in view of V and (2.11),

$$(2.20)\qquad \mu_0 \le \frac{16}{\rho} \left(2M_0 + n \log \frac{4}{\rho}\right).$$

Similarly with $t=1$, $|w| = \dfrac{1 + R_1}{2}$, (2.16) yields, in view of II and (2.11)

$$(2.21)\qquad \mu_2 \le \frac{8}{1 - R_1} \left(2M_1 + n \log(\tfrac{4}{\rho})\right).$$

Finally, using (2.7), we select some w_1 such that

$$|w_1| = \sigma_1 = R_1, \qquad \log|F(w_1)| \ge \varkappa M_1 .$$

Hence, taking (2.11) and (2.14) into account, we find

$$(2.22)\qquad \mu_1 \ge |g(w_1)| \ge \log\left|\frac{F(w_1)}{P(w_1)}\right| \ge \varkappa M_1 - M_0 - n \log \frac{4}{\rho}.$$

Assume now the lemma to be false. Then (1.7) is violated and hence

$$(2.23)\qquad n < \frac{\xi M_1}{\log(4/\rho)}.$$

An inspection of (1.5) shows that $0 < c < 1$ and hence

(2.24) $$\xi < \frac{\varkappa}{50} < \frac{1}{50}.$$

Using (2.23), (2.24) and (1.6) in (2.22), we find

(2.25) $$\mu_1 > \frac{48\,\varkappa\, M_1}{50}.$$

Similarly, from (2.20), (2.23) and (1.6)

(2.26) $$\mu_0 \leq \frac{48\,\xi\, M_1}{\rho}.$$

The analogous treatment of (2.21) yields, in view of the second inequality in (2.6),

(2.27) $$\mu_2 \leq \frac{33M_1}{(1-R)^2}.$$

By (2.25), (2.26) and (2.27)

(2.28) $$\left(\frac{\mu_0}{\mu_1}\right)^{\alpha}\left(\frac{\mu_2}{\mu_1}\right)^{\beta} < \frac{50\,\xi^{\alpha}}{\varkappa\,\rho\,(1-R)^2}, \qquad (\alpha+\beta=1).$$

To complete our proof we note that (2.17), (2.18), (2.6) and (1.5) imply

(2.29) $$1 > \alpha = \frac{\log\{(1+R_1)/2R_1\}}{\log\{2(1+R_1)/\rho\}} = \frac{\log\left(1+\frac{1-R_1}{2R_1}\right)}{\log(2(1+R_1)/\rho)} > \frac{\log\left\{1+\frac{(1-R)^2}{4}\right\}}{\log(4/\rho)} = c > 0.$$

Since $0 < \xi < 1$, (2.19), (2.28), (2.29) and (1.5) yield

$$1 < \frac{50\,\xi^{c}}{\varkappa\,\rho\,(1-R)^2} = 1.$$

This contradiction proves that (2.23) is impossible; we have thus established (1.7) and completed the proof of the lemma.

References

1. A. Edrei, Sections of the Taylor expansions of Lindelöf functions, to be published.

2. A. Edrei, E. B. Saff and R. S. Varga, Zeros of Sections of Power Series, Lecture Notes in Mathematics, vol. 1002, Springer-Verlag, Berlin, New York, 1983.

LOCATION OF POLES OF PADÉ APPROXIMANTS TO ENTIRE FUNCTIONS

J. Nuttall*
Department of Physics
University of Western Ontario
London
Ontario N6A 3K7
Canada

Abstract. We give a conjecture for a set of arcs approached asymptotically by the poles of Padé approximants to entire functions. In two examples the conjecture is shown to be correct and in a third numerical evidence supporting its validity is given.

1. Introduction

One of the most intriguing aspects of Hermite-Padé polynomials is the striking patterns that often appear when their zeros are plotted for examples of reasonably high degree. The case of the Padé approximant to exp(z), which appears on the cover of the proceedings of the previous Tampa conference [10], will be familiar to many, and other plots for this function will be found in the article by Saff and Varga [11]. For some functions with branch points, it has been predicted that most zeros and poles of high-order near-diagonal Padé approximants will lie close to an appropriate set of arcs of minimum capacity [5], and this can be proved in some cases [8]. This theory has been extended to Hermite-Padé polynomials, also for functions with branch points [7]. At the 1983 Tampa conference, Edrei and de Bruin gave further examples showing patterns of zeros of Padé and Hermite-Padé polynomials for certain entire functions.

The denominator (and also the numerator) of a Padé approximant is related to a polynomial orthogonal in a certain sense (see below),

---

*Research supported in part by Natural Sciences and Engineering Research Council Canada.

and for such a polynomial Szegö [14] gave an explicit representation as a multiple integral, observing that "the formula is not suitable in general for derivation of properties of the polynomials in question". Perhaps now Szegö would have been prepared to reconsider that remark, for it was an empirical argument based on the integral representation that first led to the idea that a set of minimum capacity might be a limit set of the zeros and poles of diagonal Padé approximants for functions with branch points [6]. The representation has been extended to Hermite-Padé polynomials [7], and a similar empirical argument applied to functions with different branch points [1]. These predictions have been verified in the special case of functions with real branch points and disjoint, real, positive discontinuities across the real axis [3].

Stimulated by the remarks of Edrei and de Bruin, we thought it might be interesting to find out whether the empirical argument could be usefully applied to the case of entire functions, and this contribution is the outcome.

2. Equation for limit set

Consider the [m/n] Padé approximant, to an entire function $F(x)$, so that

$$(1) \qquad [m/n]_F = -P_1(x)/P_2(x)$$

where

$$(2) \qquad P_1(x) + F(x)P_2(x) = O(x^{m+n+1}), \quad x \approx 0,$$

with P_1, P_2 of degree m,n respectively. If we set $p(z)=z^nP_2(z^{-1})$ we easily see that

$$(3) \qquad \int_\Gamma dz\, z^{m-n} f(z)p(z)z^k = 0 \;, \quad k=0,\ldots,n-1$$

where $f(z)=F(z^{-1})$ and the integral is taken on any curve Γ encircling the origin once.

The representation given by Szegö [14] is

(4) $$p(z)=\text{const.}\int_\Gamma dz_1\ldots\int_\Gamma dz_n\{\prod_{k=1}^{n}(z-z_k)z_k^{m-n}f(z_k)\}\ I,$$

where

(5) $$I=\prod_{i<j=1}^{n}(z_i-z_j)^2 \quad .$$

The basis of the previous argument (when m=n) was that, for a given choice of Γ, there would be a set of points $z_k^o \varepsilon\Gamma$, $k=1,\ldots,n$, unique apart from permutation, for which $|I|$ is maximum. The contour Γ is to be distorted until this maximum is smallest, and the integral evaluated approximately by expanding the integrand in the manner of the saddle point method. The prediction that results is that

(6) $$p(z) \sim \text{const}\prod_{k=1}^{n}(z-z_k^o),$$

where $\{z_k^o\}$ corresponds to the choice Γ_o that makes $|I|_{max}$ minimum.

We follow the same approach except that now the factors $z_k^{m-n}f(z_k)$ are to be included in I. Let us write

(7) $$f(z) = \exp(g(z)).$$

Then, if

(8) $$J = I\{\prod_{k=1}^{n} z_k^{m-n}f(z_k)\} ,$$

we find $\{z_i^o\}$ by solving

(9) $$\frac{\partial \log J}{\partial z_k} = 0 \ , \quad k=1,\ldots,n,$$

which gives

(10) $$2\sum_{j\neq k}(z_k-z_j)^{-1}+(m-n)z_k^{-1}+g'(z_k)=0 \ , \quad k=1,\ldots,n \quad .$$

We assume that the points $\{z_k^o\}$ are distributed smoothly on the contour Γ_o with density $\rho(z)>0$, and for large n replace (10) by

$$(11) \qquad 2\,P\int_{\Gamma}|dt|\rho(t)(z-t)^{-1}+(\sigma-1)z^{-1}+n^{-1}g'(z)=0\ , \qquad z\in\Gamma,$$

setting $\sigma=m/n$.

To solve (11) for Γ,ρ we use the techniques of Muskhelishvili [3]. Denote the interior of Γ by Γ_+, the exterior by Γ_-, and set

$$(12) \qquad \psi(z)=\int_{\Gamma}|dt|\rho(t)(t-z)^{-1}\ , \quad z\notin\Gamma\ ,$$

so that $\psi(z)$ is piecewise analytic, $z\notin\Gamma$. Then (11) may be written

$$(13) \qquad \psi_+(z)+\psi_-(z)-(\sigma-1)z^{-1}-n^{-1}g'(z)\ =\ 0\ , \quad z\in\Gamma\ ,$$

where ψ_+,ψ_- denote appropriate limits. If we define the piecewise analytic function $\chi(z)$ by

$$\chi(z)\ =\ \psi(z)\ , \quad z\in\Gamma_+$$

$$\chi(z)\ =\ -\psi(z)\ , \quad z\in\Gamma_-$$

then

$$(15) \qquad \chi_+(z)-\chi_-(z)\ =\ (\sigma-1)z^{-1}+n^{-1}g'(z)\ , \quad z\in\Gamma\ ,$$

and, since $\chi(\infty)=0$, the Plemelj formula [4] gives

$$(16) \qquad \chi(z)\ =\ (2\pi i)^{-1}\int_{\Gamma}dt\left(\ (\sigma-1)t^{-1}+n^{-1}g'(t)\ \right)(t-z)^{-1}\ , \quad z\notin\Gamma\ .$$

Now let us assume that F(x) has no zeros, so that g(z) is analytic except at z=0, and has an expansion about $z=\infty$

$$(17) \qquad g(z)\approx \text{const.}z^{-1}+O(z^{-2})\ .$$

The integral (16) then gives

$$\chi(z) = 0 , \qquad z \in \Gamma_+$$

(18)

$$\chi(z) = -(\sigma-1)z^{-1}-n^{-1}g'(z) , \quad z \in \Gamma_- .$$

If on Γ we set $dz=\exp(i\alpha(z))|dz|$, then the Plemelj formula gives

(19) $$2\pi i \exp(-i\alpha(z))\rho(z) = \psi_+(z)-\psi_-(z)$$

(20) $$= -(\sigma-1)z^{-1}-n^{-1}g'(z) \quad , z \in \Gamma.$$

There is, however, a problem with this solution, for consider

$$\int_\Gamma |dz|\rho(z) = \int_\Gamma dz \exp(-i\alpha(z))\rho(z)$$

$$= (2\pi i)^{-1}\int_\Gamma dz[-(\sigma-1)z^{-1}-n^{-1}g'(z)]$$

$$= -(\sigma-1) .$$

The integral should be unity, so that there is an inconsistency unless $\sigma=0$.

We faced a similar difficulty previously [1], and it is overcome by assuming that $\rho(z)=0$ on part of Γ, so that (11),(13) actually hold only on the other part of Γ, which we call S. Thus we now wish to solve (13) with Γ replaced by S, which consists of one or more disjoint arcs, subject to the conditions

(22) $$\rho(b_j) = 0 , \quad j=1,...,$$

where b_1, b_2,... are the ends of the arcs in S.

We illustrate with three examples.

Example 1. Suppose that $F(x)=\exp(x)$ so that $g(z)=z^{-1}$. It is simplest to assume that S consists of a single arc, with ends b_1,b_2. The solution of the modified (13) is standard [4]. Introduce the function $y(z)=[(z-b_1)(z-b_2)]^{\frac{1}{2}}$, single-valued in the complex plane cut along S, with $y(z)\simeq z$ as $z\to\infty$. Then, with

(23) $$\psi(z)=\int_S |dt|\rho(t)(t-z)^{-1} , \quad z \notin S,$$

(13) becomes

(24) $$y_+(z)\psi_+(z)-y_-(z)\psi_-(z)=y_+(z)((\sigma-1)z^{-1}-n^{-1}z^{-2}),\quad z\in S\ .$$

Thus

(25) $$y(z)\psi(z)=-1+(2\pi i)^{-1}\int_S dt\ y_+(t)((\sigma-1)t^{-1}-n^{-1}t^{-2})(t-z)^{-1}\ ,\quad z\notin S.$$

The integral in (25) is easily performed, and, with the help of (19) and (22) it is found that

(26) $$\psi(z)=-\frac{(\sigma+1)}{2}y(z)z^{-2}+\tfrac{1}{2}[(\sigma-1)z^{-1}-n^{-1}z^{-2}]$$

with b_1, b_2 being chosen so that this expression is analytic at $z=0$. This requires the conditions

$$y(0)=-N^{-1}$$

(27) $$Y_o=N^{-2}$$

$$Y_1=-2\left(\frac{\sigma-1}{\sigma+1}\right)N^{-1}$$

where we have set $N=m+n$ and

(28) $$y^2(z)=z^2+Y_1z+Y_o\ .$$

Thus

(29) $$Nb_1, Nb_2=(z^{\pm}_{\sigma-1})^{-1}$$

in the notation of [11].

The location of the arc S is found by integration from b_1, using (19) to give the slope of the curve. We now have

(30) $$2\pi i\ \exp(-i\alpha(z))\rho(z)=-(\sigma+1)y_+(z)z^{-2}\quad ,\quad z\in S\ .$$

There are three possible arcs which leave b_1 in directions differing by 120°, but only for one of them is (27) satisfied. Of the others, one passes through $z=0$ and the other turns out to be the limit curve of the zeros of the Padé approximant.

It takes little manipulation to show that the arc S predicted here is the same as that rigorously shown to be correct in [11,12].

Example 2. Consider now the function $F(x)=\exp(x^2+\mu x)$ so that $g(z)=z^{-2}+\mu z^{-1}$. We find in this case that S consists of two components and we define $y(z)=[\prod_{j=1}^{4}(z-b_j)]^{\frac{1}{2}}$. The function $\psi(z)$ is given by

$$\psi(z)=-(\frac{\sigma+1}{2})y(z)z^{-3}+\tfrac{1}{2}[(\sigma-1)z^{-1}-2n^{-1}z^{-3}-n^{-1}\mu z^{-2}] \ . \tag{31}$$

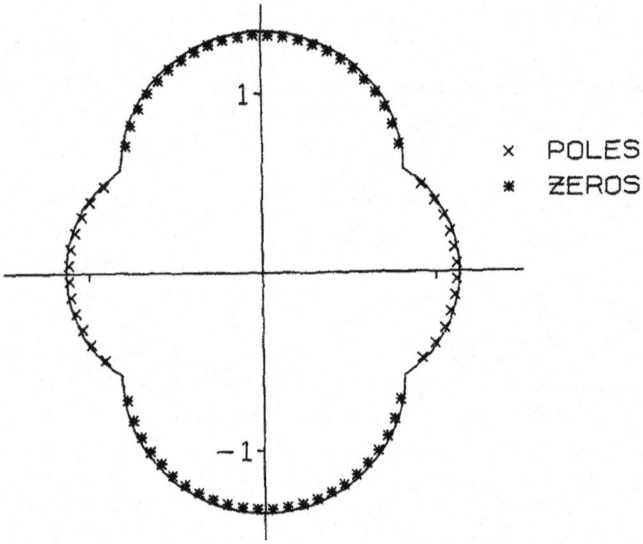

Figure 1: Zeros and poles of [48/24] for $F(x)=\exp(x^2)$, plotted in the variable $6x^{-1}$, with the corresponding theoretical limit curves.

Analyticity at z=0 means that

$$y(0)=-2N^{-1}$$

$$Y_o = 4N^{-2}$$

$$Y_1 = 4\mu N^{-2}$$

$$Y_2 =-4(\frac{\sigma-1}{\sigma+1})N^{-1}+\mu^2N^{-2},$$

where

(33) $$y^2(z)=z^4+\sum_{j=0}^{3} Y_j z^j ,$$

A study of the original argument based on (4) indicates that

(34) $$\mathrm{Re}\int_{b_1}^{b_2} dt\ y(t)t^{-3}=\mathrm{Re}\int_{b_1}^{b_3} dt\ y(t)t^{-3} = 0$$

must hold, which constitute equations for Y_3. Having found $Y_o,\ldots,Y_3$ we obtain $b_1,\ldots,b_4$ and determine the location of S by integration as before.

As $N\to\infty$ the solution will approach that for $\mu=0$, and we do not know whether it is meaningful to solve the problem with $\mu\neq 0$ - perhaps corrections that have been omitted are as important as the effect of taking $\mu\neq 0$.

We have solved the problem for $\mu=0$ and plotted the results for an example in Fig. 1. Of course, the behavior for this case can be deduced from that known for $F(x)=\exp(x)$, but the method we are presenting does not take this relation into account yet still predicts the correct result.

<u>Example 3</u>. A less trivial case is $F(x)=\exp(x^{\frac{1}{2}})+\exp(-x^{\frac{1}{2}})$, which is the Mittag-Leffler function $E_2(x)$ studied by Edrei et al [2]. The function has zeros on the negative real axis. Here we look for a solution for which S consists of a single arc and we obtain

(35) $$\psi(z)=-1+(2\pi i)^{-1}\int_S dt\ y_+(t)((\sigma-1)t^{-1}+n^{-1}g'(z)) \quad , \ z \notin S$$

where now $y(z)=[(z-b_1)(z-b_2)]^{\frac{1}{2}}$ and

(36) $$g(z)=\log(\exp(z^{-\frac{1}{2}})+\exp(-z^{-\frac{1}{2}})).$$

We expect S to lie near to $z=0$ for large N, and, provided that S is not near to the negative real axis, we may approximate $g'(z)$ in (35) by $-\frac{1}{2}z^{-3/2}$. Thus, in the case when $\sigma=1$, we use

$$(37) \qquad \psi(z)=y(z)^{-1}[-1-(4\pi in)^{-1}\int_S dt\ y_+(t)t^{-3/2}(t-z)^{-1}]\ ,\ z\notin S\ .$$

With this form of $\psi(z)$, (22) was solved numerically to yield $b_1,b_2=(-.089004\pm i.103441)n^{-2}$. The curve S was then obtained by integration. In Fig. 2 the poles of the [20/20] approximant to F(x) are compared with the calculated curve, with satisfactory agreement.

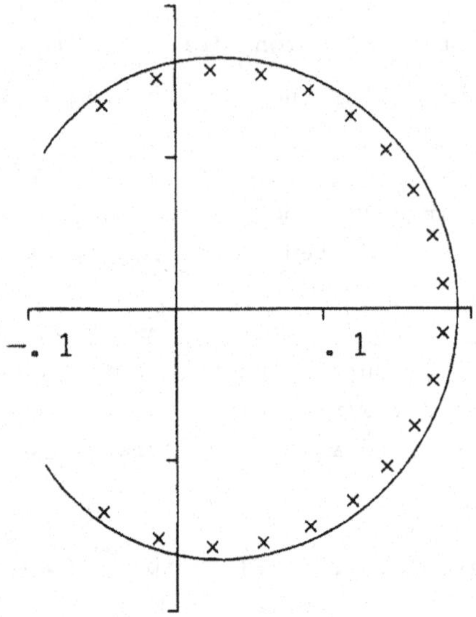

Figure 2. Poles of [20/20] for $F(x)=\exp(x^{\frac{1}{2}})+\exp(x^{-\frac{1}{2}})$, plotted in the variable $400x^{-1}$, with the theoretical limit curve.

3. Discussion

In addition to pointing towards an approach to the asymptotics of Padé approximants to entire functions, our work also suggests how the non-diagonal case for functions with branch points might be dealt with. In the diagonal case, it is thought [7] that most zeros and poles (in the z-variable) approach an appropriate set of minimum capacity. Such a set may be described as the equilibrium (unstable) location of charged flexible conducting curves ending at certain branch points and held at the same potential, using 2-dimensional electrostatics. What (13) suggests is that this specification be modified by the addition of an external field corresponding to the right-hand side of (13). Alternatively the set may be regarded as resulting for solving a max-min problem for a corresponding energy functional. Such a functional has already been introduced in the study of non-diagonal Padé approximants for functions with branch points on the real axis [13] and the related problem of incomplete polynomials. [9]

Acknowledgements. The author thanks Dr. K. Aashamar for the use of his multiple-precision code GENPREC and Mr. L.S. Luo for his help in preparing the illustrations.

References

1. Baumel, R.T., J.L. Gammel, and J. Nuttall, Asymptotic form of Hermite-Padé polynomials, IMA J.Appl.Math. 27 (1981), 335-357.
2. Edrei, A., E.B. Saff, and R.S. Varga, Zeros of sections of power series, Lect. Notes in Math. 1002 (1983).
3. Gončar, A.A., and E.A. Rahmanov, On the convergence of simultaneous Padé approximants for systems of functions of Markov type, Proc. Steklov Math. Inst. 157 (1981), 31-48.
4. Muskhelishvili, N.I., Singular Integral Equations, Noordhoff, Groningen, 1953.
5. Nuttall, J., The convergence of Padé approximants to functions with branch points, Padé and Rational Approximation (E.B. Saff and R.S. Varga, eds.) pp 101-109. Academic Press, New York, 1977.
6. Nuttall, J., Sets of minimum capacity, Padé approximants and the bubble problem, Bifurcation Phenomena in Mathematical Physics and Related Topics (C. Bardos and D. Bessis, eds.) pp 185-201. D. Reidel, Dordrecht, 1980.
7. Nuttall, J., Asymptotics of diagonal Hermite-Padé polynomials, J. Approx. Theory, to appear.
8. Nuttall, J., and S.R. Singh, Orthogonal polynomials and Padé approximants associated with a system of arcs, J. Approx. Theory 21 (1977), 1-42.
9. Saff, E.B., Incomplete and orthogonal polynomials, Proceedings of Approximation Theory Conference, Texas A & M University, 1983, to appear.
10. Saff, E.B., and R.S. Varga, eds., Padé and Rational Approximation, Academic Press, New York, 1977.
11. Saff, E.B., and R.S. Varga, On the zeros and poles of Padé approximants to e^z II, Padé and Rational Approximation (E.B. Saff and R.S. Varga, eds.), pp 195-213. Academic Press, New York, 1977.
12. Saff, E.B., and R.S. Varga, On the zeros and poles of Padé approximants to e^z III, Numer. Math. 30 (1978), 241-266.
13. Stahl, H., Beiträge zum Problem der Konvergenz von Padéapproximierenden, Dissertation, Technischen Universität Berlin (1976).
14. Szegö, G., Orthogonal Polynomials, American Mathematical Society, Providence, 1978.

APPROXIMATIONS TO e^x ARISING IN THE NUMERICAL ANALYSIS OF VOLTERRA EQUATIONS

Christopher T.H. Baker
Department of Mathematics
The University
Manchester M13 9PL, U.K.

Abstract. The study of numerical methods for Volterra integral equations yields some novel approximations to the exponential function.

1. Introduction

The study of rational approximations to the exponential function is, as is well-known, a feature of the analysis of numerical methods for initial-value problems for systems of ordinary differential equations, of the form

$$(1) \qquad y'(t) = f(y(t)), \qquad t \geq 0 \qquad (y(0) = y_0),$$

when we consider the basic scalar equation

$$(2) \qquad y'(t) = \lambda y(t), \qquad t \geq 0 \qquad (y(0) = y_0),$$

with solution $y(t) = y_0 \exp(\lambda t)$. (The theories which reveal a wider significance for (2) in the discussion of (1) are referenced in [2].)

If we write (2) in integrated form incorporating $y(0) = y_0$, we find

$$(3) \qquad y(t) - \lambda \int_0^t y(s)ds = g(t) \qquad (t \geq 0),$$

with $g(t) = y_0$; this is a basic example of the Volterra equation of the second kind:

$$(4) \qquad y(t) - \int_0^t K(t,s,y(s))ds = g(t) \qquad (t \geq 0),$$

wherein g and K are prescribed smooth functions. Equation (3) is a fruitful paradigm for (4), as we indicated briefly in [1], [2].

There is an intimate relation between certain numerical methods for (4) and numerical methods for (1); we shall suppose in both cases that the approximations $y_n \approx y(nh)$ are sought on a mesh $\{nh| \ n \geq 0\}$. The study of methods for (4) applied to (3) augments, however, the classes of problems involving the study of approximations to $\exp(\lambda h)$. It is our purpose here to consider some of the novelties which the above situation provides.

In the treatment of (2),(3) we can identify common themes. Since $y(nh) = e^{n\lambda h}$, the study involves approximations $\exp(\Lambda) \approx \mu(\Lambda)$ where $\Lambda = \lambda h$. $\mu(\Lambda)$ can be a rational function of the variable Λ; in general it is the zero of a polynomial in μ (which is the characteristic polynomial, $\Sigma(\Lambda;\mu)$, of a recurrence relation), whose coefficients are polynomials in Λ. The following concepts are of interest: $\mu(\Lambda)$ is an approximation to $\exp(\Lambda)$ of order Q if $\mu(\Lambda) = \exp(\Lambda) + O(\Lambda^{Q+1})$ as $\Lambda \to 0$. $\mu(\Lambda)$ is respectively (i) A_0-, (ii) $A(\alpha)$-, and (iii) L-*acceptable* if (i) $|\mu(\Lambda)| < 1$ for $0 > \Lambda \in R$, (ii) $|\mu(\Lambda)| < 1$ for $\mathrm{Re}(\Lambda) < 0$, $\arg(\Lambda) \in (\pi-\alpha,\pi+\alpha)$ and (iii) $\mu(\Lambda)$ is $A(\frac{\pi}{2})$-acceptable and $\mu(\Lambda) \to 0$ as $\Lambda \to -\infty$. $A(\frac{\pi}{2})$-acceptability is known as A-*acceptability*. The notions of acceptability concern quality of approximation, since $|\exp(\lambda h)| < 1$ if $\mathrm{Re}(\lambda h) < 0$; they arise in stability theories.

Suppose now that a polynomial $\Sigma(\Lambda;\mu)$ in the variable μ has coefficients depending on Λ. The associated region of *(strict) stability* S is the region of the Λ-plane for which the polynomial is a Schur polynomial, i.e. the roots $\mu = \mu(\Lambda)$ satisfy $|\mu(\Lambda)| < 1$. If $S \supseteq \{\Lambda \in R| \ \Lambda < 0\}$ then the polynomial will be called A_0-*stable*. ($A(\alpha)$-, L-*stability*, and A-*stability* are to be defined similarly.) One or more of the roots of $\Sigma(\Lambda;\mu)$ approximate $\alpha\exp(\beta\Lambda)$ (as $\Lambda \to 0$, for various α,β) including a *principal root* approximating $\exp(\Lambda)$. If there exists such a root with $\mathrm{Re}(\beta) < 0$ the polynomial (and method) are said to display *weak instability*. The utility of weakly unstable methods is very limited [9]. [4],[5],[6],[10],[13]provide background.

We summarise the relevant methods for (4) in the next section, and treat each type in turn in the following sections.

2. Underlying methods

We assume some familiarity with numerical methods for (1), in particular the consistent, zero-stable $\{\rho,\sigma\}$-linear multistep formula with first and second characteristic polynomials [7],[8],[11]:

(5) $\rho(\mu) = \alpha_0\mu^k + a_1\mu^{k-1}+\dots+\alpha_k, \quad \sigma(\mu) = \beta_0\mu^k + \beta_1\mu^{k-1}+\dots+\beta_k.$

Denoting by E the shift operator: $Ey_n = y_{n+1}$, etc., the formula for (1) is

(6) $\rho(E)y_n = h\sigma(E)f_n; \quad f_n = f(y_n).$

Simple examples with $\rho = \mu-1$ are $\sigma = 1$ (Euler), $\sigma = \mu$ (implicit Euler), $\sigma = (\mu+1)/2$ (trapezium rule) and the Adams methods, and with $\rho = \mu^2-1$, $\sigma = (\mu^2+4\mu+1)/3$, Simpson's rule. Cyclic linear multistep methods involve q such formulae, used cyclically [11] :

(7) $\rho_r(E)y_n = h\sigma_r(E)f_n \qquad (r = 0,1,\dots,q-1,\ n \equiv n_0 + r \bmod q).$

Finally, a Runge-Kutta method for (1) can be defined in terms of the elements A_{rs} $(r,s = 0,1,\dots, p)$ of a *square* generating matrix A by the formulae

(8) $y_{n+1} := y_{n,p}, \quad y_{n,r} = y_n + h\sum_s A_{rs}f(y_{n,s}) \quad (r = 0,1,\dots, p), \quad n\geq 0$

where summation over s is (here and later) for $s = 0,1,\dots,p-1,p$, and

(9) $y_{n,r} \equiv E^{\theta_r}y_n \approx E^{\theta_r}y(nh), \qquad \theta_r = \sum_s A_{rs}, \quad \theta_p = 1.$

(Thus, $E^{\theta_p} \equiv E$; $y_{n,p} \approx y((n+1)h)$.) For (1) it is *conventional* to choose A so that $A_{rp} = 0$ $(r = 0,1,\dots, p)$; see for example [11].

Methods for non-autonomous equations $x'(t) = F(t, x(t))$, $t \geq 0$ $(x(0) = x_0)$ result on deriving the equation (1) for the vector y with components $t,x(t)$. From the choice $F(t, x) = \phi(t)$, $x(0) = 0$, we thus obtain approximations x_n to $x(nh) = \int_0^{nh}\phi(s)ds$. With appropriate starting values, the multistep formulae (6),(7) yield for x_n *quadrature rule* approximations of the form

(10) $Q : \int_0^{nh} \phi(s)\,ds \approx h \sum_{j=0}^{n} w_{nj}\,\phi(jh) \qquad (n \geq k)$

and the Runge-Kutta method yields (for $E^{\theta_r}x_n$) approximations

(11) $\int_0^{nh+\theta_r h} \phi(s)\,ds \approx h \sum_{j=0}^{n-1}\sum_s A_{ps}\,\phi(jh + \theta_s h) + h\sum_s A_{rs}\phi(nh + \theta_s h).$

Formulae (10) occur naturally in their own right, but if the weights w_{nj} of a family of rules (10) can be derived from ρ,σ in this manner, the family is called $\{\rho, \sigma\}$-*reducible*; if the rules correspond to (7) they are *cyclically-reducible* (or $\{\rho_r, \sigma_r\}_0^{q-1}$-*reducible*). One may be tempted to modify (11) by the use of

$$(12) \quad \int_0^{nh+\theta_r h} \phi(s)\,ds \simeq h \sum_{j=0}^{n-1} \sum_s b_s \, \phi(jh + \theta_s h) + h \sum_s A_{rs}\phi(nh + \theta_s h)$$

for appropriate b_s ($\Sigma b_s = 1$) defining a vector b, or by the use of

$$(13) \quad \int_0^{nh+\theta_r h} \phi(s)\,ds \simeq h \sum_{j=0}^{n} w_{nj} \, \phi(jh) + h \sum_s A_{rs}\phi(nh + \theta_s h).$$

where the weights w_{nj} are those of Q in (10). The quadrature rules (11), (12), (13) are called respectively, the *extended Runge-Kutta rules*, the *mixed Runge-Kutta rules*, and the *mixed quadrature-Runge-Kutta rules*. All of these Runge-Kutta rules have the form

$$(14) \quad \int_0^{\tau} \phi(t)\,dt \simeq h \sum_k \Omega_{jk} \, \phi(\tau_k), \qquad \tau = nh+\theta_r h,$$

where the summation is over $\tau_k = jh+\theta_s h$ with $s = 0,1,\ldots,p$, and $j \le n$. We assume throughout that (10) are $\{\rho, \sigma\}$-reducible or cyclically reducible.

With the aid of the integration rules above, it is simple to discretize (4). Putting $t = nh$ in (4), the rules Q of (10) yield the equations

$$(15) \quad y_n - h \sum_{j=0}^{n} w_{nj} \, K(nh, jh, y_j) = g(nh).$$

Setting $t = nh + \theta_r h$ and using one of (11), (12), or (13), yields [2] equations of the form

$$(16) \quad y_{n,r} - h \sum_{j=-1}^{n} \sum_{s=0}^{p} \Omega_{n(p+1)+r+1,\,j(p+1)+s+1} \, K(nh+\theta_r h, jh+\theta_s h, y_{j,s})$$
$$= g(nh+\theta_r h).$$

For $j = n(p+1)+r+1$, we have either the weights, denoted $\Omega_{jk}(A)$, of the *extended RK rule* (11), or the weights $\Omega_{jk}\{b, A\}$ of the *mixed RK rule* (12), or the weights $\Omega_{jk}[Q, A]$ in (13) of a *mixed quadrature-RK rule*; (16) also admits (15) as a special case. The choices (12),(13) have but tenuous links with (1) and produce novel results.

In each of the following sections we relate a class of methods to polynomials $\Sigma(\Lambda;\mu)$, and consider common themes indicated in section 1.

3. Results via reducibility

We commence with a discussion of reducible quadrature methods. The following result indicates that *we need only appeal to the corresponding results for (cyclic) linear multistep methods* for (2) in our study. Cyclic methods correspond to matrix analogues $\{A_\ell, B_\ell\}_0^m$ of $\{\alpha_\ell, \beta_\ell\}_0^k$ in (5) (see [2] and [11]) and we have:

Theorem [3]. Suppose the weights w_{nj} in (10) are $\{\rho_r, \sigma_r\}_0^{q-1}$-reducible and $y_n - \lambda h \sum_j w_{nj} y_j = g(jh)$. Then

$$(17) \qquad \sum_{\ell=0}^{m} \{A_\ell - \lambda h B_\ell\} \phi_{n+1-\ell} = \gamma_n, \qquad \det(A_0) \neq 0,$$

where $\phi_n = [y_{nq}, y_{nq+1}, \ldots y_{(n+1)q-1}]^T$ and the matrices $\{A_\ell, B_\ell\}$ are generated by $\{\rho_r, \sigma_r\}_0^{q-1}$. γ_n depends on g and vanishes if $g(t) \equiv y_0$.

Some results of interest may not be widely known. For a set of matrices $M_0, M_1, \ldots, M_m$ (we shall take $M_\ell = A_\ell$, or $M_\ell = A_\ell - \lambda h B_\ell$) we define $M(\mu) = \sum_{\ell=0}^{m} M_\ell \mu^{m-\ell}$. Of interest is the determinant

$$(18) \qquad \Sigma(\lambda h; \mu) = \det[(A - \lambda h B)(\mu)],$$

the *characteristic polynomial* for (17). ($\Sigma(0;\mu)$ is the associated polynomial [11] for a cyclic method, $\det[A(\mu)]$.)

Theorem. Denote by $\{\mu_\ell(\lambda h)\}$ the zeros of (18), and write $\nu_\ell = \mu_\ell(0)$. If ν_ℓ is a simple zero of $\Sigma(0; \mu)$ then there exists a calculable value k_ℓ such that $\mu_\ell(\lambda h) = \nu_\ell + k_\ell \lambda h + O(\lambda h)^2$; hence

$$(19) \qquad \mu_\ell(\lambda h) = \nu_\ell \exp(k_\ell \lambda h / \nu_\ell) + O(\lambda h)^2.$$

The result (19) is of use in the study of weak instability. We can [2] show that k_ℓ can be determined from ϕ_ℓ^T, ψ_ℓ such that, for $A(\nu) = \Sigma A_\ell \nu^{m-\ell}$, $\phi_\ell^T A(\nu_\ell) = 0^T$, $A(\nu_\ell)\psi_\ell = 0$; for a particular case, in practise, we may prefer a direct scalar analysis. Let $q = 1$, and $\{\rho_0, \sigma_0\} = \{\rho, \sigma\}$; if $\nu_\ell = \mu_\ell(0)$ is a simple root of ρ, $\mu_\ell(\lambda h) = \nu_\ell + \{\sigma(\nu_\ell)/\rho'(\nu_\ell)\}\lambda h + O(\lambda h)^2$. Thus can we establish weak instability of Simpson's rule, since $\nu_1 = -1$, $\sigma(-1)/\rho'(-1) = 1/3$.

Theorem. The cyclic combination $\{\rho_r, \sigma_r\}$ may not be weakly unstable even though one of its component formulae is weakly unstable.

To illustrate: the quadrature rules (3.9) of [1] are cyclically reducible with $\{\rho_0, \sigma_0\}$ for the trapezium rule and $\{\rho_1, \sigma_1\}$ for Simpson's rule; we find $\{\mu_\ell(\lambda h)\} = \{0, 1+2\lambda h+2(\lambda h)^2+(4/3)(\lambda h)^3+(7/9)(\lambda h)^4+\cdots=\exp(2\lambda h)+O(\lambda h)^4\}$, displaying no weak instability.

Consider a linear multistep formula $\{\rho, \sigma\}$. The order of the method is Q if C_{Q+1} is the first non zero c_r where $c_0 = \rho(1)$, $c_1 = \rho'(1) - \sigma(1)$ and $c_r = [\frac{1}{r!} D_r\rho(\mu) - \frac{1}{(r-1)!} D_{r-1}\sigma(\mu)]_{\mu=1}$ $(r \geq 2)$, where D_r denotes the operator $\{(d/d\mu)\mu\}^{r-1}(d/d\mu)$, for $r \geq 1$. Given the order Q of $\{\rho, \sigma\}$ we know that $\rho(\mu) - \lambda h\sigma(\mu) = 0$ has a principal root $\mu_0(\lambda h)$ such that (see [8,p.66]):

$$(20) \quad \mu_0(\lambda h) = \exp(\lambda h) + O(\lambda h)^{Q+1}.$$

There is a corresponding result for $\Sigma(\lambda h,\mu)$ where $\{\rho_\ell,\sigma_\ell\}$ has order Q_ℓ $(\ell = 0,1,\ldots,q-1)$. The effective step is $H = qh$, and we have a principal root such that $\mu_0(\lambda h) \approx \exp(\lambda qh)$; the order of accuracy of this approximation varies with the effective order of the cyclic combination.

Concerning the region of strict stability for $\Sigma(\lambda h; \mu)$, if the formulae are *all* explicit, the region necessarily excludes a domain of small positive real λh (cf. [8, p.67]), and is bounded.

The results in this section relate to cyclic linear multistep formulae for (2) as well as the corresponding quadrature methods for (3). However, the cyclic methods for (1) do not appear to enjoy great popularity whilst the quadrature methods for (4) arise quite naturally.

4. Results for Volterra Runge-Kutta methods

Every method for (4) yields a method for (1); the Volterra Runge-Kutta methods defined by $\Omega_{jk} = \Omega_{jk}(A)$ in (16) produce our Runge-Kutta methods (8). However, the mixed methods have in general no naturally occurring analogues for treating (1).

Fortunately, *results for* $\Omega_{jk} = \Omega_{jk}\{b,A\}$ *can be related to known results for* (3). We require some notation. We denote by I the identity of order p+1, $e_0, e_1, \ldots, e_p$ its successive columns and e their sum. The corresponding identity of order p will be denoted $I^\#$, and the sum of its columns $e^\#$. The vectors ϕ_n have components $y_{n,0}$, $y_{n,1}, \ldots, y_{n,p}$ resulting from the application of a Runge-Kutta method to (4); $e_p^T \phi_n = y_{n,p}$ *is the approximation* y_{n+1} *to* $y((n+1)h)$. For the method to succeed, $I - \lambda hA$ must be invertible. Let (see [2])

(21) $\mu^*(\lambda h) = \mu^*(b,A;\lambda h) := 1 + \lambda h\, b^T d(\lambda h)$ where $d(\lambda h) = (I-\lambda hA)^{-1}e$.

Theorem. $y_{n+1} = \mu^*(\lambda h)y_n + \gamma^*_{n+1}$, where γ^*_{n+1} depends upon $g(t)$. Further, for vectors $\{g_n\}$ depending on $g(t)$,

(22) $\phi_{n+1} = (I - \lambda hA)^{-1}\{e(e_p[I - \lambda hA] - b^T)\}\phi_n + g_{n+1} - e_p^T g_n$;

this is a recurrence in which the amplification matrix has a single non-trivial eigenvalue $\mu_*(b,A;\lambda h)$, the others being zero.

Consider the choice $b^T = a_p^T := e_p^T A$. Let $\hat{\mu}(\lambda h) := \mu_*(a_p,A;\ \lambda h)$. Then $\hat{\mu}(\lambda h) = 1+\lambda h a_p^T\ d(\lambda h)$.

The value $\hat{\mu}(\lambda h)$ is significant when using the extended Runge-Kutta method with weights $\Omega_{jk}(A) = \Omega_{jk}\{a_p^T,\ A\}$. Let the method be conventional, so that $Ae_p = 0$ and the top left submatrix of A is $A^\#$ and the vector w^T sits below $A^\#$ in the full array. Then $\hat{\mu}(\lambda h) = 1 + \lambda h w^T(I^\# - \lambda hA^\#)^{-1}e^\#$; this is *the conventional result in differential equations* [11]. We have the following:

Theorem. Write $\hat{\mu}(\lambda h) = \mu(A,\lambda h)$. For a general Runge-Kutta method (10) with weights $\Omega_{jk}\{b,\ A\}$, the value $\mu^*(b,A;\lambda h)$ is the value $\mu(A^*,\lambda h)$ which corresponds to the choice $\Omega_{jk}(A^*)$ where

(23) $$A^* = \left[\begin{array}{ccc:c} & A & & 0 \\ \hdashline b_0\ldots b_{p-1} & & b_p & 0 \end{array}\right].$$

Proof. Refer to the discussion preceding the theorem (set $A^\# = A$, etc.).

Corollary. For $r = 0,1,2,\ldots,p$, $e_r^T d(\lambda h)$ is a rational approximation to $\exp(\theta_r \lambda h)$; in consequence, $\hat{\mu}(\lambda h)$ and $\mu^*(b,A;\lambda h)$ are rational approximations to $\exp(\lambda h)$.

Of course, every conventional A-stable Runge-Kutta method for (1) yields an A-acceptable $\hat{\mu}(\lambda h)$. To illustrate the existence of extended Runge-Kutta methods with unconventional A for which $\hat{\mu}(\lambda h)$ is A-acceptable and L-acceptable take $p = 1$; if $A_{11} = 5/12$, $A_{12} = -1/12$, $A_{21} = 3/4$ and $A_{22} = 1/4$, $\hat{\mu}(\lambda h)$ is a subdiagonal Padé approximant to $\exp(\lambda h)$, and is then A-acceptable and L-acceptable.

If A is lower triangular, of order p+1, then $\hat{\mu}(\lambda h)$ is a polynomial in λh of degree p+1 (p if A is conventional). We know from conventional theory that it is possible to construct explicit methods such that $\hat{\mu}(\lambda h) = t_n(\lambda h)$ is a Taylor sum for $\exp(\lambda h)$ for appropriate n:

(24) $t_n(\lambda h):=1+\lambda h+\frac{1}{2!}(\lambda h)^2+\ldots+\frac{1}{n!}(\lambda h)^n.$

Let us show that it is easy to construct methods with $\hat{\mu}(\lambda h)=1/t_n(-\lambda h)\approx \exp \lambda h$ for appropriate n. Though apparently of little interest in the treatment of (1), we shall find a role for them, later.

Lemma. Suppose that $\mu(A,\lambda h)=\hat{\mu}(\lambda h)$ corresponds to the conventional matrix A of order p+1. Then the matrix $A^-=-R_pA$, $R_p=\{\sum_0^{p-1} e_ie_i^T-ee_p^T\}$, is such that $\mu(A^-,\lambda h)=1/\mu(A,-\lambda h)\approx \exp(\lambda h)$.

Corollary. Suppose that b and A (conventional or not) are given. Let A* (of order p+2) be defined by (23) and generate $\bar{A^*}=-R_{p+1}A^*$ as above; then $\mu(\bar{A^*},\lambda h)=1/\mu^*(b,A;-\lambda h)\approx \exp(\lambda h)$.

The issue of weak instability does not arise in this section.

5. Mixed Quadrature - Runge-Kutta Methods

The mixed quadrature Runge-Kutta methods have an attraction [2] because of their apparent economy.

Theorem. Let Q be $\{\rho_r,\sigma_r\}_0^{q-1}$-reducible. Then the values y_n obtained using the mixed quadrature - Runge-Kutta method satisfy (compare (7))

(25) $\rho_r(E)y_{n+1}-\lambda h\hat{\mu}(\lambda h)\sigma_r(E)y_n=\gamma_n',\quad n\equiv r \bmod(q)+n_0$

where γ_n' vanishes if g(t) is constant.

Thus, as regards (3), *we can study the mixed quadrature-Runge-Kutta method by analyzing instead the (cyclic) linear multistep method* with $\mu\rho_r(\mu)$ replaced by $\rho_r(\mu)$ and λh replaced by $\lambda h\hat{\mu}(\lambda h)$. In particular we are interested in the auxiliary polynomials (q = 1, m = k+1)

(26) $\mu\rho(\mu)-\lambda h\hat{\mu}(\lambda h)\sigma(\mu)$,

or, when the rules Q are cyclically reducible ($q\geq 2$),

(27) $\Sigma'(\lambda h;\mu)\equiv \det\sum_{\ell=0}^{m+1}[A_\ell'-\lambda h\hat{\mu}(\lambda h)B_\ell']\mu^{m+1-\ell}$

where A_ℓ', B_ℓ' are generated by $\{\mu\rho_r,\sigma_r\}$ according to the rules for obtaining A_ℓ, B_ℓ, in (17), from $\{\rho_r,\sigma_r\}$. (26) is a special case of (27).

Theorem. When Q is weakly unstable, (27) is not necessarily weakly unstable.

Proof. Consider Simpson's rule, and note that $\mu(0) = 1$.

We shall illustrate some results of interest by concentrating upon (26). Let $\{\rho,\sigma\}$ be a linear multistep formula of order Q. Then $\rho(e^{\lambda h}) - \lambda h\sigma(e^{\lambda h}) = c_{Q+1}(\lambda h)^{Q+1} + O(\lambda h)^{Q+2}$ as $\lambda h \to 0$. As a corollary to this, we see that $\rho(\mu) - \lambda h\sigma(\mu)$ has a zero $\mu_0(\lambda h) = \exp(\lambda h) - \{c_{Q+1}/\sigma(1)\}(\lambda h)^{Q+1} + O(\lambda h)^{Q+2}$ as $\lambda h \to 0$, where $\mu_0(0) = 1$. On the other hand, $\hat{\mu}(\Lambda)$ is an approximation to $\exp(\Lambda)$ of order K, say, determined by the Runge-Kutta method. We prove the following:

Theorem. Let $\hat{\mu}(\lambda h) = \exp(\lambda h) - \gamma_K(\lambda h)^K + O(\lambda h)^{K+1}$, and let $\{\rho,\sigma\}$ be of order Q. Then $\Sigma(\lambda h;\mu) = \mu\rho(\mu) - \lambda h\hat{\mu}(\lambda h)\sigma(\mu)$ has a zero $\mu_0(\lambda h)$ with $\mu_0(0) = 1$ such that $\mu_0(\lambda h) = \exp(\lambda h) - \{c_{Q+1}/\sigma(1)\}(\lambda h)^{Q+1} - \gamma_K(\lambda h)^{K+1} + O(\lambda h)^M$, where $M = \min(K,Q)+2$.

Proof. $\exp(\lambda h)\rho(\exp \lambda h) - \lambda h\hat{\mu}(\lambda h)\sigma(\exp \lambda h) = \exp(\lambda h)[\rho(\exp \lambda h) - \lambda h\sigma(\exp\lambda h)] - \lambda h(\hat{\mu}(\lambda h) - \exp \lambda h)\sigma(\exp \lambda h) = \exp(\lambda h)c_{Q+1}(\lambda h)^{Q+1} + O(\lambda h)^{Q+2}\} + \lambda h\{\gamma_K(\lambda h)^K + O(\lambda h)^{K+1}\}\sigma(\exp \lambda h)$. In other words, $\Sigma(\lambda h;\exp \lambda h) = c_{Q+1}(\lambda h)^{Q+1} + \sigma(1)\gamma_K(\lambda h)^{K+1} + O(\lambda h)^M$. However, $\Sigma(\lambda h;\mu_0(\lambda h)) = 0$ and by the mean-value theorem $\Sigma(\lambda h;\exp \lambda h) - \Sigma(\lambda h;\mu_0(\lambda h)) = [\exp \lambda h - \mu_0(\lambda h)](d/d\mu)\Sigma(\lambda h;\zeta)$ where ζ lies between μ_0 and $\exp \lambda h$; $\zeta \to 1$ as $\lambda h \to 0$ and the value of $(d/d\mu)\Sigma(\lambda h;\zeta)$ tends smoothly to $\rho'(1) = \sigma(1) \neq 0$, as $\lambda h \to 0$. Thus, $\exp(\lambda h) - \mu_0(\lambda h) = \Sigma(\lambda h;\exp \lambda h)/\{\sigma(1) + O(\lambda h)\}$ and the result follows. Our theory can be extended to the case Q is cyclically reducible.

The following is a consequence of our earlier analysis.

Theorem. $\Sigma'(\lambda h;\mu)$ is strictly stable if and only if $\Lambda = \lambda h\hat{\mu}(\lambda h)$ is in the region of strict stability, S, of the polynomial $\Sigma(\Lambda;\mu)$ defined by the quadrature method based on quadrature rules which are $\{\mu\rho_r(\mu), \sigma_r(\mu)\}$-reducible.

The region of Λ-values for which $\Sigma(\Lambda;\mu)$ in the above statement is strictly stable is bounded due *(vide supra)* to the explicit nature of the cyclic formulae involved. Table 1 indicates, for various $\{\rho,\sigma\}$, the range of *real* Λ for which $\{\mu\rho(\mu),\sigma(\mu)\}$ is strictly stable. The values for Table 1 (where not available by a simple analysis) were computed by tracing the boundary locus [8, p. 99], using a SHARP MZ-80K computer.

Table 1

Choice of $\{\rho,\sigma\}$......	...Interval of stability for $\{\mu\rho(\mu),\sigma(\mu)\}$
Adams-Moulton formulae	Adams-Bashforth formulae
k=1 (Implicit Euler)..$S \supset (-2,0)$	k=1 (Euler)..$S \subset (-1,0)$
k=1 (Trapezium rule)..$S \supset (-2,0)$	k=2 ....$S \subset (-0.82,0)$
k=2 ....$S \supset (-1.54,0)$	k=3 ....$S \subset (-0.71,0)$
k=3 ....$S \supset (-1.40,0)$	k=4 ....$S \subset (-0.415,0)$
k=4 ....$S \supset (-1.37,0)$	

(The Adams-Moulton results apply to the weights w_{nj} of the Gregory rules.)

It does not seem to be possible to give an analytic expression for the boundary ∂S of the region $S \subset \mathbb{C}$ of strict stability for $\mu\rho(\mu) - \Lambda\sigma(\mu)$, in general. Consider, however, the particular case of the implicit Euler method. Then $\{\mu\rho = \mu^2 - \mu,\ \sigma = \mu\}$ has the same region of strict stability as does the Euler method $\{\rho = \mu-1,\ \sigma = 1\}$. Thus, the region S is here the interior of a disk: $|\Lambda - 1| < 1$.

Consider in a typical case the behaviour of $\lambda h\hat{\mu}(\lambda h)$; at $\lambda h = 0$, $\hat{\mu}(\lambda h) = 1$ and $\lambda h\hat{\mu}(\lambda h) = 0$; hence, there is an interval $(-\Lambda_0,0)$, say, where $\lambda h\hat{\mu}(\lambda h) < 0$. However, if Λ_0 is a zero of $\hat{\mu}(.)$, the roots of $\mu\rho(\mu)-\lambda h\hat{\mu}(\lambda h)\sigma(\mu)$ are those of $\mu\rho(\mu)$ and hence include the value unity.

Theorem. The region S' of strict stability of $\Sigma'(\lambda h;\mu)$ in (27) excludes those values λh for which $\hat{\mu}(\lambda h) = 0$.

It can be shown that if $\hat{\mu}(\lambda h)$ is a high order A-acceptable approximation to $\exp(\lambda h)$, then $\Sigma'(\lambda h;\mu)$ cannot be A-stable

It may be convenient to have *mixed quadrature-Runge-Kutta methods with a large region of strict stability.* We have seen the role of the zeros of $\hat{\mu}(\lambda h)$ in this discussion. Now, $\Sigma'(\lambda h;\mu)$ cannot be A_0-acceptable, and hence not A-acceptable, if $\lambda h\hat{\mu}(\lambda h)$ is unbounded as $\lambda h \to -\infty$. We therefore seek cases where $\lambda h\hat{\mu}(\lambda h) \to 0$ as $\lambda h \to -\infty$. We shall resort to the choice $\hat{\mu}(\lambda h) = 1/t_n(-\lambda h)$.

Lemma. The function $t_n(\Lambda)$ has just one negative real zero if n is odd and no real zero if n is even.

Proof. By induction, employing the facts $t_k(0) = 1$, $t_k(\infty) > 0$, $(-1)^k t_k(-\infty) > 0$, and $(d/d\Lambda)t_k(\Lambda) = t_{k-1}(\Lambda)$.

Lemma. For $n \geq 2$, $-(\sqrt{2}-1) \leq \lambda h/t_n(-\lambda h) < 0$ when $-\infty < \lambda h < 0$.

Proof. Consider the case $n = 2$ and set $\alpha(\lambda h) = \lambda h/t_2(-\lambda h)$; then

$\alpha(0) = 0$, $\alpha(-\infty) = 0$, and by our earlier lemma $\alpha(\lambda h)$ is continuous for negative λh. Further, $\alpha'(\lambda h)$ vanishes only at $-\sqrt{2}$, where α'' is positive and α assumes the value $-(\sqrt{2}-1)$; this is the minimum value of $\alpha(\lambda h)$ for negative λh. For $n \geq 3$ we observe that $t_n(-\Lambda) > t_{n-1}(-\Lambda)$ for $\Lambda < 0$, hence it follows that $\lambda h/t_n(-\lambda h)$ satisfies the given inequalities, for $n \geq 2$.

Theorem. There exist mixed quadrature Runge-Kutta methods for which $\Sigma'(\lambda h;\mu)$ is A_0- stable.

To prove our theorem, we can consult Table 1 to find the intervals of stability of various methods. Indeed, the method where Q is $\{\rho,\sigma\}$-reducible with $\rho(\mu) = \mu-1$, $\sigma(\mu) = \mu$ yields a polynomial $\mu\rho(\mu) - \Lambda\sigma(\mu)$ which is strictly stable if $-2 < \Lambda < 0$. Since $-2 < -(\sqrt{2}-1)$, the proof of the theorem is complete. We can even establish:

Theorem. There exists a mixed quadrature Runge-Kutta method for which $\Sigma'(\lambda h;\mu)$ is L-stable.

Proof. Choose $\{\rho,\sigma\}$ to correspond to the implicit Euler method, as above, and take $\hat{\mu}(\lambda h) = 1/(1-\lambda h)$. Then $\Sigma'(\lambda h;\mu) = \mu^2-\mu-\{\lambda h/(1-\lambda h)\}\mu$ has as its zeros $\mu = 0$, and $\mu = 1/(1-\lambda h)$, which are (untypically) L-acceptable.

The formula used to establish the last theorem is of low order, as is evident. We may seek $A(\alpha)$-stability if the search for A-stability combined with higher order is frustrated. We leave the answer for another occasion.

References

1. Baker, C.T.H., An introduction to the numerical treatment of Volterra and Abel-type integral equations. Lect. Notes Math 965 (1982) 1-38.
2. Baker, C.T.H., Numerical Analysis Tech Rep. 87, University of Manchester (1983).
3. Baker, C.T.H. and Wilkinson, J.C., Stability analysis of Runge-Kutta methods applied to a basic Volterra integral equation. J. Austral. Math. Soc. Ser. B 22 (1981) 515-538.
4. Brunner, H., A survey of recent advances in the numerical treatment of Volterra integral and integro-differential equations. J. Comput. Appl. Math. 8 (1982) 213-229.
5. Brunner, H., Hairer, E., and Nørsett, S.P., Runge-Kutta theory for Volterra integral equations of the second kind. Math. Comput. 39 (1982) 147-163.
6. Donelson, J.P., and Hansen, E., Cyclic composite multistep

predictor-corrector methods. SIAM J. Numer. Anal. 8 (1971) 137-157.

7. Henrici, P., Discrete Variable Methods in Ordinary Differential Equations, Wiley, New York, 1962.

8. Lambert, J.D., Computational Methods in Ordinary Differential Equations, Wiley, London, 1973.

9. Noble, B., Instability when solving Volterra integral equations of the second kind by multistep methods. Lect. Notes Math. 109 (1969) 23-29.

10. Pouzet, P. Étude en vue de leur traitement numériques des équations intégrales de type Volterra. Rev. Francaise Traitement de l'Information (Chiffres) 6 (1963) 79-112.

11. Stetter, H.J., Analysis of Discretization Methods for Ordinary Differential Equations, Springer, Berlin, 1973.

12. Wolkenfelt, P.H.M., The numerical analysis of reducible quadrature methods for Volterra integral and integro-differential equations. Academisch Proefschrift (Thesis), University of Amsterdam (1981).

13. Wolkenfelt, P.H.M., The construction of reducible quadrature rules for Volterra integral and integro-differential equations. IMA J.Numer.Anal. 2 (1982), 131-152.

ERROR EVALUATION FOR CUBIC BESSEL INTERPOLATION

Ai-Ping Bien
Institute of Applied Mathematics
National Cheng-Kung University
Tainan, Taiwan 700
R. O. C.

Fuhua Cheng
Institute of Computer and
Decision Sciences
National Tsing Hua University
Hsinchu, Taiwan 300
R. O. C.

Abstract. The convergence property of cubic Bessel interpolation is investigated and an exact error evaluation for functions of continuous third derivative when the interpolation points are uniformly spaced is given.

1. Introduction

Let $\tau = \{\tau_0, \tau_1, \ldots, \tau_n\}$ be a set of interpolation points in $[a,b]$ with $a = \tau_0 < \tau_1 < \ldots < \tau_n = b$. For a given real-valued function g defined on $[a,b]$ the cubic Bessel interpolant f_τ to g corresponding to τ is a real-valued function defined on $[a,b]$ such that on each interval $[\tau_i,\tau_{i+1}]$, $i = 0,1,\ldots,n-1$, f_τ is a polynomial P_i of degree ≤ 3 satisfying the following conditions

$$P_i(\tau_i) = g(\tau_i), \qquad P_i(\tau_{i+1}) = g(\tau_{i+1}),$$
$$P_i'(\tau_i) = s_i, \qquad P_i'(\tau_{i+1}) = s_{i+1},$$

where s_0 and s_n are free parameters, and $s_i (1 \leq i \leq n-1)$ is the slope at τ_i of the polynomial P of degree 2 which agrees with g at τ_{i-1}, τ_i and τ_{i+1} [1]. The polynomial P_i can be expressed in its Newton form

$$(1.1) \quad P_i(x) = g(\tau_i)+(x-\tau_i)s_i+(x-\tau_i)^2([\tau_i,\tau_{i+1}]g-s_i)/\Delta\tau_i \\ +(x-\tau_i)^2(x-\tau_{i+1})(s_{i+1}+s_i-2[\tau_i,\tau_{i+1}]g)/(\Delta\tau_i)^2$$

with $\Delta\tau_i = \tau_{i+1}-\tau_i$ and $[\tau_i,\tau_{i+1}]g$ the second divided difference of g at τ_i and τ_{i+1}. s_i $(1 \leq i \leq n-1)$ can be expressed as

(1.2) $s_i = (\Delta\tau_i[\tau_{i-1},\tau_i]g+\Delta\tau_{i-1}[\tau_i,\tau_{i+1}]g)/(\Delta\tau_i+\Delta\tau_{i-1})$.

When $\tau = \{\tau_0, \tau_1, \ldots, \tau_n\}$ is uniformly spaced in [a,b], we denote f_τ by f_n. It is known that if s_0 and s_n are chosen in a suitable way then cubic Bessel interpolation provides an $0(|\tau|^3)$-approximation to g [2].

The purpose of this paper is to investigate the convergence property of cubic Bessel interpolation. We shall give an exact error evaluation for cubic Bessel interpolation for functions of continuous third derivatives when τ is uniformly spaced. Since Bessel interpolation is a local method, i.e., the ith piece P_i depends only on information from, or near, the interval $[\tau_i,\tau_{i+1}]$, a different but appropriate way to choose s_0 and s_n would not affect the approximation of f_τ at interior points of [a,b] when $|\tau|$ tends to zero. We shall define s_0 and s_n the following way.

Let $g \in C^{(3)}[a,b]$. We extend g to a function $\hat{g}$ on [2a-b,ab-a] by requiring that $\hat{g}$ be a polynomial of degree three on [2a-b,a) and (b,2b-a] so that $\hat{g} \in C^{(3)}[2a-b,2b-a]$. Such a function g is uniquely defined and can be expressed as

$$\hat{g}(x) = \begin{cases} \sum_{i=0}^{3} g^{(i)}(a)(x-a)^i/i!\,, & x \in [2a-b,a) \\ g(x), & x \in [a,b] \\ \sum_{i=0}^{3} g^{(i)}(b)(x-b)^i/i!\,, & x \in (b,2b-a]. \end{cases}$$

Define $\tau_{-1} = 2\tau_0-\tau_1$ and $\tau_{n+1} = 2\tau_n-\tau_{n-1}$. Then using (1,2) with g replaced by $\hat{g}$, s_0 and s_n can be expressed as

$$\text{(1.3)}\quad \begin{aligned} s_0 &= (\Delta\tau_0[\tau_{-1},\tau_0]\hat{g}+\Delta\tau_{-1}[\tau_0,\tau_1]\hat{g})/(\Delta\tau_0+\Delta\tau_{-1}), \\ s_n &= (\Delta\tau_n[\tau_{n-1},\tau_n]\hat{g}+\Delta\tau_{n-1}[\tau_n,\tau_{n+1}]\hat{g})/(\Delta\tau_n+\Delta\tau_{n-1}). \end{aligned}$$

2. Results

Our main result is the following

Theorem 1. Let $g \in C^{(3)}[a,b]$ and $\tau = \{\tau_0, \tau_1, \ldots, \tau_n\}$ a set of uniformly spaced interpolation points in [a,b] with $a = \tau_0<\tau_1<\ldots<\tau_n=b$. If f_n is the cubic Bessel interpolant to g on [a,b] corresponding to τ such that s_0 and s_n are defined by (1.3) then

$$\lim_{n\to\infty} \|f_n-g\|\, n^3 = (\sqrt{3}/108)(b-a)^3 \|g^{(3)}\| ,$$

where $\|f\|$ is the uniform norm of f on [a,b].

Theorem 1 shows that the $O(|\tau|^3)$-approximation provided by cubic Bessel interpolation is the best possible; it cannot be improved any further. Hence it disproves a statement made by de Boor.

de Boor states in [2, problem IV.6(c)] that, in general, cubic Bessel interpolation with prescribed endslopes provides an order of $O(|\tau|^4)$-approximation if τ is uniform. However, from the above theorem and the fact that Bessel interpolation is a local method, we can tell that the above statement is not true*.

3. Proofs

The following notations will be used throughout this section, we give their definition here.

$$D = \{(\alpha,r,h)\,|\,\alpha-(1-r)h\geq 2a-b,\quad \alpha+(2-r)h\leq 2b-a,\quad h\geq 0\},$$

$$R_1(r) = -(1/2)r(1-r)^2, \qquad R_2(r) = (1/2)(1-r)(-3r^2+2r+2),$$

$$R_3(r) = (1/2)r(-3r^2+4r+1), \qquad R_4(r) = -(1/2)(1-r)r^2,$$

$$\hat{R}_1(r) = (-1-r)^3 R_1(r), \qquad \hat{R}_2(r) = (-r)^3 R_2(r),$$

$$\hat{R}_3(r) = (1-r)^3 R_3(r), \qquad \hat{R}_4(r) = (2-r)^3 R_4(r),$$

$$R(r) = \hat{R}_1(r)+\hat{R}_2(r)+\hat{R}_3(r)+\hat{R}_4(r) = r(2r-1)(r-1),$$

r_M is the value in [0,1] such that $|R(r_M)| = \max_{r\in[0,1]} |R(r)|$,

$$p(\alpha,r,h) = R_1(r)\hat{g}(\alpha-(1+r)h)+R_2(r)\hat{g}(\alpha-rh)+R_3(r)\hat{g}(\alpha+(1-r)h) +R_4(r)\hat{g}(\alpha+(2-r)h), \quad (\alpha,r,h)\in D,$$

$$\hat{P}(\alpha,r,h) = \hat{R}_1(r)\hat{g}^{(3)}(\alpha-(1+r)h)+\hat{R}_2(r)\hat{g}^{(3)}(\alpha-rh)+\hat{R}_3(r)\hat{g}^{(3)}(\alpha+(1-r)h) +\hat{R}_4(r)\hat{g}^{(3)}(\alpha+(2-r)h), \quad (\alpha,r,h)\in D.$$

---

* The authors have recently been informed by Prof. de Boor that he noticed the incorrectness of the statement, too, and that statement has been deleted from the latest printing of his book.

It is easy to see that, through a simple computation, $|R(r_M)| = \sqrt{3}/18$ with $r_M = (3\pm\sqrt{3})/6$.

We need an auxiliary result.

Lemma 1. If $\tau = \{\tau_0, \tau_1, \ldots, \tau_n\}$ is uniformly spaced in $[a,b]$ with $a = \tau_0 < \tau_1 < \ldots. < \tau_n = b$ then for any real number x in $[\tau_i, \tau_{i+1})$ $(0 \leq i < n)$ we can always find a real number s in $(0, \delta_n)$ so that

$$(f_n(x)-g(x))n^3 = (b-a)^3\hat{P}(x, r_{i,n}, s)/6$$

where $\delta_n = (b-a)/n$ and $r_{i,n} = (x-a)/\delta_n - i$.

Proof. By setting $\delta_n = (b-a)/n$, s_i can be expressed as

$$s_i = (\hat{g}(\tau_{i+1})-\hat{g}(\tau_{i-1}))/2\delta_n. \tag{3.1}$$

Let $r_{i,n} = (x-a)/\delta_n - i$. Then τ_{i-1}, τ_i, τ_{i+1}, τ_{i+2} can be expressed as

$$\begin{aligned} \tau_{i-1} &= x-(1+r_{i,n})\delta_n, & \tau_i &= x-r_{i,n}\delta_n, \\ \tau_{i+1} &= x+(1-r_{i,n})\delta_n, & \tau_{i+2} &= x+(2-r_{i,n})\delta_n. \end{aligned} \tag{3.2}$$

Now substitute (3.1) and (3.2) into (1.1) with $[\tau_i, \tau_{i+1}]g$ replaced by $(\hat{g}(\tau_{i+1})-\hat{g}(\tau_i))/\delta_n$. Then by simple algebra we have

$$f_n(x) = \hat{P}(x, r_{i,n}, \delta_n). \tag{3.3}$$

For these n, x and i, we define $D_{n,x} = \{t \mid (x, r_{i,n}, t) \in D\}$ and construct two auxiliary functions G and F on $D_{n,x}$ as follows:

$$G(t) = p(x, r_{i,n}, t) - g(x), \quad \text{and}$$
$$F(t) = G(t) - (G(u)/u^3)t^3,$$

where u is an arbitrary, fixed positive real number in $D_{n,x}$ (such a number u always exists. Actually $D_{n,x} = [0, \min\{(x+b-2a)/(r_{i,n}+1), (2b-a-x)/(2-r_{i,n})\}])$. Since $\hat{g} \in C^{(3)}[2a-b, 2b-a]$ it follows that $G \in C^{(3)}(D_{n,x})$ and $F \in C^{(3)}(D_{n,x})$. In particular, $F \in C^{(3)}[o,u]$.

We claim that $F(0) = F(u) = 0$. $F(u) = 0$ follows directly from the definition. As for $F(0)$, since $R_1(r)+R_2(r)+R_3(r)+R_4(r) = 1$ and $\hat{g}(x) = g(x)$, we have $G(0) = 0$, therefore $F(0) = 0$. Then by Roll's theorem, there exists a positive number $\xi \in (0,u)$ such that $F'(\xi) = 0$.

Next consider $F'(t) = G'(t) - (G(u)/u^3)3t^2$ on $[0,\xi]$. Since

$(-1-r)R_1(r)+(-r)R_2(r)+(1-r)R_3(r)+(2-r)R_4(r) = 0$ it follows that $G'(0) = 0$ and so $F'(0) = 0$. Therefore, by Roll's theorem again, there exists a positive number $\theta \in (0,\xi)$ such that $F^{(2)}(\theta) = 0$.

Finally consider $F^{(2)}(t) = G^{(2)}(t)-(G(u)/u^3)6t$ on $[0,\theta]$. Since $(-1-r)^2R_1(r)+(-r)^2R_2(r)+(1-r)^2R_3(r)+(2-r)^2R_4(r) = 0$, we have $G^{(2)}(0)=0$ and so $F^{(2)}(0) = 0$. Applying Rolle's theorem one more time, we can find a positive number $s \in (0,\theta)$ such that $F^{(3)}(s) = 0$ or

$$G(u)/u^3 = G^{(3)}(s)/6.$$

Since $G^{(3)}(s)$ is nothing but $\hat{p}(x,r_{i,n},s)$, therefore, from the definition of $G(t)$ we see that

$$(3.4) \qquad (p(x,r_{i,n},u)-g(x))/u^3 = \hat{p}(x,r_{i,n},s)/6,$$

where u is an arbitrarily chosen positive number in $D_{n,x}$. Simply let $u = \delta_n$ and substitute (3.3) into (3.4). Then the lemma follows (under the assumption $x \in [\tau_i,\tau_{i+1}) \subset [a,b]$, it is easy to see that $\delta_n \leq \min\{(x+b-2a)/(r_{i,n}-1),(2b-a-x)/(2-r_{i,n})\}$).

Now the proof of Theorem 1. Since $f_n \in C^{(1)}[a,b]$ and $g \in C^{(3)}[a,b]$ we may choose, for each positive integer n, a real number $x_n \in [a,b)$ such that

$$(3.5) \qquad |f_n(x_n)-g(x_n)| = \|f_n-g\|,$$

where $\|f\|$ is the uniform norm of f on [a,b]. For each of these x_n let i_n be the integer such that $x_n \in [\tau_{i_n},\tau_{i_n+1})$. Then by Lemma 1 there exists a positive number $s_n \in (0,\delta_n)$ such that

$$(3.6) \qquad |f_n(x_n)-g(x_n)|n^3 = (b-a)^3|\hat{p}(x_n,r_n,s_n)|/6$$

with $r_n = (x_n-a)/\delta_n-i_n$. Reorganize $\hat{p}(x_n,r_n,s_n)$

$$\begin{aligned}\hat{p}(x_n,r_n,s_n) = {} & \hat{R}_1(r_n)\hat{g}^{(3)}(x_n-(1+r_n)s_n)+\hat{R}_2(r_n)\hat{g}^{(3)}(x_n-r_ns_n)\\ & +\hat{R}_3(r_n)\hat{g}^{(3)}(x_n+(1-r_n)s_n)+\hat{R}_4(r_n)\hat{g}^{(3)}(x_n+(2-r_n)s_n)\end{aligned}$$

so that

$$\begin{aligned}\hat{p}(x_n,r_n,s_n) = {} & R(r_n)\hat{g}^{(3)}(x_n)+\hat{R}_1(r_n)(\hat{g}^{(3)}(x_n-(1+r_n)s_n)-\hat{g}^{(3)}(s_n))\\ & +\hat{R}_2(r_n)(\hat{g}^{(3)}(x_n-r_ns_n)-\hat{g}^{(3)}(x_n))\end{aligned}$$

$$+\hat{R}_3(r_n)(\hat{g}^{(3)}(x_n+(1-r_n)s_n)-\hat{g}^{(3)}(x_n))$$
$$+\hat{R}_4(r_n)(\hat{g}^{(3)}(x_n+(2-r_n)s_n)-\hat{g}^{(3)}(x_n))$$
$$= R(r_n)\hat{g}^{(3)}(x_n)+\varepsilon_1(n).$$

Then we have

$$|\hat{p}(x_n,r_n,s_n)| \le |R(r_n)| \; \|g^{(3)}\| + |\varepsilon_1(n)|.$$

Since $r_n = (x_n-a)/\delta_n-i_n \in [0,1]$ and the maximum of $|R(r)|$ on $[0,1]$ is $\sqrt{3}/18$, with the fact that $\|\hat{g}^{(3)}\| = \|g^{(3)}\|$ on $[a,b]$, it follows that

$$\text{(3.7)} \qquad \|\hat{p}(x_n,r_n,s_n)\| \le (\sqrt{3}/18)\,\|g^{(3)}\| + |\varepsilon_1(n)|.$$

From the facts that $g^{(3)}$ is uniformly continuous on $[2a-b,2b-a]$ and $\lim_{n\to\infty} s_n=0$ we have $\lim_{n\to\infty} \varepsilon_1(n)=0$. Therefore, from (3.5), (3.6) and (3.7),

$$\text{(3.8)} \qquad \lim_{n\to\infty} \|f_n-g\|\, n^3 \le (\sqrt{3}/108)(b-a)^3 \|g^{(3)}|.$$

On the other hand, let y_0 be in $[a,b]$ such that $|g^{(3)}(y_0)|=\|g^{(3)}\|$. For any positive integer n, there exists a nonnegative integer j_n such that $y_0 \in [\tau_{j_n},\tau_{j_n+1})$. In each of these $[\tau_{j_n},\tau_{j_n+1}]$ we choose a real number y_n satisfying the equation $(y_n-a)/\delta_n-j_n=r_M$. By Lemma 1 we can find a real number $\theta_n \in (0,\delta_n)$ such that

$$|f_n(y_n)-g(y_n)|\,n^3 = (b-a)^3|\hat{p}(y_n,r_M,\theta_n)|/6.$$

If we reorganize the right hand side of $\hat{p}$ to the following from

$$\hat{p}(y_n,r_M,\theta_n) = R(r_M)\hat{g}^{(3)}(y_n)$$
$$+\hat{R}_1(r_M)(\hat{g}^{(3)}(y_n-(1+r_M)\theta_n)-\hat{g}^{(3)}(y_n))$$
$$+\hat{R}_2(r_M)(\hat{g}^{(3)}(y_n-r_M\theta_n)-\hat{g}^{(3)}(y_n))$$
$$+\hat{R}_3(r_M)(\hat{g}^{(3)}(y_n+(1-r_M)\theta_n)-\hat{g}^{(3)}(y_n))$$
$$+\hat{R}_4(r_M)(g^{(3)}(y_n+(2-r_M)\theta_n)-\hat{g}^{(3)}(y_n))$$
$$= \hat{R}(r_M)\hat{g}^{(3)}(y_n)+\varepsilon_2(n)$$

then we have

$$|\hat{p}(y_n, r_M, \theta_n)| \geq (\sqrt{3}/18)\,|\hat{g}^{(3)}(y_n)| - |\varepsilon_2(n)|,$$

or

(3.9) $$|f_n(y_n) - g(y_n)|\, n^3 \geq ((\sqrt{3}/108)\,|\hat{g}^{(3)}(y_n)| - |\varepsilon_2(n)|/6)\,(b-a)^3.$$

Since $|y_n - y_0| \leq \delta_n$ for all n implies that $\lim_{n\to\infty} y_n = y_0$ and $\hat{g}^{(3)}$ is continuous on [a,b], we have

(3.10) $$\lim_{n\to\infty} \hat{g}^{(3)}(y_n) = \hat{g}^{(3)}(y_0) = g^{(3)}(y_0) = \|g^{(3)}\|.$$

By using the same argument as we did for $\varepsilon_1(n)$, we can show that

(3.11) $$\lim_{n\to\infty} |\varepsilon_2(n)| = 0.$$

Therefore, from (3.9), (3.10), (3.11) and the fact that $\|f_n - g\| \geq |f_n(Y_n - g(y_n)|$,

(3.12) $$\lim_{n\to\infty} \|f_n - g\|\, n^3 \geq (\sqrt{3}/108)(b-a)^3\, \|g^{(3)}\|,$$

and theorem 1 follows from (3.8) and (3.12). □

References

1. G. Birkhoff and C. de Boor, Piecewise polynomial interpolation and approximation, in "Approximation of Functions", H. L. Garabedian ed., Elsevier Pub. Co., Armsterdam, 1965.

2. C. de Boor, A Practical Guide to Splines, Springer-Verlag, Berlin/New York, 1978.

Extended Numerical Computations on the "1/9" Conjecture In Rational Approximation Theory

*A. J. Carpenter, A. Ruttan, and R. S. Varga**

Institute for Computational Mathematics

Kent State University

Kent, Ohio 44242, U.S.A.

Abstract. The behavior of the constants $\lambda_{n,n}(e^{-x})$, denoting the errors of best uniform approximation to e^{-x} on the interval $[0,+\infty)$ by real rational functions having numerator and denominator polynomials of degree at most n, has generated much recent interest in the approximation theory literature. Based on high-precision calculations, we present here the table of constants $\{\lambda_{n,n}(e^{-x})\}_{n=0}^{30}$, rounded to forty significant digits, and we discuss their significance to related conjectures in this area.

1. Introduction

It is well-known (cf. [15, Chapter 8]) that rational approximations to e^{-x} permeate the numerical analysis and approximation theory literature, in that these approximations arise quite naturally in the numerical solution of heat-conduction problems and in the study of numerical methods for ordinary differential equations. Our interest here centers on the following specific problem. If $\pi_{m,n}$ denotes the class of rational functions $p_m(x)/q_n(x)$, where $p_m(x)$ is a real polynomial of degree at most m and $q_n(x)$ is a real polynomial of degree at most n, then what can be said about the quantities

$$\lambda_{m,n}(e^{-x}) := \min\left\{ \| e^{-x} - r_{m,n} \|_{L_\infty[0,+\infty)} : r_{m,n} \in \pi_{m,n} \right\},\ m \le n \tag{1.1}$$

and

$$\Lambda_1 := \varliminf_{n\to\infty} \lambda_{n,n}^{1/n}(e^{-x})\ ;\ \Lambda_2 := \varlimsup_{n\to\infty} \lambda_{n,n}^{1/n}(e^{-x})\ ? \tag{1.2}$$

The history of that question has its beginnings in a 1969 paper of Cody, Meinardus, and Varga [4], where they showed that

* Research supported by the Air Force Office of Scientific Research and by the Department of Energy.

$$\lim_{n\to\infty}\left\{ \| e^{-x} - \frac{1}{s_n(x)} \|_{L_\infty[0,+\infty)} \right\}^{1/n} = \frac{1}{2} < 1 \ , \tag{1.3}$$

where $s_n(x) := \sum_{k=0}^{n} x^k/k!$ is the n-th partial sum of e^x. Thus (cf. (1.1)),

$$\overline{\lim_{n\to\infty}} \ \lambda_{0,n}^{1/n}(e^{-x}) \le \frac{1}{2} \ . \tag{1.4}$$

In fact, a slightly better result than (1.3) was established in [4], namely

$$\overline{\lim_{n\to\infty}} \ \lambda_{0,n}^{1/n}(e^{-x}) \le K \doteqdot \frac{1}{2.298} \ . \tag{1.5}$$

Since it is obvious from (1.1) that

$$\lambda_{0,n}(e^{-x}) \ge \lambda_{1,n}(e^{-x}) \ge \cdots \ge \lambda_{n,n}(e^{-x}) \ , \ n = 0, 1, \cdots , \tag{1.6}$$

it is evident that (1.5) implies, with (1.2) and (1.6), that

$$0 \le \Lambda_1 \le \Lambda_2 \le \frac{1}{2.298} \ , \tag{1.7}$$

so that best rational approximation to e^{-x} on $[0,+\infty)$ exhibits *geometric convergence* to zero. It is this geometric convergence which has fascinated so many researchers.

In [4], a tabulation of the computed values $\{\lambda_{0,n}(e^{-x})\}_{n=0}^{9}$ and $\{\lambda_{n,n}(e^{-x})\}_{n=0}^{14}$ was also given. The numbers in these tabulations exhibited a striking regularity and added to the interest in the problem, despite the relatively low accuracy (about 4 significant digits) of these numbers computed. Two subsequent papers by Schönhage [12] in 1973 and Newman [7] in 1974 added significantly to this interest. First, Schönhage [12] obtained the very precise estimates

$$\frac{1}{6\sqrt{(4n+4)\log 3 + 2\log 2}} \le 3^n \lambda_{0,n}(e^{-x}) \le \sqrt{2} \ , \ n = 0, 1, \cdots , \tag{1.8}$$

so that in fact

$$\lim_{n\to\infty} \lambda_{0,n}^{1/n} = \frac{1}{3} \ , \tag{1.9}$$

whence (cf. (1.2) and (1.6))

$$\Lambda_2 \le \frac{1}{3} \ . \tag{1.10}$$

Then, Newman [7] showed that the convergence of $\lambda_{n,n}(e^{-x})$ to zero was *at most* geometric, i.e., (cf. (1.2))

$$0 < \frac{1}{1280} \le \Lambda_1 \ . \tag{1.11}$$

On the other hand, as the determination of $\lambda_{n,n}(e^{-x})$ depends on asymptotically twice as many coefficients (in $r_{n,n}(x)$ of (1.1)) as does the determination of $\lambda_{0,n}(e^{-x})$, one could wildly guess from (1.9) that

$$\lambda_{n,n}^{1/n} \sim \left(\frac{1}{3}\right)^2 = 1/9 \ . \tag{1.12}$$

But, as the computed values $\{\lambda_{n,n}(e^{-x})\}_{n=0}^{14}$ of [4] indeed seemed to *roughly* agree with (1.12), the following conjecture was born:

Conjecture 1 (cf.[11]). $\lim_{n\to\infty} \lambda_{n,n}^{1/n}(e^{-x}) = 1/9$, i.e., $\Lambda_1 = \Lambda_2 = 1/9$.

The race was then on to improve upon the bounds for Λ_1 of (1.11) and Λ_2 of (1.10). We list, in chronological order, the successive refinements for Λ_1 and Λ_2:

1969	Cody/Meinardus/Varga [4]		$\Lambda_2 \le \frac{1}{2.298}$
1973	Schönhage [12]		$\Lambda_2 \le \frac{1}{3}$
1974	Newman [7]	$\frac{1}{1280} \le \Lambda_1$	
1978	Rahman/Schmeisser [9]	$\frac{1}{380} \le \Lambda_1$	
1978	Rahman/Schmeisser [10]		$\Lambda_2 < \frac{1}{4.091}$
1980	Blatt/Braess [1]	$\frac{1}{52} < \Lambda_1$	
1981	Németh [6]		$\Lambda_2 < \frac{1}{6.475}$
1982	Schönhage [13]	$\frac{1}{13.928} < \Lambda_1$	
1984	Opitz/Scherer [8]		$\Lambda_2 < \frac{1}{9.037}$

It is clear that this last result of Opitz and Scherer [8] rigorously *disproves* Conjecture 1; the geometrical convergence rate of $\lambda_{n,n}^{1/n}(e^{-x})$ is actually *better* than 1/9. What is curious is that Schönhage [13] and Trefethen and Gutknecht [14] simultaneously observed (by examining the ratios $\lambda_{n-1,n-1}(e^{-x}) / \lambda_{n,n}(e^{-x})$ of the numbers tabulated in [4]) that

Conjecture 1 was surely ***numerically*** false, even before this was rigorously established later by Opitz and Scherer [8]. In fact, Schönhage made the following conjecture:

Conjecture 2 ([13]): $\lim_{n\to\infty} \lambda_{n,n}^{1/n}(e^{-x}) = \frac{3}{2}\left(2-\sqrt{3}\right)^2 \doteq 1/9.285469$,

while Trefethen and Gutknecht [14] instead conjectured

Conjecture 3 ([14]): $\lim_{n\to\infty} \lambda_{n,n}^{1/n}\left(e^{-x}\right) \doteq 1/9.28903$.

Obviously, Conjectures 2 and 3 are very close, so close in fact that the tabulation given in [4] is simply not accurate enough to settle either of these conjectures.

2. Statement of Numerical Results

The numbers $\{\lambda_{n,n}(e^{-x})\}_{n=0}^{14}$ of [4] from 1969 no doubt contributed auxiliary interest to the theoretical problem of either settling Conjecture 1 or determining improved bounds for Λ_1 and Λ_2 of (1.2). But, in the intervening fifteen years since [4] appeared, we knew of no further ***direct*** computation of the numbers $\lambda_{n,n}(e^{-x})$ to either ***improve*** the accuracies of the calculations of [4], or to extend the ***length*** of the table $\{\lambda_{n,n}(e^{-x})\}_{n=0}^{14}$ of [4]. Thus, it was felt that a numerical ***up-date*** for such calculations was in order! (We note, however, that upper estimates of $\{\lambda_{n,n}(e^{-x})\}_{n=0}^{18}$, based on the Carathéodory-Fejér method, appear in [14].)

Based on calculations done on a VAX-11/780 in the Department of Mathematical Sciences at Kent State University, using Richard Brent's MP (multiple precision) package [2] with 230 decimal digits, the numbers $\{\lambda_{n,n}(e^{-x})\}_{n=0}^{30}$ were determined to an accuracy of about 200 decimal digits. These numbers are given below, rounded to forty significant digits.

n	$\lambda_{n,n}\,(e^{-x})$
0	5.0000000000000000000000000000000000000 (-01)
1	6.6831042161850463470611623827115147261 45 (-02)
2	7.3586701695805292800125541630806037567 45 (-03)
3	7.9938063633568782880811900971119616897 66 (-04)
4	8.6522406952888523482243458254146735250 07 (-05)
5	9.3457131530266464767536568207923979896 09 (-06)
6	1.0084543748996707079345287764100020604 07 (-06)
7	1.0874974913752479608665313072729334784 85 (-07)
8	1.1722652116334907177954323039388804735 11 (-08)
9	1.2632924833223141460949321009097283343 34 (-09)
10	1.3611205233454477498707881615368423764 73 (-10)
11	1.4663111949374871406681261995577526903 48 (-11)
12	1.5794568370512387714867567328183815746 85 (-12)
13	1.7011870763403529664164865499450815333 37 (-13)
14	1.8321743782540412751555017565131565305 59 (-14)
15	1.9731389966128034286256658020822992417 70 (-15)
16	2.1248537104952237487996344364187178090 45 (-16)
17	2.2881485632478919604052208612692419494 72 (-17)
18	2.4639157377651692748310829623232282977 74 (-18)
19	2.6531146580633127669264550346953305434 63 (-19)
20	2.8567773835490937066908938449300680288 30 (-20)
21	3.0760143495057905069144218639753086839 48 (-21)
22	3.3120205005513186907513737108226141460 29 (-22)
23	3.5660818606364245847698227997651372597 24 (-23)
24	3.8395825821681321269364868473011895629 43 (-24)
25	4.1340125172853630062707580554526301970 56 (-25)
26	4.4509753557304246897932636072797330395 12 (-26)
27	4.7921973758889041899314199978855209710 52 (-27)
28	5.1595368582571326546650112912554530364 11 (-28)
29	5.5549942137516226746420079038791349910 28 (-29)
30	5.9807228828496954372714270071247982846 42 (-30)

<u>Table 1:</u> $\{\ \lambda_{n,n}\,(e^{-x})\ \}_{n=0}^{30}$

A short description of how the actual computations for $\lambda_{n,n}\,(e^{-x})$ were performed, will be given in §3.

As the ratios $\lambda_{n-1,n-1}\,/\,\lambda_{n,n}\,(e^{-x})$ figured into the formulation of Conjectures 2 and 3 of §1, we next tabulate these ratios below, rounded to twenty digits.

n	ρ_n
1	7.4815532397221509829 (+00)
2	9.0819455991000169709 (+00)
3	9.2054646248528427538 (+00)
4	9.2390013695637342229 (+00)
5	9.2579780201008948071 (+00)
6	9.2673633886078728002 (+00)
7	9.2731650684028757880 (+00)
8	9.2768895688704833336 (+00)
9	9.2794442071765347804 (+00)
10	9.2812683495309755120 (+00)
11	9.2826170054814049413 (+00)
12	9.2836420758101343366 (+00)
13	9.2844394306651793616 (+00)
14	9.2850718606898552365 (+00)
15	9.2855819149043751996 (+00)
16	9.2859992519340952302 (+00)
17	9.2863450591648612312 (+00)
18	9.2866347991400778935 (+00)
19	9.2868799705918004397 (+00)
20	9.2870892682832631480 (+00)
21	9.2872693653333554026 (+00)
22	9.2874254522088778824 (+00)
23	9.2875616152014957848 (+00)
24	9.2876811067903443961 (+00)
25	9.2877865418013514399 (+00)
26	9.2878800417598657237 (+00)
27	9.2879633425048853224 (+00)
28	9.2880378753756707951 (+00)
29	9.2881048291364217039 (+00)
30	9.2881651976905816378 (+00)

Table 2: $\{ \rho_n := \lambda_{n-1,n-1}(e^{-x}) / \lambda_{n,n}(e^{-x}) \}_{n=1}^{30}$

Now, as the ratios $\rho_n := \lambda_{n-1,n-1}(e^{-x}) / \lambda_{n,n}(e^{-x})$ of Table 2, appear to be converging linearly in $1/n^2$ (an observation already used in [14]), i.e.,

$$\rho_n \doteqdot \rho + \frac{K_1}{n^2} + \cdots, \text{ as } n \to \infty , \tag{2.1}$$

where K_1 is independent of n, we have applied Richardson's extrapolation (cf. Brezinski [3, p.6]), with $x_n = 1/n^2$, to the last eleven entries in Table 2, to accelerate the convergence of the entries of Table 2. These are given in Table 3 - 10 below.

9.2890264097244993503 (+00)
9.2890262501648177328 (+00)
9.2890261238332083354 (+00)
9.2890260227584489782 (+00)
9.2890259411144138332 (+00)
9.2890258745847956720 (+00)
9.2890258199319277514 (+00)
9.2890257746993546055 (+00)
9.2890257370035920984 (+00)
9.2890257053863190015 (+00)

Table 3: 1st Richardson's extrapolation.

9.2890254903568100301 (+00)
9.2890254907395748783 (+00)
9.2890254910173236639 (+00)
9.2890254912217618362 (+00)
9.2890254913741950640 (+00)
9.2890254914892118820 (+00)
9.2890254915769523220 (+00)
9.2890254916445664944 (+00)
9.2890254916971628979 (+00)

Table 4: 2nd Richardson's extrapolation.

9.2890254919264426247 (+00)
9.2890254919246363634 (+00)
9.2890254919235212362 (+00)
9.2890254919227472919 (+00)
9.2890254919222163736 (+00)
9.2890254919218439885 (+00)
9.2890254919215797099 (+00)
9.2890254919213896706 (+00)

Table 5: 3rd Richardson's extrapolation.

9.2890254919205312241 (+00)
9.2890254919208485672 (+00)
9.2890254919207963074 (+00)
9.2890254919208120946 (+00)
9.2890254919208127682 (+00)
9.2890254919208150150 (+00)
9.2890254919208161591 (+00)

Table 6: 4th Richardson's extrapolation.

9.2890254919214127327	(+00)
9.2890254919206982369	(+00)
9.2890254919208432825	(+00)
9.2890254919208141655	(+00)
9.2890254919208198985	(+00)
9.2890254919208187594	(+00)

Table 7: 5th Richardson's extrapolation.

9.2890254919196627357	(+00)
9.2890254919210653837	(+00)
9.2890254919207671899	(+00)
9.2890254919208296190	(+00)
9.2890254919208167344	(+00)

Table 8: 6th Richardson's extrapolation.

9.2890254919227707308	(+00)
9.2890254919203837979	(+00)
9.2890254919209142567	(+00)
9.2890254919207983626	(+00)

Table 9: 7th Richardson's extrapolation.

9.2890254919178974094	(+00)
9.2890254919214990876	(+00)
9.2890254919206635242	(+00)

Table 10: 8th Richardson's extrapolation.

It is interesting to see from these Richardson extrapolations that the last 32 entries from Tables 5 - 10 *all* agree, when rounded to twelve digits, to the number $\rho = 9.28902549192$. It appears that the best extrapolated value of ρ of (2.1) comes from Table 6, which thus yields numerically, to fifteen decimals, that

$$1 / \rho = \lim_{n \to \infty} \lambda_{n,n}^{1/n} (e^{-z}) \doteqdot \frac{1}{9.289\,025\,491\,920\,81} . \tag{2.2}$$

This would appear, numerically, to *refute* Conjecture 2 of §1. In addition, as the constant in (2.2) is distinctly different from $1/9.037$, it also appears that the claim of Opitz and Scherer [8], that their method might produce "optimal" results, is numerically surely false! Despite the general numerical agreement of the extrapolations of Tables 3 - 10 in estimating a common value (2.2) for Λ_1 and Λ_2, it must be emphasized, however, that we have presented here only numerical results.

As a second method for estimating the quantity ρ of (2.2), assume that, as $n \to +\infty$,

$$\lambda_{n,n} (e^{-z}) = \frac{1}{\rho^n} \left\{ \gamma_0 + \frac{\gamma_1}{n} + \frac{\gamma_2}{n^2} + \cdots \right\} + \text{lower order terms} , \tag{2.3}$$

so that

$$\left(\lambda_{n,n} (e^{-z}) \right)^{1/n} = \frac{1}{\rho} + \frac{c_1}{n} + \frac{c_2}{n^2} + \cdots + \text{lower order terms} , \tag{2.4}$$

where the constants c_j of (2.4) depend on ρ and the γ_j's. The form of (2.4) suggests that the convergence of $\left(\lambda_{n,n} (e^{-x}) \right)^{1/n}$ to $1/\rho$ can be similarly accelerated by Richardson's extrapolation, now with $x_n = 1/n$. The numerical results of this acceleration, applied to $\left\{ \left(\lambda_{n,n} (e^{-x}) \right)^{1/n} \right\}_{n=1}^{30}$, produced essentially the *same* value for ρ as that given in (2.2). (For brevity, we have not included here the analogs of Tables 3 - 10 for these extrapolations.)

With the assumption of (2.3), it follows that

$$\rho^n \lambda_{n,n} (e^{-x}) = \gamma_0 + \frac{\gamma_1}{n} + \frac{\gamma_2}{n^2} + \cdots + \text{lower order terms} , \tag{2.5}$$

which further suggests that the constants γ_j in (2.5) can be successively determined from the known values of $\lambda_{n,n} (e^{-x})$, the assumed values of ρ of (2.2), and Richardson's extrapolations, again with $x_n = 1/n$. The resulting estimates for the constants γ_j are given in Table II.

Table II:
Numerical estimates for the γ_j of (2.5)

j	γ_j
0	+0.656 213 133 75
1	-0.054 684 427 8
2	+0.029 620 072 8
3	-0.016 012 6
4	+0.008 627 4
5	-0.005 74

The approximation

$$\tilde{\lambda}_{n,n} := \frac{1}{\rho^n} \left\{ \gamma_0 + \frac{\gamma_1}{n} + \cdots + \frac{\gamma_5}{n^5} \right\}, \tag{2.6}$$

based on the numbers of (2.2) and Table II, give excellent approximations of $\left\{ \lambda_{n,n} (e^{-x}) \right\}_{n=1}^{30}$, except for very small values of n.

To round out our discussion of the "1/9" Conjecture, we list, as in [4], the coefficients of the extremal polynomials $p_n (x)$ and $q_n (x)$ (with $q_n (0) = 1$) for which

$$\lambda_{n,n} (e^{-x}) = \| e^{-x} - \frac{p_n(x)}{q_n(x)} \|_{L_\infty[0,+\infty)} , \tag{2.7}$$

for $n = 1,2, \cdots , 30$, these coefficients being rounded to twenty places to conserve space. These can be found in §4.

3. Description of the Numerical Computations.

Initially, our computations were done on $[0, +\infty)$ with an essentially standard Remez algorithm (cf. Meinardus [5]) using Brent's MP package [2] to handle the high-precision computations. The values $\{\lambda_{n,n}(e^{-x})\}_{n=1}^{13}$ were calculated in this fashion, using 43 decimal-digit arithmetic. However, $\lambda_{n+1,n+1}(e^{-x})$ had approximately 3 digits *less* accuracy than $\lambda_{n,n}(e^{-x})$, indicating that the method used in these initial computations was *highly* ill-conditioned.

To achieve a better conditioning, our original approximation problem (1.1) was restated in the form

$$\lambda_{n,n}(e^{-x}) = \min\left\{ \| e^{-c_n(1+t)/(1-t)} - \tilde{r}_{n,n}(t) \|_{L_\infty[-1,+1)} : \tilde{r}_{n,n} \in \pi_{n,n} \right\}, \quad (3.1)$$

resulting from the change of variables ·

$$x = c_n \left(\frac{1+t}{1-t} \right), c_n > 0 \ , \text{ where } x \in [0, +\infty) \ , \ t \in [-1,1) \ . \quad (3.2)$$

Ideally, the constant c_n should be chosen so as to distribute the set of $2n+2$ alternant points (cf. [5]), associated with the interval $[-1, +1]$ of (3.1), as uniformly as possible in $[-1, +1]$.

The reformulated problem (3.1) was solved by the following implementation of the Remez algorithm:

1) Obtain an estimate for the alternants $\{t_j\}_{j=0}^{2n+1}$ (where $-1 = t_0 < t_1 < \cdots < t_{2n+1} = 1$), and for a value for the constant c_n of (3.1), using previous data.

2) Find real polynomials $p_n(t)$ and $q_n(t)$ (with $q_n(0) := 1$), each of degree n, and a positive constant λ which satisfy

$$e^{-c_n(1+t_k)/(1-t_k)} - \frac{p_n(t_k)}{q_n(t_k)} - (-1)^k \lambda = 0 \ , \ \left(k = 0,1, \cdots, 2n+1\right) \ , \quad (3.3)$$

on the current alternants $\{t_k\}_{k=0}^{2n+1}$ in $[-1, +1]$. A Newton's method, involving $2n+2$ parameters consisting of the $2n+1$ coefficients of $p_n(t)$ and $q_n(t)$ and the constant λ, was used to solve the nonlinear problem of (3.3). To add stability to these calculations, the polynomials $p_n(t)$ and $q_n(t)$ were expressed in terms of the Chebyshev polynomial basis $\{T_k(t)\}_{k=0}^{n}$.

3) A new estimate of the alternants was then found by finding a set of local extrema, with alternating signs, of the function

$$F(t) := e^{-c_n(1+t)/(1-t)} - p_n(t)/q_n(t) \ , \text{ defined on } [-1, +1] \ .$$

If the new alternants were sufficiently close to the old alternants, the algorithm was terminated. Otherwise, step 2) above was repeated, etc.

With a sufficiently good estimate for the constants c_n, the new algorithm was *significantly* more stable than the standard Remez algorithm applied on $[0, +\infty)$: the converged value $\lambda_{n+1,n+1}(e^{-x})$ of this new algorithm had approximately *one* digit less accuracy than the previous converged value $\lambda_{n,n}(e^{-x})$. This is about as much as can be expected since $\lambda_{n+1,n+1}(e^{-x})$ is roughly $1/9.29$ times $\lambda_{n,n}(e^{-x})$ with increasing n (cf. Table 2)!

The most time-consuming computer portion of our modified algorithm occurred in step 2) above. Now, each Newton step in 2) requires solving a $(2n+2) \times (2n+2)$ matrix equation, and this is clearly compounded by the extra computer time necessary in carrying out all calculations in very high precision. On starting the above algorithm with 20 digit accuracy in the associated parameters and on using 230 digit arithmetic from Brent's MP package, this algorithm only needed at most 8 Newton updates to achieve a final 200 digit accuracy in the associated parameters. But for $n = 30$, this required, for example, 15 cpu hours on our VAX-11/780 to determine $\lambda_{30,30}(e^{-x})$!

We remark that these costly computing times occurred, despite the fact that our initial estimates used in step 1) of the algorithm above were surprisingly good. If $\{t_j\}_{j=0}^{2n+1}$ denoted the alternants in the interval $[-1, +1]$ for our problem (3.1) and if $\{x_j\}_{j=0}^{2n+1}$ were the images, under the transformation in (3.2), of the alternants in the interval $[0, +\infty)$, then on choosing $c_n \cong \sqrt{x_n \cdot x_{n+1}}$, we found that the associated alternants $\{t_j\}_{j=0}^{2n+1}$ in $[-1, +1]$ became *unexpectedly* similar to the extrema $\{\tilde{t}_j := \cos/\pi(1 - \frac{j}{2n+1})/\}_{j=0}^{2n+1}$ of the Chebyshev polynomial $T_{2n+1}(t)$ on the interval $[-1, +1]$. More precisely, the ratios of these alternants,

$$u_j := \frac{\tilde{t}_j}{t_j} \quad (j = 0,1, \cdots, 2n+1) , \tag{3.4}$$

formed a nearly symmetric inverted bell-shaped curve on $[0,2n+1]$, i.e., these ratios were nearly one for j small or j near $2n+1$, and these ratios decreased slowly to about 0.76 as j approached the center point of the interval $[0,2n+1]$. This observation led us to the following estimate

$$\hat{t}_j := \cos\left[\pi\left(-\frac{j}{2n+1}\right)\right]\left\{1 - 3.36\left(\frac{j}{2n+1}\right)^2\left(1 - \frac{j}{2n+1}\right)^2\right\}, \tag{3.5}$$

$$j = 0,1, \cdots ,2n+1 ,$$

of the alternants $\{t_j\}_{j=0}^{2n+1}$ in $[-1, +1]$, which numerically achieved a relative deviation of at most 6% from the actual alternants $\{t_j\}_{j=0}^{2n+1}$, even when we used the estimates

$$\hat{c}_n := c_{n-1}^2 / c_{n-2} \tag{3.6}$$

for c_n where c_{n-1} and c_{n-2} were determined from previously run cases.

Summarizing, using the estimates of (3.5) and (3.6), using the transformed problem of (3.1), and using the Chebyshev polynomial basis $\{T_k(x)\}_{k=0}^{n}$, resulted in a significantly better-conditioned computation for the values of $\lambda_{n,n}(e^{-x})$. (We stopped our computations with the case $n = 30$ from cpu-time considerations, rather than from accuracy considerations!)

4. Coefficients of $p_n(x)$ and $q_n(x)$.

Tables of coefficients for best approximants to e^{-x}		
i	q	p
	n = 1	
0	1.0000000000000000000 (+00)	1.0668310421618504635 (+00)
1	1.7271172505820169235 (+00)	-1.1542504579210602494 (-01)
	n = 2	
0	1.0000000000000000000 (+00)	9.9264132983041947072 (-01)
1	6.6930154271087127186 (-01)	-1.8833350198927415815 (-01)
2	5.7224957904836489341 (-01)	4.2109959068982177855 (-03)
	n = 3	
0	1.0000000000000000000 (+00)	1.0007993806363356878 (+00)
1	7.9829357089752213329 (-01)	-2.2365742718351887787 (-01)
2	2.2040971161511489626 (-01)	1.2499601545398984435 (-02)
3	1.2485918642725863159 (-01)	-9.9810015898578281854 (-05)
	n = 4	
0	1.0000000000000000000 (+00)	9.9991347759304711148 (-01)
1	7.5668306888329708214 (-01)	-2.4025402432545953884 (-01)
2	2.9175397633746512345 (-01)	1.8400562307678039215 (-02)
3	4.5750548404322635677 (-02)	-4.4981502907081176448 (-04)
4	1.9376829538777680730 (-02)	1.6765299308108737248 (-06)
	n = 5	
0	1.0000000000000000000 (+00)	1.0000093457131530266 (+00)
1	7.5017443629508826484 (-01)	-2.5023100706418111745 (-01)
2	2.6991013134417674897 (-01)	2.2480613306965212876 (-02)
3	6.7668626041566587102 (-02)	-8.3363085734239059333 (-04)
4	6.9346135560032124409 (-03)	1.0779810679092561383 (-05)
5	2.3446790106210413736 (-03)	-2.1912697469186570498 (-08)
	n = 6	
0	1.0000000000000000000 (+00)	9.9999899154562510033 (-01)
1	7.4317310793725353126 (-01)	-2.5677508985594545088 (-01)
2	2.6898234032991615456 (-01)	2.5389670322537157467 (-02)
3	6.1593026160813895476 (-02)	-1.1769059339745022305 (-03)
4	1.1364907743793030262 (-02)	2.4820964613817624866 (-05)
5	8.2567981485296730024 (-04)	-1.9070014316258941354 (-07)
6	2.3230231175265367241 (-04)	2.3426628258627078732 (-10)

Tables of coefficients for best approximants to e^{-x}		
i	q	p
	n = 7	
0	1.0000000000000000000 (+00)	1.0000001087497491375 (+00)
1	7.3860755265403652073 (-01)	-2.6139890245157325374 (-01)
2	2.6609542167331571699 (-01)	2.7548737236512233353 (-02)
3	6.2210380540681505225 (-02)	-1.4675743675443654044 (-03)
4	1.0229633036518400366 (-02)	4.0604885270256007598 (-05)
5	1.4878817819751908909 (-03)	-5.3705891418335754085 (-07)
6	8.0883914233407339765 (-05)	2.6538677816935889717 (-09)
7	1.9484208914619273525 (-05)	-2.1189028316079702843 (-12)
	n = 8	
0	1.0000000000000000000 (+00)	9.9999998827734788367 (-01)
1	7.3516490200874711006 (-01)	-2.6483430847822992853 (-01)
2	2.6438044169339228008 (-01)	2.9207044273505478870 (-02)
3	6.1718734777325241574 (-02)	-1.7107715899069427972 (-03)
4	1.0520815163731983883 (-02)	5.6307821021754562350 (-05)
5	1.3283453064064323585 (-03)	-1.0147775758029899384 (-06)
6	1.5910254290684040602 (-04)	9.0013440975771190466 (-09)
7	6.7271456842176719503 (-06)	-3.0312488609110424332 (-11)
8	1.4167507178008938403 (-06)	1.6608075800347647028 (-14)
	n = 9	
0	1.0000000000000000000 (+00)	1.0000000012632924833 (+00)
1	7.3251419564078598353 (-01)	-2.6748589953942883104 (-01)
2	2.6303062535807977162 (-01)	3.0517580603500653821 (-02)
3	6.1530784916685721148 (-02)	-1.9147812452801917483 (-03)
4	1.0492608831391450658 (-02)	7.1103869037310644849 (-05)
5	1.3950029476562314910 (-03)	-1.5678080149581536439 (-06)
6	1.4116049483486023948 (-04)	1.9535758505225041515 (-08)
7	1.4351266620345007596 (-05)	-1.2209626369889029705 (-10)
8	4.8597568414012272418 (-07)	2.9287304416715685077 (-13)
9	9.0914553514881083438 (-08)	-1.1485167207995354803 (-16)
	n = 10	
0	1.0000000000000000000 (+00)	9.9999999986388794767 (-01)
1	7.3040628740509415862 (-01)	-2.6959370125162598429 (-01)
2	2.6198433793450755661 (-01)	3.1577898264638956869 (-02)

Tables of coefficients for best approximants to e^{-x}		
i	q	p
3	6.1359835762604515656 (-02)	-2.0872336141098597946 (-03)
4	1.0522214938304485349 (-02)	8.4694800522056656479 (-05)
5	1.4004400136031827693 (-03)	-2.1529632009223805112 (-06)
6	1.5162908445438365994 (-04)	3.3595495648065426068 (-08)
7	1.2670597578885658604 (-05)	-3.0243887922221152380 (-10)
8	1.1179771406674077683 (-06)	1.3835175675810288679 (-12)
9	3.1024563729347710543 (-08)	-2.4479596835781524301 (-15)
10	5.2207134645030587538 (-09)	7.1060202430410289851 (-19)
	n = 11	
0	1.0000000000000000000 (+00)	1.0000000000146631119 (+00)
1	7.2869093514903338347 (-01)	-2.7130906619020039085 (-01)
2	2.6114358806002462135 (-01)	3.2452672671798064591 (-02)
3	6.1230547379183840216 (-02)	-2.2343529914673730280 (-03)
4	1.0537271208026105008 (-02)	9.7033279144419487693 (-05)
5	1.4150097974152344401 (-03)	-2.7417852073123794696 (-06)
6	1.5333091232704136771 (-04)	5.0236657702890298313 (-08)
7	1.3934767796316549369 (-05)	-5.7755502193955032327 (-10)
8	9.8315547981894275422 (-07)	3.8862322897518799981 (-12)
9	7.6575875007520006935 (-08)	-1.3413459199227922075 (-14)
10	1.7740700295278932878 (-09)	1.8010050564587997692 (-17)
11	2.7127219457634251348 (-10)	-3.9776945578255130913 (-21)
	n = 12	
0	1.0000000000000000000 (+00)	9.9999999999842054316 (-01)
1	7.2726794442587404691 (-01)	-2.7273205541723861212 (-01)
2	2.6045423235897757189 (-01)	3.3186285409181941015 (-02)
3	6.1125881342176750711 (-02)	-2.3610296684603661423 (-03)
4	1.0551656317596859800 (-02)	1.0818210684852625702 (-04)
5	1.4248371121782907054 (-03)	-3.3170690870773473397 (-06)
6	1.5627772051583776268 (-04)	6.8564050007593925822 (-08)
7	1.4198481758017543395 (-05)	-9.4025623763914245637 (-10)
8	1.1076204551770729043 (-06)	8.2159261557706839823 (-12)
9	6.7122345531815556515 (-08)	-4.2460559325998987934 (-14)
10	4.6772292551801273223 (-09)	1.1335755380226379804 (-16)
11	9.1859907864599662598 (-11)	-1.1824226782945190429 (-19)
12	1.2870807304176959932 (-11)	2.0328884594951322377 (-23)

Tables of coefficients for best approximants to e^{-x}		
i	q	p
	n = 13	
0	1.0000000000000000000 (+00)	1.0000000000001701187 (+00)
1	7.2606856648187456846 (-01)	-2.7393143353638373887 (-01)
2	2.5987868966049413531 (-01)	3.3810123496675834027 (-02)
3	6.1040005208020200371 (-02)	-2.4710700386539540993 (-03)
4	1.0563659959887015315 (-02)	1.1824617529896343260 (-04)
5	1.4333960951724789950 (-03)	-3.8691771493804491498 (-06)
6	1.5836802154851644352 (-04)	8.7846766808947538873 (-08)
7	1.4609445035018602359 (-05)	-1.3766793766013695399 (-09)
8	1.1372692221789108095 (-06)	1.4546096489288275848 (-11)
9	7.7482794810543576337 (-08)	-9.9041130338816606894 (-14)
10	4.0886476353685167018 (-09)	4.0201482804353265611 (-16)
11	2.5766613699837803265 (-10)	-8.4753849754741511891 (-19)
12	4.3455473513268560364 (-12)	7.0050690177811160395 (-22)
13	5.6184959576874667073 (-13)	-9.5581127116884329757 (-26)
	n = 14	
0	1.0000000000000000000 (+00)	9.9999999999998167826 (-01)
1	7.2504400105794982801 (-01)	-2.7495599893993723129 (-01)
2	2.5939097352682523362 (-01)	3.4346972429252957162 (-02)
3	6.0968196803506752125 (-02)	-2.5674425707902795730 (-03)
4	1.0574051936471402715 (-02)	1.2734060684477834084 (-04)
5	1.4405532019175131023 (-03)	-4.3932761847806353276 (-06)
6	1.6019218139205361453 (-04)	1.0753187241762202746 (-07)
7	1.4908242237501135777 (-05)	-1.8710223403465319988 (-09)
8	1.1820298688854366697 (-06)	2.2849466635625509298 (-11)
9	8.0164052804759260531 (-08)	-1.9038591551025381666 (-13)
10	4.8361843519832766389 (-09)	1.0310120249635897805 (-15)
11	2.2472046467801857194 (-10)	-3.3490171642170337489 (-18)
12	1.2922822504487751142 (-11)	5.6743670690330211959 (-21)
13	1.8921825619816040160 (-13)	-3.7794532322670955969 (-24)
14	2.2710727625899900737 (-14)	4.1610013267680029528 (-28)
	n = 15	
0	1.0000000000000000000 (+00)	1.0000000000000019731 (+00)
1	7.2415866907620477911 (-01)	-2.7584133092403855215 (-01)

Tables of coefficients for best approximants to e^{-x}		
i	q	p
2	2.5897242355314705843 (-01)	3.4813754481829550920 (-02)
3	6.0907277733570110800 (-02)	-2.6524779865394277857 (-03)
4	1.0583087170044096379 (-02)	1.3557652521034370441 (-04)
5	1.4466872665940473451 (-03)	-4.8874813823443484833 (-06)
6	1.6173222871264890572 (-04)	1.2722046758585241530 (-07)
7	1.5171907890884190088 (-05)	-2.4081661405899631864 (-09)
8	1.2151064697266154881 (-06)	3.2966530346210070367 (-11)
9	8.4196487972429069905 (-08)	-3.2068470095050508776 (-13)
10	5.0409480029770692861 (-09)	2.1476966393883189321 (-15)
11	2.7233683230693980458 (-10)	-9.4109689260605595636 (-18)
12	1.1248069393139919485 (-11)	2.4856294581031545360 (-20)
13	5.9468225157795131984 (-13)	-3.4361705671901568590 (-23)
14	7.6317957265269706634 (-15)	1.8714400301425160987 (-26)
15	8.5472063683108394376 (-16)	-1.6864826197411413312 (-30)
	n = 16	
0	1.0000000000000000000 (+00)	9.9999999999999978751 (-01)
1	7.2338602292194837605 (-01)	-2.7661397707802372040 (-01)
2	2.5860931332113989163 (-01)	3.5223290398593890975 (-02)
3	6.0854946127264699182 (-02)	-2.7280223945541603972 (-03)
4	1.0591015301175281646 (-02)	1.4305532540830578363 (-04)
5	1.4519917779283776247 (-03)	-5.3516653691319551221 (-06)
6	1.6306180422515751551 (-04)	1.4663511678787708208 (-07)
7	1.5396764745302603497 (-05)	-2.9748521541610146711 (-09)
8	1.2446986743617001063 (-06)	4.4667074036373546096 (-11)
9	8.7213007002183174707 (-08)	-4.9124750188063887407 (-13)
10	5.3517844906526438896 (-09)	3.8755282930914110147 (-15)
11	2.8593027098738583509 (-10)	-2.1199503237701174494 (-17)
12	1.3963572572460278486 (-11)	7.6277421016130829045 (-20)
13	5.1671893800279173586 (-13)	-1.6607979818792513091 (-22)
14	2.5276802007963606269 (-14)	1.8979708477544314608 (-25)
15	2.8667698160212141620 (-16)	-8.5606377701670908334 (-29)
16	3.0093104224231407541 (-17)	6.3943444171177598099 (-33)

Tables of coefficients for best approximants to e^{-x}		
i	q	p
	n = 17	
0	1.0000000000000000000 (+00)	1.0000000000000000229 (+00)
1	7.2270585462136953087 (-01)	-2.7729414537863365694 (-01)
2	2.5829131995630046565 (-01)	3.5585465335003458111 (-02)
3	6.0809507390072969545 (-02)	-2.7955519228564120076 (-03)
4	1.0598023489381217368 (-02)	1.4986697693060245133 (-04)
5	1.4566258090943959608 (-03)	-5.7867107664997448463 (-06)
6	1.6421910387421514909 (-04)	1.6558969231502529775 (-07)
7	1.5592669068294270482 (-05)	-3.5600331752523462572 (-09)
8	1.2701896118955897646 (-06)	5.7694610687453000688 (-11)
9	8.9952033533504590178 (-08)	-7.0111135292795387525 (-13)
10	5.5866285320057224635 (-09)	6.3018857979045448319 (-15)
11	3.0690410653109898959 (-10)	-4.0934546068922329038 (-17)
12	1.4763544085912781412 (-11)	1.8540997178336935538 (-19)
13	6.5692146982166977577 (-13)	-5.5478662451290864179 (-22)
14	2.1929711332561360156 (-14)	1.0077964568000473113 (-24)
15	9.9803140529347369232 (-16)	-9.6311299616434657123 (-28)
16	1.0076443559489818370 (-17)	3.6381008264169349244 (-31)
17	9.9533336444995010637 (-19)	-2.2774706078188437603 (-35)
	n = 18	
0	1.0000000000000000000 (+00)	9.9999999999999999754 (-01)
1	7.2210251162213760760 (-01)	-2.7789748837786202942 (-01)
2	2.5801053189666607896 (-01)	3.5908020274519729805 (-02)
3	6.0769684666293474499 (-02)	-2.8562580858878249220 (-03)
4	1.0604260828765657181 (-02)	1.5609017337127136364 (-04)
5	1.4607084457312575111 (-03)	-6.1940471597288549472 (-06)
6	1.6523598870223731736 (-04)	1.8396482187296904295 (-07)
7	1.5764526871377682191 (-05)	-4.1548085790428176917 (-09)
8	1.2926132501100778041 (-06)	7.1794537028167480115 (-11)
9	9.2334736432550722847 (-08)	-9.4773868843647724074 (-13)
10	5.8031481469953154253 (-09)	9.4758123475451790323 (-15)
11	3.2288400362818480622 (-10)	-7.0617684389507100553 (-17)
12	1.6023022340045240158 (-11)	3.8258499143063765432 (-19)
13	6.9925714572325997863 (-13)	-1.4520088830640599985 (-21)

Tables of coefficients for best approximants to e^{-x}		
i	q	p
14	2.8542945098191319824 (-14)	3.6534688613148991179 (-24)
15	8.6471483697462350423 (-16)	-5.5958102671944549554 (-27)
16	3.6787450097487522068 (-17)	4.5177277417421400742 (-30)
17	3.3278408185195843503 (-19)	-1.4434969618996796174 (-33)
18	3.1040251435758619580 (-20)	7.6480564016753553998 (-38)
	n = 19	
0	1.0000000000000000000 (+00)	1.0000000000000000003 (+00)
1	7.2156368259201663937 (-01)	-2.7843631740798340183 (-01)
2	2.5776078409791612555 (-01)	3.6197101505900533482 (-02)
3	6.0734497954567183445 (-02)	-2.9111115140177481016 (-03)
4	1.0609846263465004573 (-02)	1.6179325924143951781 (-04)
5	1.4643324939311864714 (-03)	-6.5753674125082918706 (-06)
6	1.6613650526094958487 (-04)	2.0168905169819757417 (-07)
7	1.5916587025359119477 (-05)	-4.7521805561279600593 (-09)
8	1.3124457578828000080 (-06)	8.6730006775887139868 (-11)
9	9.4451373237383608747 (-08)	-1.2276354018264591744 (-12)
10	5.9933092623793340401 (-09)	1.3411822822603597433 (-14)
11	3.3784614979453334011 (-10)	-1.1186004966950956669 (-16)
12	1.6989658352357050158 (-11)	6.9970429220569463847 (-19)
13	7.6742363437140323820 (-13)	-3.1972725111092105805 (-21)
14	3.0580043687699495419 (-14)	1.0273733186333559896 (-23)
15	1.1518577416976816152 (-15)	-2.1951396392431808895 (-26)
16	3.1835397068991986730 (-17)	2.8615408787772833807 (-29)
17	1.2713653178005351833 (-18)	-1.9694708209381382654 (-32)
18	1.0364344279541674044 (-20)	5.3703307695262059488 (-36)
19	9.1570582958910789228 (-22)	-2.4294725589568881359 (-40)
	n = 20	
0	1.0000000000000000000 (+00)	9.9999999999999999997 (-01)
1	7.2107955165146436895 (-01)	-2.7892044834853562639 (-01)
2	2.5753720158407837071 (-01)	3.6457649932613876971 (-02)
3	6.0703182600751914030 (-02)	-2.9609098242596261130 (-03)
4	1.0614875834295449959 (-02)	1.6703541699804974713 (-04)
5	1.4675710228275782849 (-03)	-6.9324562692610725077 (-06)
6	1.6693957290448895787 (-04)	2.1872483964129629603 (-07)
7	1.6052079646727331527 (-05)	-5.3467565092977333192 (-09)
8	1.3301225337360296890 (-06)	1.0228954715465658016 (-10)

Tables of coefficients for best approximants to e^{-x}		
i	q	p
9	9.6338964603806016053 (-08)	-1.5368313511438283042 (-12)
10	6.1639056613443707964 (-09)	1.8096461988824864400 (-14)
11	3.5110926854287101220 (-10)	-1.6580745736734816354 (-16)
12	1.7908832457629530219 (-11)	1.1672198605029734819 (-18)
13	8.2008117841834923615 (-13)	-6.1936889537571481080 (-21)
14	3.3938864156618761186 (-14)	2.4105681922839642535 (-23)
15	1.2417688637946327705 (-15)	-6.6185196281694492006 (-26)
16	4.3385159620628243371 (-17)	1.2113724541256915285 (-28)
17	1.0990484330810003653 (-18)	-1.3552946153009542146 (-31)
18	4.1354019630164880823 (-20)	8.0169605314221382434 (-35)
19	3.0538918819268542585 (-22)	-1.8805470199361703461 (-38)
20	2.5628928552136086178 (-23)	7.3216143452337990707 (-43)
	n = 21	
0	1.0000000000000000000 (+00)	1.0000000000000000000 (+00)
1	7.2064219697569183005 (-01)	-2.7935780302430817048 (-01)
2	2.5733587895330020783 (-01)	3.6693681977608392584 (-02)
3	6.0675133266582940643 (-02)	-3.0063138655381823795 (-03)
4	1.0619427824397480596 (-02)	1.7186787185029200536 (-04)
5	1.4704823869739844457 (-03)	-7.2670890316381390456 (-06)
6	1.6766020704056002618 (-04)	2.3505832195838590627 (-07)
7	1.6173579847685692566 (-05)	-5.9344580351910059567 (-09)
8	1.3459762326494999013 (-06)	1.1828954830323715653 (-10)
9	9.8033984897602339980 (-08)	-1.8712232943943080645 (-12)
10	6.3173038544853450718 (-09)	2.3495449807922150045 (-14)
11	3.6312508159412913565 (-10)	-2.3313838537000271146 (-16)
12	1.8730981404230082919 (-11)	1.8111884218044142747 3 (-18)
13	8.7093081612820652481 (-13)	-1.0875663018513674485 (-20)
14	3.6548676757025251170 (-14)	4.9414360113574654171 (-23)
15	1.3936848498319246143 (-15)	-1.6523102799180557777 (-25)
16	4.7051618127896484209 (-17)	3.9082049594935053126 (-28)
17	1.5317669769113303944 (-18)	-6.1754498246162749622 (-31)
18	3.5714624755160836554 (-20)	5.9746919108510846444 (-34)
19	1.2703379067912114642 (-21)	-3.0598920596277695902 (-37)
20	8.5380603270749883481 (-24)	6.2192394025759794712 (-41)
21	6.8232509117901788034 (-25)	-2.0988417714945058814 (-45)

Tables of coefficients for best approximants to e^{-x}		
i	q	p
	n = 22	
0	1.0000000000000000000 (+00)	1.0000000000000000000 (+00)
1	7.2024515522426885778 (-01)	-2.7975484477573114216 (-01)
2	2.5715365085971926369 (-01)	3.6908495635450404166 (-02)
3	6.0649864512193881866 (-02)	-3.0478754020575993647 (-03)
4	1.0623566628151799912 (-02)	1.7633500843928321830 (-04)
5	1.4731137404802299197 (-03)	-7.5809738495069090674 (-06)
6	1.6831048917843324445 (-04)	2.5069191557238854811 (-07)
7	1.6283152234777024234 (-05)	-6.5122619757180812495 (-09)
8	1.3602761130470247007 (-06)	1.3457379364976452568 (-10)
9	9.9564343590539204149 (-08)	-2.2268026983155948583 (-12)
10	6.4560913902503564972 (-09)	2.9560245331729836665 (-14)
11	3.7401859742633337134 (-10)	-3.1410724352013750719 (-16)
12	1.9483106529080606242 (-11)	2.6553887718291344662 (-18)
13	9.1680999305772174791 (-13)	-1.7676644378244532847 (-20)
14	3.9107766743159333128 (-14)	9.1278680283845964971 (-23)
15	1.5123507567108144027 (-15)	-3.5805166566755439172 (-25)
16	5.3401899140075195528 (-17)	1.0365955917689346517 (-27)
17	1.6707425727208528640 (-18)	-2.1277784397335996581 (-30)
18	5.0886706159921823345 (-20)	2.9231399336791845047 (-33)
19	1.0961475446293379047 (-21)	-2.4623396257008074572 (-36)
20	3.6965545855001175030 (-23)	1.0991176527978871486 (-39)
21	2.2708806892898350258 (-25)	-1.9484008877459215375 (-43)
22	1.7320983079796653153 (-26)	5.7367451049989032790 (-48)
	n = 23	
0	1.0000000000000000000 (+00)	1.0000000000000000000 (+00)
1	7.1988310082459591732 (-01)	-2.8011689917540408269 (-01)
2	2.5698792471734873844 (-01)	3.7104823892752821325 (-02)
3	6.0626982524584947958 (-02)	-3.0860584471325009762 (-03)
4	1.0627345664071083966 (-02)	1.8047536072786826345 (-04)
5	1.4755035992707838480 (-03)	-7.8757210411818256606 (-06)
6	1.6890024843251790692 (-04)	2.6563900673673161074 (-07)
7	1.6382473993464912301 (-05)	-7.0779812242602691625 (-09)
8	1.3732404247572612432 (-06)	1.5101143424472028548 (-10)
9	1.0095302286038609121 (-07)	-2.5997957789837943232 (-12)

Tables of coefficients for best approximants to e^{-x}		
i	q	p
10	6.5822413182372985325 (-09)	3.6233532714776076309 (-14)
11	3.8394889629746827515 (-10)	-4.0861034248600722572 (-16)
12	2.0170542366772364610 (-11)	3.7137851529683374768 (-18)
13	9.5918906854811887872 (-13)	-2.7001985390889146513 (-20)
14	4.1436204320974462060 (-14)	1.5528939801402053295 (-22)
15	1.6304901228232053668 (-15)	-6.9530886094237998157 (-25)
16	5.8386383707137038815 (-17)	2.3720625335812422216 (-27)
17	1.9175029935153725110 (-18)	-5.9872927305549306726 (-30)
18	5.5808542257176158674 (-20)	1.0736346677944520471 (-32)
19	1.5960495536735968784 (-21)	-1.2905749729715581657 (-35)
20	3.1871496109206449820 (-23)	9.5241140961329381142 (-39)
21	1.0217421583117155147 (-24)	-3.7278671871533691988 (-42)
22	5.7595314217244562516 (-27)	5.7982136100082272330 (-46)
23	4.2015926344320482739 (-28)	-1.4983223279431735600 (-50)
	n = 24	
0	1.0000000000000000000 (+00)	1.0000000000000000000 (+00)
1	7.1955160628060490012 (-01)	-2.8044839371939509988 (-01)
2	2.5683655680548874312 (-01)	3.7284950524883842970 (-02)
3	6.0606164481744409458 (-02)	-3.1212558501085499598 (-03)
4	1.0630809594431807553 (-02)	1.8432246866428426565 (-04)
5	1.4776837304194045942 (-03)	-8.1528290297762335410 (-06)
6	1.6943756353916543523 (-04)	2.7992015745235962935 (-07)
7	1.6472921423980865550 (-05)	-7.6300845409641694341 (-09)
8	1.3850483386300453833 (-06)	1.6749426980810398258 (-10)
9	1.0221886364132138740 (-07)	-2.9867405776081327288 (-12)
10	6.6974106945354655796 (-09)	4.3453522989129354348 (-14)
11	3.9303598949948647010 (-10)	-5.1625886299634749395 (-16)
12	2.0801904728762698738 (-11)	4.9974970771678653853 (-18)
13	9.9823495547390858172 (-13)	-3.9209132343147921877 (-20)
14	4.3607684640656254187 (-14)	2.4722281164775052693 (-22)
15	1.7388589608575720948 (-15)	-1.2376529579577131695 (-24)
16	6.3423901723117428381 (-17)	4.8389987731625314556 (-27)
17	2.1120571959754931088 (-18)	-1.4452601356069529738 (-29)
18	6.4766664595094796687 (-20)	3.2004921170130934113 (-32)
19	1.7596348598140616683 (-21)	-5.0438750863387113193 (-35)

Tables of coefficients for best approximants to e^{-x}

i	q	p
20	4.7406321676460227043 (-23)	5.3360424355118835887 (-38)
21	8.8030256384366476434 (-25)	-3.4694587438114214128 (-41)
22	2.6892325986479079123 (-26)	1.1974216241920071748 (-44)
23	1.3959597685006867882 (-28)	-1.6430816452663319539 (-48)
24	9.7582905119403701973 (-30)	3.7467762281382790571 (-53)
	n = 25	
0	1.0000000000000000000 (+00)	1.0000000000000000000 (+00)
1	7.1924696060030549152 (-01)	-2.8075303939969450848 (-01)
2	2.5669775918887669273 (-01)	3.7450798588571201206 (-02)
3	6.0587143251270532876 (-02)	-3.1538023041200807920 (-03)
4	1.0633996033260288963 (-02)	1.8790560971052078695 (-04)
5	1.4796805670782497197 (-03)	-8.4136803468272165599 (-06)
6	1.6992913597947736280 (-04)	2.9356040915848232932 (-07)
7	1.6555634671491244526 (-05)	-8.1675511125608142608 (-09)
8	1.3958482728461237761 (-06)	1.8393384178735986993 (-10)
9	1.0337750836440238611 (-07)	-3.3845209609800166958 (-12)
10	6.8029738899716716771 (-09)	5.1157161639534070020 (-14)
11	4.0138303634825116937 (-10)	-6.3644923388319778193 (-16)
12	2.1383556797380942966 (-11)	6.5117237108438679018 (-18)
13	1.0343638215272956605 (-12)	-5.4595347139888024431 (-20)
14	4.5624243518361710181 (-14)	3.7273813754702834958 (-22)
15	1.8408841519514984572 (-15)	-2.0530724455614180911 (-24)
16	6.8081267555534785131 (-17)	9.0077109606214317012 (-27)
17	2.3116269500240241604 (-18)	-3.0948366861501914604 (-29)
18	7.1858776322636732020 (-20)	8.1407882328111744921 (-32)
19	2.0647764426693640793 (-21)	-1.5906727254440811555 (-34)
20	5.2528760787303468994 (-23)	2.2153039233678690245 (-37)
21	1.3370915947538380277 (-24)	-2.0736074544841591910 (-40)
22	2.3154233915482175675 (-26)	1.1940545565030128827 (-43)
23	6.7551774268161192942 (-28)	-3.6523509679158950883 (-47)
24	3.2397074059310773266 (-30)	4.4437519314004684085 (-51)
25	2.1738952552926457167 (-31)	-8.9869101966470571763 (-56)

Tables of coefficients for best approximants to e^{-x}		
i	q	p
	n = 26	
0	1.0000000000000000000 (+00)	1.0000000000000000000 (+00)
1	7.1896603004635407628 (-01)	-2.8103396995364592372 (-01)
2	2.5657002887003749352 (-01)	3.7603998823683417241 (-02)
3	6.0569695881738960595 (-02)	-3.1839846317881614457 (-03)
4	1.0636936873433653069 (-02)	1.9125041898775982050 (-04)
5	1.4815162820564590937 (-03)	-8.6595435825426075478 (-06)
6	1.7038057158576408119 (-04)	3.0658737713813768797 (-07)
7	1.6631566345893463369 (-05)	-8.6897545335970724597 (-09)
8	1.4057642432742099133 (-06)	2.0025861713950916330 (-10)
9	1.0444204490518077866 (-07)	-3.7903724436935358376 (-12)
10	6.9000857608018811274 (-09)	5.9282411666625107531 (-14)
11	4.0907662030930223269 (-10)	-7.6842585097056414965 (-16)
12	2.1921125846523355118 (-11)	8.2570666493171204585 (-18)
13	1.0678718214314360999 (-12)	-7.3392007703057431530 (-20)
14	4.7503981805777539255 (-14)	5.3712708293521833434 (-22)
15	1.9363701772525106571 (-15)	-3.2137958886540163074 (-24)
16	7.2507250514592475300 (-17)	1.5565082769644566430 (-26)
17	2.4975442158093370662 (-18)	-6.0214447480232786212 (-29)
18	7.9241460923961745271 (-20)	1.8284175028861242452 (-31)
19	2.3072848547429883203 (-21)	-4.2589594095915223042 (-34)
20	6.2318770493623148075 (-23)	7.3811490698898474272 (-37)
21	1.4887308305062634160 (-24)	-9.1298836604138566248 (-40)
22	3.5900145501058512309 (-26)	7.5983415064438853460 (-43)
23	5.8126365856240802587 (-28)	-3.8935547750075112746 (-46)
24	1.6227799580300107200 (-29)	1.0604684624677904373 (-49)
25	7.2122228957098244519 (-32)	-1.1493650186177443495 (-53)
26	4.6529920015189527423 (-33)	2.0710352729171641461 (-58)
	n = 27	
0	1.0000000000000000000 (+00)	1.0000000000000000000 (+00)
1	7.1870615013494964452 (-01)	-2.8129384986505035548 (-01)
2	2.5645209321774275677 (-01)	3.7745943082793112249 (-02)
3	6.0553634833924406861 (-02)	-3.2120499830101943170 (-03)
4	1.0639659327332069240 (-02)	1.9437941312076668187 (-04)
5	1.4832096123430519539 (-03)	-8.8915786943714812065 (-06)

Tables of coefficients for best approximants to e^{-x}		
i	q	p
6	1.7079659574127093742 (-04)	3.1902991377590222703 (-07)
7	1.6701518602605391310 (-05)	-9.1963709811680089503 (-09)
8	1.4149006878178168423 (-06)	2.1641140277115697685 (-10)
9	1.0542351021349550139 (-07)	-4.2018705007566813448 (-12)
10	6.9897228455967626327 (-09)	6.7769798099866907653 (-14)
11	4.1619044516026066335 (-10)	-9.1133406626101867879 (-16)
12	2.2419392381250479060 (-11)	1.0230254833623119193 (-17)
13	1.0990305718118410865 (-12)	-9.5764061479699012137 (-20)
14	4.9259005778171551594 (-14)	7.4512740901471078481 (-22)
15	2.0260314782637897578 (-15)	-4.7931707838444243896 (-24)
16	7.6681358771735639629 (-17)	2.5293126309095483331 (-26)
17	2.6758680762905260877 (-18)	-1.0834661026569340573 (-28)
18	8.6169847303890952011 (-20)	3.7161403305502155654 (-31)
19	2.5634235918361219624 (-21)	-1.0024877187066737147 (-33)
20	7.0127193297565789436 (-23)	2.0780918351231215531 (-36)
21	1.7855742180920933932 (-24)	-3.2097160980790598559 (-39)
22	4.0156816304849403139 (-26)	3.5424654128735486313 (-42)
23	9.1964363393424183436 (-28)	-2.6331524240973874715 (-45)
24	1.3955671627874349543 (-29)	1.2060058776911720646 (-48)
25	3.7352134986647930254 (-31)	-2.9375861296063709736 (-52)
26	1.5427084484811426672 (-33)	2.8484004094090161063 (-56)
27	9.5834724343683731142 (-35)	-4.5925891451883766222 (-61)
	n = 28	
0	1.0000000000000000000 (+00)	1.0000000000000000000 (+00)
1	7.1846504099977450858 (-01)	-2.8153495900022549142 (-01)
2	2.5634286747132473766 (-01)	3.7877826471550229077 (-02)
3	6.0538801220168057895 (-02)	-3.2382124179360921395 (-03)
4	1.0642186749988000749 (-02)	1.9731243218656025214 (-04)
5	1.4847764994109842628 (-03)	-9.1108440565059548173 (-06)
6	1.7118121976301399562 (-04)	3.3091718065250326410 (-07)
7	1.6766171748975058231 (-05)	-9.6873069289091329257 (-09)
8	1.4233461897756964635 (-06)	2.3234704839756243162 (-10)
9	1.0633127738462594514 (-07)	-4.6169087559105242168 (-12)
10	7.0727167212993401677 (-09)	7.6563385687590018588 (-14)
11	4.2278753905749988235 (-10)	-1.0642631862580357359 (-15)
12	2.2882484070110947837 (-11)	1.2424881652802987399 (-17)

i	q	p
Tables of coefficients for best approximants to e^{-x}		
13	1.1280736630891817950 (-12)	-1.2181315591198636789 (-19)
14	5.0900982136752396390 (-14)	1.0008004746340252594 (-21)
15	2.1102989402066160773 (-15)	-6.8628244877211120581 (-24)
16	8.0629352903914881994 (-17)	3.9041630685993974915 (-26)
17	2.8453127951750166544 (-18)	-1.8273116597117645173 (-28)
18	9.2876348488845996777 (-20)	6.9600084535166773350 (-31)
19	2.8055162407493957644 (-21)	-2.1270994500225483862 (-33)
20	7.8493728999571373376 (-23)	5.1221768334261579751 (-36)
21	2.0231282677112118888 (-24)	-9.4924644132126223194 (-39)
22	4.8688630300634044605 (-26)	1.3124217192547875882 (-41)
23	1.0332584543810187690 (-27)	-1.2979704782618576593 (-44)
24	2.2522930190976109514 (-29)	8.6529029938315506542 (-48)
25	3.2105491540280196138 (-31)	-3.5567748438917412531 (-51)
26	8.2519544489273895218 (-33)	7.7792379820555470653 (-55)
27	3.1755253582770512366 (-35)	-6.7753365272078992201 (-59)
28	1.9020809159366749124 (-36)	9.8138565931627609198 (-64)
n = 29		
0	1.0000000000000000000 (+00)	1.0000000000000000000 (+00)
1	7.1824074043079310993 (-01)	-2.8175925956920689007 (-01)
2	2.5624142131497740435 (-01)	3.8000680884184294416 (-02)
3	6.0525059535231118368 (-02)	-3.2626582310163976817 (-03)
4	1.0644539295547595500 (-02)	2.0006701267299431847 (-04)
5	1.4862305915618925334 (-03)	-9.3183042499982610991 (-06)
6	1.7153787091780660524 (-04)	3.4227801419691581593 (-07)
7	1.6826106556886635018 (-05)	-1.0162642462709301594 (-08)
8	1.4311763809427150035 (-06)	2.4803044888644507226 (-10)
9	1.0717335838264462378 (-07)	-5.0336720080283779322 (-12)
10	7.1497799310138340033 (-09)	8.5611335339544449737 (-14)
11	4.2892213438580353073 (-10)	-1.2262802013924015446 (-15)
12	2.3313973576581319597 (-11)	1.4832100877490882467 (-17)
13	1.1552060390438050984 (-12)	-1.5158313990467185641 (-19)
14	5.2440069076448366312 (-14)	1.3074657548508829771 (-21)
15	2.1896191080064318492 (-15)	-9.4902885995221209857 (-24)
16	8.4364220579657309921 (-17)	5.7698403519637535860 (-26)
17	3.0067321387130736625 (-18)	-2.9188924120576619426 (-28)
18	9.9296151586572481448 (-20)	1.2179775468499820477 (-30)

Tables of coefficients for best approximants to e^{-x}

i	q	p
19	3.0419849906975770270 (-21)	-4.1449405346182376688 (-33)
20	8.6456065781235580902 (-23)	1.1339292479243459054 (-35)
21	2.2812963979843922853 (-24)	-2.4481138035452179148 (-38)
22	5.5537857813866255467 (-26)	4.0730588289709086731 (-41)
23	1.2663351963339644273 (-27)	-5.0614377161525805088 (-44)
24	2.5413421113019371806 (-29)	4.5033487174180149912 (-47)
25	5.2836220838809921666 (-31)	-2.7029531453596257962 (-50)
26	7.0894030065454535655 (-33)	1.0009296137524799591 (-53)
27	1.7525818103023344647 (-34)	-1.9731092335931916942 (-57)
28	6.2991336260625001585 (-37)	1.5493207620467243315 (-61)
29	3.6427133054769394938 (-38)	-2.0235251334280446010 (-66)
	n = 30	
0	1.0000000000000000000 (+00)	1.0000000000000000000 (+00)
1	7.1803155042772127612 (-01)	-2.8196844957227872388 (-01)
2	2.5614695234169101695 (-01)	3.8115401913969740830 (-02)
3	6.0512293508462864581 (-02)	-3.2855502860341809776 (-03)
4	1.0646734444855619090 (-02)	2.0265870261805029904 (-04)
5	1.4875836417649791494 (-03)	-9.5148379859906376412 (-06)
6	1.7186949502113754284 (-04)	3.5314050182780261684 (-07)
7	1.6881821841067465703 (-05)	-1.0622586963885646562 (-08)
8	1.4384562284711296335 (-06)	2.6343483216755972319 (-10)
9	1.0795664283018752186 (-07)	-5.4506073294691798165 (-12)
10	7.2215265693946799983 (-09)	9.4866156508756075292 (-14)
11	4.3464114384271304094 (-10)	-1.3964554205647248956 (-15)
12	2.3716969565318967420 (-11)	1.7441252036441660423 (-17)
13	1.1806074393066802850 (-12)	-1.8506688650091489877 (-19)
14	5.3885346845345913909 (-14)	1.6676820706603970154 (-21)
15	2.2643856170276607118 (-15)	-1.2737164571814720528 (-23)
16	8.7901123390735385090 (-17)	8.2159993047662503208 (-26)
17	3.1604309408297432480 (-18)	-4.4525490040349927562 (-28)
18	1.0545543403758406495 (-19)	2.0130724366066200126 (-30)
19	3.2699924326595110152 (-21)	-7.5239822024090788005 (-33)
20	9.4303280463436687932 (-23)	2.2978810217829931996 (-35)
21	2.5286339977569998853 (-24)	-5.6508242227847947169 (-38)
22	6.3086412701711342899 (-26)	1.0982155012464389275 (-40)

Tables of coefficients for best approximants to e^{-x}		
i	q	p
23	1.4540002113970188434 (-27)	-1.6467595005222406689 (-43)
24	3.1480227761306722217 (-29)	1.8462045089123337587 (-46)
25	5.9861194489977205340 (-31)	-1.4832272593982169255 (-49)
26	1.1893082101510933116 (-32)	8.0441303745414552340 (-53)
27	1.5049933964043840759 (-34)	-2.6930919888571651980 (-56)
28	3.5836255739795333870 (-36)	4.8015689571472672268 (-60)
29	1.2057358269368541653 (-38)	-3.4109451463472630706 (-64)
30	6.7397798848782102402 (-40)	4.0308755782861197983 (-69)

Acknowledgement

We wish to thank Stephen Friedl and Craig Mohrman of Kent State University for their considerable help in producing the displays of the paper. In addition, we are indebted to Ava D. Logsdon for the considerable effort she expended in preparing this manuscript.

References

1. H.-P. Blatt and D. Braess, "Zur rationalen Approximation von e^{-x} auf $[0, \infty)$", J. Approxiation Theory 30(1980), 169 - 172.
2. Richard Brent, "A FORTRAN multiple-precision arithmetic package", Assoc. Comput. Mach. Trans. Math. Software 4(1978), 57 - 70.
3. C. Brezinski, Algorithmes d'Accélération de la Convergence , Éditions Technip, Paris, 1978.
4. W. J. Cody, G. Meinardus, and R. S. Varga, "Chebyshev rational approximation to e^{-x} in $[0, +\infty)$ and applications to heat-conduction problems", J. Approximation Theory 2(1969), 50 - 65.
5. G. Meinardus, Approximation of Functions: Theory and Numerical Methods , Springer-Verlag, New York, 1967.
6. G. Németh, "Notes on generalized Padé approximation", in Approximation and Function Spaces (Z. Ciesielski, ed.), pp. 484 - 508, North-Holland Publishing Co., Amsterdam, 1981.
7. D. J. Newman, "Rational approximation to e^{-x} ", J. Approximation Theory 10(1974), 301 - 303.

8. H.-U. Opitz and K. Scherer, "On the rational approximation of e^{-x} on $[0, \infty)$", Constructive Approximation (to appear).

9. Q. I. Rahman and G. Schmeisser, "Rational approximation to e^{-x} ", J. Approximation Theory 23(1978), 146 - 154.

10. Q. I. Rahman and G. Schmeisser, "Rational approximation to e^{-x}. II", Trans. Amer. Math. Soc. 235(1978), 395 - 402.

11. E. B. Saff and R. S. Varga, "Some open questions concerning polynomials and rational functions", in Padé and Rational Approximation (E. B. Saff and R. S. Varga, eds.), pp. 483 - 488, Academic Press, Inc., New York, 1977.

12. A. Schönhage, "Zur rationalen Approximierbarkeit von e^{-x} über $[0, \infty)$, J. Approximation Theory 7(1973), 395 - 398.

13. A. Schönhage, "Rational approximation to e^{-x} and related L^2– problems", SIAM J. Numer. Anal. 19(1982), 1067 - 1082.

14. L. N. Trefethen and M. H. Gutknecht, "The Carathéodory-Fejér method for real rational approximation", SIAM J. Numer. Anal. 20(1983), 420 - 436.

15. R. S. Varga, Matrix Iterative Analysis, Prentice-Hall, Inc., Englewood Cliffs, N.J., 1962.

COMPUTING WITH THE FABER TRANSFORM

S. W. Ellacott
Department of Mathematics
Brighton Polytechnic
Brighton BN2 4GJ
England

E. B. Saff[†]
Department of Mathematics
University of South Florida
Tampa, Florida 33620
U.S.A.

Abstract Some theoretical and computational aspects of Faber-Padé approximants are discussed. In particular, a Montessus type theorem is proved and a new method for computing the approximants is presented. Results of numerical tests for the latter are included.

1. Introduction

The purpose of this paper is to further discuss the Faber-Padé (FP) approximants introduced in [4]. In this section we review some of their basic properties and, in Section 2, we prove a Montessus type theorem. A new method for computing the FP approximants is presented in Section 3.

Let E be a closed bounded point set (not a single point) in the z-plane whose complement K is simply connected on the sphere. By the Riemann mapping theorem, there exists a conformal map $w = \phi(z)$ of K onto $|w| > 1$ with the property that, in a neighborhood of infinity,

$$\phi(z) = \frac{z}{c} + a_0 + \frac{a_1}{z} + \frac{a_2}{z^2} + \dots , \quad c > 0 . \tag{1.1}$$

If $F(w)$ is analytic on $|w| \le 1$, then the Faber transform of F is defined by

$$f(z) = T(F)(z) := \frac{1}{2\pi i}\int_{\Gamma_\rho} \frac{F(\phi(\xi))}{\xi - z}\, d\xi , \quad z \text{ inside } \Gamma_\rho , \tag{1.2}$$

where $\Gamma_\rho := \{\xi : |\phi(\xi)| = \rho\}$ and $\rho(>1)$ is chosen so that $F(w)$ is analytic on $|w| \le \rho$. When E is a Jordan region bounded by

[†]Research supported, in part, by the National Science Foundation.

a rectifiable Jordan curve C, then $T(F)$ is further defined for any F analytic in $|w| < 1$ and continuous on $|w| \leq 1$ by replacing Γ_ρ by C in (1.2). (Further extensions are discussed in the paper [1] by Anderson in this volume.)

Denoting by ψ the inverse of the mapping ϕ, it is straightforward to prove that the Faber transform has the following "singularity preserving property."

Lemma 1.1. *Let F be analytic on the closed disk $|w| \leq 1$, and let $f = T(F)$. Then $F(w) - f(\psi(w))$ can be extended analytically to $|w| > 1$, including the point at infinity.*

As a consequence of the above property we have

Proposition 1.2. *Let $R(w)$ be a type (m,n) rational function with all its poles in $|w| > 1$. Then $r(z) := T(R)(z)$ is a type $(\tilde{m},n)$ rational function, where $\tilde{m} := \max(m,n-1)$, and the poles of $r(z)$ are the images under ψ of those of $R(w)$, with corresponding multiplicities.*

Proof. It follows from (1.2) and Lemma 1.1 that r is meromorphic in the extended plane and hence is rational. The second part of the proposition is also an easy consequence of Lemma 1.1. (Different proofs of this result are given in [1] and [5].) □

We observe also that $\phi_n(z) := T(w^n)(z)$, $n = 0,1,\ldots,$ is a polynomial of degree n, the so-called *Faber polynomial*. If

$$F(w) = \sum_{k=0}^{\infty} a_k w^k ,$$

then $T(F)$ has the expansion

$$f(z) = T(F)(z) = \sum_{k=0}^{\infty} a_k \phi_k(z) .$$

Indeed, this property is often taken as the definition of the Faber transform. In practice, of course, we will be given the function f rather than F, but provided the mapping function ψ is known the coefficients a_n can easily be computed from the former (see [3]).

Observe that if $R(w)$ is a normal type (m,n) Padé approximant to $F(w)$, and has all its poles in $|w| > 1$, then $r = T(R)$ satisfies

$$(1.3) \qquad f(z) - r(z) = \sum_{k=m+n+1}^{\infty} b_k \phi_k(z)$$

for suitable coefficients b_k. This (with an obvious modification if R is not normal) is the Faber-Padé approximant of $f(z)$ as introduced in [4] (see also [5]). In the special case when E is the real interval $[-1,1]$, the FP approximant reduces to the Chebyshev-Padé approximant. For arbitrary point sets E, the FP approximant has two apparent drawbacks. First, it need not be of the "correct" type if $m < n-1$; second, the associated rational $R(w)$ is required to have no poles in the unit disk. Although, as is well-known, the first difficulty can be overcome in the special case of Chebyshev-Padé approximation, there appears to be no simple technique to extend this to the general setting. The second problem is, for the case of meromorphic functions F, addressed in the next section.

2. A Montessus Theorem

The following theorem guarantees the existence of the Faber-Padé under certain conditions and also shows that it behaves in the expected manner. The proof is a straightforward application of the singularity preserving property (Lemma 1.1) and it is possible to generalize other properties of the classical Padé approximants in a similar manner. With E and ϕ as described in the introduction, we have

Theorem 2.1. Let f be analytic on E and meromorphic with precisely n poles (counting multiplicities) in the Jordan region E_ρ bounded by the level curve $|\phi(z)| = \rho$, $\rho > 1$. Then for each m sufficiently large, the type (m,n) Faber-Padé approximant $r_{m,n}$ exists satisfying (1.3) on E. The $r_{m,n}$ have precisely n finite poles, and as $m \to \infty$, these poles approach, respectively, the n poles of f in E_ρ. Moreover, the sequence $r_{m,n}$ converges uniformly to f on every compact subset of E_ρ which excludes the poles of f.

Proof. $F = T^{-1}(f)$ exists since f is analytic on the closed set E and hence has a Faber series expansion that converges in an open set containing E. In view of Lemma 1.1, the function F is analytic on $|w| \leq 1$ and meromorphic with n poles in $|w| < \rho$ (these are the images of the poles of f under the map $w = \phi(z)$). From the classical Montessus theorem (see e.g. [2, p. 246]), it follows that there exists a sequence $R_{m,n}$, $m \geq m_0$, of type (m,n) Padé approximants to F with the following properties:

(A) For each $m \geq m_0$,

$$F(w) - R_{m,n}(w) = O(w^{m+n+1}) \quad \text{as} \quad w \to 0 \; ;$$

(B) For each $m \geq m_0$, $R_{m,n}$ has precisely n finite poles which approach the n poles of F in $|w| < \rho$ (with corresponding multiplicities);

(C) $\lim_{m \to \infty} R_{m,n}(w) = F(w)$ uniformly on every compact subset of $|w| < \rho$ which contains no poles of F.

From property (B), we see that for each m large, $R_{m,n}$ is analytic on $|w| \leq 1$ and hence its Faber transform exists. With

$$r_{m,n} := T(R_{m,n}) \quad ,$$

we note from Proposition 1.2 that $r_{m,n}$ is a type (m,n) rational for each m large. In view of property (A) we have

$$(2.1) \qquad f(z) - r_{m,n}(z) = \sum_{k=m+n+1}^{\infty} b_k^{(m)} \phi_k(z) \quad , \quad z \in E \; ,$$

and, since the poles of $r_{m,n}$ are the images under ψ of the poles of $R_{m,n}$, the assertion of the theorem regarding the poles of $r_{m,n}$ follows immediately.

To prove convergence, observe that from (1.2) we have

$$(2.2) \qquad f(z) - r_{m,n}(z) = \frac{1}{2\pi i} \int_{\Gamma_\sigma} \frac{F(\phi(\xi)) - R_{m,n}(\phi(\xi))}{\xi - z} \, d\xi \quad ,$$

for z inside $\Gamma_\sigma : |\phi(\xi)| = \sigma$, where $\sigma (>1)$ is suitably chosen. Equation (2.2) is valid for z on any compact set $K \subset E_\rho \setminus \{n$ poles of $f\}$ provided m is sufficiently large and Γ_σ is replaced by the curve $\Gamma_\tau : |\phi(\xi)| = \tau$, with $\rho - \tau > 0$ sufficiently small, together with small circles around the poles of f . Since, from property (C), the sequence $R_{m,n}(\phi(\xi))$ converges uniformly to $F(\phi(\xi))$ on these curves, the convergence of $r_{m,n}(z)$ to $f(z)$ on K follows. □

3. Computing the Faber Transform of a Rational Function

A crucial stage in computing Faber-Padé or Faber-CF approximants is the computation of the transform of a rational function R analytic on the unit disk. In [4] this was carried out by computing the poles of R and applying Proposition 1.2 in an obvious fashion. We describe here an alternative and much faster method based on the integral representation (1.2) of the transform. In so doing, we suppose that E is bounded by a rectifiable Jordan curve C and that the origin lies interior to C . Then, for any $f = T(F)$, we have

$$\frac{f^{(k)}(0)}{k!} = \frac{-1}{4\pi^2} \int_{|z|=\delta} \frac{1}{z^{k+1}} \int_C \frac{F(\phi(\xi))}{\xi - z} \, d\xi dz \quad ,$$

where $\delta > 0$ is sufficiently small. By interchanging the order of integration and computing the integral with respect to z we obtain

$$\frac{f^{(k)}(0)}{k!} = \frac{1}{2\pi i} \int_C \frac{F(\phi(\xi))}{\xi^{k+1}} \, d\xi \quad . \tag{3.1}$$

If F is entire, we may replace the curve C by a circle and evaluate as many of the integrals as we require simultaneously by the trapezium rule and the fast Fourier transform. However, this is not in general possible for the case required here where $F = R$ is a type (m,n) rational function. Instead, we make the substitution $w = \phi(\xi)$ in (3.1) and obtain, for $r = T(R)$,

$$\frac{r^{(k)}(0)}{k!} = \frac{1}{2\pi i} \int_{|w|=\rho} \frac{R(w)\psi'(w)}{[\psi(w)]^{k+1}} \, dw \quad , \qquad k = 0,1,\dots \quad . \tag{3.2}$$

In (3.2), the constant $\rho > 1$ is chosen sufficiently small to ensure that the circle $|w| = \rho$ does not enclose any poles of R . Note that in practice this may be easily checked by evaluating (to the nearest integer) the integral

$$\frac{1}{2\pi i} \int_{|w|=\rho} \frac{Q'(w)}{Q(w)}\, dw \quad ,$$

where Q is the denominator of R , since this integral gives the number of zeros of Q inside the circle.

To calculate the transform of a type (m,n) rational function R (with $m \geq n-1$) we first evaluated the integrals (3.2) for $k = 0,1,\ldots,m+n+1$ using the 512 point trapezium rule on a VAX 11 using double precision (about 16 decimal digits). These $m+n+2$ values uniquely determine the type (m,n) rational $r = T(R) = p/q$, and p,q can be computed from the Padé equations. Some of our numerical results are given below. Although these refer only to real poles (the correct position of the pole is easier to calculate in this case), the method has also been used successfully with conjugate pairs of poles.

Example 1: $\psi(w) = w + 1/4w$ (an ellipse).
Since this curve is analytic we may choose $\rho = 1$ here.

i) R is type (2,2) with a pole at 1.1. The corresponding pole of the transformed rational r was calculated to be 1.327272727272730 which is correct to 16 figures.

ii) R is type (4,5), near degenerate, with a pole at 2.0 and a zero at 2.01. The corresponding pole of the transformed rational r was calculated to be 2.124999998829719 which is correct to 10 figures.

iii) R is type (4,4), degenerate with a pole and zero at 2.0. The spurious pole and zero of the transformed rational r agreed to 15 figures, but they were inside the ellipse. This suggests that it would be advisable to check for degeneracy before making use of these approximants.

iv) R is type (2,2), with a double pole at 2. The poles of the transformed rational r were only calculated correct to eight figures, but this turned out to be due to the ill-conditioning inherent in the determination of multiple zeros of a polynomial; examination of the coefficients of the rational function revealed them to be correct to 16 figures. Thus this is another reason for preferring this method of calculating the approximants over that given in [4].

Example 2. $\psi(w) = \int (1+w^{-4})^{1/2}\, dw$ (a square).

In this example we chose $\rho = 1.1$ and evaluated $\psi(w)$ on the circle $|w| = \rho$ by expanding as a series which was then summed using the fast Fourier transform. For an example where R has poles at 2 and 5, the poles of the transformed rational r agreed with the true values as accurately as the latter could be computed, which was to about eight figures.

The above examples therefore indicate that the method described here is an effective way to evaluate the Faber transform of a rational function when computing Faber-Padé and Faber-CF approximants.

References

1. J. M. Anderson, "The Faber operator." This volume.
2. G. A. Baker and P. Graves-Morris, "Padé approximants. Part 1: Basic Theory." Encyclopedia of Mathematics and its Applications, vol. 13. Addison-Wesley. Massachusetts, 1981.
3. S. W. Ellacott, "Computation of Faber series with application to numerical polynomial approximation in the complex plane." Math. Comp. vol. 40, no. 162, 1984, pp. 575-587.
4. S. W. Ellacott, "On the Faber Transform and efficient numerical rational approximation." SIAM J. Numer. Anal., vol. 20, no. 5, October 1983, pp. 989-1000.
5. T. Ganelius, Degree of rational approximation. In: Lectures on Approximation and Value Distribution, Les Presses de l'Université de Montréal, Montréal, Canada (1982).

A-STABLE METHODS FOR SECOND ORDER DIFFERENTIAL SYSTEMS AND THEIR RELATION TO PADÉ APPROXIMANTS

I. Gladwell
Department of Mathematics
University of Manchester
Manchester M13 9PL
U.K.

R.M. Thomas
Department of Mathematics
U.M.I.S.T.
Manchester M60 1QD
U.K.

Abstract. We discuss a number of different methods for second order differential systems $\underset{\sim}{x}'' = \underset{\sim}{F}(t, \underset{\sim}{x})$ each of which reduce to two-step methods for linear homogeneous systems $\underset{\sim}{x}'' + K\underset{\sim}{x} = \underset{\sim}{g}(t)$. It is shown that some apparently unconnected methods are closely related when applied to this simple problem.

1. Implicit Runge-Kutta Methods

For the first order initial value problem of order N

$$\underset{\sim}{y}' = \underset{\sim}{f}(\underset{\sim}{y}), \quad \underset{\sim}{y}(0) = \underset{\sim}{y}_0, \quad t > 0 \tag{1.1}$$

implicit Runge-Kutta (IRK) methods have been proposed for cases where the requirement of numerical stability imposes a restriction on the increment h in t. The s-stage IRK method may be represented by the tableau

$$\begin{array}{c|c} \underset{\sim}{b} & A \\ \hline & \underset{\sim}{c}^T \end{array} \tag{1.2}$$

where $b_i = \sum_{j=1}^{s} a_{ij}$, $A = (a_{ij})$, Butcher (1976). With this notation, a step of the IRK method may be written

$$\underset{\sim}{y}_{n+1} = \underset{\sim}{y}_n + h \sum_{j=1}^{s} c_j \underset{\sim}{f}(\underset{\sim}{Y}_j), \tag{1.3a}$$

where

$$\underset{\sim}{Y}_j = \underset{\sim}{y}_n + h \sum_{k=1}^{s} a_{jk}\, \underset{\sim}{f}(\underset{\sim}{Y}_k), \quad j = 1, \ldots, s. \tag{1.3b}$$

(The coefficients $\underset{\sim}{b}$ do not come into this formula but are needed to define the method for the inhomogeneous problem $\underset{\sim}{y}' = \underset{\sim}{f}(t, \underset{\sim}{y})$.)

The equations (1.3) constitute a system of $(s+1)N$ nonlinear equations for the unknown $\underset{\sim}{y}_{n+1}$ and the intermediate values $\underset{\sim}{Y}_j$, $j = 1, \ldots, s$, given the value $\underset{\sim}{y}_n$. Techniques for solving these equations have received some attention (Cooper and Butcher, 1983) but these will not concern us here where we study their properties for linear systems

$$\underset{\sim}{y}' = H\underset{\sim}{y}, \quad \underset{\sim}{y}(0) = \underset{\sim}{y}_0 \tag{1.4}$$

and an explicit solution of equations (1.3) may be obtained. In fact a general form of the solution of equations (1.3) applied to the linear system (1.4) is available but this does not seem to provide insight into the properties of the methods and so we content ourselves with three well-known examples of maximal (2s-th) order methods:

<u>s = 1</u>

The tableau is

$$\begin{array}{c|c} \frac{1}{2} & \frac{1}{2} \\ \hline & 1 \end{array} ; \tag{1.5}$$

that is

$$\underset{\sim}{y}_{n+1} = \underset{\sim}{y}_n + h\underset{\sim}{f}(\underset{\sim}{Y}_1) \tag{1.6a}$$

where

$$\underset{\sim}{Y}_1 = \underset{\sim}{y}_n + \tfrac{1}{2}h\underset{\sim}{f}(\underset{\sim}{Y}_1). \tag{1.6b}$$

For equation (1.4), the method (1.6) immediately gives

$$\underset{\sim}{y}_{n+1} = \left(I - \frac{hH}{2}\right)^{-1} \left(I + \frac{hH}{2}\right)\underset{\sim}{y}_n \tag{1.7}$$

that is

$$\underset{\sim}{y}_{n+1} = R_{11}(hH)\underset{\sim}{y}_n \tag{1.8}$$

where $R_{kk}(z)$ is the (k,k)-Padé approximant to e^{-z}.

s = 2

In this case the tableau (1.2) is

$$\begin{array}{c|cc} \frac{1}{2}-\frac{\sqrt{3}}{6} & \frac{1}{4} & \frac{1}{4}-\frac{\sqrt{3}}{6} \\ \frac{1}{2}+\frac{\sqrt{3}}{6} & \frac{1}{4}+\frac{\sqrt{3}}{6} & \frac{1}{4} \\ \hline & \frac{1}{2} & \frac{1}{2} \end{array} \tag{1.9}$$

and it can be shown that using this method to solve equation (1.4) gives

$$\underset{\sim}{y}_{n+1} = R_{22}(hH)\underset{\sim}{y}_n. \tag{1.10}$$

s = 3

Finally for $s = 3$ the tableau is

$$\begin{array}{c|ccc} \frac{1}{2}-\frac{\sqrt{15}}{10} & \frac{5}{36} & \frac{2}{9}-\frac{\sqrt{15}}{15} & \frac{5}{36}-\frac{\sqrt{15}}{30} \\ \frac{1}{2} & \frac{5}{36}+\frac{\sqrt{15}}{24} & \frac{2}{9} & \frac{5}{36}-\frac{\sqrt{15}}{24} \\ \frac{1}{2}+\frac{\sqrt{15}}{10} & \frac{5}{36}+\frac{\sqrt{15}}{30} & \frac{2}{9}+\frac{\sqrt{15}}{15} & \frac{5}{36} \\ \hline & \frac{5}{18} & \frac{4}{9} & \frac{5}{18} \end{array} \tag{1.11}$$

which when applied to equation (1.4) gives

$$\underset{\sim}{y}_{n+1} = R_{33}(hH)\underset{\sim}{y}_n. \tag{1.12}$$

The first of these cases ($s = 1$) has been included to illustrate the conversion of the tableau to a Padé approximant method for linear systems (Varga, 1961). We will be concerned only with the cases $s = 2,3$ below.

2. Stability

Our concern in this paper will be with methods for second order systems

$$\underset{\sim}{x}'' = \underset{\sim}{F}(\underset{\sim}{x}),\ \underset{\sim}{x}(0) = \underset{\sim}{x}_0,\ \underset{\sim}{x}'(0) = \underset{\sim}{x}_0',\quad t > 0 \tag{2.1}$$

where $\partial \underset{\sim}{F}/\partial \underset{\sim}{x}$ is symmetric and negative definite for all $\underset{\sim}{x}$, and with their counterparts for linear homogeneous problems

$$\underset{\sim}{x}'' + K\underset{\sim}{x} = \underset{\sim}{0},\ \underset{\sim}{x}(0) = \underset{\sim}{x}_0,\ \underset{\sim}{x}'(0) = \underset{\sim}{x}_0',\quad t > 0 \tag{2.2}$$

where K is symmetric and positive definite. It is for this reason that our examples in the previous section are of methods which are A-stable (Lambert, 1973) and are I-stable (Nφrsett and Wanner, 1979), that is $|R_{k,k}(it)| \leq 1$ for all real t, (Birkhoff and Varga, 1965).

For methods designed directly for second·order systems (2.2) the test equation

$$x'' + \omega^2 x = 0,\quad \omega \text{ a real constant}, \tag{2.3}$$

is used and, in this context, we say a numerical method is P-stable (Lambert and Watson, 1976) if the principal roots of the difference scheme have modulus one for all values ωh. A method designed for second order systems (2.2) with the property of P-stability has the property of I-stability if the method can be considered as one for first order systems.

3. Equivalent Two-Step Methods

When applied to the equation (2.2) written in the form of the equation (1.4), namely with

$$H = \begin{pmatrix} 0 & I \\ -K & 0 \end{pmatrix},\quad \underset{\sim}{y} = \begin{pmatrix} \underset{\sim}{x} \\ \underset{\sim}{x}' \end{pmatrix}, \tag{3.1}$$

the (s, s)-Padé approximant method has the form

$$P(-hH)\underset{\sim}{y}_{n+1} = P(hH)\underset{\sim}{y}_n. \tag{3.2}$$

<u>Theorem 3.1</u> The Padé approximant method (3.2) is a two-step method for $\underset{\sim}{x}_{n+1}$.

<u>Proof</u> Let

$$P(hH) = I + a_1 hH + a_2 (hH)^2 + \ldots + a_s (hH)^s.$$

It may be shown, by induction, that

$$H^{2i} = \begin{bmatrix} (-K)^i & 0 \\ 0 & (-K)^i \end{bmatrix}, \quad H^{2i-1} = \begin{bmatrix} 0 & (-K)^{i-1} \\ (-K)^i & 0 \end{bmatrix}.$$

for $i = 1,2,\ldots, [(s+1)/2]$. Hence

$$P(hH) = \begin{bmatrix} p_1(h^2K) & hp_2(h^2K) \\ -hKp_2(h^2K) & p_1(h^2K) \end{bmatrix} \tag{3.3}$$

where

$$p_1(h^2K) = I - a_2h^2K + a_4h^4K^2 + \ldots + a_mh^m(-K)^{m/2}$$

$$p_2(h^2K) = a_1I - a_3h^2K + a_5h^4K^2 + \ldots + a_\ell h^{\ell-1}(-K)^{(\ell-1)/2}$$

and $m = 2[s/2]$, $\ell = 2[(s+1)/2] - 1$. Hence (3.2) may be written in terms of (3.3). Now premultiplying by

$$\begin{bmatrix} p_1(h^2K) & hp_2(h^2K) \\ 0 & I \end{bmatrix}$$

we obtain

$$\begin{bmatrix} p_1(h^2K)^2 + h^2Kp_2(h^2K)^2 & 0 \\ hKp_2(h^2K) & p_1(h^2K) \end{bmatrix} \begin{bmatrix} \underset{\sim}{x}_{n+1} \\ \underset{\sim}{x}'_{n+1} \end{bmatrix}$$

$$= \begin{bmatrix} p_1(h^2K)^2 - h^2Kp_2(h^2K)^2 & 2hp_1(h^2K)p_2(h^2K) \\ -hKp_2(h^2K) & p_1(h^2K) \end{bmatrix} \begin{bmatrix} \underset{\sim}{x}_n \\ \underset{\sim}{x}'_n \end{bmatrix}. \tag{3.4}$$

The first of these equations gives

$$2hp_1(h^2K)p_2(h^2K)\underset{\sim}{x}'_n = \{p_1(h^2K)^2 + h^2Kp_2(h^2K)^2\}\underset{\sim}{x}_{n+1}$$
$$-\{p_1(h^2K)^2 - h^2Kp_2(h^2K)^2\}\underset{\sim}{x}_n . \tag{3.5}$$

This equation may now be used to substitute for $\underset{\sim}{x}'_n$ and $\underset{\sim}{x}'_{n+1}$ in the second of equations (3.4) to give

$$\{p_1(h^2K)^2 + h^2Kp_2(h^2K)^2\}(\underset{\sim}{x}_{n+1} + \underset{\sim}{x}_{n-1})$$

$$+ \{2h^2Kp_2(h^2K)^2 - 2p_1(h^2K)^2\}\underset{\sim}{x}_n = \underset{\sim}{0}. \quad \square \tag{3.6}$$

We now consider two examples from section 1. First, the (2,2)-Padé approximant has

$$P(hH) = I + \tfrac{1}{2}hH + \frac{1}{12}h^2H^2 \tag{3.7}$$

and so

$$p_1(h^2K) = I - \frac{1}{12}h^2K, \quad p_2(h^2K) = \tfrac{1}{2}I.$$

Substituting in (3.6) we obtain

$$(I + \frac{1}{12}h^2K + \frac{1}{144}h^4K^2)(\underset{\sim}{x}_{n+1} + \underset{\sim}{x}_{n-1})$$

$$+ (-2I + \frac{5}{6}h^2K - \frac{1}{72}h^4K^2)\underset{\sim}{x}_n = \underset{\sim}{0}. \tag{3.8}$$

Finally, consider the (3,3)-Padé approximant for which

$$P(hH) = I + \tfrac{1}{2}hH + \frac{1}{10}h^2H^2 + \frac{1}{120}h^3H^3 \tag{3.9}$$

and hence

$$p_1(h^2K) = I - \frac{1}{10}h^2K, \quad p_2(h^2K) = \tfrac{1}{2}I - \frac{1}{120}h^2K.$$

Again substituting in (3.6) we obtain

$$\{I + \frac{1}{20}h^2K + \frac{1}{600}h^4K^2 + \frac{1}{14400}h^6K^3\}(\underset{\sim}{x}_{n+1} + \underset{\sim}{x}_{n-1})$$

$$+ \{-2I + \frac{9}{10}h^2K - \frac{11}{300}h^4K^2 + \frac{1}{7200}h^6K^3\}\underset{\sim}{x}_n = \underset{\sim}{0}. \tag{3.10}$$

We emphasize that the formulae (3.8) and (3.10) have been derived for comparison purposes; it is likely that in practice a one-step form such as (3.5) (for $\underset{\sim}{x}_{n+1}$) would be used.

4. Direct Hybrid Methods

Cash (1981) and Chawla (1981) have introduced a class of direct hybrid methods for second order systems (2.1). These have been extended and tested for linear problems (2.2) by Thomas (1983) and their implementation and efficiency for nonlinear problems are being investigated currently by the authors. The basic direct hybrid method has the form

$$\underset{\sim}{x}_{n+1} - 2\underset{\sim}{x}_n + \underset{\sim}{x}_{n-1} = h^2\{\beta_0(\underset{\sim}{x}''_{n+1} + \underset{\sim}{x}''_{n-1}) + \gamma\underset{\sim}{x}''_n + \beta_1(\underset{\sim}{x}''_{n+\alpha_1} + \underset{\sim}{x}''_{n-\alpha_1})\} \tag{4.1a}$$

where the off-step values are defined by

$$\underset{\sim}{x}_{n+\alpha_1} = \hat{A}\underset{\sim}{x}_{n+1} + \hat{B}\underset{\sim}{x}_n + \hat{C}\underset{\sim}{x}_{n-1} + h^2\{\hat{s}\underset{\sim}{x}''_{n+1} + \hat{t}\underset{\sim}{x}''_n + \hat{u}\underset{\sim}{x}''_{n-1}\}, \tag{4.1b}$$

$$\underset{\sim}{x}_{n-\alpha_1} = \tilde{A}\underset{\sim}{x}_{n+1} + \tilde{B}\underset{\sim}{x}_n + \tilde{C}\underset{\sim}{x}_{n-1} + h^2\{\tilde{s}\underset{\sim}{x}''_{n+1} + \tilde{t}\underset{\sim}{x}''_n + \tilde{u}\underset{\sim}{x}''_{n-1}\}$$

and

$$\underset{\sim}{x}''_n = \underset{\sim}{F}(t_n, \underset{\sim}{x}_n),\ \underset{\sim}{x}''_{n\pm1} = \underset{\sim}{F}(t_n \pm h, \underset{\sim}{x}_{n\pm1}), \underset{\sim}{x}''_{n\pm\alpha_1} = \underset{\sim}{F}(t_n \pm \alpha_1 h, \underset{\sim}{x}_{n\pm\alpha_1}). \tag{4.1c}$$

The parameters in (4.1) may be chosen freely but to achieve <u>fourth-order accuracy</u> they must satisfy

$$\begin{aligned}
&\beta_0 = \frac{1}{12} - \beta_1\alpha_1^2,\ \gamma = \frac{5}{6} + 2\beta_1(\alpha_1^2 - 1),\\
&\hat{B} = 1 + \alpha_1 - 2\hat{A},\ \tilde{B} = 1 - \alpha_1 - 2\tilde{A},\\
&\hat{C} = \hat{A} - \alpha_1,\ \tilde{C} = \tilde{A} + \alpha_1,\\
&\hat{s} = \hat{u} + \frac{1}{6}(\alpha_1^3 - \alpha_1),\ \tilde{s} = \tilde{u} - \frac{1}{6}(\alpha_1^3 - \alpha_1),\\
&\hat{t} = -\hat{A} + \frac{2}{3}\alpha_1 + \frac{1}{2}\alpha_1^2 - \frac{1}{6}\alpha_1^3 - 2\hat{u},\ \tilde{t} = -\tilde{A} - \frac{2}{3}\alpha_1 + \frac{1}{2}\alpha_1^2 + \frac{1}{6}\alpha_1^3 - 2\tilde{u}.
\end{aligned} \tag{4.2}$$

[Methods satisfying these equations also have fourth order phase lag, and are in phase (Gladwell and Thomas, 1983 and Thomas, 1983).] In addition to fourth order accuracy, the other requirement which we impose and which is shared with the diagonal Padé approximants is P-stability which property is obtained if and only if

$$\begin{aligned}
&\omega^2h^2 + \beta_1(\hat{A} + \tilde{A} - \alpha_1^2)\omega^4h^4 \geq 0,\\
&4 + \omega^2h^2\{4\beta_1(\hat{A} + \tilde{A} - \alpha_1^2) - \frac{2}{3}\} + \omega^4h^4\{\beta_1(\alpha_1^2 - \hat{A} - \tilde{A}) - 4\beta_1(\hat{u} + \tilde{u})\} \geq 0
\end{aligned} \tag{4.3}$$

for all ωh.

The conditions (4.3) are satisfied by Cash's choice of parameters

$$\alpha_1 = \frac{1}{2},\ \hat{A} + \tilde{A} = \frac{1}{4},\ \beta_1 = \frac{41}{90},\ \tilde{u} + \hat{u} = -\frac{1}{64} \tag{4.4}$$

leaving the choice of two free parameters, $\tilde{A}$ and $\tilde{u}$ say. In contrast Chawla imposes a certain degree of symmetry by setting

$$\hat{B} = \tilde{B},\ \tilde{C} = \hat{A},\ \tilde{A} = \hat{C},\ \hat{t} = \tilde{t},\ \hat{u} = \tilde{s},\ \hat{s} = \tilde{u}$$

$$\hat{A} + \hat{B} + \hat{C} = 1,\ \hat{A} - \hat{C} = \alpha_1,\ \hat{s} + \hat{u} = \frac{1}{2}(\alpha_1^2 - \hat{A} - \hat{C}) - \hat{t}, \tag{4.5}$$

$$\beta_0 = \frac{1}{12} - \beta_1\alpha_1^2,\ \gamma = \frac{5}{6} - 2\beta_1(1 - \alpha_1^2).$$

These conditions are sufficient to make the method fourth order accurate. To obtain P-stability, Chawla also requires that

$$\hat{B} = 0,\ \hat{t} = 0,\ \gamma = 0,\quad \alpha_1 \in (0, 1). \tag{4.6}$$

Consider now the linear inhomogeneous system

$$\underset{\sim}{x}'' + K\underset{\sim}{x} = \underset{\sim}{g}(t),\quad \underset{\sim}{x}(0) = \underset{\sim}{x}_0,\ \underset{\sim}{x}'(0) = \underset{\sim}{x}_0',\quad t > 0. \tag{4.7}$$

For this system the hybrid method (4.1) yields

$$\begin{aligned}
&\{I + [\beta_0 + \beta_1(\hat{A} + \tilde{A})]h^2K - \beta_1(\hat{s} + \tilde{s})h^4K^2\}\underset{\sim}{x}_{n+1}\\
&+ \{-2I + [\gamma + \beta_1(\hat{B} + \tilde{B})h^2K - \beta_1(\hat{t} + \tilde{t})h^4K^2\}\underset{\sim}{x}_n\\
&+ \{I + [\beta_0 + \beta_1(\hat{C} + \tilde{C})h^2K - \beta_1(\hat{u} + \tilde{u})h^4K^2\}\underset{\sim}{x}_{n-1}\\
&\quad = h^2\{[\beta_0 - \beta_1(\hat{s} + \tilde{s})h^2K]\underset{\sim}{g}(t_{n+1}) + [\beta_0 - \beta_1(\hat{u} + \tilde{u})h^2K]\underset{\sim}{g}(t_{n-1})\\
&\qquad + [\gamma - \beta_1(\hat{t} + \tilde{t})h^2K]\underset{\sim}{g}(t_n) + \beta_1\ [\underset{\sim}{g}(t_{n+\alpha_1}) + \underset{\sim}{g}(t_{n-\alpha_1})]]\}\ .
\end{aligned} \tag{4.8}$$

Observe that for these linear problems it is only linear combinations of those constants associated with the $n \pm 1$ points which concern us, hence simplifying the earlier conditions (4.2) considerably.

If we set $\underset{\sim}{g} \equiv \underset{\sim}{0}$ in (4.7), and hence (4.8), we may compare this version of the direct hybrid method with the (2,2)-Padé approximant formula (3.8) (which is also of fourth order) and we observe that in this case (4.8) can only be equivalent to (3.8) if

$$\beta_1(\hat{u} + \tilde{u}) = -\frac{1}{144}\ ,\quad \hat{A} + \tilde{A} - \alpha_1^2 = 0 \tag{4.9}$$

in addition to the fourth order conditions (4.2). We see immediately that Cash's choice of parameters satisfies the first of these conditions but not the second whereas Chawla's choice satisfies neither. However, it is clear that there are fourth-order hybrid methods which are equivalent in this sense to (2,2)-Padé approximant methods and indeed we will meet one in the next section.

When $\underset{\sim}{g} \not\equiv \underset{\sim}{0}$ it is interesting to ask whether the direct hybrid method can ever give the same formula as would be obtained by applying the implicit Runge-Kutta method given by (1.9) to equation (4.7). The technique of derivation of the two-step method equivalent to this Runge-Kutta method for equation (5.3) with (3.1) implies immediately that the corresponding formula for equation (4.7) has too many off-step points to be equivalent to the fourth-order direct hybrid methods. Indeed for the same order, the direct hybrid methods will always involve less function evaluations.

We turn now to sixth order methods. Cash (1981) suggested a method for achieving sixth-order accuracy whereby he introduced two further off-step points, $\underset{\sim}{x}_{n\pm\alpha_2}$. Details are given in Cash (1981) and in Thomas (1983). Here we content ourselves with the following comments which correspond directly to those for fourth-order methods:

(i) There exist sixth-order P-stable direct hybrid methods with four off-step points;

(ii) A subset of these methods is equivalent to the (3,3)-Padé approximant method (3.10) when applied to equation (2.2);

(iii) Cash's particular choice of parameters does not correspond to a member of this subset;

(iv) The implicit Runge-Kutta method (1.11) when applied to equation (4.7) is not equivalent to a direct hybrid method.

5. Gellert's Method

In this final section we consider a method derived by entirely different means but which turns out to be equivalent for linear problems to some of the fourth order methods discussed above. Gellert (1978) derived his methods for the implicit linear second order system

$$M\underset{\sim}{x}'' + C\underset{\sim}{x}' + K\underset{\sim}{x} = \underset{\sim}{g}(t). \tag{5.1}$$

He approximates the vector $\underset{\sim}{x}$ by a cubic Hermite polynomial and $\underset{\sim}{g}$ by a quadratic Lagrangian polynomial. He then differentiates equation (5.1), and, separately, integrates it once and twice giving four equations (including (5.1)) and into them he substitutes the approximations for $\underset{\sim}{x}$ and $\underset{\sim}{g}$.

Finally he takes two linear combinations of the four formulae which result. Full details are given in Thomas and Gladwell (1982); here we give the final result of this complicated derivation when applied to the equation (4.7):

$$\begin{bmatrix} I + \frac{5}{12}h^2K & \frac{h}{2}I - \frac{h^3}{12}K \\ \frac{h}{2}K & I - \frac{h^2}{12}K \end{bmatrix} \begin{bmatrix} \underset{\sim}{x}_{n+1} \\ \underset{\sim}{x}'_{n+1} \end{bmatrix} = \begin{bmatrix} I - \frac{7}{12}h^2K & \frac{3h}{2}I - \frac{h^3}{12}K \\ -\frac{h}{2}K & I - \frac{h^2}{12}K \end{bmatrix} \begin{bmatrix} \underset{\sim}{x}_n \\ \underset{\sim}{x}'_n \end{bmatrix}$$

$$+ \begin{bmatrix} \frac{h^2}{4}\underset{\sim}{g}_n + \frac{2}{3}h^2\underset{\sim}{g}_{n+\frac{1}{2}} + \frac{h^2}{12}\underset{\sim}{g}_{n+1} \\ \frac{h}{6}\underset{\sim}{g}_n + \frac{2}{3}h\underset{\sim}{g}_{n+\frac{1}{2}} + \frac{h}{6}\underset{\sim}{g}_{n+1} \end{bmatrix} \qquad (5.2)$$

and this formula is fourth order accurate and P-stable. After some manipulation it can be seen that this method is, in fact, equivalent to a method for the first order system

$$\underset{\sim}{y}' = H\underset{\sim}{y} + \underset{\sim}{G}(t) \qquad (5.3a)$$

with

$$H = \begin{bmatrix} 0 & I \\ -K & 0 \end{bmatrix}, \quad \underset{\sim}{G} = \begin{bmatrix} \underset{\sim}{0} \\ \underset{\sim}{g}(t) \end{bmatrix}, \qquad (5.3b)$$

of the form

$$\left(I - \frac{h}{2}H + \frac{h^2}{12}H^2\right)\underset{\sim}{y}_{n+1} = \left(I + \frac{h}{2}H + \frac{h^2}{12}H\right)\underset{\sim}{y}_n$$

$$+ \frac{h}{6}\left(\underset{\sim}{G}_{n+1} + 4\underset{\sim}{G}_{n+\frac{1}{2}} + \underset{\sim}{G}_n\right) - \frac{h^2}{12}H\left(\underset{\sim}{G}_{n+1} - \underset{\sim}{G}_n\right). \qquad (5.4)$$

It is obvious immediately that for the homogeneous problem, $\underset{\sim}{g} \equiv \underset{\sim}{0}$, Gellert's method is the (2,2)-Padé approximant method (1.10). Writing the method (5.4) in partitioned form we have

$$\begin{bmatrix} I - \frac{h^2}{12}K & -\frac{h}{2}I \\ \frac{h}{2}K & I - \frac{h^2}{12}K \end{bmatrix} \begin{bmatrix} \underset{\sim}{x}_{n+1} \\ \underset{\sim}{x}'_{n+1} \end{bmatrix} = \begin{bmatrix} I - \frac{h^2}{12}K & \frac{h}{2}I \\ -\frac{h}{2}K & I - \frac{h^2}{12}K \end{bmatrix} \begin{bmatrix} \underset{\sim}{x}_{n} \\ \underset{\sim}{x}'_{n} \end{bmatrix}$$

$$+ \begin{bmatrix} -\frac{h^2}{12}\{\underset{\sim}{g}_{n+1} - \underset{\sim}{g}_n\} \\ \frac{h}{6}\{\underset{\sim}{g}_{n+1} + 4\underset{\sim}{g}_{n+\frac{1}{2}} + \underset{\sim}{g}_n\} \end{bmatrix}. \tag{5.5}$$

Now we proceed in a similar way to the proof of Theorem 3.1 (in fact for the homogeneous equation this is just a special case of the proof) to eliminate the terms $\underset{\sim}{x}'_n$ and $\underset{\sim}{x}'_{n+1}$ so as to produce a two-step method. After routine manipulation we obtain

$$(I + \frac{h^2}{12}K + \frac{h^4}{144}K^2)(\underset{\sim}{x}_{n+1} + \underset{\sim}{x}_{n-1}) - (2I - \frac{5}{6}h^2K + \frac{h^4}{72}K^2)\underset{\sim}{x}_n =$$

$$\frac{h^4}{144}(\underset{\sim}{g}_{n+1} + \underset{\sim}{g}_{n-1}) + \frac{h^2}{3}(\underset{\sim}{g}_{n+\frac{1}{2}} + \underset{\sim}{g}_{n-\frac{1}{2}}) + \frac{h^2}{12}(4I - \frac{h^2}{6}K)\underset{\sim}{g}_n. \tag{5.6}$$

This is equivalent to the fourth order direct hybrid method (4.8) (with order conditions (4.2)) when

$$\alpha_1 = \frac{1}{2}, \quad \beta_1 = \frac{1}{3}, \quad \hat{A} + \tilde{A} = \frac{1}{4}, \quad \hat{u} + \tilde{u} = -\frac{1}{48}. \tag{5.7}$$

6. Conclusion

We have shown that a variety of methods for first and second order systems are equivalent to Padé approximant methods when applied to linear homogeneous differential systems. Also, we have shown how the methods relate when applied to inhomogeneous systems.

References

1. Birkhoff, G. and Varga, R. S. Discretization errors for well set Cauchy problems. I. J. Math. and Physics, 44, 1-23 (1965).

2. Butcher, J.C. Runge-Kutta Methods, ch. 5 of G. Hall and J.M. Watt (eds.) Modern Numerical Methods for Ordinary Differential Equations, Oxford, (1976).

3. Cash, J.R. High order, P-stable formulae for the numerical integration of periodic initial value problems. Numer. Math., 37, 355-370 (1981).

4. Chawla, M.M. Two-step fourth order P-stable methods for second order differential equations. BIT, 21, 190-193 (1981).

5. Cooper, G.J. and Butcher, J.C. An iteration scheme for implicit Runge-Kutta methods. IMA J. Numer. Anal., 3, 127-140 (1983).

6. Gellert, M. A new algorithm for integration of dynamic systems. Computers and Structures, 9, 401-408 (1978).

7. Gladwell, I. and Thomas, R.M. Damping and phase analysis for some methods for solving second order ordinary differential equations. Int. J. Numer. Meth. Engng., 19, 495-503 (1983).

8. Lambert, J.D. and Watson, I.A. Symmetric multistep methods for periodic initial value problems. JIMA, 18, 189-202 (1976).

9. Nørsett, S.P. and Wanner, G. The real-pole sandwich for rational approximations and oscillation equations. BIT, 19, 79-94 (1979).

10. Thomas, R.M. and Gladwell, I. Extensions to methods of Gellert and of Brusa and Nigro. NA Report 75, University of Manchester (1982), to appear in Int. J. Numer. Meth Engng. (in press)

11. Thomas, R.M. Phase properties of high order, P-stable formulae. Report 176, Department of Computer Studies, University of Leeds (1983), to appear in BIT.

12. Varga, R. S. On higher-order stable implicit methods for solving parabolic partial differential equations, J. Math. and Physics, 40, 220-231 (1961).

SHAPE PRESERVING RATIONAL SPLINE INTERPOLATION

John A. Gregory
Department of Mathematics and Statistics
Brunel University
Uxbridge UB8 3PH
England.

Abstract A rational cubic function is presented which has shape preserving interpolation properties. It is shown that the rational cubic can be used to construct C^2 rational spline interpolants to monotonic or convex sets of data which are defined on a partition $x_1 < x_2 < \ldots < x_n$ of the real interval $[x_1, x_n]$.

1. INTRODUCTION

Let (x_i, f_i), $i = 1,\ldots,n$, be a set of real data, where $x_1 < x_2 < \ldots < x_n$ and let d_i, $i = 1,\ldots,n$, denote first derivative values defined at the knots x_i, $i = 1,\ldots,n$. We assume that the data satisfy either monotonic or convex constraints and we seek a monotonic or convex function $s \in C^1[x_1, x_n]$, piecewise defined on the partition $x_1 < x_2 < \ldots < x_n$, which is such that

(1.1) $s(x_i) = f_i$ and $s^{(1)}(x_i) = d_i$, $i = 1,\ldots,n$.

Moreover, given an appropriate definition of $s(x)$ we seek values for the d_i such that $s \in C^2[x_1, x_n]$, i.e. $s(x)$ is *twice continuously differentiable*.

The constraints on the data are assumed to have one of the following forms:

(1.2) $f_1 < f_2 < \ldots < f_n$ *(monotonic increasing data)*,

(1.3) $\Delta_1 < \Delta_2 < \ldots < \Delta_{n-1}$ *(convex data)* ,

where

$$(1.4) \qquad \Delta_i = (f_{i+1} - f_i)/h_i \ , \quad h_i = x_{i+1} - x_i \ , \quad i = 1,\ldots,n-1 \ .$$

Necessary conditions on the derivative parameters, for a strictly monotonic or convex interpolant, are then given respectively by

$$(1.5) \qquad d_i > 0 \ , \quad i = 1,\ldots,n \quad \textit{(for monotonicity)} \ ,$$

$$(1.6) \qquad d_1 < \Delta_1 < d_2 < \ldots < \Delta_{i-1} < d_i < \Delta_i < \ldots < d_n \quad \textit{(for convexity)} \ .$$

Also, in the case of monotonic increasing and convex data, where $\Delta_1 > 0$ in (1.3), then $d_1 > 0$ in (1.6) and any convex interpolant is then necessarily monotonic.

In the absence of monotonic or convex constraints, a familiar solution to the interpolation problem (1.1), with $s \in C^2[x_1,x_n]$, is the cubic interpolating spline. However, the cubic interpolating spline is not necessarily shape preserving, see Section 5, and hence we find it appropriate to generalize to a rational cubic form introduced in a recent paper by Delbourgo and Gregory [2]. The rational cubic also includes, as a special case, a rational quadratic form, used for the interpolation of monotonic data, in [4] and [1]. We begin with a review of some of the work of these earlier papers, set in the context of the piecewise rational cubic interpolant. The rational C^2 spline interpolation of shaped data is then discussed. In particular a new analysis of the solution of the spline equations for monotonic interpolation given in [1] is presented. The study of the spline equations for convex interpolation is not yet complete and we only report on some preliminary results in this case.

Rational splines have been studied by a number of authors using a definition due to Schaback [6] and [7]. However, the constraint of shape preservation leads us to consider a rather different approach to the subject here.

2. THE SHAPE PRESERVING RATIONAL CUBIC

For $x \in [x_i,x_{i+1}]$, $i = 1,\ldots,n-1$, a piecewise rational cubic is defined by

$$(2.1) \quad s(x) = \frac{f_{i+1}\theta^3 + (r_i f_{i+1} - h_i d_{i+1})\theta^2(1-\theta) + (r_i f_i + h_i d_i)\theta(1-\theta)^2 + f_i(1-\theta)^3}{1 + (r_i - 3)\theta(1-\theta)}$$

where

$$(2.2)\qquad \theta = (x - x_i)/h_i\ ,\quad h_i = x_{i+1} - x_i\ .$$

Here, r_i denotes a parameter which is such that

$$(2.3)\qquad r_i > -1$$

and this ensures a strictly positive denominator in (2.1). The interpolation properties (1.1) are easily verified and when $r_i = 3$ the rational cubic clearly degenerates to the standard cubic Hermite polynomial.

Error Bound. Given $f \in C^4[x_i,x_{i+1}]$ and $f_i = f(x_i)$, $f_{i+1} = f(x_{i+1})$, then an error bound for the rational cubic on $[x_i,x_{i+1}]$ is given by

$$(2.4)\qquad |f(x) - s(x)| \le \frac{h_i}{4c_i} \max\{|f_i^{(1)} - d_i|, |f_{i+1}^{(1)} - d_{i+1}|\}$$
$$+ \frac{1}{384c_i}\{(1+|r_i-3|/4)h_i^4\|f^{(4)}\| + 4|r_i-3|(h_i^3\|f^{(3)}\| + 3h_i^2\|f^{(2)}\|)\},$$

where

$$c_i = \min\{1,(1+r_i)/4\} \quad \text{and} \quad \|f\| = \max_{x_i \le x \le x_{i+1}} |f(x)|\ .$$

A proof of this result is given in [2], where it is also shown that the rational cubic has shape preserving properties with the following choices of the parameters r_i.

Monotonic Case. If the data satisfy the monotonicity condition (1.2) and the necessary derivative condition (1.5) holds, then

$$(2.5)\qquad r_i = 1 + (d_i + d_{i+1})/\Delta_i$$

ensures that $s^{(1)}(x) > 0$ on $[x_i,x_{i+1}]$. Hence $s(x)$ is monotonic increasing.

Convex Case. If the data satisfy the convexity condition (1.3) and the necessary derivative condition (1.6) holds, then

$$(2.6)\qquad r_i = 1 + (d_{i+1} - \Delta_i)/(\Delta_i - d_i) + (\Delta_i - d_i)/(d_{i+1} - \Delta_i)$$

ensures that $s^{(2)}(x) > 0$ on $[x_i,x_{i+1}]$. Hence $s(x)$ is convex.

The values of r_i defined by (2.5) and (2.6) are not the only ones which ensure shape preservation. However, they are distinguished by the fact that

$$(2.7)\qquad r_i - 3 = O(h_i^2)$$

for the appropriate choice of r_i, where f is either strictly monotonic or strictly convex in the error bound (2.4) and where

$$(2.8)\qquad d_i - f_i^{(1)} = O(h_i^2) = d_{i+1} - f_{i+1}^{(1)} .$$

In particular, if

$$(2.9)\qquad d_i - f_i^{(1)} = O(h_i^3) = d_{i+1} - f_{i+1}^{(1)}$$

then an optimal $O(h_i^4)$ error bound (2.4) is achieved.

The monotonic case (2.5) is also distinguished by the fact that the rational cubic on $x_i \leq x \leq x_{i+1}$ then degenerates to the rational quadratic form

$$(2.10)\quad s(x) = \frac{f_{i+1}\theta^2+\Delta_i^{-1}(f_{i+1}d_i + f_i d_{i+1})\theta(1-\theta) + f_i(1-\theta)^2}{\theta^2 + \Delta_i^{-1}(d_i + d_{i+1})\theta(1-\theta) + (1-\theta)^2}$$

and it is this form that is investigated in references [4] and [1].

3. MONOTONIC RATIONAL SPLINE EQUATIONS

Let the data set (x_i, f_i), $i = 1,\ldots,n$, satisfy the monotonicity constraint (1.2) and let s(x) be the piecewise rational cubic with r_i defined by (2.5). (Thus s(x) has the piecewise rational quadratic form (2.10)). Given $d_1 > 0$ and $d_n > 0$ we seek derivative values d_i, $i = 2,\ldots,n-1$, which satisfy the C^2 consistency constraints

$$(3.1)\qquad s^{(2)}(x_{i-}) = s^{(2)}(x_{i+}) , \qquad i = 2,\ldots,n-1$$

and which are such that the necessary condition (1.5) holds, i.e. $d_i > 0$, $i = 2,\ldots,n-1$. The C^2 consistency constraint gives the non-linear system

$$(3.2)\qquad d_i[-c_i + a_{i-1}d_{i-1} + (a_{i-1} + a_i)d_i + a_i d_{i+1}] = b_i , \quad i = 2,\ldots,n-1 ,$$

where

(3.3) $a_i = 1/(h_i\Delta_i)$, $b_i = \Delta_{i-1}/h_{i-1} + \Delta_i/h_i$, $c_i = 1/h_{i-1} + 1/h_i$,

are all positive values. We then have the following theorem.

Theorem 3.1 *For strictly increasing data and given end conditions* $d_1 > 0$, $d_n > 0$, *there exist unique values* d_i, $i = 2,\dots,n-1$, *satisfying the non-linear consistency equations* (3.2) *and the monotonicity conditions* $d_i > 0$, $i = 2,\dots,n-1$.

It can also be shown that $d_i - f_i^{(1)} = O(h^3)$ in (2.4), where $h = \max\{h_i\}$. Hence an $O(h^4)$ error bound is achieved for the rational spline.

A proof of Theorem 3.1 is a consequence of the study of the iterative solution of (3.2) and this we now consider in detail.

Equation (3.1) is a quadratic in the unknown d_i. The monotonicity condition $d_i > 0$ means that the positive root of the quadratic is the only one of interest and this gives the non-linear system

(3.4) $d_i = G_i(d_{i-1}, d_{i+1})$, $i = 2,\dots,n-1$,

where

(3.5) $$\begin{cases} G_i(\xi,\eta) \equiv \dfrac{1}{2(a_{i-1}+a_i)}\{\zeta + [\zeta^2 + 4(a_{i-1}+a_i)b_i]^{1/2}\} , \\ \zeta \equiv c_i - a_{i-1}\xi - a_i\eta . \end{cases}$$

Thus

(3.6) $$\begin{cases} \dfrac{\partial G_i}{\partial \xi} = \dfrac{-a_{i-1}}{2(a_{i-1}+a_i)}\{1 + \zeta/[\zeta^2 + 4(a_{i-1}+a_i)b_i]^{1/2}\} , \\ \dfrac{\partial G_i}{\partial \eta} = \dfrac{-a_i}{2(a_{i-1}+a_i)}\{1 + \zeta/[\zeta^2 + 4(a_{i-1}+a_i)b_i]^{1/2}\} . \end{cases}$$

Equations (3.4) suggest the two iterative methods

(3.7) $d_i^{(k+1)} = G_i(d_{i-1}^{(k)}, d_{i+1}^{(k)})$, $i = 2,\dots,n-1$, *(Jacobi iteration)* ,

(3.8) $d_i^{(k+1)} = G_i(d_{i-1}^{(k+1)}, d_{i+1}^{(k)})$, $i = 2,\dots,n-1$, *(Gauss-Seidel iteration)* ,

where $d_1^{(k+1)} = d_1^{(k)} = d_1$ and $d_n^{(k+1)} = d_n^{(k)} = d_n$ are the given end conditions.

Our convergence analysis of the non-linear Jacobi and Gauss-Seidel methods requires the following lemma, which is a result of the monotonic increasing nature of $G_i(\xi,\eta)$, considered as a function of ζ.

Lemma 3.1 *Let $\alpha_i, \beta_i > 0$ be defined by*

$$(3.9) \qquad \alpha_i = G_i(\beta_{i-1},\beta_{i+1}) \ , \quad \beta_i = G_i(0,0) \ , \quad i = 2,\ldots,n-1 \ ,$$

where $\beta_1 = d_1$ and $\beta_n = d_n$ in (3.9). Then

$$(3.10) \qquad \alpha_i \le G_i(d_{i-1},d_{i+1}) \le \beta_i \ , \ i = 2,\ldots,n-1 \ , \quad \forall \ \alpha_i \le d_i \le \beta_i \ , \quad i = 2,\ldots,n-1 \ .$$

The Lemma states that G_i, $i = 2,\ldots,n-1$, maps the closed interval $I = [\alpha_2,\beta_2] \times \ldots \times [\alpha_{n-1},\beta_{n-1}]$ into itself. Thus there exist $d_i > 0$, $i = 2,\ldots,n-1$, in this interval, which satisfy the non-linear system (3.4) and hence the consistency equations (3.2) (Schauder's Fixed Point Theorem). Also, given any $d_i^{(0)}$, $i = 2,\ldots,n-1$, it is easily shown that the Jacobi and Gauss-Seidel iterates lie in the interval I within three iterative steps. Thus we consider the mapping G_i, $i = 2,\ldots,n-1$, defined on I and we then have the following convergence theorem.

Theorem 3.2 *The Jacobi and Gauss-Seidel iterations converge to the (unique) positive solution of the non-linear consistency equations.*

Proof Let $d_i \in [\alpha_i,\beta_i]$, $i = 2,\ldots,n-1$, satisfy the fixed point equation (3.4) and consider the Jacobi iteration (3.7) defined on $[\alpha_i,\beta_i]$, $i = 2,\ldots,n-1$. The Mean Value Theorem gives

$$\begin{aligned} |d_i - d_i^{(k+1)}| &= |G_i(d_{i-1},d_{i+1}) - G_i(d_{i-1}^{(k)},d_{i+1}^{(k)})| \\ &\le |d_{i-1} - d_{i-1}^{(k)}| \left|\frac{\partial G_i(\xi_i,\eta_i)}{\partial \xi}\right| + |d_{i+1} - d_{i+1}^{(k)}| \left|\frac{\partial G_i(\xi_i,\eta_i)}{\partial \eta}\right| \ , \end{aligned}$$

where (ξ_i,η_i) is some point on the open line segment joining (d_{i-1},d_{i+1}) and $(d_{i-1}^{(k)},d_{i+1}^{(k)})$. Now, from (3.6),

$$(3.11) \qquad \left|\frac{\partial G_i}{\partial \xi}\right| \le \frac{K\,a_{i-1}}{a_{i-1}+a_i} \ , \qquad \left|\frac{\partial G_i}{\partial \eta}\right| \le \frac{K\,a_i}{a_{i-1}+a_i} \ ,$$

where it can be shown that

$$\begin{aligned} (3.12) \qquad K &\le \max_{2\le i\le n-1} \tfrac{1}{2}\{1 + 1/[1 + 4(a_{i-1}+a_i)b_i/c_i^2]^{1/2}\} \\ &\le \tfrac{1}{2}\,(1 + 1/\sqrt{5}) \ . \end{aligned}$$

Thus

$$\| \underline{d} - \underline{d}^{(k+1)} \|_\infty \le K \| \underline{d} - \underline{d}^{(k)} \|_\infty \ ,$$

where

$$\| \underline{d} \|_\infty = \max_{2 \le i \le n-1} |d_i| .$$

Since $0 < K < 1$ it follows that the Jacobi iteration converges. An analysis of the Gauss-Seidel iteration gives

$$\| \underline{d} - \underline{d}^{(k+1)} \|_\infty \le K' \| \underline{d} - \underline{d}^{(k)} \|_\infty ,$$

where

$$K' = K \max_{2 \le i \le n-1} \left\{ \frac{a_i}{a_i + (1-K) a_{i-1}} \right\} .$$

Since $0 < K' < K$ it follows that the Gauss-Seidel iteration also converges.

Asymptotic Rates of Convergence. Consider the Jacobi and Gauss-Seidel error iterates respectively defined by

$$e_i^{(k+1)} \equiv d_i - d_i^{(k+1)} = G_i(d_{i-1}, d_{i+1}) - G_i(d_{i-1}^{(k)}, d_{i+1}^{(k)}) , \quad i = 2, \ldots, n-1 ,$$

$$e_i^{(k+1)} \equiv d_i - d_i^{(k+1)} = G_i(d_{i-1}, d_{i+1}) - G_i(d_{i-1}^{(k+1)}, d_i^{(k)}) , \quad i = 2, \ldots, n-1 .$$

Then a Taylor expansion analysis gives

$$(3.13) \quad \begin{cases} e_i^{(k+1)} = l_i e_{i-1}^{(k)} + u_i e_{i+1}^{(k)} + \rho_i^{(k)} , & i = 2, \ldots, n-1 , \quad \text{(Jacobi)} \\ e_i^{(k+1)} = l_i e_{i-1}^{(k+1)} + u_i e_{i+1}^{(k)} + \sigma_i^{(k)} , & i = 2, \ldots, n-1 , \quad \text{(Gauss-Seidel)} \end{cases}$$

where

$$l_i = \frac{\partial G_i(d_{i-1}, d_{i+1})}{\partial \xi} \qquad u_i = \frac{\partial G_i(d_{i-1}, d_{i+1})}{\partial \eta}$$

and $\rho_i^{(k)}$, $\sigma_i^{(k)}$ are second order terms. (It should be noted that $e_1^{(k)} = e_n^{(k)} = 0$ in (3.13).) These error equations can be expressed as

$$(3.14) \quad \begin{cases} \underline{e}^{(k+1)} = (L + U)\underline{e}^{(k)} + \underline{\rho}^{(k)} , & \text{(Jacobi)} \\ \underline{e}^{(k+1)} = (I - L)^{-1} U \underline{e}^{(k)} + (I - L)^{-1} \underline{\sigma}^{(k)} , & \text{(Gauss-Seidel)} \end{cases}$$

where it can be shown that

$$\| \underline{\rho}^{(k)} \|_\infty \le C_1 \| \underline{e}^{(k)} \|_\infty^2 , \quad \| \underline{\sigma}^{(k)} \|_\infty \le C_2 \| \underline{e}^{(k)} \|_\infty^2 .$$

It can also be shown, using (3.4) and (3.2), that the non-zero off-diagonal coefficients of the strictly lower and upper $(n-2) \times (n-2)$ matrices L and U can be expressed as

$$\text{(3.15)} \quad \begin{cases} l_i = \dfrac{-a_{i-1}}{a_{i-1} + a_i + b_i/d_i^2} , & i = 3,\ldots,n-1 , \\[2ex] u_i = \dfrac{-a_i}{a_{i-1} + a_i + b_i/d_i^2} , & i = 2,\ldots,n-2 . \end{cases}$$

Equations (3.14) show that the asymptotic convergence rates are dependent on the spectral radii $\rho(L+U)$, for the Jacobi method, and $\rho((I-L)^{-1}U)$, for the Gauss-Seidel method. These spectral radii give the R_1 convergence factors of the methods, as defined by Ortega and Rheinboldt [5,Chap.9 and 10]. From (3.11) and (3.12) it follows that

$$\text{(3.16)} \quad \rho(L+U) \le \| L+U \|_\infty \le \tfrac{1}{2}(1 + 1/\sqrt{5}) < 1$$

and since $I - L - U$ is a tri-diagonal matrix with non-zero diagonal, the 2-cyclic case of linear successive over-relaxation (SOR) applies, see Varga [8,Chap.4]. Thus

$$\text{(3.17)} \quad \rho((I-L)^{-1}U) = \rho^2(L+U) ,$$

It can also be shown that the spectral radii are non-zero, since $L + U$ has non-zero off-diagonal elements all of the same sign. Thus the iterative methods are linearly order convergent and in view of (3.17) and (3.16) we have the following theorem.

Theorem 3.3 *The Gauss-Seidel iteration converges asymptotically faster than the Jacobi iteration.*

Linear SOR theory also suggests the accelerated iteration

$$\text{(3.18)} \quad x_i^{(k+1)} = (1-\omega)x_i^{(k)} + \omega G_i(x_{i-1}^{(k+1)}, x_{i+1}^{(k)}) , \quad i = 2,\ldots,n-1 , \quad \text{(SOR } \textit{iteration}) ,$$

where

$$\omega = 2/\{1 + [1 - \rho^2(L+U)]^{1/2}\} .$$

This value of ω minimizes the R_1 convergence factor of the iteration, i.e. it minimizes the spectral radius $\rho(I - \omega L)^{-1}(\omega U + (1-\omega)I))$. The SOR iteration may be worthy of further investigation, although the numerical results in [1] indicate that the Gauss-Seidel method converges well in practice. Second order convergent methods might also be investigated, but the Gauss-Seidel iteration is simple, robust, and the global convergence of the algorithm makes it particularly attractive.

4. CONVEX RATIONAL SPLINE EQUATIONS

Assume that the data satisfy the convex constraint (1.3) and let $s(x)$ be the piecewise rational cubic with r_i defined by (2.6). Given $d_1 < \Delta_1$ and $d_n > \Delta_n$ we seek derivative values d_i, $i = 2,\ldots,n-1$, which satisfy the C^2 consistency constraints

$$s^{(2)}(x_{i-}) = s^{(2)}(x_{i+}) \ , \quad i = 2,\ldots,n-1 \ ,$$

and the necessary convexity condition (1.6). The consistency constraints give the non-linear system

$$(4.1) \qquad \left(\frac{\Delta_i - d_i}{d_i - \Delta_{i-1}}\right)^2 = \frac{h_i}{h_{i-1}} \left(\frac{d_{i+1} - \Delta_i}{\Delta_{i-1} - d_{i-1}}\right) , \quad i = 2,\ldots,n-1 \ ,$$

where, for d_i satisfying the convexity conditions, we must have $(\Delta_i - d_i)/(d_i - \Delta_{i-1}) > 0$ and $(d_{i+1} - \Delta_i)/(\Delta_{i-1} - d_{i-1}) > 0$. Thus we consider the positive square root of (4.1) and then

$$(4.2) \qquad d_i = G_i(d_{i-1},d_{i+1}) \ , \quad i = 2,\ldots,n-1 \ ,$$

where

$$(4.3) \qquad G_i(\xi,\eta) = \frac{h_{i-1}{}^{1/2}(\Delta_{i-1} - \xi)^{1/2}\Delta_i + h_i{}^{1/2}(\eta - \Delta_i)^{1/2}\Delta_{i-1}}{h_{i-1}{}^{1/2}(\Delta_{i-1} - \xi)^{1/2} + h_i{}^{1/2}(\eta - \Delta_i)^{1/2}} \ .$$

Now, for $\Delta_{i-1} < d_i < \Delta_i$, $i = 2,\ldots,n-1$, $G_i(d_{i-1},d_{i+1})$ is a convex combination of Δ_{i-1} and Δ_i, and hence $\Delta_{i-1} < G_i(d_{i-1},d_{i+1}) < \Delta_i$, $i = 2,\ldots,n-1$. More precisely, it can be shown that there exists $\varepsilon > 0$ such that

$$(4.4) \qquad \Delta_{i-1} + \varepsilon \leq G_i(d_{i-1},d_{i+1}) \leq \Delta_i - \varepsilon \ , \quad i = 2,\ldots,n-1 \ ,$$
$$\forall \ \Delta_{i-1} + \varepsilon \leq d_i \leq \Delta_i - \varepsilon \ , \quad i = 2,\ldots,n-1 \ .$$

Since (4.4) defines a mapping of closed intervals into themselves, it follows that there exist d_i, $i = 2,\ldots,n-1$, contained in these intervals such that the non-linear system (4.2) is satisfied.

The above discussion assures the existence of a solution satisfying both the consistency and convexity conditions. The uniqueness of the solution can also be shown but an iterative method which guarantees convergence to this solution has yet to be found. This problem is now being investigated, as is also the error analysis of the quantities $d_i - f_i^{(1)}$.

5. EXAMPLES

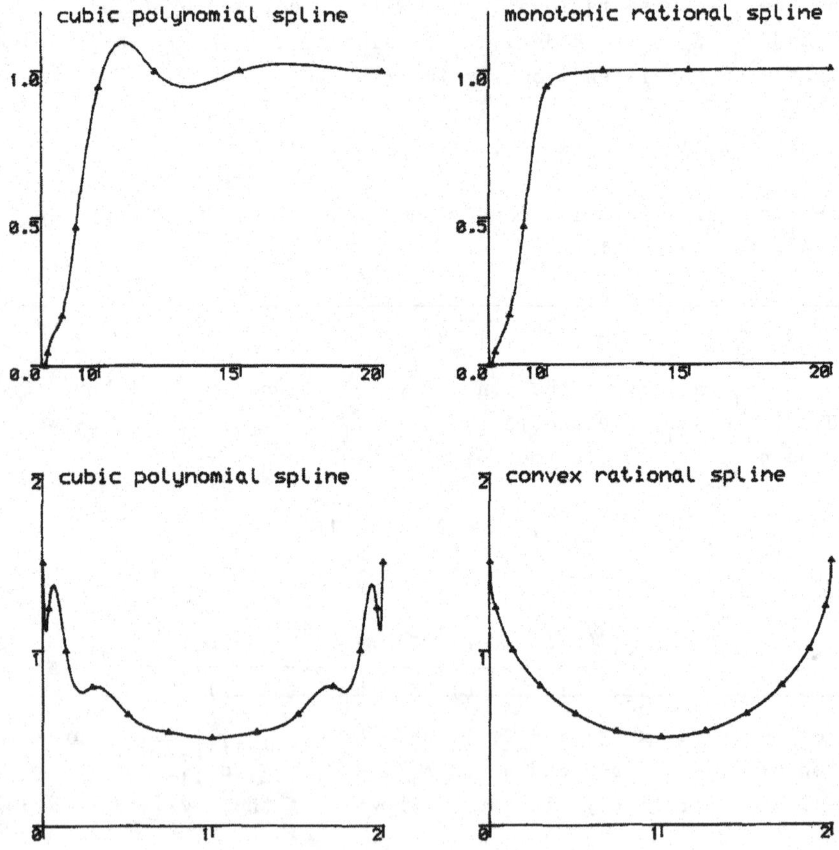

FIGURES. Results for a monotonic and a convex data set

Remarks The monotonic data set is one used by Fritsch and Carlson [3], with zero derivative end conditions. The convex data set is obtained from a semi-circle of radius unity, with the derivative end conditions -50 and +50 replacing the infinite gradients. The convex rational spline equations are solved by a Gauss-Seidel iteration based on equations (4.2).

REFERENCES

1. Delbourgo, R. and Gregory, J.A., C^2 rational quadratic spline interpolation to monotonic data, IMA J. Num. Analysis 3 (1983), 141-152.
2. Delbourgo, R. and Gregory, J.A., Shape preserving piecewise rational interpolation, to appear in SIAM J. Sci. Stat. Comput.
3. Fritsch, F.N. and Carlson, R.E., Monotone piecewise cubic interpolation, SIAM J. Num. Analysis 17 (1980), 235-246.
4. Gregory, J.A. and Delbourgo, R., Piecewise rational quadratic interpolation to monotonic data, IMA J. Num. Analysis 2 (1982), 123-130.
5. Ortega J.M. and Rheinboldt, W.C., Iterative Solution of Nonlinear Equations in Several Variables, Academic Press, New York, 1970.
6. Schaback, R., Spezielle rationale splinefunktionen, J. Approx. Theory 7 (1973), 281-292.
7. Schaback, R., Interpolation mit nichtlinearen klassen von splinefunktionen, J. Approx. Theory 8 (1973), 173-188.
8. Varga, R., Matrix Iterative Analysis, Prentice-Hall, Englewood Cliffs, 1962.

AN APPLICATION OF GAUSSIAN ELIMINATION TO INTERPOLATION BY GENERALIZED RATIONAL FUNCTIONS

T. Håvie
UNIT/NTH
Alfred Getz vei 1
N7034 Trondheim
Norway

M.J.D. Powell
DAMTP
Silver Street
Cambridge CB3 9EW
England

Abstract. We consider the calculation of $r(\xi)$, where ξ is a given number, and where $\{r(x)=p(x)/q(x);\ x\in\mathbb{R}\}$ is a generalized rational function whose coefficients should satisfy some interpolation conditions. We study a procedure that obtains $r(\xi)=p(\xi)/q(\xi)$ by applying Gaussian elimination to remove the unknown coefficients from a system of linear equations. It is shown that the procedure breaks down only if $r(\xi)$ or the coefficients of the rational function are not properly defined. It is proved that the intermediate equations of Gaussian elimination are related to rational interpolating functions that depend on subsets of the coefficients and data. A numerical example demonstrates the procedure.

1. Introduction

We consider interpolation by rational functions of the form

$$r(x) = p(x)/q(x)$$
$$= \sum_{j=0}^{m} a_j p_j(x) \Big/ \sum_{j=0}^{n} b_j q_j(x)\ , \qquad x\in\mathbb{R}\ , \tag{1}$$

where $p(.)$ and $q(.)$ belong to general linear spaces of dimensions $(m+1)$ and $(n+1)$, that are spanned by the functions $\{p_j(.);\ j=0,1,\ldots,m\}$ and $\{q_j(.);\ j=0,1,\ldots,n\}$ respectively. Due to homogeneity the coefficients $\{a_j;\ j=0,1,\ldots,m\}$ and $\{b_j;\ j=0,1,\ldots,n\}$ give only $(m+n+1)$ independent degrees of freedom in the rational function. Therefore in the interpolation problem $r(.)$ is usually determined by the conditions

$$r(x_t) = f(x_t)\ ; \qquad t=0,1,\ldots,m+n\ , \tag{2}$$

where the interpolation points $\{x_t;\ t=0,1,\ldots,m+n\}$ and the function values $\{f(x_t);\ t=0,1,\ldots,m+n\}$ are data. These conditions imply the equations

$$\sum_{j=0}^{m} a_j p_j(x_t) - f(x_t)\sum_{j=0}^{n} b_j q_j(x_t) = 0,\quad t=0,1,\ldots,m+n, \tag{3}$$

which are linear constraints on the coefficients of $r(.)$. A frequently used procedure for calculating $r(.)$, which is not always successful, sets $b_o=1$, and then obtains the remaining coefficients of $r(.)$ by solving the linear system (3).

If, instead of the coefficients of $r(.)$, one requires the value of $r(\xi)$, where ξ is a prescribed number that is not a data point, then there are some advantages in augmenting the system (3) by the equations

$$\sum_{j=0}^{m} a_j p_j(\xi) = p(\xi) \tag{4}$$

and

$$\sum_{j=0}^{n} b_j q_j(\xi) = q(\xi). \tag{5}$$

One may apply Gaussian elimination to form a nontrivial linear combination of the equations (3) - (5) whose left hand side is zero, which usually determines $r(\xi) = p(\xi)/q(\xi)$ without the calculation of $\{a_j;\ j=0, 1,\ldots,m\}$ and $\{b_j;\ j=0,1,\ldots,n\}$. However, if values of these coefficients are required, they can be found by back-substitution in the triangular system of equations that is given by Gaussian elimination.

We study this method for calculating $r(\xi)$ and find that it has some interesting and useful properties. Section 2 considers some details of implementation, and gives particular attention to a pivoting procedure and possible causes of failure. In Section 3 it is shown that the equations that arise during Gaussian elimination are related to rational interpolating functions that depend on subsets of the coefficients and data. Finally Section 4 gives an example of the procedure and discusses some of its features.

Because our method requires $O([m+n]^3)$ computer operations, it is inefficient when $p(.)$ and $q(.)$ are algebraic polynomials of degrees at most m and n respectively. Suitable algorithms for the polynomial case are proposed by Graves-Morris [1], Larkin [3], Werner [4] and Wuytack [6].

It is shown by Håvie [2] that the calculation of $r(\xi)$ may be organized in a manner similar to the one used in the classical algorithms for linear iterative polynomial interpolation due to Neville and Aitken. One computes a lower triangular table of rational approximations of increasing order, each element in the table being a rational function interpolating f at a subset of the interpolation points defined in (2), the last element in the main diagonal being the rational approximation defined by (3) - (5).

The advantage of this approach is that computing the table row by row one may, by inspection in the table, terminate the calculation when a wanted accuracy is obtained. The calculation may also be organized in such a way that the row and column elements in the system (3) - (5) are not computed before needed in the Gaussian elimination process. This approach may give a considerable saving of computing time as compared to solving the complete system (3) - (5). A disadvantage of this procedure is, however, that since the pivoting strategy is fixed, the calculation may be more unstable than the procedure that is discussed in the next section.

2. The calculation of $r(\xi)$

In order to preserve accuracy and to prevent unnecessary failure during the elimination of the coefficients $\{a_j;\ j=0,1,\ldots,m\}$ and $\{b_j;\ j=0,1,\ldots,n\}$ from equations (3) - (5), it may be important to include a pivoting procedure in the calculation of $r(\xi)$. We recommend, therefore, that the calculation begin by scaling equations and coefficients so that the matrix on the left hand side of the system (3) - (5) satisfies the following conditions. The modulus of every matrix element is at most one, and in each row and column of the matrix there is at least one element of modulus one. Further, by letting expressions (4) and (5) be the first two equations of the system, and by re-ordering columns, one can ensure that the matrix has the form

$$\begin{pmatrix} \pm 1 & 0 & x & \cdots & x \\ 0 & \pm 1 & x & \cdots & x \\ x & x & x & \cdots & x \\ \vdots & \vdots & \vdots & & \vdots \\ x & x & x & \cdots & x \end{pmatrix}, \tag{6}$$

where x denotes an element that may be zero or non-zero.

It is now suitable to apply Gaussian elimination with column pivoting. We note that this procedure chooses the first two pivots to be the elements that are shown as ±1 in the matrix (6). Therefore, after the two coefficients that correspond to the first two columns of the matrix (6) have been eliminated, it is usual for non-zero multiples of $p(\xi)$ and $q(\xi)$ to occur on the right hand side of every equation, which is important to the observations of Section 3. As noted in Section 1, at the end of the elimination calculation one has an equation whose left hand side is identically zero, so usually the right hand side is a linear combination of $p(\xi)$ and $q(\xi)$ that determines the required ratio $r(\xi) = p(\xi)/q(\xi)$.

If every pivot is non-zero, and if the final equation is the expression

$$0 = \bar{\rho}p(\xi) + \bar{\sigma}q(\xi) \quad , \tag{7}$$

where $\bar{\rho}$ and $\bar{\sigma}$ are numbers that are computed by the elimination calculation, then it can be deduced from the backward error analysis of Wilkinson [5] that the contribution to $\bar{\rho}$ and $\bar{\sigma}$ from computer rounding errors can be accounted for by small changes to the values of the functions $\{p_j(.); j=0,1,\ldots,m\}$ and $\{q_j(.); j=0,1,\ldots,n\}$ that occur in equations (3) - (5). Therefore, if these function values are found to normal accuracy by a subroutine that is provided by the user of the interpolation procedure, and if one has no additional information about the numerator and denominator of $r(.)$, then the accuracy of the Gaussian elimination method is similar to the accuracy of the data. It follows that there is little point in choosing new basis functions for the linear spaces that contain $p(.)$ and $q(.)$. However, if it is known that $p(.)$ and $q(.)$ are algebraic polynomials, then, to make full use of this extra information, it may be necessary to work with orthogonal basis functions. Thus an algorithm for general rational interpolation can be much simpler than an algorithm for a special case.

The remainder of this section considers possible failures of the given procedure due to zero matrix elements. If a row or column of the matrix of the system (3) - (5) were identically zero, then the calculation would be terminated by the scaling procedure that normally gives the matrix of form (6). Otherwise, because computer rounding errors tend to prevent the occurrence of numbers that are exactly zero, it is usually possible to complete the calculation. However, the column pivoting procedure may fail to find a non-zero pivot during the Gaussian elimination, or $\bar{\rho}$ and $\bar{\sigma}$ may both be zero in equation (7). In all of these cases the calculation of $r(\xi)$ breaks down, but, remembering the remarks of the previous paragraph, we can defend our procedure by the comment that failure would occur in exact arithmetic for data that is close to the actual data. We do not regard the case when $\bar{\rho}=0$ and $\bar{\sigma}\neq 0$ as unsuccessful, because it indicates that $r(.)$ has a pole at $x = \xi$.

Difficulties should occur if the interpolation conditions (3) do not define the rational function (1), or if $r(.)$ is defined almost everywhere but $p(\xi) = q(\xi) = 0$. We ask whether in these cases exact arithmetic would lead to one of the failures that are mentioned in the previous paragraph. The interpolation conditions (3) do not define $r(.)$ if and only if the last $(m+n+1)$ rows of the $(m+n+3) \times (m+n+2)$ matrix (6) are linearly dependent; moreover a zero pivot occurs during Gaussian elimination if and only if the rank of this matrix is less than $(m+n+2)$. Therefore insufficient interpolation conditions coincide with the completion of Gaussian elimination if and only if the rank of the matrix (6) is exactly $(m+n+2)$, but the last $(m+n+1)$ rows of this matrix span a space of dimension only $(m+n)$. In this case, except for a constant scaling factor, only one non-trivial linear combination of the equations (3) - (5) has a zero left hand side, and the multipliers of equations (4) and (5) are both zero, which is the condition $\bar{\rho}=\bar{\sigma}=0$. Alternatively, if $r(.)$ is defined almost everywhere but $p(\xi)=q(\xi)=0$, then the last $(m+n+1)$ rows of the matrix (6) are linearly independent, but all rows of this matrix are orthogonal to the vector of coefficients that is defined by the interpolation conditions. Hence the columns of the matrix are linearly dependent, so a zero pivot occurs during Gaussian elimination. Thus it is not possible to complete our calculation in exact arithmetic whenever a difficulty should arise.

Conversely, if Gaussian elimination is completed but $\bar{\rho}=\bar{\sigma}=0$, then the interpolation conditions (3) are linearly dependent. Alternatively, if a zero pivot occurs during elimination, then either there is dependence in the interpolation conditions or the first two rows of the matrix (6) are each in the linear space that is spanned by the last $(m+n+1)$ rows, which implies $p(\xi)=q(\xi)=0$. Therefore, assuming that no row or column of the matrix of the system (3) - (5) is identically zero, we claim that our procedure breaks down only when a failure should occur, which is a strong advantage over many other algorithms that have been proposed for rational interpolation.

So far, however, we have not considered two other exceptional cases. One is the possibility that $r(\xi)=p(\xi)/q(\xi)$ is defined by the data but the interpolation conditions do not determine $r(.)$ almost everywhere. It occurs, for example, when ξ is one of the interpolation points $\{x_t;\ t=0,1,\ldots,m+n\}$, but the conditions (3) are linearly dependent. In this case, if all computer arithmetic were exact, one could find a suitable extension to Gaussian elimination, but it may be no more successful than the method that has been described already when computer rounding errors occur. Secondly, the coefficients that give $r(\xi)=p(\xi)/q(\xi)$ may also imply $p(x_t)=q(x_t)=0$ for some t, which does not violate condition (3) but one should doubt whether equation (2) holds. The example of Section 4 shows that this case may escape notice during the calculation of $r(\xi)=p(\xi)/q(\xi)$.

3. The intermediate equations of Gaussian elimination

In this section we study the coefficients of a typical equation that is generated during the Gaussian elimination procedure. It is shown that they are related to rational functions that interpolate the data that are used to derive the coefficients.

Without loss of generality we assume that the intermediate equation has the form

$$\sum_{j=u+1}^{m} a_j\alpha_j + \sum_{j=v+1}^{n} b_j\beta_j = \rho p(\xi) + \sigma q(\xi) , \tag{8}$$

where the numbers $\{\alpha_j;\ j=u+1,u+2,\ldots,m\}$, $\{\beta_j;\ j=v+1,v+2,\ldots,n\}$, ρ and σ are calculated by the Gaussian elimination procedure. Further, we assume that condition (8) is derived by eliminating the coefficients $\{a_j;\ j=0,1,\ldots,u\}$ and $\{b_j;\ j=0,1,\ldots,v\}$ from the equations

$$\left.\begin{aligned} &\sum_{j=0}^{m} a_j p_j(\xi) = p(\xi) \\ &\sum_{j=0}^{n} b_j q_j(\xi) = q(\xi) \\ &\sum_{j=0}^{m} a_j p_j(x_t) - f(x_t)\sum_{j=0}^{n} b_j q_j(x_t) = 0,\ t=0,1,\ldots,k \end{aligned}\right\} , \tag{9}$$

where $k=u+v$. Thus any set of coefficients that satisfies expression (9) also satisfies condition (8). Because a zero pivot causes termination, we also assume without loss of generality that the $(k+2)\times(k+2)$ matrix

$$\begin{pmatrix} p_o(\xi) & \ldots & p_u(\xi) & 0 & \ldots & 0 \\ 0 & \ldots & 0 & q_o(\xi) & \ldots & q_v(\xi) \\ p_o(x_o) & \ldots & p_u(x_o) & -f(x_o)q_o(x_o) & \ldots & -f(x_o)q_v(x_o) \\ \vdots & & \vdots & \vdots & & \\ p_o(x_{k-1}) & \ldots & p_u(x_{k-1}) & -f(x_{k-1})q_o(x_{k-1}) & \ldots & -f(x_{k-1})q_v(x_{k-1}) \end{pmatrix} \tag{10}$$

is nonsingular.

The main result of this section is implied by the following lemma. It is that, if the coefficients of the rational function of form

$$\bar{r}(x) = \sum_{j=0}^{u} \bar{a}_j p_j(x) / \sum_{j=0}^{v} \bar{b}_j q_j(x), \quad x \in \mathbb{R}, \tag{11}$$

are defined by the interpolation conditions of expression (9), then usually the value of $\bar{r}(\xi)$ is $-\sigma/\rho$.

<u>Lemma 1</u> The equation

$$\rho \sum_{j=0}^{u} \bar{a}_j p_j(\xi) + \sigma \sum_{j=0}^{v} \bar{b}_j q_j(\xi) = 0 \tag{12}$$

holds if $\{\bar{a}_j;\ j=0,1,\dots,u\}$ and $\{\bar{b}_j;\ j=0,1,\dots,v\}$ are any coefficients that satisfy the homogeneous conditions

$$\sum_{j=0}^{u} \bar{a}_j p_j(x_t) - f(x_t) \sum_{j=0}^{v} \bar{b}_j q_j(x_t) = 0, \quad t=0,1,\dots,k. \tag{13}$$

<u>Proof</u> The numbers

$$\left.\begin{aligned} &a_j = \begin{cases} \bar{a}_j, & 0 \le j \le u \\ 0, & u < j \le m \end{cases} \qquad b_j = \begin{cases} \bar{b}_j, & 0 \le j \le v \\ 0, & v < j \le n \end{cases} \\ &p(\xi) = \sum_{j=0}^{u} \bar{a}_j p_j(\xi) \qquad q(\xi) = \sum_{j=0}^{v} \bar{b}_j q_j(\xi) \end{aligned}\right\} \tag{14}$$

satisfy expression (9). Therefore they also satisfy equation (8), which gives the required result. □

In order to interpret the multiplier α_ℓ of equation (8), where ℓ is any integer in $[u+1, m]$, we consider a rational function of the form

$$\bar{r}(x) = [p_\ell(x) + \sum_{j=0}^{u} \bar{a}_j p_j(x)] / \sum_{j=0}^{v} \bar{b}_j q_j(x), \quad x \in \mathbb{R}. \tag{15}$$

Because we wish to choose the coefficients so that this $\bar{r}(.)$ not only satisfies the interpolation conditions of expression (9) but also has a pole at $x = \xi$, we require the matrix

$$\begin{pmatrix} 0 & \dots & 0 & q_o(\xi) & \dots & q_v(\xi) \\ p_o(x_o) & \dots & p_u(x_o) & -f(x_o)q_o(x_o) & \dots & -f(x_o)q_v(x_o) \\ \vdots & & \vdots & \vdots & & \vdots \\ p_o(x_k) & \dots & p_u(x_k) & -f(x_k)q_o(x_k) & \dots & -f(x_k)q_v(x_k) \end{pmatrix} \tag{16}$$

to be nonsingular. Therefore the following lemma is useful.

Lemma 2 The matrix (16) is nonsingular if $\rho\neq 0$.

Proof The construction of equation (8) implies that, if $\rho\neq 0$, the vector $(p_o(\xi)\ \ldots\ p_u(\xi)\ 0\ \ldots\ 0)$ is in the row space of the matrix (16). Hence, due to the nonsingularity of the matrix (10), not only are the first (k+1) rows of the matrix (16) linearly independent, but also the last row of expression (16) is linearly independent of the other rows. Therefore this matrix is nonsingular. □

This lemma allows us to define the coefficients of the rational function (15) by the equations

$$\left.\begin{aligned} &\sum_{j=0}^{v} \bar{b}_j q_j(\xi) = 0 \\ &p_\ell(x_t) + \sum_{j=0}^{u} \bar{a}_j p_j(x_t) - f(x_t) \sum_{j=0}^{v} \bar{b}_j q_j(x_t) = 0, \quad t=0,1,\ldots,k \end{aligned}\right\}, \tag{17}$$

and the next lemma relates the numerator of this rational function at $x=\xi$ to α_ℓ.

Lemma 3 The equation

$$\alpha_\ell/\rho = p_\ell(\xi) + \sum_{j=0}^{u} \bar{a}_j p_j(\xi) \tag{18}$$

holds if $\rho\neq 0$ and if the coefficients $\{\bar{a}_j;\ j=0,1,\ldots,u\}$ and $\{\bar{b}_j;\ j=0,1,\ldots,v\}$ are defined by expression (17).

Proof The numbers

$$\left.\begin{aligned} &a_j = \begin{cases} \bar{a}_j, & 0\leq j\leq u \\ 0, & u<j\leq m,\ j\neq\ell \\ 1, & j=\ell \end{cases} \qquad b_j = \begin{cases} \bar{b}_j, & 0\leq j\leq v \\ 0, & v<j\leq n \end{cases} \\ &p(\xi) = p_\ell(\xi) + \sum_{j=0}^{u} \bar{a}_j p_j(\xi) \qquad q(\xi) = 0 \end{aligned}\right\} \tag{19}$$

satisfy expression (9). It follows from equation (8) that the lemma is true. □

The analogous interpretation of β_ℓ, where ℓ is any integer in $[v+1, n]$, is given in Lemma 4, but we omit the proof of this lemma because it is similar to the proofs of Lemmas 2 and 3.

Lemma 4 If $\sigma \neq 0$, then the conditions

$$\left.\begin{aligned} &\sum_{j=0}^{u} \bar{a}_j p_j(\xi) = 0 \\ &\sum_{j=0}^{u} \bar{a}_j p_j(x_t) - f(x_t)[q_\ell(x_t) + \sum_{j=0}^{v} \bar{b}_j q_j(x_t)] = 0, \quad t=0,1,\dots,k \end{aligned}\right\} \quad (20)$$

define the coefficients $\{\bar{a}_j;\ j=0,1,\dots,u\}$ and $\{\bar{b}_j;\ j=0,1,\dots,v\}$ uniquely. Further, the equation

$$\beta_\ell/\sigma = q_\ell(\xi) + \sum_{j=0}^{v} \bar{b}_j q_j(\xi) \quad (21)$$

is satisfied. □

Lemma 1 may be useful for choosing automatically the values of m and n when many function values $\{f(x_t);\ t=0,1,2,\dots\}$ are available, but at present we cannot suggest any applications of Lemmas 3 and 4.

4. An example

The procedure and some of the given results are illustrated by calculating the value at $x = 2$ of the generalized rational function

$$r(x) = [a_o + a_1 x + a_2 (x-2)_+]/[b_o + b_1 x], \quad x \in \mathbb{R}, \quad (22)$$

that interpolates the data $f(0) = 0$, $f(1) = 1$, $f(3) = 3$ and $f(4) = \theta$, where θ is a parameter. Therefore the procedure eliminates a_o, a_1, a_2, b_o and b_1 from the equations

$$\left.\begin{aligned} a_o + 2a_1 \qquad\qquad &= p(2) \\ b_o + 2b_1 \qquad\qquad &= q(2) \\ a_o \qquad\qquad &= 0 \\ (a_o + a_1) - (b_o + b_1) &= 0 \\ (a_o + 3a_1 + a_2) - 3(b_o + 3b_1) &= 0 \\ (a_o + 4a_1 + 2a_2) - \theta(b_o + 4b_1) &= 0 \end{aligned}\right\} . \quad (23)$$

We omit the recommended pre-scaling and pivoting and we let the order of the coefficients be a_o, b_o, a_1, b_1 and a_2. It follows that the intermediate equations

$$\left.\begin{aligned} -2a_1 \qquad\quad &= -p(2) \\ -a_1 + b_1 \qquad &= -p(2) + q(2) \\ a_1 + a_2 - 3b_1 &= -p(2) + 3q(2) \\ 2a_1 + 2a_2 - 2\theta b_1 &= -p(2) + \theta q(2) \end{aligned}\right\} \quad (24)$$

,

$$\left.\begin{aligned} b_1 \qquad &= -\tfrac{1}{2}p(2) + q(2) \\ a_2 - 3b_1 &= -1\tfrac{1}{2}p(2) + 3q(2) \\ 2a_2 - 2\theta b_1 &= -2p(2) + \theta q(2) \end{aligned}\right\} \quad (25)$$

and

$$\left.\begin{aligned} a_2 &= -3p(2) + 6q(2) \\ 2a_2 &= (-2-\theta)p(2) + 3\theta q(2) \end{aligned}\right\} \quad (26)$$

are generated by Gaussian elimination, which leads to the identity

$$0 = (4-\theta)p(2) + (3\theta - 12)q(2) . \quad (27)$$

Therefore the required value of the rational function is $r(2)=$ $=p(2)/q(2)=3$, except that the calculation breaks down when $\theta=4$.

It is remarkable that, for $\theta\neq 4$, the coefficients of $r(.)$ and $r(2)$ are independent of θ. The generalized rational interpolant is

$$r(x) = [3x - 6(x-2)_+]/[4-x] , \quad x\in\mathbb{R} , \quad (28)$$

but at $x=4$ a 0/0 singularity occurs so $r(x)$ is defined by continuity. Thus the coefficients $a_o=0$, $a_1=3$, $a_2=-6$, $b_o=4$ and $b_1=-1$ satisfy the interpolation conditions (3) for all values of θ. Unlike the polynomial case, the 0/0 singularity does not allow the degrees of the numerator and denominator of $r(.)$ to be reduced. Further, as mentioned at the end of Section 2, it is unlikely that the singularity will be noticed during the calculation of $r(2)$.

The example also illustrates the remark in Section 2 that, if the Gaussian elimination is completed but $\bar\rho=\bar\sigma=0$ in equation (7), then the interpolation conditions (3) are linearly dependent. Specifically expression (23) shows that the interpolation conditions are dependent if the rank of the matrix

$$\begin{pmatrix} 1 & 0 & 0 & 0 & 0 \\ 1 & 1 & 0 & -1 & -1 \\ 1 & 3 & 1 & -3 & -9 \\ 1 & 4 & 2 & -\theta & -4\theta \end{pmatrix} \quad (29)$$

is only 3, which happens when $\theta=4$.

Of course one may also use the example to check the lemmas of Section 3. For instance, Lemma 1 and expression (26) imply that the ratios $p(2)/q(2)=2$ and $p(2)/q(2)=3\theta/(2+\theta)$ are the values at $x=2$ of the 1 - 1 polynomial rational functions that interpolate the first, second and third data values and the first, second and fourth data values respectively. In fact these rational functions are $\{r(x) = x; x \in \mathbb{R}\}$ and $\{r(x) = 3\theta x/[4\theta - 4 + (4 - \theta)x]; x \in \mathbb{R}\}$, which agree with Lemma 1.

There is no need to include the first k interpolation conditions when one eliminates (k + 2) coefficients from (k + 3) equations of the system (3) - (5). Instead one may prefer, for example, to use equations (4) and (5) and (k + 1) consecutive interpolation conditions, which could give an algorithm that is analogous to one that is proposed by Larkin [3]. Further consideration of different algorithms that come from different combinations of interpolation conditions can be found in Håvie [2].

Acknowledgement

The work of T. Håvie was supported by a NATO Research Grant (No. 027-81).

References

1. Graves-Morris, P.R., Practical reliable rational interpolation, J. Inst. Math. Applic., 25 (1980), 267-286.

2. Håvie, T., Two algorithms for iterative interpolation and extrapolation using generalized rational functions, Math. and Comp. Report 4/83, ISBN 82-7151-056-8, University of Trondheim, 1983.

3. Larkin, F.M., Some techniques for rational interpolation, Comput. J., 10 (1967), 178-187.

4. Werner, H., A reliable method for rational interpolation, in Padé Approximation and its Applications, Lecture Notes in Mathematics No. 765, ed. Wuytack, L., Springer-Verlag, Berlin, 1979.

5. Wilkinson, J.H., The Algebraic Eigenvalue Problem, Oxford University Press, Oxford, 1965.

6. Wuytack, L., An algorithm for rational interpolation similar to the q-d algorithm, Numer. Math., 20 (1973), 418-424.

PROBLEMS IN NUMERICAL CHEBYSHEV APPROXIMATION BY INTERPOLATING RATIONALS

B. Nelson and Jack Williams,
Department of Mathematics
University of Manchester,
Manchester M13 9PL,
England.

Abstract. We describe some theory and practice for the problem of real Chebyshev approximation to a continuous function $f(x)$ whose zeros (if any) in the range of interest are known. Our typical approximant is of the form $B(x)(P(x)/Q(x))^p$, where $B(x)$ is a specified continuous function having the same zeros as $f(x)$, p is specified and $P(x)$, $Q(x)$ are polynomials.

1. Introduction

In Williams [9] the practical problem of approximating oscillatory decay-type functions was considered as a particular case of the following general problem. Let $D[0, b]$, $0 < b < \infty$, denote the class of real continuous functions of the form $f(x) = B(x)g(x)$, where B, $g \in C[0, b]$, $g > 0$ $\forall x \in [0, b]$ and where $B(x)$ has finitely many zeros $x_1, x_2, \ldots, x_s$ in $[0, b]$. Given $f \in D[0, b]$, n and p, $0 < p < \infty$, let $V(n, p)$ denote the class of interpolating rationals of the form

$$F(A, x) = \frac{B(x)}{L(A,x)^p}, \quad x \in [0, b],$$

in which $L(A, x) = \sum_{r=1}^{n} a_r\phi_r(x)$, and the continuous functions $\phi_r(x)$, $r = 1, 2, \ldots, n$; $\phi_1 \equiv 1$, form a Chebyshev set on $[0, b]$; also $L(A, x) > 0$ on $[0, b]$. With $\|.\|$ the Chebyshev norm on $[0, b]$, $F(A^*, x) \in V(n, p)$ is a best Chebyshev approximation to f if

$$\|f - F(A^*, x)\| \leq \|f - F(A, x)\| \; \forall F(A, x) \in V(n, p).$$

Existence of $F(A^*, x)$ is not straightforward. Useful sufficient conditions expressed in terms of $B(x)$ are given by Taylor and Williams [8]. They show that if $\phi_r(x)$, $r = 2, 3, \ldots, n$ are analytic in open discs with centres at the interior zeros $x_\nu \in (0, b)$ and

i) $\lim_{x \to x_\nu \pm} |B(x)|/|x - x_\nu|^p = \infty$ if $x_\nu = 0$ or b,

ii) $\lim_{x \to x_\nu} |B(x)|/|x - x_\nu|^{2p} = \infty$ if $x_\nu \in (0, b)$,

then a best approximation exists. Less useful conditions on g are also given in [8].

Characterization of a best approximation in terms of a modified alternation property, along with uniqueness are also given in [9]. The characterization is then used as the basis for a multiple point exchange algorithm (Remes algorithm) for computing a best approximation. Here we describe some work which extends the algorithm described in [9] to the case of approximation from the more general class of rationals

$$V(m,n,p) := \{F(A,x) = B(x)\left(\frac{P(x)}{Q(x)}\right)^p, \; P(x) \geq 0, \; Q(x) > 0, \; x \in [0, b]\}.$$

Here $p > 0$ is specified and the parameters A are the coefficients of P and Q, *ordinary* polynomials of specified degrees m and n respectively. The case $m = 0$ with the normalisation $P(x) \equiv 1$ corresponds to the Williams problem with $\phi_r(x) \equiv x^{r-1}$. We note here that P/Q (understood to be irreducible) is a *nonnegative rational*, so that when $B(x) \equiv 1$, $p = 1$, the corresponding approximation problem is not equivalent to conventional rational approximation in which the numerator polynomial may take negative values. In this case, however, $f \equiv g > 0$, and the problems, at least for high enough m, n, will actually be equivalent. This follows since continuous functions on $[0, b]$ are uniformly approximable by ordinary rationals; so ultimately a best approximation must be positive.

Best Chebyshev approximation by $V(m, n, p)$ was introduced by Schmidt [7], who generalised the existence theorem (involving $B(x)$ above) of Taylor and Williams. Schmidt also showed that the conditions on $B(x)$ are actually essential when $m \geq 2$ and $n \geq 2$, in the sense that, if they are not satisfied there exists an $f \in D[0, b]$, that does not have a best approximation in $V(m, n, p)$. In addition the characterization (by modified alternation) and uniqueness results of Williams [9] are also generalised.

It therefore seems quite natural to try to extend the algorithm of [9] to the family $V(m, n, p)$. In Section 2 the exchange algorithm for this problem is outlined, a key part of which is the numerical solution of the levelling equations. Section 3 discusses important questions of the existence and uniqueness of solutions of these equations; their numerical solution is then treated in Section 4. Finally in Section 5 we present some illustrative numerical examples.

2. The Exchange Algorithm

We now assume throughout that best approximations exist. The numerical method for computing a best approximation $F(A^*, x) = B(x)(P^*(x)/Q^*(x))^p$ is based on the following characterization theorem.

Theorem 1 (Schmidt [7]). $F(A^*, x) \in V(m, n, p)$ is a best approximation to $f \in D[0, b]$ if and only if there exist $d + 1$ points $0 \leq x_1 < x_2 < \dots x_{d+1} \leq b$ such that

$$|f(x_r) - F(A^*, x_r)| = \|f - F(A^*, x)\|, \quad r = 1, 2, \dots, d + 1,$$

and

$$\operatorname{sign}(f(x_r))[f(x_r) - F(A^*, x_r)] = -\operatorname{sign}(f(x_{r+1})) \times [f(x_{r+1}) - F(A^*, x_{r+1})], \qquad r = 1, 2, \dots, d,$$

where $d \equiv d(A^*) = 1 + \max\{m + \deg Q^*, n + \deg P^*\}$.

It is clear that the above set of points (the extremal set) cannot contain a point at which $f(x)$ vanishes. In order to apply the exchange algorithm it is necessary to assume that $f \in D[0, b]$ is a normal point, that is, P^*/Q^* is not degenerate, so the number of points in the reference is given by $N = 1 + d(A^*) = m + n + 2$. Following Williams [9] an initial reference $X^{(0)} := \{x_1^{(0)}, x_2^{(0)}, \dots, x_N^{(0)}\} \subset [0, b]$ is selected which represents an approximation to the extremal set. The kth iteration of the algorithm is as follows.

Stage 1 Solution of the levelling equations on the reference $X^{(k)}$. An approximation $F(A_k, x) \in V(m, n, p)$ is computed which satisfies

$$E_k(x_r^{(k)}) \equiv \mathrm{sign}(f(x_r^{(k)}))[f(x_r^{(k)}) - F(A_k, x_r^{(k)})] = (-)^r \lambda_k,$$

$$r = 1, 2, \ldots, N$$

Stage 2 Update reference $X^{(k)}$. The extrema of $E_k(x)$ are found from which $X^{(k+1)}$ is selected, satisfying

i) $E_k(x)$ alternates in sign on $X^{(k+1)}$,

ii) $\min_{x \in X^{(k+1)}} |E_k(x)| \geq |\lambda_k|$ and $X^{(k+1)}$ contains a point at which $|E_k(x)| = \|E_k\|$.

Here we wish to concentrate on Stage 1; Stage 2 considerations are briefly discussed in [9]. The convergence of the algorithm rests essentially on whether for each k the levelling equations have a solution contained in $V(m, n, p)$. If this can be guaranteed, then it is clear from the fundamental paper of Barrar and Loeb [2], (also see Burke [4]), that for f a normal point of $D[0, b]$, the algorithm yields $\{F(A_k, x)\}$ satisfying $\|f - F(A_k, x)\| \to \|f - F(A^*, x)\|$ with a geometric rate of convergence. In [2] the proofs apply to those approximating functions for which the best approximation is characterised by standard Chebyshev alternation. The modified alternation property for the family $V(m, n, p)$ does not affect the arguments. The convergence proof requires that $X^{(0)}$ is sufficiently close to the extremal set. As numerical examples show, in practice a poor approximation $X^{(0)}$ can lead immediately (or subsequently) to an approximation with a pole in $[0, b]$.

An important device in monitoring the convergence of the exchange algorithm is the following de la Vallée Poussin result; the algorithm can therefore be terminated when the error curve $E_k(x)$ is sufficiently levelled. The result is implicit in the work of Schmidt [7].

Lemma 1 If for $F(A, x) \in V(m, n, p)$ there exist $s = d(A) + 1$ points, $0 \leq x_1 < x_2 < \ldots < x_s \leq b$, such that $\mathrm{sign}(f(x_r))[f(x_r) - F(A, x_r)] = -\mathrm{sign}(f(x_{r+1}))[f(x_{r+1}) - F(A, x_{r+1})]$, $r = 1, 2, \ldots, s-1$, then

$$\min_{1 \leq r \leq s} |f(x_r) - F(A, x_r)| \leq \|f - F(A^*, x)\| \leq \|f - F(A, x)\| .$$

3. Solution of the Levelling Equations, Existence and Uniqueness.

If now $0 \leq x_1 < x_2 < \ldots < x_N \leq b$ denotes a typical reference, the levelling equations are of the general form

$$(1)\quad \operatorname{sign}(f(x_r))\left[f(x_r) - B(x_r)\left(\frac{P(x_r)}{Q(x_r)}\right)^p\right] = (-)^r\lambda, \quad r = 1, 2,\dots,N,$$

which can be written,

$$(2)\quad P(x_r) - \alpha_r(\lambda)^{-1}Q(x_r) = 0, \quad \alpha_r(\lambda)^{-1} = \left(\frac{|f(x_r)| - (-)^r\lambda}{|B(x_r)|}\right)^{1/p}.$$

We are now concerned with questions of existence and uniqueness of solutions of (2). There appear to be two approaches. Firstly, with the normalisation $Q(0) = 1$ (appropriate since $x \in [0, b]$), the system (2) can be treated as N algebraic equations

$$(3)\quad G_r(\underline{a}, \underline{b}; \lambda) = 0, \quad r = 1, 2,\dots,N$$

where $\underline{a} = (a_0, a_1,\dots,a_m)^T$, $\underline{b} = (b_1, b_2,\dots,b_n)^T$ are the unknown coefficients of $P(x)$ and $Q(x)$ respectively. Secondly, we can regard (2) as a homogeneous system

$$(4)\quad B\begin{pmatrix}\underline{a}\\ b_0\\ \underline{b}\end{pmatrix} = \operatorname{diag}(\alpha_r(\lambda)^{-1})C\begin{pmatrix}\underline{a}\\ b_0\\ \underline{b}\end{pmatrix}$$

where $B = (V_m|0)$, $C = (V_n|0)$ are square matrices of size N with $V_t = (x_r^{j-1})$ having $t + 1$ columns.

To solve (2) numerically we propose to apply Newton's method to (3), but use (4) to investigate existence and uniqueness of solutions. We seek solutions with $P(x_r) \geq 0$, $Q(x_r) > 0$, therefore λ must satisfy $\lambda \in [-\rho_1, \rho_2]$ where

$$\rho_1 = \min_{r\ \text{odd}} |f(x_r)|, \quad \rho_2 = \min_{r\ \text{even}} |f(x_r)|.$$

In fact this is necessary for certain p in order for $\alpha_r(\lambda)^{-1}$ to be real-valued. Also, from a practical point of view, if a solution satisfies $\lambda = -\rho_1$ or $\lambda = \rho_2$ then $P(x_r) = 0$ for some r. It therefore seems probable that $P(x)$ changes sign in $[0, b]$ which corresponds to $F(A, x) \notin V(m, n, p)$. We therefore seek solutions that satisfy $\lambda \in (-\rho_1, \rho_2)$ and hence $\alpha_r(\lambda)^{-1} > 0$, $r = 1, 2,\dots,N$. For a nontrivial solution of (4), λ must now satisfy

$$\det[B - \mathrm{diag}(\alpha_r(\lambda)^{-1})C] = 0,$$

or equivalently,

$$(5) \quad F(\lambda) \equiv \det[-\mathrm{diag}(\alpha_r(\lambda))B + C] = 0.$$

The case $m = 0$ has been treated by Williams [9, Thm. 4.1] by a different approach. Here (5) gives for $m = 0$,

$$F(\lambda) = \begin{vmatrix} -\alpha_1(\lambda) & 1 & x_1 & x_1^2 & .. & x_1^n \\ -\alpha_2(\lambda) & 1 & x_2 & x_2^2 & .. & x_2^n \\ . & . & . & . & & . \\ . & . & . & . & & . \\ -\alpha_N(\lambda) & 1 & x_N & x_N^2 & .. & x_N^n \end{vmatrix}$$

$$= \sum_{r=1}^{N} (-)^r \alpha_r(\lambda) v_r$$

where v_r is a Vandermonde determinant of size $N - 1$. Since the reference points are ordered, $v_r > 0$. For $\lambda \in (-\rho_1, \rho_2)$, $F(\lambda)$ is continuously differentiable and $F'(\lambda) > 0$. Since

$$\lim_{\lambda_+ \to -\rho_1} F(\lambda) = -\infty, \qquad \lim_{\lambda_- \to \rho_2} F(\lambda) = +\infty,$$

there exists a unique $\lambda^* \in (-\rho_1, \rho_2)$ satisfying $F(\lambda^*) = 0$.

For general $m > 0$, $n > 0$ and an *arbitrary reference*, we are unable to make any conclusive statements about the existence of a solution of (5). For uniqueness we have the following

Theorem 2 Let $(P_1(x)/Q_1(x), \lambda_1)$, $\lambda_1 \in (-\rho_1, \rho_2)$ be a solution of (2) with P_1/Q_1 nondegenerate, $Q_1(0) = 1$, which satisfies $Q_1(x_r) > 0$, $r = 1, 2, \ldots, N$. Then there are no other normalised solutions which are positive on the reference $\{x_r\}$.

Proof Assume $(P_2/Q_2, \lambda_2)$ is also a solution with $\lambda_2 \in (-\rho_1, \rho_2)$ $Q_2(0) = 1$, $Q_2(x_r) > 0$, $r = 1, 2, \ldots, N$. For $r = 1, 2, \ldots, N$,

$$\frac{P_1(x_r)}{Q_1(x_r)} = \alpha_r(\lambda_1)^{-1}, \quad \frac{P_2(x_r)}{Q_2(x_r)} = \alpha_r(\lambda_2)^{-1}.$$

With $\lambda \in (-\rho_1, \rho_2)$, $\alpha_r(\lambda) > 0$ is monotonically increasing (decreasing) for r even (odd) and hence $P_1Q_2 - P_2Q_1$ has $N - 1$ zeros in $[0, b]$. There follows

$$P_1(Q_2 - Q_1) - Q_1(P_2 - P_1) \equiv 0$$

which is an element of the linear space with basis

$$xP_1,\ x^2P_1,\ldots,x^nP_1;\quad Q_1,\ xQ_1,\ldots,x^mQ_1.$$

Since P_1/Q_1 is nondegenerate this is a Chebyshev set (Cheney [5]) and so $Q_2 \equiv Q_1$, $P_2 \equiv P_1$; $\lambda_2 \equiv \lambda_1$ via $\alpha_r(\lambda)$ being one to one.

We conclude that if such a solution $(P_1/Q_1, \lambda_1)$ has been computed, there are no other acceptable solutions. We cannot, however, guarantee that $B.(P_1/Q_1)^p \in V(m, n, p)$; this must be checked numerically by showing that Q_1 has no zeros in $[0, b]$. Numerical examples certainly show that there can be cases in which a solution of the levelling equations is not contained in $V(m, n, p)$.

4. Numerical Solution of the Levelling Equations

We have successfully applied Newton's method to the levelling equations in the form (3). Convergence depends of course on the use of sufficiently good starting approximations for $\underline{a}$, $\underline{b}$ ($\lambda \approx 0$ is satisfactory). This can be a difficulty with the initial reference $X^{(0)}$, but the equations for subsequent references can usually be successfully solved with previously computed solutions as starting values. We are unable to offer a general device for generating starting approximations for the case $X^{(0)}$. In some numerical experiments with $X^{(0)}$ in which starting approximations were all unsuccessful, the only way to proceed was by actually changing $X^{(0)}$ on the basis that it must be a poor approximation to the extremal set.

It is necessary to adapt the convergence criterion in Newton's method to the actual character of the original levelling equations. For the exchange algorithm to proceed, the error curve $E_k(x)$ must alternate in sign on the current reference, with approximately equal magnitude. Thus for a reference $0 \leqslant x_1 < x_2 < \ldots x_N \leqslant b$, if $\underline{a}^{(\nu)}$, $\underline{b}^{(\nu)}$, $\lambda^{(\nu)}$ denote a current Newton iterate the convergence test is based on residuals of the original system (1). If we define ε_r by

$$\varepsilon_r = |f(x_r)| - |B(x_r)|\left|\frac{P^{(\nu)}(x_r)}{Q^{(\nu)}(x_r)}\right|^p,\quad r = 1, 2,\ldots,N,$$

then the iteration is terminated when the following conditions are satisfied.

i) $\operatorname{sign}(\varepsilon_r) = (-)^r \operatorname{sign}(\lambda^{(\nu)})$,

ii) $\dfrac{\left| |\lambda^{(\nu)}| - |\varepsilon_r| \right|}{|\varepsilon_r|} < \mathrm{EPS}.$

In practice we found EPS = .001 satisfactory, and it could be reduced.

Numerical experiments have indicated that as m, n increase, the Jacobian $J(\underline{a}, \underline{b}, \lambda)$ of the system (3) can become increasingly ill-conditioned. This means that the Newton corrections to $\underline{a}^{(\nu)}$, $\underline{b}^{(\nu)}$, $\lambda^{(\nu)}$ can be (and usually are) poorly determined, so that ultimately the convergence test cannot be satisfied; in fact it becomes impossible to determine $E_k(x)$ with the correct sign changing properties. The algorithm then fails. Changing the basis of $P(x)$, $Q(x)$ from $\{x^r\}$ to scaled powers $\{(x/b)^r\}$ and the Chebyshev polynomials $\{T_r(x)\}$ leads to no improvement in general. This has been noted previously in the case $m = 0$ by Almacany et al [1]. Increased machine precision is necessary to proceed further. We emphasise here that illconditioning is not associated with $\|E_k\|$ being very small. It can occur with $\|f\|$ of order 1 and $\|E_k\|$ of order 0.1. We found that some Jacobians of size 10 × 10 could have condition numbers of order 10^{12}; also for $m = n = 2$ condition numbers of order 10^8 were possible. We found that with a machine precision of 14 decimal digits, although some problems had condition numbers of order 10^{10}, the algorithm determined the best approximation with an accuracy of three correct significant figures in the norm. The illconditioning is associated with *corrections* to the coefficients; high accuracy in this context is not always necessary. For problems with low m, n the algorithm is perfectly satisfactory; this is often the case of approximating functions defined as data on a discrete set. See [1] for examples of discrete approximation with $m = 0$.

5. Numerical Examples

1. We treat the problem of approximating the reciprocal Gamma function, see Williams [9] for the case $m = 0$. Here $f(x) = -1/\Gamma(x - 3)$, $x \in [0, 14]$, which has simple zeros at $x = 0, 1, 2$ and 3; we take

$$F(A, x) = x(1-x)(2-x)(3-x)\left(\frac{\sum_{r=0}^{m} a_r x^r}{1+\sum_{r=1}^{n} b_r x^r}\right)^p .$$

The behaviour of the exchange algorithm is shown in Table 1 for the case $m = 3$, $n = 5$ and $p = 4$. Each row corresponds to a reference point and each column to an iteration; the process is terminated, via Lemma 1, when the error curve is levelled to an accuracy of three significant figures. Our working precision was 14 decimal digits.

x_1	.1	.191	.139	.136	.136
x_2	.3	.442	.405	.397	.397
x_3	.6	.783	.761	.758	.758
x_4	1.3	2.396	1.666	1.673	1.673
x_5	3.3	3.712	3.620	3.615	3.615
x_6	4.0	4.539	4.562	4.570	4.571
x_7	5.0	5.519	5.594	5.631	5.682
x_8	6.0	6.552	6.835	6.876	6.876
x_9	7.0	8.040	8.440	8.430	8.430
x_{10}	9.0	10.460	10.813	10.760	10.760
λ	$4.692,10^{-7}$	$-9.633,10^{-6}$	$-1.076,10^{-5}$	$-1.078,10^{-5}$	levelled

Table 1.

Some best approximation errors for $p = 4$ are shown in Table 2.

	n = 6	n = 8
m = 0	$6.67.10^{-4}$	$2.02.10^{-4}$
1	$5.72.10^{-4}$	$5.17.10^{-5}$
2	$5.45.10^{-5}$	$2.94.10^{-6}$
3	$8.40.10^{-6}$	$1.34.10^{-6}$

Table 2

2. Here we consider a simple example of simultaneous best rational approximation and interpolation at non zeros. We require an approximation to $f(x) = e^x$, $x \in [0, 1]$, which interpolates the function at $x = 0$, $\frac{1}{2}$ and 1. Arbitrarily close uniform approximation by rational functions which simultaneously interpolate at a finite number of points, is possible by a theorem of J.L. Walsh (see Davies [6, p.121]).

Let $p_2(x)$ be the unique quadratic which interpolates e^x at 0, $\frac{1}{2}$ and 1. The standard error formula yields

$$E(x) = e^x - p_2(x) = \frac{x}{6}(x - \tfrac{1}{2})(x - 1)e^{\xi_x},$$

for some $\xi_x \in (0, 1)$, hence $E \in D[0, 1]$ where $B(x) = x(x-\frac{1}{2})(x-1)$. A best approximation to $E(x)$ of the form

$$F(A, x) = B(x)\left(\frac{P(x)}{Q(x)}\right)^p,$$

can now be obtained using the exchange algorithm. There results a "best" interpolating rational approximation to e^x of the form $p_2(x) + F(A, x)$. With $p = 1$ and $\deg P = \deg Q = 2$, the details of the computation are as follows. With

$$X^{(0)} := \{.1, .25, .4, .6, .75, .9\}$$

and corresponding Newton starting values,

$$\underline{a}^{(0)} = (1, .5, .25)^T, \ \underline{b}^{(0)} = (1, .5, .1)^T, \ \lambda^{(0)} = 0,$$

convergence was achieved after four steps of the algorithm, yielding,

$$X^{(4)} := \{.032, .155, .344, .688, .862, .971\},$$

numerator coefficients (.49678315, .05531536, .00188762),
denominator coefficients (1, -.22687520, .01476809).
The resulting interpolating rational (numerator degree 5, denominator degree 2), $p_2(x) + F(A^*, x)$, approximates e^x with error $5.38{,}10^{-11}$.

3. Finally we give an example in which the algorithm is applied to approximation on a discrete set $X := \{x_1, x_2, \ldots, x_M\}$. We assume that best approximations exist which are continuous on $[x_1, x_M]$. See Almacany et al [1] for a treatment of discrete approximation in the case $m = 0$.

The example is the approximation of the Titanium heat data, stud-

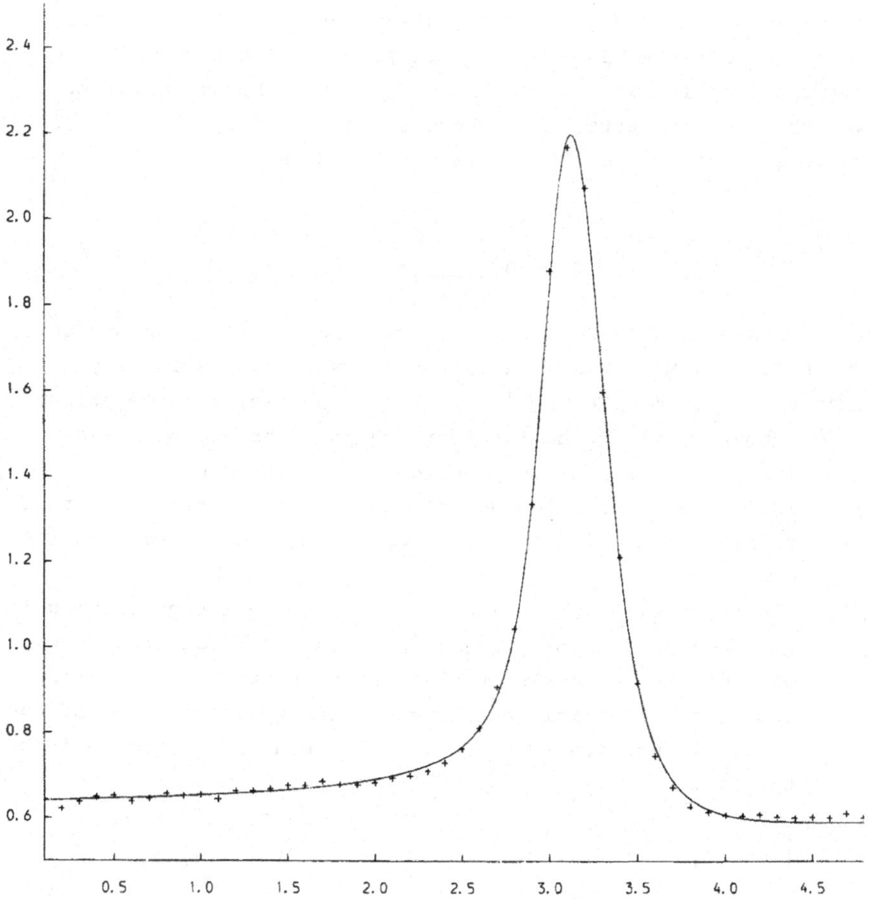

Figure 1

ied by de Boor [3] in relation to spline approximation. The data are plotted in Figure 1 along with the best approximation ($B(x) \equiv 1$) of the form

$$F(A, x) = \left(\sum_{r=0}^{4} a_r x^r \Big/ \sum_{r=0}^{4} b_r x^r \right)^{1/3} .$$

The error of the best approximation is approximately .02, which was the smallest error possible ranging over several runs with various m, n and p. We have selected $p = 1/3$ here merely to show the possible curve shapes. Higher m, n always produced poles in the range. See [1] for the case $m = 0$. In [3, p.273] a plot of a least squares quartic spline with seven knots (five variable interior knots) is shown, the greatest error throughout the range is about .04; this seems to be down to the level of noise in the data.

6. Conclusions

On the basis of this work we feel that Chebyshev approximation by the form $F(A, x) = B(x)(P(x)/Q(x))^p$ is a practical possibility using the exchange algorithm. The choice of $B(x)$ allows, to some extent, information about $f(x)$ to be incorporated into the approximating form. The inhibiting factors in practice are as follows.

i) The presence of poles during the algorithm which in turn is related to the difficulty of producing a good starting reference $X^{(0)}$.

ii) The onset with increasing m and n of illconditioning in the Newton equations, ultimately leading to an error curve without the correct sign changing properties. This arises from the polynomial bases used in the representation of the rational function $P(x)/Q(x)$. The resulting Jacobian is of the form,

$$\begin{bmatrix} 1 & x_1 & x_1^2 & \cdots & x_1^m & \alpha_1^{-1}x_1 & \alpha_1^{-1}x_1^2 & \cdots & \alpha_1^{-1}x_1^n & k_1 \\ 1 & x_2 & x_2^2 & \cdots & x_2^m & \alpha_2^{-1}x_2 & \alpha_2^{-1}x_2^2 & \cdots & \alpha_2^{-1}x_2^n & k_2 \\ \cdot & \cdot & \cdot & & \cdot & \cdot & \cdot & & \cdot & \cdot \\ \cdot & \cdot & \cdot & & \cdot & \cdot & \cdot & & \cdot & \cdot \\ \cdot & \cdot & \cdot & & \cdot & \cdot & \cdot & & \cdot & \cdot \\ 1 & x_N & x_N^2 & \cdots & x_N^m & \alpha_N^{-1}x_N & \alpha_N^{-1}x_N^2 & \cdots & \alpha_N^{-1}x_N^n & k_N \end{bmatrix}$$

and from the well known behaviour of Vandermondes, illconditioning is apparent with increasing m as columns 1 to $m + 1$ tend to become linearly dependent. Similarly with increasing n where approaching linear dependence of columns $m + 2$ to $N - 1$ is observed by first premultiplying by the diagonal matrix, $\text{diag}(\alpha_i^{-1})$. For a general reference $\{x_i\}$, a change in basis has no significant affect on this behaviour. A completely different representation of

$P(x)/Q(x)$ is needed to overcome this problem, but it is not clear that such a suitable representation exists.

Finally, the introduction by Schmidt [7] of the numerator polynomial $P(x)$ in the form introduced by Williams [9], allows greater flexibility in the type of functions to be approximated.

References

1. Almacany, M., Dunham, C.B., and Williams, J., Discrete Chebyshev approximation by interpolating rationals, IMA J. num. Analysis, to appear.

2. Barrar, R.B., and Loeb, H.L., On the Remez algorithm for non-linear families, Numer. Math. 15 (1970), 382-391.

3. de Boor, C., A Practical Guide to Splines, Springer-Verlag, 1978.

4. Burke, M.E., Nonlinear best approximation on discrete sets, J. Approx. Theory 16 (1976), 133-141.

5. Cheney, E.W., Introduction to Approximation Theory, McGraw-Hill, New York, 1966.

6. Davies, P.J., Interpolation and Approximation, Blaisdell, 1963.

7. Schmidt, D., An existence theorem for Chebyshev approximation by interpolating rationals, J. Approx. Theory 27 (1979), 146-152.

8. Taylor, G.D., and Williams, J., Existence questions for the problem of Chebyshev approximation by interpolating rationals, Math. Comp. 28 (1974), 1097-1104.

9. Williams, J., Numerical Chebyshev approximation by interpolating rationals, Math. Comp. 26 (1972), 199-206.

EXPONENTIAL FITTING OF RESTRICTED RATIONAL APPROXIMATIONS TO THE EXPONENTIAL FUNCTION*

Syvert P. Norsett**
Department of Computer Science

Stewart R. Trickett
Department of Applied Mathematics

University of Waterloo
Waterloo, Ontario
Canada N2L 3G1

Abstract

Let $R_{n/m}(z,\gamma) = P_n(z;\gamma)/(1-\gamma z)^m$ be a restricted rational approximation to $\exp(z)$, $z \in \mathbb{C}$, of order n for all real γ. In this paper we discuss how γ can be used to obtain fitting at a real non-positive point z_1. It is shown that there are exactly $\min(n + 1, m)$ different positive values of γ with this property.

1. Introduction

Solving systems of ordinary differential equations numerically is an old field of mathematics. The first proper method is due to Euler [2, p. 271], and is the basic member for practically every class of methods known today. Since Euler's time a wide variety of solvers have been constructed, analyzed, and used as step-advancers for modern production codes.

The usual approach is to approximate the unknown solution by easily computable functions such as polynomials and rationals. For the differential equation

$$y' = \lambda y, \quad y(a) = y_0, \quad \lambda \in \mathbb{C} \text{ constant}$$

we find that the function

$$u(x) = \left[\sum_{j=0}^{m} N^{(j)}((x - a)/h) z^{m-j} \Big/ \sum_{j=0}^{m} N^{(j)}(0) z^{m-j} \right] y_0$$

$$=: \; S(x;z)\, y_0, \quad z = \lambda h,$$

with

---

* This research was partially supported by the Natural Sciences and Engineering Research Council of Canada under grant A1244.

** Visiting Professor from NTH, Trondheim, Norway.

$$N(t) = \frac{1}{m!} \prod_{i=1}^{m} (t - c_i), \quad c_i \in \mathbb{R}$$

has the property

$$u'(a + c_i h) = \lambda u(a + c_i h), \quad i = 1, \ldots, m.$$

In other words, $u \in \pi_m$ satisfies the differential equation at m points (with derivatives of the equation being satisfied if some of the points coincide). Since $y_0 = y(a)$, it follows that $y_1 = u(a+h) = S(a+h;z)y_0$ is an approximation to $y(a+h)$, and as a consequence

$$R(z) = S(a + h;z) = \sum_{j=0}^{m} N^{(j)}(1)z^{m-j} / \sum_{j=0}^{m} N^{(j)}(0)z^{m-j}$$

is an approximation to $\exp(z)$. By way of construction, $R(z)$ is an approximation of at least order m, i.e. $R(z) = e^z + O(z^{m+1})$. Higher order is possible by choosing $c_1, \ldots, c_m$ in a proper manner. The highest order 2m is obtained when $c_1, \ldots, c_m$ are the zeros of the Legendre polynomial, scaled to [0,1].

Apart from this very natural occurrence, rational approximations to the exponential have been found useful for solving ordinary differential equations which are known to pose some rather serious stability problems. See, for example, Norsett [6], Lawson [5], Lau [4], and Swayne [10].

In the model equation $y' = \lambda y$, λ is generally considered to represent an eigenvalue of the Jacobian for a more general non-linear system. Hence we are interested in R(A), where A is a matrix. If A has special structure (for example, if it is sparse or banded), the computation of R(A)y, y a vector, requires in general the "inversion" of a matrix having a different structure than A itself. We avoid this difficulty by considering rationals whose denominators have the form

$$\prod_{i=1}^{m} (1 - \gamma_i z), \quad \gamma_i \in \mathbb{R}\,. \tag{1}$$

These approximations were studied in Norsett [8], Norsett and Wolfbrandt [9], and Wanner, Hairer, and Norsett [12]. A surprising result is that the maximum order at z = 0 is n + 1. Although the number of free parameters is m + n + 1, the restriction to real poles is a demanding property.

From a computational viewpoint the most efficient approach is to choose $\gamma_1 = \gamma_2 = \ldots = \gamma_m = \gamma \in \mathbb{R}$. This lead Norsett [7] to study approximations of the form

$$R_{n/m}(z;\gamma) = \sum_{i=0}^{n} (-1)^m L_m^{(m-i)}(1/\gamma)(\gamma z)^i/(1-\gamma z)^m \quad , \tag{2}$$

where $L_m^{(i)}(x)$ is the i'th derivative of the Laguerre polynomial

$$L_m(x) = \sum_{j=0}^{m} \frac{(-1)^j}{j!}\binom{m}{j} x^j \quad . \tag{3}$$

The order is n for all $\gamma \in \mathbb{R}$, and n + 1 when

$$\begin{cases} L_{m+1}^{(1)}(1/\gamma) = 0 & \text{if } n = m, \text{ and} \\ L_m^{(m-n-1)}(1/\gamma) = 0 & \text{if } n < m. \end{cases} \tag{4}$$

When the order is n+1 we have the restricted Padé approximations.

<u>Definition 1</u> Let $R(z) \in \pi_n/\pi_m$ be a rational approximation to exp(z). Then R(z) is of order $p \geq 0$ at $z_0 \in \mathbb{C}$ if

$$R(z) = \exp(z) + O(|z-z_0|^{p+1}).$$

R(z) is exponentially fitted of degree $q \geq 1$ at $z_0 \in \mathbb{C}$ if it is of order $q-1$ at z_0.

When R(z) is of the restricted type

$$R(z) = P_n(z)/\prod_{i=1}^{m}(1-\gamma_i z), \quad P_n(z) \in \pi_n, \tag{5}$$

we remarked that the order at $z_0 = 0$ is at most n + 1. The natural question is then, is it possible to use the freedom in choosing the poles to perform exponential fitting at other real points? This is of more than theoretical interest; by basing numerical ODE methods on approximations which are fitted at the largest (and hence, more persistent) real eigenvalues of the Jacobian, unusual accuracy can be achieved during the early stages of solution. Numerical examples illustrating this point for restricted rational approximations are given by Trickett [11].

From the stability point of view we are particularly interested in the case of positive poles and negative fitting points. However, from the order star theory one finds:

Theorem 2 (Iserles [3]) Let R(z) be as in (5) with $\gamma_i > 0$. Then the maximum sum of exponentially fitted degrees at non-positive real points is n + 2. □

We therefore assume that all poles are equal, and that the order at $z_0 = 0$ is n. Hence R(z) is of the form (2). From Theorem 2 we now have the possibility of a single additional fitting point $z_1 \in \mathbb{R}$, $z_1 \le 0$. In the next section we discuss this possibility, and find that for each such point there are exactly $q = \min(n+1, m)$ positive values of γ where

$$E_{n/m}(z_1;\gamma) := R_{n/m}(z_1;\gamma) - \exp(z_1) = 0 . \qquad (6)$$

2. Exponential Fitting

We will first give some preliminary results for the error function $E_{n/m}$ and the zeros of Laguerre polynomials.

Lemma 3

$$\frac{\partial}{\partial\gamma} E_{n/m}(z;\gamma) = \begin{cases} -\dfrac{1}{\gamma^2}\left[\dfrac{-\gamma z}{1-\gamma z}\right]^{m+1} L_m^{(1)}(1/\gamma) & \text{if } n = m, \\[2ex] (-1)^{m+1} mz \dfrac{(\gamma z)^n}{(1-\gamma z)^{m+1}} L_{m-1}^{(m-n-1)}(1/\gamma) & \text{if } n < m. \end{cases}$$

Proof For fixed $z \in \mathbb{C}$

$$\frac{\partial}{\partial\gamma} E_{n/m}(z;\gamma) = \frac{\partial}{\partial\gamma} R_{n/m}(z;\gamma)$$

$$= (-1)^m zm(1-\gamma z)^{-m-1} \sum_{i=0}^{n} L_m^{(m-i)}(1/\gamma)(\gamma z)^i$$

$$+ (-1)^m (1-\gamma z)^{-m} \sum_{i=0}^{n} \left[L_m^{(m-i+1)}(1/\gamma)(-\gamma^2)(\gamma z)^i + L_m^{(m-i)}(1/\gamma)\, iz(\gamma z)^{i-1} \right]$$

$$= (-1)^m z(1-\gamma z)^{-m-1} \times$$

$$\sum_{i=0}^{n} \left[m\gamma z L_m^{(m-i)}(1/\gamma) - 1/\gamma(1-\gamma z) L_m^{(m-i+1)}(1/\gamma) + i(1-\gamma z) L_m^{(m-i)}(1/\gamma) \right] (\gamma z)^{i-1} .$$

Now the following recurrence relations hold for Laguerre polynomials:

$$xL_m^{(k+1)}(x) = (m-k)L_m^{(k)}(x) - mL_{m-1}^{(k)}(x), \quad k = 0, \ldots, m, \tag{7a}$$

$$L_m^{(k+1)}(x) = L_{m-1}^{(k+1)}(x) - L_{m-1}^{(k)}(x), \quad k = 0, \ldots, m. \tag{7b}$$

Hence (defining $L_{m-1}^{(-1)}(1/\gamma) := \frac{1}{m\gamma} L_m^{(1)}(1/\gamma)$)

$$\frac{\partial}{\partial\gamma} E_{n/m}(z;\gamma) = (-1)^m mz(1-\gamma z)^{-m-1} \sum_{i=0}^{n} \left[L_{m-1}^{(m-i)}(1/\gamma) - \gamma z L_{m-1}^{(m-i-1)}(1/\gamma) \right] (\gamma z)^{i-1}$$

$$= (-1)^m mz(1-\gamma z)^{-m-1} \left[L_{m-1}^{(m)}(1/\gamma)(\gamma z)^{-1} - L_{m-1}^{(m-n-1)}(1/\gamma)(\gamma z)^n \right]$$

from which the result follows immediately. □

<u>Lemma 4</u> For $0 \le n \le m$, let $\alpha_{m,i}^n > 0$, $i=1, \ldots, n$ be the zeros of $L_m^{(m-n)}(x)$, numbered in increasing order. Then

$$\alpha_{m,i}^{n+1} < \alpha_{m,i}^n < \alpha_{m,i+1}^{n+1} \quad \text{and} \tag{8}$$

$$\alpha_{m,i}^{n+1} < \alpha_{m-1,i}^n < \alpha_{m,i}^n, \tag{9}$$

for $i = 1, \ldots, n$ and $n < m$.

<u>Proof</u> Rolle's theorem and relation (7a). □

Let us first treat the case where $n = m$.

<u>Theorem 5</u> For each $z_1 < 0$ there exists exactly m different values of $\gamma > 0$ such that $E_{m/m}(z_1;\gamma) = 0$. Denote these values of γ by $u_i(z_1)^{-1}$, $i = 1, \ldots, m$, ordered so that $u_i(z_1) < u_{i+1}(z_1)$. Then

$$u_i(z_1) \in I_i := (\alpha_{m,i}^m, \alpha_{m+1,i}^m), \quad i = 1, \ldots, m.$$

Furthermore, $\lim_{z_1 \to 0^-} u_i(z_1) = \alpha_{m+1,i}^m$ and $\lim_{z_1 \to -\infty} u_i(z_1) = \alpha_{m,i}^m$.

<u>Proof</u> Let $z_1 < 0$ be fixed. Then $E_{m/m}(z_1;\gamma) = 0$ is equivalent to an m'th degree polynomial in γ, and hence has at most m solutions. By Theorem 4.3 of Norsett [8]

$$E_{m/m}(z;\gamma) = e^{z}\int_0^{-z}\left(\frac{\gamma y}{1+\gamma y}\right)^{m+1}\left[L_m(1/\gamma) - \frac{1}{\gamma y}L_{m+1}^{(1)}(1/\gamma)\right] dy \qquad (10)$$

giving

$$\text{sign } E_{m/m}(z_1;1/\alpha_{m+1,i}^{m}) = \text{sign } L_m(\alpha_{m+1,i}^{m}) = \text{sign } L_m^{(1)}(\alpha_{m+1,i}^{m}), \qquad (11)$$

where the second equality is from (7b), and similarly

$$\text{sign } E_{m/m}(z_1;1/\alpha_{m,i}^{m}) = -\text{ sign } L_{m+1}^{(1)}(\alpha_{m,i}^{m}) = -\text{ sign } L_m^{(1)}(\alpha_{m,i}^{m}). \qquad (12)$$

From (8) and (9) of Lemma 4 with different values of m and n

$$\alpha_{m,i-1}^{m-1} < \alpha_{m,i}^{m} < \alpha_{m+1,i}^{m} < \alpha_{m,i}^{m-1}, \quad i = 1, \ldots, m \quad ,$$

where we define $\alpha_{m,0}^{m-1}:=0$ and $\alpha_{m,m}^{m-1}:=\infty$. It now follows that

$$L_m^{(1)}(\alpha_{m+1,i}^{m})\cdot L_m^{(1)}(\alpha_{m,i}^{m}) > 0.$$

By (11), (12), and the continuity of $E_{m/m}(z_1;\gamma)$, there must now exist a zero $1/\gamma = u_i(z_1)$ in each of the intervals I_i.

By definition of $\alpha_{m,i}^{m}$ and $\alpha_{m+1,i}^{m}$, the last two statements on the limiting values of u_i follow easily. □

<u>Theorem 6</u> Let m and I_i, $i = 1,\ldots,m$ be as in Theorem 5, and define $I = \bigcup_{i=1}^{m} I_i$. Then for each $1/\gamma \in I$ there exists one and only one real z_1 such that $E_{m/m}(z_1;\gamma) = 0$.

<u>Proof</u> From Theorem 3.1 of Norsett [8],

$$E_{m/m}(z;\gamma) = -\frac{1}{\gamma}L_{m+1}^{(1)}(1/\gamma)(-\gamma z)^{m+1} + O(z^{m+2}),$$

and from (2)

$$E_{m/m}(z;\gamma) \to L_m(1/\gamma) \quad \text{as } z \to -\infty.$$

Now from Lemma 4

$$L_{m+1}^{(1)}(1/\gamma)\cdot L_m(1/\gamma) > 0 \quad \text{for } 1/\gamma \in I .$$

Hence sign $E_{m/m}(z;\gamma)$ as $z \to -\infty$ is opposite to that as $z \to 0^-$, and it follows that there exists at least one solution $z_1 < 0$ such that $E_{m/m}(z_1;\gamma) = 0$. By Theorem 2, no more fitting is possible. □

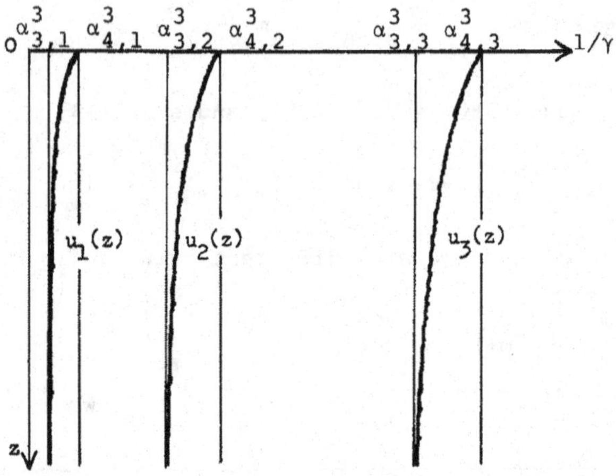

Figure 1. The curves $1/\gamma = u_i(z)$ for $R_{3/3}$.

Example 7 In the case of m = 2 we find

$$E_{2/2}(z;\gamma) = \left[(1 + (1-2\gamma)z + (1/2-2\gamma+\gamma^2)z^2 \right] / (1-\gamma z)^2 - e^z .$$

$E_{2/2}(z;\gamma) = 0$ gives

$$u_i(z) = \frac{z(1+z-e^z) \pm (z(1+e^z)(z/2-\tanh(z/2)))^{1/2}}{(1+z+z^2/2-e^z)},$$

where i = 2 corresponds to the plus sign and i = 1 to the minus sign. From Theorem 5

$$0{\cdot}58579 \le u_1(z) \le 1.26794,$$

$$3.41425 \le u_2(z) \le 4.73216.$$

Further, by a result of Burrage [1], $R_{2/2}(z;\gamma)$ is A-acceptable iff $1/\gamma \in (0,4]$. Hence the fitted approximation $R_{2/2}(z;1/u_1(z_1))$ is A-acceptable for all $z_1 < 0$, whereas $R_{2/2}(z;1/u_2(z_1))$ is A-acceptable only when

$z_1 \in (-\infty, -4.79872)$. □

When $n < m$ the approach is very similar.

Theorem 8 Let $0 \le n < m$ and fix $z_1 < 0$. Then there exists exactly $n+1$ different values of $\gamma > 0$ such that $E_{n/m}(z_1;\gamma) = 0$. Denote these values of γ by $u_i(z_1)^{-1}$, $i = 0,\dots,n$, ordered so that $u_i(z_1) < u_{i+1}(z_1)$. Then

$$u_i(z_1) \in I_i := (\alpha^n_{m,i}, \alpha^{n+1}_{m,i+1}), \quad i = 0,\dots,n,$$

where we define $\alpha^n_{m,0} = 0$. Furthermore, $\lim_{z_1 \to 0^-} u_i(z_1) = \alpha^{n+1}_{m,i+1}$ and $\lim_{z_1 \to -\infty} u_i(z_1) = \alpha^n_{m,i}$, $i = 0,\dots,n$.

Proof The proof in this case is somewhat different than that of Theorem 5. We could have used a similar idea, but will instead take the approach indicated by Trickett [11].

Let us first consider $E_{n/m}(z_1;\gamma)$ with $1/\gamma \in I_i$ for i between 1 and n. From Norsett [8] the error has the form

$$\pi_1 := E_{n/m}(z;1/\alpha^{n+1}_{m,i+1}) \tag{13}$$

$$= \left(\frac{-z}{\alpha^{n+1}_{m,i+1}}\right)^{n+2} \left\{ \begin{array}{ll} -\dfrac{\alpha^n_{m,i+1}}{m+1} L^{(1)}_m(\alpha^m_{m,i+1}) & \text{for } n=m-1 \\ (-1)^{n+m+3} L^{(m-n-2)}_m(\alpha^{n+1}_{m,i+1}) & \text{for } n<m-1 \end{array} \right\} + O(z^{n+3})$$

and

$$\pi_2 := E_{n/m}(z,1/\alpha^n_{m,i}) = (-1)^{n+m+2} L^{(m-n-1)}_m(\alpha^n_{m,i}) \left(\frac{-z}{\alpha^n_{m,i}}\right)^{n+1} + O(z^{n+2}) . \tag{14}$$

By the location of the zeros $\alpha^n_{m,i}$ and $\alpha^{n+1}_{m,i}$, and the relation

$$xL^{(k)}_m(x) + (k-1-x)L^{(k-1)}_m(x) + (m-k+2)L^{(k-2)}_m(x) = 0$$

we have as $z \to 0^-$

$$\operatorname{sign}(\pi_1\cdot\pi_2) = \operatorname{sign}\begin{Bmatrix} L_m^{(1)}(\alpha_{m,i+1}^{m})\cdot L_m(\alpha_{m,i}^{m-1}), & n=m-1 \\ L_m^{(m-n-2)}(\alpha_{m,i+1}^{n+1})\cdot L_m^{(m-n-1)}(\alpha_{m,i}^{n}), & n<m-1 \end{Bmatrix} < 0.$$

But $\gamma = 1/\alpha_{m,i+1}^{n+1}$ and $1/\alpha_{m,i}^{n}$ both give rise to methods of optimal order at $z = 0$, implying that the corresponding errors are of one sign for all $z < 0$. Hence sign $(\pi_1\cdot\pi_2) < 0$ for all real $z < 0$. By continuity there is at least one value of $1/\gamma\epsilon I_i$ such that $E_{n/m}(z_1;\gamma)=0$

In particular, we have for $n < m$ and z real and negative

$$E_{n/m}(z;1/\alpha_{m,1}^{n+1}) > 0$$

and

$$E_{n/m}(z;\gamma) \to -e^z < 0 \quad \text{as } \gamma \to \infty .$$

Hence there exists some value $1/\gamma\epsilon(0,\alpha_{m,1}^{n+1})=I_0$ such that $E_{n/m}(z_1;\gamma) = 0$.

For a given $z_1 < 0$ we have now found n+1 zeros of $E_{n/m}(z_1;\gamma) = 0$. From Lemma 3 this is the maximum number. □

<u>Theorem 9</u> Let n, m, and I_i, i=0,...,n be as in Theorem 8, and let $I = \bigcup_{i=0}^{n} I_i$. Then for each $1/\gamma\epsilon I$ there exists one and only one real $z_1 < 0$ such that $E_{n/m}(z_1;\gamma) = 0$.

<u>Proof</u> As in (14) we have from Norsett [8]

$$E_{n/m}(z;\gamma) = (-1)^{n+m+2}L_m^{(m-n-1)}(1/\gamma)(-\gamma z)^{n+1} + 0(z^{n+2}),$$

and from (2)

$$(1-\gamma z)^{n-m}E_{n/m}(z;\gamma) \to (-1)^{n+m}L_m^{(m-n)}(1/\gamma) \quad \text{as } z \to \infty.$$

By Lemma 4

$$L_m^{(m-n-1)}(1/\gamma)\cdot L_m^{(m-n)}(1/\gamma) < 0, \quad 1/\gamma\epsilon I.$$

It follows that for each $1/\gamma\epsilon I$ there exists at least one solution $z_1<0$ such that $E_{n/m}(z_1;\gamma)=0$. Again by Theorem 2, no more fitting is possible.

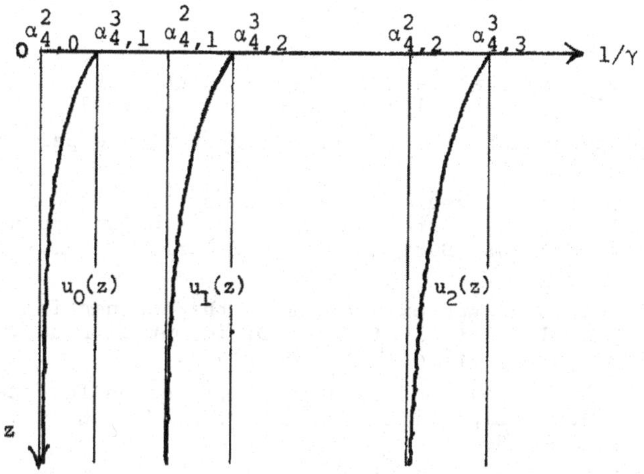

Figure 2. The curves of $1/\gamma = u_i(z)$ for $R_{2/4}$.

Example 10 In the case $m = 2$, $n = 0$ we find

$$E_{0,2}(z;\gamma) = \frac{1}{(1-\gamma z)^2} - e^z.$$

Setting $E_{0,2}(z_1;\gamma) = 0$ gives

$$\gamma = \frac{1 \pm \exp(-z_1/2)}{z_1}.$$

Choosing the positive solution, we get

$$u_0(z_1) = \frac{z_1}{1 - \exp(-z_1/2)}$$

with $u_0(0) = \alpha^1_{2,1} = 2$ and $\lim_{z_1 \to -\infty} u_0(z_1) = 0$. The approximation is A-acceptable for all $z_1 < 0$. □

References

[1] Burrage, K., "A special family of Runga-Kutta methods for solving stiff differential equations", BIT 18 (1978), pp. 22-41.

[2] Euler, L., Opera Omnia, Series Prima, Vol. 11, Leipzig and Berlin, 1913.

[3] Iserles, A., "Generalized order star theory", Padé Approximation and its Applications, Amsterdam 1980 (ed. M.G. De Bruin and H. van Rossum), LNiM 888, Springer-Verlag, Berlin, 1981, pp. 228-238.

[4] Lau, T., "A Class of Approximations to the Exponential Function for the Numerical Solution of Stiff Differential Equations", Ph.D. Thesis, University of Waterloo, 1974.

[5] Lawson, J.D., "Generalized Runga-Kutta processes for stable systems with large Lipschitz constants", SIAM J. Numer. Anal. 4 (1967), pp. 372-380.

[6] Norsett, S.P., "An A-stable modification of the Adams-Bashforth methods", Conf. on the Numerical Solution of Differential Equations (ed. A. Dold and B. Eckmann), LNiM 109, Springer-Verlag, Berlin, 1969, pp. 214-219.

[7] ___________, "One-step methods of Hermite type for numerical integration of stiff systems", BIT 14 (1974), pp. 63-77.

[8] ___________, "Restricted Padé-approximations to the exponential function", SIAM J. Numer. Anal. 15 (1978), pp. 1008-1029.

[9] Norsett, S.P., and Wolfbrandt, A., "Attainable order of rational approximations to the exponential function with only real poles", BIT 17 (1977), pp. 200-208.

[10] Swayne, D.A., "Computation of Rational Functions with Matrix Argument with Application to Initial-Value Problems", Ph.D. Thesis, University of Waterloo, 1975.

[11] Trickett, S.R., "Rational Approximations to the Exponential Function for the Numerical Solution of the Heat Conduction Problem", Master's Thesis, University of Waterloo, 1984.

[12] Wanner, G., Hairer, E., and Norsett, S.P., "Order stars and stability theorems", BIT 18 (1978), pp. 475-489.

QUADRATURE FORMULAE AND MOMENT PROBLEMS

W. J. Thron*
Department of Mathematics
Campus Box 426
University of Colorado
Boulder, CO 80309
U.S.A.

Abstract. Quadrature formulae can play a role in the solution of the moment problem. Formulae with degree of precision $2n - 1 - k$, $k = 0,1$, have previously been used. Here we study when, for $k \geq 2$, the node polynomials have only real, simple zeros and when the weights are non-negative.

1. Introduction and Brief Review

An excellent survey of Gauss-Christoffel quadrature formulae was recently compiled by Gautschi [4]. Here we give a brief review of those parts of the theory which are pertinent for this article. The following notation will be used.

Π is the set of all polynomials with coefficients in $\mathbb{C}$,

Π_n consists of all $P \in \Pi$ of degree $\leq n$,

Π^R is the subset of those $P \in \Pi$ whose coefficients are in $\mathbb{R}$,

$\Pi^+ = [P \in \Pi^R, P \not\equiv 0, P(x) \geq 0, x \in \mathbb{R}]$.

A linear functional γ on Π will be called positive definite if

$$P \in \Pi^+ \Rightarrow \gamma(P) > 0 .$$

A quadrature formula of order n is an expression

$$F_n(P) = \sum_{\nu=1}^{n} \lambda_\nu^{(n)} P(x_\nu^{(n)}) .$$

If $\gamma(P) = F_n(P)$ for all $P \in \Pi_{2n-1-k}$, then F_n is called a quadrature formula for γ with degree of precision $d(F_n, \gamma) \geq 2n-1-k$. We shall be interested only in values $0 \leq k \leq n$ for k. The distinct numbers $x_1^{(n)}, \ldots, x_n^{(n)}$ will be called the nodes and the numbers $\lambda_1^{(n)}, \ldots, \lambda_n^{(n)}$ the weights of the formula F_n. The node

*Research supported in part by the U.S. National Science Foundation under grant no. MCS-8202230.

polynomial of F_n is defined to be

$$M^{(n)}(x) = \prod_{v=1}^{n} (x-x_v^{(n)}) .$$

With every positive definite linear functional γ on π there is associated a unique sequence $\{R_n\}$ of monic polynomials

$$R_n(x) = x^n + g_1^{(n)}x^{n-1} + \ldots + g_n^{(n)} ,$$

which are orthogonal with respect to γ , that is

$$\gamma(R_nR_m) = \begin{cases} 0 , & \text{for } n \neq m , \\ N_n > 0 , & \text{for } n = m . \end{cases}$$

The sequence $\{R_n\}$ is known to satisfy a three term recurrence relation

$$R_n = (x-a_n)R_{n-1} - k_nR_{n-2} , \ n \geq 1 , \ R_0 = 1 , \ R_{-1} = 0 ;$$

here $a_n \in \mathbb{R}$, $n \geq 1$, and $k_n = N_{n-1}/N_{n-2}$, $n \geq 2$.

In terms of the sequence $\{R_n\}$ one can define sets of polynomials

$$\Theta_{n,k} = [P: P \in \Pi^R , P = R_n + \beta_1 R_{n-1} + \ldots + \beta_k R_{n-k}] .$$

In 1826 (in a special case) Jacobi established the connection between quadrature formulae and orthogonal sequences of polynomials. He showed that a formula F_n for γ with $d(F_n,\gamma) = 2n-1-k$ exists iff

$$M^{(n)}(x) \in \Theta_{n,k}$$

and has real, simple zeros.

Next, introduce $\Lambda_v^{(n)} \in \Pi_{n-1}$ by

$$\Lambda_v^{(n)} = \prod_{v \neq \mu} \left(\frac{x-x_\mu^{(n)}}{x_v^{(n)}-x_\mu^{(n)}}\right) .$$

Then, if $d(F_n,\gamma) \geq n-1$, or $0 \leq k \leq n$,

$$\lambda_v^{(n)} = \gamma(\Lambda_v^{(n)})$$

so that the weights are completely determined by the nodes.

In 1884 Stieltjes gave an ingenious argument to show that for $d(F_n,\gamma) \geq 2n-2$ one must have $\lambda_v^{(n)} > 0$. Note that $\Psi_{n,v} = (\Lambda_v^{(n)})^2 - \Lambda_v^{(n)} = M^{(n)}K_v^{(n)}$, where $K_v^{(n)} \in \Pi_{n-2}$ and that $\Psi_{n,v} \in \Pi_{2n-2}$. Hence

$$\lambda_v^{(n)} = \gamma(\Lambda_v^{(n)}) = \gamma((\Lambda_v^{(n)})^2) - \gamma(M^{(n)}K_v^{(n)}) > 0 .$$

Since n is usually fixed throughout a given argument, we shall from now on frequently omit the superscript n and write simply, M, Λ_v, λ_v, x_v.

Quadrature formulae with all nodes preassigned were first studied by Newton and Cotes for

$$\gamma(f) = \int_a^b f(x)dx .$$

One then has a degree of precision $d(F_n,\gamma) = n-1$. Gauss, in 1814, allowed all nodes to be dependent on γ, where

$$\gamma(f) = \int_{-1}^{+1} f(x)dx , \tag{1.1}$$

in order to maximize $d(F_n,\gamma)$. He obtained $d(F_n,\gamma) = 2n-1$. Christoffel, in 1858, still for the special case (1.1) considered the situation where k nodes are preassigned. With suitable restrictions on the preassigned nodes one then has $d(F_n,\gamma) = 2n-1-k$. We shall discuss some aspects of this problem in Section 4.

Much of the work on quadrature formulae was originally done for special cases of γ. It is surprising how many of the methods and results carry over to the general case of a positive definite functional defined only on Π, which we study here. Nevertheless in a number of cases, in particular if the functional γ has finite support and some or all of the nodes lie outside this set, arguments can be used which are not available in the general case. Thus it is desirable to have new and occasionally more transparent or simpler proofs for the general case even when proofs for more special cases already exist. It would only be tedious to state explicitly for each earlier result for what type of γ it was proved. So we shall, in general, not do so.

As we shall show in Section 2, quadrature formulae comprise one approach to the solution of the moment problem. So far formulae with $d(F_n,\gamma) = 2n-1-k$, where $k = 0$ or 1, have been used. However, larger values of k, in particular $k = 2$, can also give solutions of the problem, provided the nodes are real and simple and the weights are non-negative. It thus becomes desirable to determine sufficient conditions for these properties to hold. This is done in Section 3 and 5 , respectively.

This article was motivated by joint work with O. Njåstad [7]. In particular most of the theorems in Section 4 resulted from this cooperation.

2. Connection with Moment Problems

A linear functional γ on π is determined by its moments

$$c_n = \gamma((-x)^n) , \quad n = 0,1,2,\ldots$$

Similarly, if γ is given, the moments are determined. One can thus associate a sequence of orthogonal polynomials directy with a sequence of moments. This was done in [12], [5], [1] and [2] and possibly earlier by other authors.

The moment problem is the following: For a given sequence $\{c_n\}$ of real numbers one is to determine necessary and sufficient conditions so that the functional γ on Π can be extended to a functional

$$\Gamma(f) = \int_{-\infty}^{\infty} f(x)d\psi(x) ,$$

where ψ is non-decreasing and bounded and has an infinite number of points of increase on $(-\infty,\infty)$. Since we demand $\Gamma(P) = \gamma(P)$, $P \in \Pi$ we would have

$$c_n = \int_{-\infty}^{\infty} (-x)^n d\psi(x) , \quad n = 0,1,2,\ldots ,$$

so that the c_n are the moments with respect to ψ . Besides the question of existence the moment problem is concerned with the question of uniqueness of the function ψ .

One approach to the solution of the moment problem is as follows: Consider the "moment generating" function

$$F(z) = \int_{-\infty}^{\infty} \frac{z d\psi(x)}{z+x} = \Gamma\left(\frac{z}{z+x}\right) .$$

It is "formally equal" to

$$\sum_{n=0}^{\infty} \Gamma((\frac{-x}{z})^n) = \sum_{n=0}^{\infty} \frac{\gamma((-x)^n)}{z^n} = \sum_{n=0}^{\infty} \frac{c_n}{z^n} .$$

This suggests looking at

$$\sum_{v=0}^{n} \frac{c_v}{z^v} = \gamma\left(\frac{1-(\frac{-x}{z})^{n+1}}{1-(\frac{-x}{z})}\right) .$$

If there exists a quadrature formula F_n for γ and for every $n \in \mathbb{N}$ then we can write

$$\sum_{v=0}^{n} \frac{c_v}{z^v} = \sum_{v=1}^{n} \lambda_v^{(n)} \frac{z(1-(\frac{-x_v^{(n)}}{z})^{n+1})}{z + x_v^{(n)}}$$

$$= \sum_{v=1}^{n} \frac{z\lambda_v^{(n)}}{z+x_v^{(n)}} - \frac{1}{z^n} \sum_{v=1}^{n} \frac{\lambda_v^{(n)}(-x_v^{(n)})^{n+1}}{z + x_v^{(n)}} .$$

The sequence $\{G_n\}$ of rational functions

$$G_n(z) = \sum_{v=1}^{n} \frac{z\lambda_v^{(n)}}{z+x_v^{(n)}}$$

then "corresponds" to $F(z)$ and one can hope to get one (or more) valid F as limits of subsequences of $\{G_n\}$. Here the Helly theorems can be used. To obtain ϕ one inverts the Stieltjes transform

$$F(z) = \int_{-\infty}^{\infty} \frac{z d\phi(x)}{z + x} .$$

3. Elements of $\Theta_{n,k}$ with Real, Simple Zeros

The following lemmas can be proved.

Lemma 3.1. *Let*

$$T_{n,k} = R_n + \beta_1 R_{n-1} + \dots + \beta_k R_{n-k} \in \Theta_{n,k} .$$

Then the number m *of real zeros of odd order of* $T_{n,k}$ *satisfies*

$$m \geq \begin{cases} n-k, & \text{if } k \text{ is even} , \\ n-k+1 , & \text{if } k \text{ is odd} . \end{cases}$$

Lemma 3.2. *If* $T_{n,k}$ *in Lemma 3.1 satisfies the additional condition* $\beta_k < 0$, *then*

$$m \le \begin{cases} n-k+2 , & \text{if } k \text{ is even} , \\ n-k+1 , & \text{if } k \text{ is odd} . \end{cases}$$

Proof of Lemma 3.2. Let $\xi_1,\dots,\xi_m$ be the real zeros of odd order of $T_{n,k}$. Set

$$P_m(x) = \prod_{\mu=1}^{m} (x-\xi_\mu) = R_m(x) + \delta_1 R_{m-1}(x) + \dots + \delta_m R_o(x) .$$

Then, if $m \le n-k$,

$$\gamma(T_{n,k}P_m) = 0 \quad \text{or} \quad \gamma(\beta_k R^2_{n-k}) = \beta_k N_{n-k} ,$$

because of the orthogonality of the sequences $\{R_n\}$. Hence $\gamma(T_{n,k}P_m) \le 0$. On the other hand, since $T_{n,k}(x)P_m(x)$ $= \prod_k (x-\omega_k)^{2m_k} \cdot \prod_\sigma ((x+\alpha_\sigma)^2 + \beta_\sigma^2) \prod_{\mu=1} (x-\xi_\mu)^{2\mu_\mu}$, where $\omega_k, \alpha_\sigma, \beta_\sigma \in \mathbb{R}$, γ is positive definite and $T_{n,k}P_m \in \Pi^+$ we must have $\gamma(T_{n,k}P_m) > 0.$ This is a contradiction and hence $m > n-k+1$. The sum s of the multiplicities of the zeros ξ_μ is odd if m is odd and even if m is even. However $n-s$, the number (counting multiplicity) of the remaining zeros is even, since they come in pairs. Thus $n-m$ is even and the lemma is proved.

Lemma 3.1 can be proved along similar lines.

An immediate corollary of these lemmas is:

Corollary 3.3. *All zeros of* $T_{n,0}$ *and* $T_{n,1}$ *are real and simple. If* $\beta_k < 0$, *then* $T_{n,2}$ *has only real, simple zeros.*

Zeros of polynomials $T_{n,2}$ were studied by Micchelli and Rivlin [6]. See also references in Section 5. The results for $T_{n,2}$ can be somewhat improved using the recurrence relations for the sequence $\{R_n\}$. We have

$$T_{n,2} = (x-a_n+\beta_1)R_{n-1} + (\beta_2-k_n)R_{n-2} .$$

It is known that the zeros of R_{n-2} separate the zeros of R_{n-1} . From this it is easy to deduce that the zeros of $T_{n,2}$ are all real and simple provided $\beta_2 - k_n < 0$. Thus:

Theorem 3.4. *The zeros of*

$$R_n + \beta_1 R_{n-1} + \beta_2 R_{n-2} \in \Theta_{n,2}$$

are all real and simple if $\beta_2 < k_n$.
Using the recurrence relations one can also prove:

Lemma 3.5. $P \in \Theta_{n-1,k-2} \Rightarrow (x-\alpha)P \in \Theta_{n,k}$ *for* $\alpha \in \mathbb{R}$. *In particular*

$$R_n + aR_{n-1} + bR_{n-2} + cR_{n-3} = (x-\alpha)(R_{n-1}+\beta R_{n-2})$$

if a *and* c *are arbitrary real numbers and*

$$b = k_n - \frac{c}{k_{n-1}} \left(\frac{c}{k_{n-1}} - a - a_n + a_{n-1}\right) .$$

Note however that $(x-\alpha)P \in \Theta_{n,k}$ does not in general imply $P \in \Theta_{n-1,k-2}$.

That multiple zeros and complex conjugate zeros can actually occur for polynomials in $\Theta_{n,k}$ is easily illustrated. The first Legendre polynomials are:

$$R_0 = 1 , \quad R_1 = x , \quad R_2 = x^2 - \frac{1}{3} , \quad R_3 = x^3 - \frac{3}{5} x ,$$

$$R_4 = x^4 - \frac{6}{7} x^2 + \frac{3}{35} , \ldots$$

and hence

$$R_3 + \frac{8}{5} R_1 = x(x^2+1) ,$$

$$R_3 + \frac{3}{5} R_1 = x^3 ,$$

$$R_4 + \frac{6}{7} R_2 = x^4 - \frac{1}{5} .$$

Note also ,

$$R_3 + \frac{4}{15} R_1 = xR_2 , \quad R_4 - \frac{1}{2} R_3 + \frac{9}{35} R_2 = (x - \frac{1}{2})R_3 .$$

4. Preassigned Nodes

Let $\zeta_1,\ldots,\zeta_k$ be distinct real numbers. We investigate here under what conditions on these numbers there will exist an M satisfying

$$(4.1) \qquad M \in \Theta_{n,k} , \quad M(\zeta_\mu) = 0 , \quad \mu = 1,\ldots,k .$$

The ζ_μ can be thought of as being preassigned nodes for a quadrature formula of degree of precision $2n-1-k$, except that we are leaving aside the question whether all zeros of M , obtained in this manner, are real and simple.

The polynomial

(4.2) $M = R_n + \alpha_1 R_{n-1} + \dots + \alpha_k R_{n-k}$

will satisfy (4.1) if the α_k satisfy the system of equations

$$R_n(\zeta_\mu) + \sum_{k=1}^{k} \alpha_k R_{n-k}(\zeta_\mu) = 0 \ , \quad \mu = 1,\dots,k \ .$$

This system has a unique solution $\alpha_1,\dots,\alpha_k$ iff the determinant

$$D_{n-1}(\zeta_1,\dots,\zeta_k) = \begin{vmatrix} R_{n-1}(\zeta_1) & \dots & R_{n-k}(\zeta_1) \\ R_{n-1}(\zeta_k) & \dots & R_{n-k}(\zeta_k) \end{vmatrix}$$

does not vanish. The resulting M is then given by

$$M(x) = D_n(x,\zeta_1,\dots,\zeta_k)/D_{n-1}(\zeta_1,\dots,\zeta_k)$$

as was known to Christoffel in a special case.

Next, we explore in more detail what happens if there exists a $T \in \Theta_{n-1,k-2}$ so that

$$T(\zeta_1) = T(\zeta_2) = \dots = T(\zeta_{k-1}) = 0 \ .$$

If $T(\zeta_k) \neq 0$ and $D_{n-2}(\zeta_1,\dots,\alpha_{k-1}) \neq 0$, then

$$|D_{n-1}(\zeta_1,\dots,\zeta_k)| = |T(\zeta_k)D_{n-2}(\zeta_1,\dots,\zeta_{k-1})| \neq 0$$

so that there is exactly one M which satisfies (4.1). Since $(x-\zeta_k)T(x)$ satisfies these conditions (using Lemma 3.5), it must be the desired M .

If $T(\zeta_k) = 0$, then $D_{n-1}(\zeta_1,\dots,\zeta_k) = 0$ and $M(x) = (x-\alpha)T(x)$ with arbitrary $\alpha \in \mathbb{R}$, are all solutions of (4.1). We thus have proved.

<u>Theorem</u> 4.1. <u>Let</u> $\zeta_1,\dots,\zeta_k$ <u>be a set of distinct real numbers. Further, assume that there exists a</u> $T \in \Theta_{n-1,k-2}$ <u>such that</u> $T(\zeta_\mu) = 0$ <u>for</u> $\mu = 1,\dots,k-1$. <u>Then</u>

(A) <u>if</u> $T(\zeta_k) \neq 0$ <u>and</u> $D_{n-2}(\zeta_1,\dots,\zeta_{k-1}) \neq 0$, <u>there is exactly one</u> M , <u>namely</u> $M(x) = (x-\zeta_k)T(x)$, <u>which satisfies</u> (4.1);

(B) <u>if</u> $T(\zeta_k) = 0$ <u>and</u> $\alpha \in \mathbb{R}$, <u>then every</u> $M(x) = (x-\alpha)T(x)$ <u>satisfies</u> (4.1).

If $D_{n-1}(\zeta_1,\dots,\zeta_k) = 0$, then there may be no solutions or, as we have already seen, there may be an infinite number of solutions. A case for which there is no solution of (4.1) is the following.

<u>Theorem</u> 4.2. <u>If</u> $\zeta_1,\ldots,\zeta_k$ <u>are chosen so that</u> $R_{n-k}(\zeta_\mu) = 0$, $\mu = 1,\ldots,k$, <u>then there is no quadrature formula</u> F_n <u>for</u> γ <u>with</u> $d(F_n,\gamma) = 2n-1-k$ <u>which has the</u> ζ_μ <u>as preassigned nodes</u>.

<u>Proof</u>. Since the node polynomial M must be in $\Theta_{n,k}$, it is of the form (4.2) so that, if we set $\alpha_o = 1$, the α_μ must satisfy the system of equations

$$\sum_{\sigma=0}^{k-1} \alpha_\sigma R_{n-\sigma}(\zeta_\mu) = 0 , \quad \mu = 1,\ldots,k .$$

For this system to have a nontrivial solution $1,\alpha_1,\ldots,\alpha_k$ it is necessary that

$$D = D_n(\zeta_1,\ldots,\zeta_k) = 0 .$$

Using the following notation for D

$$D = \,]R_n(\zeta_\mu) , \ldots , R_{n-k+1}(\zeta_\mu)[,$$

that is by displaying only a typical row of the determinant, we can evaluate D by using the recurrence relations for $\{R_n\}$ and elementary properties of determinants as follows.

$$D = \,]\zeta_\mu R_{n-1}(\zeta_\mu),\ldots,\zeta_\mu R_{n-k+2}(\zeta_\mu), R_{n-k+2}(\zeta_\mu),R_{n-k+1}(\zeta_\mu)[$$

$$= \,]\zeta_\mu^{k-2} R_{n-k+2}(\zeta_\mu),\zeta_\mu^{k-3} R_{n-k+2}(\zeta_\mu),\ldots \zeta_\mu^{\sigma} R_{n-k+2}(\zeta_\mu,R_{n-k+1}(\zeta_\mu)[$$

From the assumption $R_{n-k}(\zeta_\mu) = 0$, $\mu = 1,\ldots,k$ one deduces $R_{n-k+2}(\zeta_\mu) = (\zeta_\mu-\alpha_{n-k+2})R_{n-k+1}(\zeta_\mu)$ and hence

$$D = \,]\zeta_\mu^{k-1} R_{n-k+1}(\zeta_\mu), \ldots, \zeta_\mu R_{n-k+1}(\zeta_\mu),R_{n-k+1}(\zeta_\mu)[$$

$$= \prod_{\mu=1}^{k} R_{n-k+1}(\zeta_\mu)\,]\zeta_\mu^{k-1},\ldots,\zeta_\mu^{o}[$$

$$= \prod_{\mu=1}^{k} R_{n-k+1}(\zeta_\mu) \prod_{1\le\kappa\le\lambda\le k} (\zeta_\kappa-\zeta_\lambda) .$$

Since R_{n-k} and R_{n-k+1} have no common zeros and since the ζ_μ are all distinct, the theorem follows.

Thus we have shown that a unique M, no M and an infinite number of M can all occur. Since in the first and second case the M may have non-real or multiple zeros, it may not be a node polynomial for a quadrature formula F_n.

A sufficient condition for a unique M, already known to Christoffel in case γ is of the form (1.1), is that all $|\zeta_\mu| > 1$. That is all preassigned nodes are outside the interval of support of γ. A modification of an argument of Gautschi [4] yields the following for slightly more general functionals.

Theorem 4.3. Let γ be a positive definite linear functional and let $Z = (x-\zeta_1) \dots (x-\zeta_k)$, $\zeta_\mu \in \mathbb{R}$, $\zeta_\mu \neq \zeta_k$, be such that $\gamma(PZ) > 0$ for all $P \in \Pi^+$. Then $\zeta_1,\dots,\zeta_k$ can be the set of preassigned nodes for a quadrature formula F_n with $d(F_n,\gamma) = 2n-1-k$, with all real, simple nodes.

Proof. Define γ^* on Π by

$$\gamma^*(P) = \gamma(PZ) .$$

Then γ^* is a positive definite linear functional on Π and hence there exists a sequence $\{R_n^*\}$ of monic polynomials orthogonal with respect to γ^*. Let $M = ZN \in \Theta_{n,k}$. Then $\gamma(M\cdot H) = 0$ for all $H \in \Pi_{n-k-1}$. It follows that $\gamma^*(NH) = 0$ for all $H \in \Pi_{n-k-1}$. Since $N \in \Pi_{n-k}$, one then has $N = R_{n-k}^*$ so that all of its zeros are real and simple.

5. Formulae with Non-Negative Weights

The question of when a formula has positive weights (and how many) was, after Stieltjes' initial results, taken up again in the 1930s by Pólya [9], Fejér [3] and Shohat [10] and very recently by Peherstorfer [8] and Sottas and Wanner [11]. We shall here only consider an extension of a result of Fejér and secondly the situation when $M = (x-\alpha)T$, $T(\zeta_\mu) = 0$, $\mu = 1,\dots,k-1$.

Theorem 5.1. Let $M = R_n + \alpha_1 R_{n-1} + \alpha_2 R_{n-2}$ be the node polynomial of a formula F_n with $d(F_n,\gamma) = 2n-3$. If $\alpha_1,\alpha_2 \in \mathbb{R}$ and $\alpha_2 < k_n$, then the weights of the formula are all positive.

Proof. We recall that $\lambda_v = \gamma(\Lambda_v)$ and, following Stieltjes, take a look at $\Psi_v = \Lambda_v - \Lambda_v^2$. We can write

$$\Psi_v = M(-\delta_1^2(v)R_{n-2} + H) , \quad H \in \Pi_{n-3} ,$$

if $\Lambda_v = \sum_{\mu=1}^{n} \delta_\mu(v)R_{n-\mu}$.

The leading coefficient in Ψ_v is then $-\delta_1^2(v)$. Further, recalling that $\gamma(R_\mu^2) = N_\mu$, we have

$$\gamma(\Psi_v) = -\delta_1^2(v)\alpha_2 N_{n-2} \quad \text{and} \quad \gamma(\Lambda_v^2) = \sum_{\mu=1}^{n} \delta_\mu^2(v) N_{n-\mu} .$$

Hence

$$\begin{aligned} \lambda_v = \gamma(\Lambda_v) &= \gamma(\Psi_v) + \gamma(\Lambda_v^2) \\ &= -\delta_1^2(v)\alpha_2 N_{n-2} + \delta_1^2(v) N_{n-1} + \ldots + \delta_n^2(v) N_o \\ &\geq \delta_1^2(v) N_{n-1}\left(1 - \alpha_2 \frac{N_{n-2}}{N_{n-1}}\right) . \end{aligned}$$

Since $k_n = N_{n-1}/N_{n-2}$, it follows that $\lambda_v > 0$ if $k_n - \alpha_2 > 0$ (since $\delta_1(v) \neq 0$). This completes the proof of the theorem.

Combining this result with Theorem 3.4 we arrive at

Theorem 5.2. Let $M = R_n + \alpha_2 R_{n-1} + \alpha_2 R_{n-2}$, $\alpha_1 \in \mathbb{R}$, $\alpha_2 < k_n$. Then M has real, simple zeros and thus is the node polynomial for a quadrature formula F_n , with $d(F_n,\gamma) = 2n-3$, all of whose weights are positive.

Our last result is the following.

Theorem 5.3. Let $\zeta_1,\ldots,\zeta_k$ be preassigned nodes for a quadrature formula F_n and assume that $T \in \Theta_{n-1,k-2}$ has only real, simple zeros and satisfies $T(\zeta_\mu) = 0$, $\mu = 1,\ldots,n-1$, $T(\zeta_k) \neq 0$ and $D_{n-2}(\zeta_1,\ldots,\zeta_{k-1}) \neq 0$ and $1 \leq k \leq n-1$. Then, if we set $x_n = \zeta_k$, so that the node polynomial of F_n becomes $M = (x-\zeta_k)T$, we have

$$\lambda_\mu^{(n)} = \lambda_\mu^{(n-1)}, \quad \mu = 1,\ldots,n-1; \quad \lambda_n^{(n)} = 0 .$$

Here $\lambda^{(n-1)}$ are the weights of the formula F_{n-1} defined by the node polynomial T .

Proof. Let all zeros of T be $x_1,\ldots,x_{n-1}$. Then we have on the one hand

$$\gamma(P) = \sum_{\mu=1}^{n-1} \lambda_\mu^{(n-1)} P(x_\mu) , \quad P \in \Pi_{2(n-1)-(k-1)-1} = \Pi_{2n-2-k} .$$

On the other hand, for the node polynomial $M = (x-\zeta_k)T$,

$$\gamma(P) = \sum_{\mu=1}^{n-1} \lambda_\mu^{(n)} P(x_\mu) + \lambda_n^{(n)} P(\zeta_k) , \quad P \in \Pi_{2n-1-k} .$$

Then for all $P \in \Pi_{2n-2-k}$,

$$(5.1) \qquad 0 = \sum_{\mu=1}^{n-1} (\lambda_\mu^{(n)} - \lambda_\mu^{(n-1)})P(x_\mu) + \lambda_n^{(n)}P(\zeta_k) .$$

Set $P = T \in \Pi_{n-1}$. If $k \leq n-1$, then $T \in \Pi_{2n-2-k}$. Thus

$$0 = \lambda_n^{(n)}T(\zeta_k)$$

and it follows that $\lambda_n^{(n)} = 0$. Substituting $\Lambda_\mu^{(n-1)}$, $\mu = 1,\ldots,n-1$ in (5.1) one obtains $0 = \lambda_\mu^{(n)} - \lambda_\mu^{(n-1)}$. This completes the proof of the theorem.

References

1. Brezinski, C. Padé-type approximation and general orthogonal polynomials, ISNM 50, Birkhäuser Verlag (1980).

2. Draux, A. Polynômes Orthogonaux Formels - Applications, Lecture Notes in Mathematics No.974, Springer-Verlag, Berlin (1983).

3. Fejér, L. Mechanische Quadraturen mit positiven Cotesschen Zahlen, Math. Zeitschr. 37 (1933), 287-309.

4. Gautschi, W. A survey of Gauss-Christoffel quadrature formulae, E. B. Christoffel (P.L. Butzer and F. Fehér, Eds.), Birkhäuser Verlag (1981), 72-147.

5. Gragg, W. B. Matrix interpretations and applications of the continued fraction algorithm, Rocky Mtn. J. Math. 4 (1974), 213-225.

6. Micchelli, C. A. and Rivlin, T. J. Numerical integration rules near Gaussian quadrature, Israel J. Math. 16 (1973), 287-299.

7. Njåstad, O. and Thron, W. J. The theory of sequences of orthogonal L-polynomials, Kgl. Norske Vidensk. Selsk. Skrifter (1983), No. 1, 54-91.

8. Peherstorfer F. Characterization of positive quadrature formulas, SIAM J. Math. Anal. 12 (1981), 935-942.

9. Pólya, G. Über die Konvergenz von Quadraturverfahren, Math. Zeitschr. 37 (1933), 264-286.

10. Shohat, J. On mechanical quadratures, in particular, with positive coefficients, Trans. AMS 42 (1937), 461-496.

11. Sottas, G. and Wanner, G. The number of positive weights of a quadrature formula, BIT 22 (1982), 339-352.

12. Wall, H. S. Analytic Theory of Continued Fractions, Van Nostrand, New York, 1948.

DISCRETE ℓ_p APPROXIMATION BY RATIONAL FUNCTIONS

G.A. Watson
Department of Mathematical Sciences
University of Dundee
Dundee DD1 4HN
Scotland

Abstract. The numerical solution of rational discrete ℓ_p approximation problems is considered. For the cases $1 < p < \infty$, Gauss-Newton and separated Gauss-Newton methods are developed, and convergence results established. An algorithm for the problem with $p = 1$ is also outlined.

1. Introduction

For given functions $g_j : R^N \to R$, $j = 0,1,\dots,m$ and $h_j : R^N \to R$, $j = 0,1,\dots,n$, and points $x_i \in R^N$, $i = 1,2,\dots,t$, consider the family of generalized rational functions

$$\mathcal{R} = \{P/Q : P = \sum_{j=0}^{m} a_{j+1} g_j(x) \ , \ Q = h_0(x) + \sum_{j=1}^{n} b_j h_j(x) \ , \ Q_i > 0 \ , \ i = 1,2,\dots,t\}$$

where Q_i (and subsequently P_i) denotes $Q(x_i,b)(P(x_i,a))$. (The explicit dependence on parameters will normally be suppressed when no confusion can arise.) Let B be the subset of R^n defined by

$$B = \{b \in R^n \ , \ P/Q \in \mathcal{R}\} .$$

Then of interest here is the numerical solution of the following problem:

$$\text{given } f \in R^t, \text{ find } a \in R^{m+1}, \ b \in B \ \text{ to minimize } \ \|r\|_p \tag{1.1}$$

where $r \in R^t$ has i^{th} component

$$r_i = f_i - P_i/Q_i \ , \quad i = 1,2,\dots,t \tag{1.2}$$

and the norm is the ℓ_p norm

$$\|r\|_p = \left(\sum_{i=1}^{t} |r_i|^p\right)^{1/p}, \quad 1 \le p < \infty . \tag{1.3}$$

This problem, in particular when $1 \le p \le 2$, is a central problem in data fitting, when the correct model of the data is a function from R. In theory, existence of a best approximation from R cannot be guaranteed, although this will not be regarded as a practical difficulty. Lack of convexity means that local minima may be expected, and the intention here will merely be to determine such a point. This paper is primarily concerned with the treatment of (1.1) when $1 < p < \infty$, and the next two sections are devoted to this. Finally, Section 4 deals briefly with the important case $p = 1$.

The limiting case $p = \infty$ of (1.3), when (1.1) is a discrete rational Chebyshev approximation problem, will not be considered here. A number of algorithms are available, with the differential correction algorithm often recommended (see [1],[9] [10]).

2. The Cases $1 < p < \infty$

From an algorithmic point of view, it is important to be able to define the following diagonal matrix

$$D = \mathrm{diag}\{|r_i|^{p-2},\ i = 1,2,\ldots,t\} \tag{2.1}$$

at all points $a \in R^{m+1}$, $b \in B$ which are of interest. This matrix is defined for all $p \ge 2$, and for $1 < p < 2$ provided that $r_i \ne 0$ for every i : it will be assumed that this is so. When $p = 1$, a solution to (1.1) is characterized by certain zero components of r (see Section 4) so it is the case that some elements of D will become increasingly large as p gets close to 1 ; however, for reasonable values of p, it is possible to work with D except in pathological cases. The problem (1.1) may then be rewritten as the minimization of

$$\phi = \frac{1}{p} r^T D r \tag{2.2}$$

over all $a \in R^{m+1}$, $b \in B$. Let ϕ' denote the vector in R^{m+n+1} whose components are the partial derivatives of ϕ with respect to its parameters, and let ϕ'' be the corresponding $(m+n+1) \times (m+n+1)$ matrix of second partial derivatives. Then it is easily seen that

$$\phi' = C^T D r \tag{2.3}$$

$$\phi'' = (p-1)C^T DC + \sum_{i=1}^{t} r_i |r_i|^{p-2} T_i \ , \tag{2.4}$$

where C is the $t \times (m+n+1)$ matrix of partial derivatives of r with respect to its parameters, and T_i is the $(m+n+1) \times (m+n+1)$ matrix of corresponding second derivatives of r_i, $i = 1,2,\ldots,t$. If $a \in R^{m+1}$, $b \in B$ solves (1.1) then $\phi' = 0$, and the Newton step δ for the solution of this system of equations satisfies

$$((p-1)C^T DC + \sum_{i=1}^{t} r_i |r_i|^{p-2} T_i)\delta = -C^T Dr \ . \tag{2.5}$$

2.1 The Gauss-Newton Method

For reasons which are now well understood (2.5) as it stands is not suitable as the basis of an iterative procedure for solving (1.1). For the special case $p = 2$, the approximation of the Hessian matrix of ϕ given by replacing the second term of (2.5) by μI, where $\mu \geq 0$ is adaptively fixed at each iteration, gives a method of Levenberg-Marquardt type. When $C^T C$ remains positive definite, it is possible to set $\mu = 0$ on each iteration, and this is the basis of the familiar Gauss-Newton method for nonlinear least squares problems. This interpretation of the Gauss-Newton method suggests the following iterative scheme for the solution of (1.1), valid when $C^T DC$ is (uniformly) positive definite.

(1) calculate δ to satisfy

$$(p-1)C^T DC\ \delta = -C^T Dr \tag{2.6}$$

(2) set

$$\begin{pmatrix} a \\ b \end{pmatrix} := \begin{pmatrix} a \\ b \end{pmatrix} + \gamma\delta \tag{2.7}$$

where γ is chosen to reduce ϕ subject to the restriction that the new $b \in B$.

The solution δ to (2.6) may be efficiently obtained by solving the ℓ_2 problem

$$\underset{\delta}{\text{minimize}} \ \| (p-1)D^{\frac{1}{2}}C\delta + D^{\frac{1}{2}}r \|_2 \ . \tag{2.8}$$

The requirement that the new $b \in B$ means that γ must not exceed the value

$$\bar{\gamma} = \min_i \{-\frac{Q_i}{\delta Q_i}, \quad \delta Q_i < 0\}$$

where $\delta Q_i = \sum_{j=1}^{n} \delta_{m+j+1} h_j(x_i)$, $i = 1,2,\ldots,t.$

A different form of generalization of the Gauss-Newton method is given in [12] and [14] motivated by the fact that the resulting algorithm is norm-independent: in the ℓ_p case, δ is calculated by solving the linear ℓ_p approximation problem

$$\underset{\delta}{\text{minimize}} \; \|C\delta + r\|_p . \tag{2.9}$$

This is no longer a finite calculation when $p \neq 2$, although asymptotically methods using (2.6) and (2.9) (with $\gamma = 1$) are equivalent.

It is well-known that when $p = 2$, the iteration (2.6), (2.7) with γ set to 1 is frequently convergent. It is also well-known that for certain problems associated with $\|r\|_2$ large at the solution, a reduced step length is required, but even if it is not the rate of convergence can be very slow. For such problems attention has been directed to better approximations of the Hessian matrix, and a number of attempts have been made to incorporate the first term of (2.4) into an approximate Hessian (see, for example, [2],[3],[4]). It is clear that such ideas are not confined to the case $p = 2$, so that current research on methods for nonlinear least squares problems will have implications for other values.

It may also be argued that unsatisfactory performance of the Gauss-Newton method means that the wrong model is being used. In any event, the intention here is to concentrate on this method, and its generalizations based on (2.6), (2.7). Clearly the basic iteration applies to any nonlinear ℓ_p approximation problem. One way in which advantage can be taken of the special structure of the rational functions is by exploiting the fact that r is an affine function of a for fixed b. For the case $p = 2$, the idea of treating the variables a and b separately has been pursued by a number of authors (see, for example, [6],[7],[8],[11],[13]). The generalization of some of the ideas developed in the review paper [13] to values of p other than 2 is now considered.

2.2 Separation of the Variables

Differentiating ϕ separately with respect to a and b gives

$$\phi' = \begin{bmatrix}\phi_a\\ \phi_b\end{bmatrix} = \begin{bmatrix}G^T Dr\\ H^T Dr\end{bmatrix}$$

where $C = [G \vdots H]$. Further

$$\phi'' = \begin{bmatrix}\phi_{aa} & \phi_{ab}\\ \phi_{ba} & \phi_{bb}\end{bmatrix}$$

$$= \begin{bmatrix}(p-1)G^T DG & (p-1)G^T DH + \sum_{i=1}^{t} r_i|r_i|^{p-2}R_i\\ (p-1)H^T DG + \sum_{i=1}^{t} r_i|r_i|^{p-2}R_i^T & (p-1)H^T DH + \sum_{i=1}^{t} r_i|r_i|^{p-2}S_i\end{bmatrix}$$

where

$$R_i = \frac{1}{Q_i^2} g_i h_i^T ,$$

$$S_i = -\frac{2P_i}{Q_i^3} h_i h_i^T ,$$

with

$$g_i = [g_0(x_i),\dots,g_m(x_i)]^T$$

$$h_i = [h_1(x_i),\dots,h_n(x_i)]^T .$$

Let

$$A = (p-1)C^T DC ,$$

$$E = \sum_{i=1}^{t} r_i|r_i|^{p-2}T_i .$$

Then

$$\phi'' = A + E = \begin{bmatrix}A_{11} & A_{12} + E_{12}\\ A_{21} + E_{21} & A_{22} + E_{22}\end{bmatrix}$$

partitioning in an obvious way. Let

$$\phi_a = 0 \tag{2.10}$$

define $a = a(b)$, so that $\phi_b = 0$ becomes

$$F(b) = \phi_b(a(b),b) = 0 . \tag{2.11}$$

Then

$$F' = \phi_{bb} - \phi_{ba}(\phi_{aa})^{-1}\phi_{ab}$$

provided that G^TDG is positive definite, and the Newton step s in the variables b satisfies

$$Ns = -F$$

where

$$N = [(A_{22} + E_{22}) - (A_{12} + E_{12})^T A_{11}^{-1}(A_{12} + E_{12})] . \qquad (2.12)$$

There are two convenient ways in which the Hessian matrix N can be approximated, and where the resulting systems of equations can be readily solved as linear least squares problems. The approximations are

$$N_I = A_{22} - A_{12}^T A_{11}^{-1} A_{12} + E_{12}^T A_{11}^{-1} E_{12}$$

and

$$N_{II} = A_{22} - A_{12}^T A_{11}^{-1} A_{12} .$$

Let the QR factorization of $D^{\frac{1}{2}}G$ be

$$D^{\frac{1}{2}}G = [Q_1 \vdots Q_2]\begin{bmatrix} U \\ 0 \end{bmatrix}$$

where $[Q_1 \vdots Q_2]$ is a $t \times t$ orthogonal matrix, and U is an $(m+1) \times (m+1)$ upper triangular matrix. Then an s satisfying

$$N_{II}\, s = -F$$

may be obtained as the ℓ_2 solution of the system of $(t-m-1)$ equations in n unknowns

$$(p-1)Q_2^T D^{\frac{1}{2}} Hs = -Q_2^T D^{\frac{1}{2}} r . \qquad (2.13)$$

Now let the $(m+1) \times n$ matrix V satisfy

$$U^T V = E_{12} . \qquad (2.14)$$

Then

$$V^T V = E_{12}^T (G^T DG)^{-1} E_{12}$$
$$= (p-1) E_{12}^T A_{11}^{-1} E_{12}$$

and an s satisfying

$$N_I s = -F$$

may be obtained as the ℓ_2 solution of the system of t equations in n unknowns

$$\begin{bmatrix} (p-1) Q_2^T D^{\frac{1}{2}} H \\ V \end{bmatrix} s = \begin{bmatrix} -Q_2^T D^{\frac{1}{2}} r \\ 0 \end{bmatrix} . \tag{2.15}$$

When $p = 2$, the solution of (2.10) for a is a finite problem, and may be efficiently obtained by solving the linear least squares problem

$$\underset{a}{\text{minimize}} \; || Ga + f ||_2 . \tag{2.16}$$

An apparent drawback in exploiting separability in other cases is that (2.10) loses this finite property, and so iterative methods must be used. Newton's method (with line search if necessary) may be readily applied to this problem as a sequence of linear least squares problems (for example [14]), and in particular when b is close to a solution, few steps would be expected to be necessary starting from the current value of a. In addition, far from a solution to (1.1), it may be sufficient to work with a fairly crude estimate of the solution to (2.10), for although this means that the problems (2.13) and (2.15) no longer provide the required directions, progress may still be possible. It remains to be seen whether these ideas can give rise to methods which give an improvement in efficiency over the direct application of the unseparated Gauss-Newton method. Further investigations, and in particular numerical experiments, are obviously needed. Meantime some local convergence results relevant to all these methods are developed.

3. Local Convergence

The Gauss-Newton method (2.6), (2.7) and its separated variants described in the previous section when used without a line search can be interpreted as simple iterative methods with iteration functions, respectively

$$g_G(z) = z - A^{-1}\phi' \qquad \text{where} \quad z = \begin{pmatrix} a \\ b \end{pmatrix} \tag{3.1}$$

$$g_I(b) = b - N_I^{-1} F \tag{3.2}$$

$$g_{II}(b) = b - N_{II}^{-1} F \tag{3.3}$$

Local convergence therefore depends on the size of the eigenvalues of g' at fixed points of these iterations. If the smallest intervals containing these eigenvalues are $[\alpha_G,\beta_G]$, $[\alpha_I,\beta_I]$ and $[\alpha_{II},\beta_{II}]$ respectively, then the analysis of Ruhe and Wedin [13] gives the following relationships between these intervals.

Theorem 1 (i) If $\alpha_I < 0$ and $\beta_I > 0$, then

$$\alpha_G \le \alpha_I \ , \ \beta_I \le \beta_G \ .$$

(ii) If $\alpha_{II} < 0$ and $0 < \beta_{II} \le 1$, then

$$\alpha_G \le \alpha_{II} \ , \ \beta_{II} \le \beta_G \ .$$

(iii) If $\alpha_I \le 0$ and $0 \le \beta_{II} \le 1$, then

$$\alpha_{II} \le \alpha_I \ , \ \beta_{II} \le \beta_I \ .$$

Attention will be focussed therefore on the unseparated Gauss-Newton method, for which at a point such that $\phi'(z) = 0$,

$$g_G'(z) = -A^{-1}E \quad . \tag{3.4}$$

It is interesting that (3.4) is also true for the iteration function arising from the interpretation as simple iteration of the Gauss-Newton method based on (2.9). Some of the results obtained in [12] (from a different standpoint) for the latter method are therefore relevant here; however some improvements in both the complexity and the quality of these results is obtained by working directly with (3.4).

Let the maximum modulus eigenvalue of (3.4) be $\rho_p (=\rho_p(f))$. Then

$$\rho_p = \max_{z \neq 0} h(z)$$

where

$$h(z) = \frac{\left|\sum_{i=1}^{t} r_i |r_i|^{p-2} z^T T_i z\right|}{(p-1) z^T C^T D C z} \quad . \tag{3.5}$$

In order to bound the denominator of (3.5) away from zero, it is convenient to consider two separate cases. The proofs of the following theorems are straightforward.

Theorem 2 Let $1 < p \le 2$, and let r be a minimum for (1.1) with

(i) $r_i \neq 0$, $i = 1,2,\ldots,t$, if $p < 2$,

(ii) $\min_{\|z\|_2=1} z^T C^T C z \geq \Delta_p > 0$.

Then

$$z^T C^T D C z \geq \|r\|_p^{p-2} \Delta_p .$$

Theorem 3 Let $2 < p < \infty$, and let r be a minimum for (1.1) with

(i) $r \neq 0$

(ii) $\min_j \min_{\|z\|_2=1} z^T C_j^T C_j z \geq \delta_p > 0$

where $\{C_j\}$ is the set of $n \times n$ submatrices of C,

(iii) $\|r\|_p / |r_n| \leq K_p$

where (without loss of generality)

$$|r_1| \geq |r_2| \geq \ldots\ldots \geq |r_t| .$$

Then

$$z^T C^T D C z \geq \|r\|_p^{p-2} K_p^{2-p} \delta_p .$$

The condition (ii) of Theorem 2 is a consequence of C having full rank. The condition (ii) of Theorem 3 is a consequence of C satisfying the considerably stronger Haar condition. The latter condition, together with (i) and the fact that $\phi' = 0$, implies that at least n components of r must be non-zero, so that the existence of K_p follows. Notice that usually $\|r\|_p / |r_n| \to 1$ as $p \to \infty$.

Theorem 4 Let the conditions of Theorem 2 or Theorem 3 be satisfied. Then

$$\rho_p \leq \frac{1}{p-1} c_p \|r\|_p , \qquad (3.6)$$

where

(i) $c_p = \alpha_p / \Delta_p$, $1 < p \leq 2$

(ii) $c_p = \alpha_p K_p^{p-2} / \delta_p$, $2 < p < \infty$

and $\|\lambda\|_p \leq \alpha_p$, with $\lambda \in R^t$ having ith component the maximum modulus eigenvalue of T_i, $i = 1,2,\ldots,t$.

Proof Let $\|z\|_2 = 1$ and let $v = \nabla_r \|r\|_p$. Then

$$\left|\sum_{i=1}^{t} r_i |r_i|^{p-2} z^T T_i z\right| = \|r\|_p^{p-1} \left|\sum_{i=1}^{t} v_i z^T T_i z\right| \le \alpha_p \|r\|_p^{p-1}$$

using the fact that $\|v\|_q \le 1$, where $1/p + 1/q = 1$. The result follows using Theorems 2 and 3. □

The constants occurring in these theorems clearly depend on f as well as p. However the pointwise conditions can be strengthened in an obvious way to give uniform results valid over a range of data vectors f. Rather more can in fact be said about the constant α_p. The matrix T_i is given by

$$T_i = \begin{bmatrix} 0 & R_i \\ R_i^T & S_i \end{bmatrix} = \frac{1}{Q_i^3} \begin{bmatrix} 0 & Q_i g_i h_i^T \\ Q_i h_i g_i^T & -2P_i h_i h_i^T \end{bmatrix}.$$

This is a rank-2 matrix whose non-zero eigenvalues may be readily shown to be

$$\lambda = -\frac{h_i^T h_i}{Q_i^2} \left\{ \frac{P_i}{Q_i} \pm \sqrt{\frac{P_i^2}{Q_i^2} + \frac{g_i^T g_i}{h_i^T h_i}} \right\}.$$

Let

$$Q_i \ge \varepsilon > 0 \;, \; i = 1,2,\ldots,t. \tag{3.7}$$

Then a suitable α_p is given by

$$\alpha_p = t^{1/p} \max_{1 \le i \le t} \left\{ \frac{h_i^T h_i}{\varepsilon^2} \left\{ \|f\| + \sqrt{\|f\|^2 + \frac{g_i^T g_i}{h_i^T h_i}} \right\} \right\}.$$

Clearly a strengthening of the pointwise inequalities (3.7) is necessary to give a uniform result.

4. The ℓ_1 Problem

Let S denote the set

$$S = \{i : r_i = 0\}$$

and let

$$\theta_i = \text{sign}\ (r_i)\ ,\ i \in S^c\ ,$$

where S^c denotes the indices from $\{1,2,...,t\}$ not in S.

Theorem 5 Let $a \in R^{m+1}$, $b \in B$ solve (1.1) when $p = 1$, and let

$$\Lambda = \{\lambda \in R^t : |\lambda_i| \le 1\ ,\ \lambda_i = \theta_i\ ,\ i \in S^c\}.$$

Then there exists $\lambda \in \Lambda$ such that

$$C^T\lambda = 0\ . \tag{4.1}$$

Proof See, for example, [5]. □

Let $z = \binom{a}{b}$ be given, where $a \in R^{m+1}$, $b \in B$, and let $s \in R^{m+n+1}$ be a given non-zero vector. Let

$$\delta P = \sum_{j=0}^{n} s_{j+1} g_j(x)$$

$$\delta Q = \sum_{j=1}^{n} s_{m+j+1} h_j(x).$$

Then for scalar $\gamma > 0$

$$\| r(z+\gamma s)\|_1 = \sum_{i=1}^{t} \left|f_i - \frac{P_i + \gamma\delta P_i}{Q_i + \gamma\delta Q_i}\right|$$

$$= \|r\|_1 + \sum_{i\in S^c}\theta_i\rho_i(C)s(\gamma-\gamma^2\frac{\delta Q_i}{Q_i} + O(\gamma^3))$$

$$+ \sum_{i\in S}|\rho_i(C)s(\gamma-\gamma^2\frac{\delta Q_i}{Q_i} + O(\gamma^3))| \tag{4.2}$$

where γ is small enough, say $\gamma \le \hat{\gamma}$, that no component of $r_i(z + \gamma s)$, $i \in S^c$ changes sign, and where $\rho_i(C)$ denotes the ith row of C. It follows from (4.2) that s is a descent direction for $\|r\|_1$ at z provided that

$$\sum_{i\in S^c}\theta_i\rho_i(C)s + \sum_{i\in S}|\rho_i(C)s| < 0\ . \tag{4.3}$$

It also follows from (4.2) that if S contains $k < m+n+1$ indices, it is possible to find a descent direction s which has the property that

$$r_i(z + \gamma s) = 0 \ , \quad i \in S \tag{4.4}$$

for all $\gamma, 0 \le \gamma \le \hat{\gamma}$, by choosing s so that

$$\rho_i(C)s = 0 \ , \quad i \in S \ ,$$

$$\sum_{i \in S^c} \theta_i \rho_i(C)s < 0 \ ,$$

unless the coefficient of s in the last inequality is zero.

If S contains exactly $m+n+1$ indices, then the conditions of Theorem 5 may be tested by regarding λ_i, $i \in S$ as unknowns in (4.1). If $|\lambda_j| > 1$, for some $j \in S$, it is again possible to find a descent direction s, with the property that

$$r_i(z + \gamma s) = 0 \ , i \in S - \{j\} \tag{4.5}$$

for all γ, $0 \le \gamma \le \hat{\gamma}$, by choosing s so that

$$\rho_i(C)s = 0 \ , \ i \in S - \{j\}$$

$$|\rho_j(C)s| + \sum_{i \in S^c} \theta_i \rho_i(C)s < 0 \quad .$$

A method for calculating a suitable s based on these ideas is developed in [15]. The situation when S contains $k > m+n+1$ indices corresponds to a form of degeneracy which would not be expected to arise with genuinely error-contaminated data, although a theoretical resolution of this problem is also given. Usually, it is possible to move to a new rational approximation from R with smaller $\|r\|_1$ which is such that S contains $(k+1)$ indices, or regains $(m+n+1)$ indices. The process is repeated until the conditions of Theorem 5 are satisfied. If such a move is not possible at any time, then second derivative information is incorporated into the algorithm by exploiting the fact that if $\rho_i(C)s = 0$, $i \in S$, then the right hand side of (4.2) is just a power series in γ. Therefore either convergence is obtained in a finite number of steps or a second order convergence rate is possible. Full details of the proposed algorithm, together with numerical results of its application to polynomial rational approximation of a number of data sets, are given in [15].

References

1. Barrodale, I., Powell, M.J.D. and Roberts, F.D.K., The differential correction algorithm for rational L_∞ approximation, SIAM J. Num. Anal. 9 (1972), 493-504.

2. Dennis, J.E., Gay, D.M. and Welsch, R.E., An adaptive nonlinear least-squares algorithm, ACM Trans. Math. Software 7 (1981), 348-368.

3. Dennis, J.E. and Schnabel, R.B., Least change secant updates for quasi-Newton methods, SIAM Rev. 21 (1979), 443-459.

4. Dennis, J.E. and Walker, H.F., Convergence theorems for least change secant update methods, SIAM J. Num. Anal. 18 (1981), 949-987.

5. Fletcher, R. and Watson, G.A., First and second order conditions for a class of nondifferentiable optimization problems, Math. Prog. 18 (1980), 291-307.

6. Golub, G.H. and Pereyra, V., The differentiation of pseudo-inverses and nonlinear least squares problems whose variables separate, SIAM J. Num. Anal. 10 (1973), 413-432.

7. Golub, G.H. and Pereyra, V., Differentiation of pseudo-inverses, separable nonlinear least squares problems and other tales, Generalized Inverses and Applications, Academic Press, New York, 1976.

8. Kaufman, L., Variable projection method for solving separable nonlinear least squares problems, BIT 15 (1975), 49-57.

9. Kaufman, E.H., McCormick, S.F. and Taylor, G.D., An adaptive differential-correction algorithm, J. Approx. Th. 37 (1983), 197-211.

10. Lee, C.M. and Roberts, F.D.K., A comparison of algorithms for rational approximation, Math. of Comp. 27 (1973), 111-121.

11. Osborne, M.R., Some special nonlinear least squares problems, SIAM J. Num. Anal. 12 (1975), 571-592.

12. Osborne, M.R. and Watson, G.A., Nonlinear approximation problems in vector norms, in Numerical Analysis, Dundee 1977, ed. G.A. Watson, Springer-Verlag, Berlin, 1978.

13. Ruhe, A. and Wedin, P.Å., Algorithms for separable nonlinear least squares problems, SIAM Rev. 22 (1980), 318-337.

14. Watson, G.A., Approximation Theory and Numerical Methods, Wiley, Chichester, 1980.

15. Watson, G.A., Discrete ℓ_1 approximation by rational functions, University of Dundee, Department of Mathematical Sciences Report NA/67, 1983; IMAJ Num. Anal. 4, in press.

WHAT IS BEYOND SZEGÖ'S THEORY OF ORTHOGONAL POLYNOMIALS?

Attila Máté
Department of Mathematics
Brooklyn College of the
City University of New York
Brooklyn, New York 11210
U.S.A.

Paul Nevai
Department of Mathematics
Ohio State University
Columbus, Ohio 43210
U.S.A.

and

Vilmos Totik
Bolyai Institute
University of Szeged
6720 Szeged, Hungary

Abstract. Consider a system $\{\phi_n\}$ of polynomials orthonormal on the unit circle with respect to the measure $d\mu$ with $\mu' > 0$ almost everywhere. Then

$$\lim_{n\to\infty} \int_{-\pi}^{\pi} ||\phi_n(e^{i\theta})|\sqrt{\mu'(\theta)} - 1|^2 \, d\theta = 0 .$$

This result enables one to extend many results of Szegö's theory to the case $\mu' > 0$.

The purpose of this paper is to extend results of Szegö's theory of orthogonal polynomials to orthogonal polynomials corresponding to measures whose derivatives are positive almost everywhere.

Let $d\mu$ be a nonnegative measure on $[-\pi, \pi]$ such that its support,

---

This material is based upon work supported by the National Science Foundation under Grant Nos. MCS 8100673 (first author) and MCS-83-00882 (second author) and by the PSC-CUNY Research Award Program of the City University of New York under Grant No. 662043 (first author). The third author made his contributions to the paper while visiting the Ohio State University.

supp($d\mu$), is infinite, and let $\phi_n(d\mu, z) = \kappa_n(d\mu)z^n + \dots$, $\kappa_n > 0$ denote the corresponding orthonormal polynomials, that is

$$\frac{1}{2\pi}\int_{-\pi}^{\pi} \phi_n(e^{i\theta})\,\overline{\phi_m(e^{i\theta})}\,d\mu(\theta) = \delta_{nm} .$$

Similarly, if $d\alpha$ is a nonnegative measure on $[-1, 1]$ and supp($d\alpha$) is infinite, then the corresponding orthonormal polynomials are denoted by $p_n(d\alpha, x) = \gamma_n(d\alpha)x^n + \dots$, $\gamma_n > 0$.

If $g \geq 0$ is such that $\log g \in L^1$ then the Szegö function $D(g)$ is defined by

$$D(g, z) = \exp\{\frac{1}{4\pi}\int_{-\pi}^{\pi} \log g(t)\,\frac{1 + ze^{-it}}{1 - ze^{-it}}\,dt\} , \qquad |z| < 1 .$$

It is well known that $D(g) \in H^2(|z| < 1)$, and

$$\lim_{r \to 1-0} D(g, re^{i\theta}) = D(g, e^{i\theta})$$

exists for almost every θ and $|D(g, e^{i\theta})|^2 = g(\theta)$ a.e. One of the main results of Szegö's theory [2], [3], [13] is that

$$(1) \qquad \lim_{n \to \infty}\int_{-\pi}^{\pi} |\phi_n(d\mu, z)z^{-n}\,\overline{D(\mu', z)} - 1|^2\,d\theta = 0 , \qquad z = e^{i\theta}$$

whenever $\log \mu' \in L^1$. We can prove the following

Theorem 1. *Let $\mu' > 0$ almost everywhere. Then*

$$\lim_{n \to \infty}\int_{-\pi}^{\pi} |\,|\phi_n(d\mu, z)|\sqrt{\mu'(\theta)} - 1|^2\,d\theta = 0 , \qquad z = e^{i\theta} ,$$

and for every $f \in L^2$

$$\lim_{n \to \infty}\int_{-\pi}^{\pi} f(\theta)|\phi_n(z)|^{-1}\,d\theta = \int_{-\pi}^{\pi} f(\theta)\sqrt{\mu'(\theta)}\,d\theta , \qquad z = e^{i\theta} ,$$

holds. If, in addition, μ is absolutely continuous then

$$\lim_{n \to \infty}\int_{-\pi}^{\pi} |\,|\phi_n(z)|^{-1} - \sqrt{\mu'(\theta)}|^2\,d\theta = 0 , \qquad z = e^{i\theta} .$$

Szegö [3], [13] proves (1) (see also Kolmogorov [4] and Krein [5]) by solving an extremal problem which can be stated as

$$(2) \qquad \lim_{n\to\infty} \kappa_n(d\mu_1)/\kappa_n(d\mu_2) = D(\mu_1'/\mu_2', 0)$$

if $\log \mu_1' \in L^1$ and $\log \mu_2' \in L^1$. The corresponding result for $\mu' > 0$ a.e. is

Theorem 2. Let $\mu_1' > 0$ a.e. and let $d\mu_2 = \rho d\mu_1$, where the function $\rho > 0$ is such that $R\rho^{\pm 1} \in L^\infty(d\mu_1)$ for a trigonometric polynomial R . Then (2) holds. Moreover,

$$\lim_{n\to\infty} \phi_n(d\mu_2, z)/\phi_n(d\mu_1, z) = \overline{D(\rho^{-1}, \bar{z}^{-1})} , \qquad |z| > 1 ,$$

and

$$\lim_{n\to\infty} \int_{-\pi}^{\pi} |\phi_n(d\mu_2, z)\, \overline{D(\rho, z)} - \phi_n(d\mu_1, z)|^2\, \mu_1'(\theta)\, d\theta = 0 , \quad z = e^{i\theta} ,$$

are satisfied.

As a typical pointwise asymptotic result on the unit circle we mention

Theorem 3. Let $\mu_1' > 0$ a.e. and let $d\mu_2 = \rho d\mu_1$, where the function $\rho > 0$ is such that $R\rho^{\pm 1} \in L^\infty(d\mu_1)$ for some trigonometric polynomial R . Suppose that at at a point t

$$(3) \quad \rho(t) > 0 \ \text{ and } \ |\rho(t) - \rho(\theta)| \le K|t - \theta|$$

are satisfied for $|t - \theta| < \delta$. Then

$$(4) \quad \lim_{n\to\infty} \phi_n(d\mu_2, e^{it})/\phi_n(d\mu_1, e^{it}) = \overline{D(\rho^{-1}, e^{it})} .$$

If (3) is uniformly satisfied on a set E (with the same δ for every $t \in E$) then the convergence in (4) is also uniform on E .

For orthogonal polynomials on the real line we can prove

<u>Theorem 4. Let</u> $\alpha' > 0$ <u>a.e. in</u> $[-1, 1]$ <u>and let</u> ℓ <u>be a fixed integer. Then for every</u> $f \in L^{\infty}$

$$\lim_{n \to \infty} \int_{-1}^{1} f(x)p_n(d\alpha, x)p_{n+\ell}(d\alpha, x)d\alpha(x) =$$

$$\lim_{n \to \infty} \int_{-1}^{1} f(x)p_n(d\alpha, x)p_{n+\ell}(d\alpha, x)\alpha'(x)dx = \frac{1}{\pi}\int_{-1}^{1} f(x)\, T_{|\ell|}(x)(1 - x^2)^{-1/2}\, dx ,$$

<u>where</u> $T_{|\ell|}$ <u>denotes the</u> $|\ell|$<u>-th Chebyshev polynomial, and thus</u>

$$\lim_{n \to \infty} \int_{-1}^{1} p_n^2(d\alpha, x)d\alpha_s(x) = 0 ,$$

<u>where</u> α_s <u>is the singular component of</u> α .

The Turán determinant $D_n(d\alpha)$ is defined by

$$D_n(d\alpha, x) = p_n^2(d\alpha, x) - p_{n+1}(d\alpha, x)p_{n-1}(d\alpha, x) .$$

<u>Theorem 5. Let</u> $\alpha' > 0$ <u>a.e. in</u> $[-1, 1]$ <u>and let</u> $\Delta \subset (-1, 1)$ <u>be a fixed closed interval. Then there exists</u> $N = N(\Delta, \alpha)$ <u>such that</u> $D_n(x) > 0$, $x \in \Delta$, <u>for</u> $n > N$. <u>If</u> $f \in L^2$ <u>then</u>

$$\lim_{n \to \infty} \int_{\Delta} f(x)\, D_n(x)^{-1/2}\, dx = \sqrt{\frac{\pi}{2}} \int_{\Delta} f(x) \sqrt{\alpha'(x)}\, (1 - x^2)^{-1/4}\, dx .$$

<u>If</u> $f \in C$ <u>then</u>

$$\lim_{n \to \infty} \int_{\Delta} f(x)\, D_n(x)^{-1}\, dx = \frac{\pi}{2} \int_{\Delta} f(x)(1 - x^2)^{-1/2}\, d\alpha(x) .$$

<u>If</u> α <u>is absolutely continuous then</u>

$$\lim_{n \to \infty} \int_{\Delta} |(1 - x^2)^{1/2}\, D_n(x)^{-1} - \frac{\pi}{2}\alpha'(x)|dx = 0 .$$

<u>Theorem 6. If</u> $\alpha' > 0$ <u>a.e. in</u> $[-1, 1]$ <u>then</u>

$$\lim_{n \to \infty} \int_{-1}^{1} |D_n(x)\, \alpha'(x) - \frac{2}{\pi}(1 - x^2)^{1/2}|dx = 0 .$$

These limit relations show the significance of Turán's determinants. For example, one can recover both the absolutely continuous and singular parts of the measure from them.

The proofs of our theorems depend on results obtained by Szegö [13], Rahmanov [11], [12] (cf. Máté-Nevai [6], where an error in a key result of [11] was pointed out; a corrected proof appears in [12], and a simpler proof is given in Máté-Nevai-Totik [7]) and Nevai [8], [10], and their details will be published elsewhere together with some applications which include pointwise asymptotics for orthogonal polynomials on the support of the measure, weighted mean convergence of orthogonal series and Lagrange interpolation, Müntz-Szász type approximations and Wiener-Ingham-Turán type inequalities [14]. Finally, the results about Turán's determinant (Theorems 5 and 6) follow from the complex case and the relation

$$|\phi_n(z)|^2 \sin^2\theta = \frac{\pi}{2} D_n(x) + o(|\phi_n(z)|^2) , \qquad x = \cos\theta,\ z = e^{i\theta} ,$$

which can be proved to hold uniformly for $-1 \le x \le 1$ as $n \to \infty$.

As an illustration, we show how Theorem 4 can be used for obtaining the asymptotic distribution of the eigenvalues of certain Toeplitz matrices. Let $d\alpha$ be as above, $f \in L^\infty(d\alpha)$, and let

$$A_n(f) = A_n(d\alpha, f) = \{\int_{-1}^{1} f(x)\, p_k(d\alpha, x)\, p_m(d\alpha, x)\, d\alpha(x)\}_{k,m=0}^{n-1}$$

be the corresponding truncated Toeplitz matrix. The problem we treat here is to find the distribution of the eigenvalues of $A_n(d\alpha, f)$ as $n \to \infty$. We denote by $\Lambda_{kn}(f)$, $0 \le k < n$, $n = 1, 2,\ldots$ these eigenvalues and prove

Theorem 7. *Assume that* $\mathrm{supp}(d\alpha) = [-1, 1]$, $\alpha' > 0$ *almost everywhere in* $[-1, 1]$ *and* $f \in L^\infty(d\alpha)$. *If* G *is a continuous function defined on a closed interval containing the essential range of* f *then*

$$(5) \qquad \lim_{n\to\infty} \frac{1}{n} \Sigma_{k=0}^{n-1} G(\Lambda_{kn}(f)) = \frac{1}{\pi} \int_{-1}^{1} \frac{G(f(x))}{(1-x^2)^{1/2}}\, dx .$$

This result was found by the second author in [9] who proved it by a different method. The special case $f(x) = x$ is equivalent to a remarkable result of Erdös and Turán [1] about the distribution of zeros of orthogonal polynomials.

Proof of Theorem 7. We divide the proof into several steps.

Step 1. Let $g \in L^\infty(d\alpha)$. We claim that

$$\limsup_{n\to\infty} \frac{1}{n} |\mathrm{tr}\, A_n(g)| \le \left(\frac{1}{\pi}\int_{-1}^{1} g^2 V\right)^{1/2} ,$$

where $\mathrm{tr}\, A$ denotes the trace of the matrix A and $V(x) = (1 - x^2)^{-1/2}$ is the Chebyshev weight. Indeed, according to Theorem 4 we have

$$\lim_{n\to\infty} \frac{1}{n} \mathrm{tr}\, A_n(g) = \lim_{n\to\infty} \frac{1}{n} \sum_{k=0}^{n-1} \int_{-1}^{1} g\, p_k^2\, d\alpha = \frac{1}{\pi}\int_{-1}^{1} gV ,$$

and the absolute value of the right-hand side can be estimated according to Schwarz's inequality as being

$$\le \frac{1}{\pi}\int_{-1}^{1} |gV^{1/2}| \cdot V^{1/2} \le \frac{1}{\pi}\left(\int_{-1}^{1} g^2 V \cdot \int_{-1}^{1} V\right)^{1/2} = \left(\frac{1}{\pi}\int_{-1}^{1} g^2\, V\right)^{1/2} ,$$

establishing the claim above.

Step 2. Let $g_1, g_2 \in L^\infty(d\alpha)$. We will show that

$$\limsup_{n\to\infty} \frac{1}{n}|\mathrm{tr}(A_n(g_1)\, A_n(g_2))| \le \frac{1}{\pi}\left(\int_{-1}^{1} g_1^2\, V \cdot \int_{-1}^{1} g_2^2\, V\right)^{1/2}$$

To this end, we use Schwarz's inequality for sums to obtain the estimate

$$\limsup_{n\to\infty} \frac{1}{n} |\mathrm{tr}(A_n(g_1)\, A_n(g_2))| =$$

$$\limsup_{n\to\infty} \frac{1}{n} \left| \sum_{k=0}^{n-1} \sum_{m=0}^{n-1} \int_{-1}^{1} g_1 p_k p_m\, d\alpha \cdot \int_{-1}^{1} g_2 p_m p_k\, d\alpha \right| \le$$

$$\limsup_{n\to\infty} \frac{1}{n} \sum_{k=0}^{n-1} \left(\sum_{m=0}^{n-1} \left(\int_{-1}^{1} g_1 p_k p_m\, d\alpha\right)^2\right)^{1/2} \cdot \left(\sum_{m=0}^{n-1} \left(\int_{-1}^{1} g_2 p_k p_m\, d\alpha\right)^2\right)^{1/2} .$$

By Bessel's inequality, we have $(i = 1, 2)$

$$\sum_{m=0}^{n-1} \left(\int_{-1}^{1} (g_i p_k) p_m\, d\alpha\right)^2 \le \int_{-1}^{1} (g_i p_k)^2\, d\alpha .$$

Thus, using Schwarz's inequality for sums again, and then applying Theorem 4, we obtain that the above lim sup on the right-hand side is

$$\leq \limsup_{n\to\infty} \frac{1}{n} \sum_{k=0}^{n-1} \left(\int_{-1}^{1} (g_1 p_k)^2 \, d\alpha\right)^{1/2} \cdot \left(\int_{-1}^{1} (g_2 p_k)^2 \, d\alpha\right)^{1/2} \leq$$

$$\limsup_{n\to\infty} \left(\frac{1}{n} \sum_{k=0}^{n-1} \int_{-1}^{1} g_1^2 \, p_k^2 \, d\alpha\right)^{1/2} \cdot \left(\frac{1}{n} \sum_{k=0}^{n-1} \int_{-1}^{1} g_2^2 \, p_k^2 \, d\alpha\right)^{1/2} =$$

$$\frac{1}{\pi} \left(\int_{-1}^{1} g_1^2 \, V \cdot \int_{-1}^{1} g_2^2 \, V\right)^{1/2} ,$$

verifying the claim of Step 2.

<u>Step 3</u>. Let g_1 and g_2 be polynomials. We claim that

$$\lim_{n\to\infty} \frac{1}{n} \left(\mathrm{tr}\, A_n(g_1 g_2) - \mathrm{tr}(A_n(g_1)\, A_n(g_2))\right) = 0 .$$

Indeed, we have

$$\limsup_{n\to\infty} \left|\frac{1}{n} \left(\mathrm{tr}\, A_n(g_1 g_2) - \mathrm{tr}(A_n(g_1)\, A_n(g_2))\right)\right| =$$

$$\limsup_{n\to\infty} \left|\frac{1}{n} \sum_{k=0}^{n-1} \left(\int_{-1}^{1} g_1 g_2 p_k^2 \, d\alpha - \sum_{m=0}^{n-1} \int_{-1}^{1} g_1 p_k p_m \, d\alpha \int_{-1}^{1} g_2 p_m p_k \, d\alpha\right)\right| .$$

The first integral on the right can be rewritten by using the expansion $(i = 1, 2)$

$$g_i(x)\, p_k(x) = \sum_{m=0}^{\infty} p_m(x) \int_{-1}^{1} g_i(t)\, p_k(t)\, p_m(t)\, d\alpha(t)$$

(only finitely many terms are non-zero, as g_i is a polynomial), and it becomes

$$\sum_{m=0}^{\infty} \int_{-1}^{1} g_1 p_k p_m \, d\alpha \int_{-1}^{1} g_2 p_k p_m \, d\alpha .$$

Thus the above lim sup equals

$$\limsup_{n\to\infty} \frac{1}{n} \left| \sum_{k=0}^{n-1} \sum_{m=n}^{\infty} \int_{-1}^{1} g_1 p_k p_m \, d\alpha \int_{-1}^{1} g_2 p_k p_m \, d\alpha \right| .$$

Let ℓ be the maximum of the degrees of g_1 and g_2. Then all terms here with $|k - m| > \ell$ are zero in view of the orthogonality relations; thus there are fewer than ℓ^2 non-zero terms between the absolute value signs. Since we have $(i = 1, 2)$

$$\left|\int_{-1}^{1} g_i p_k p_m \, d\alpha\right| \le \|g_i\|_{L^\infty(d\alpha)} \int_{-1}^{1} |p_k p_m| d\alpha \le \|g_i\|_{L^\infty(d\alpha)} ,$$

we can see that the last expression involving lim sup is

$$\le \limsup_{n \to \infty} \left(\frac{\ell^2}{n} \|g_1\|_{L^\infty(d\alpha)} \|g_2\|_{L^\infty(d\alpha)}\right) = 0 ,$$

which establishes the claim made at the beginning of the present Step.

We can complete the proof of Theorem 7 as follows. Let $g_1, g_2 \in L^\infty(d\alpha)$. By choosing two sequences of polynomials $\{g_{i\ell}\}_{\ell=1}^{\infty}$ $(i = 1, 2)$ with $g_{i\ell} \to g_i$ in $L^2(Vdx)$, it follows from the claims established in Step 1-3 that

$$\lim_{n \to \infty} \frac{1}{n} (\mathrm{tr}\, A_n(g_1 g_2) - \mathrm{tr}(A_n(g_1)\, A_n(g_2))) = 0 .$$

Now let $f \in L^\infty_{d\alpha}$ and let G be an arbitrary polynomial. It follows by an iteration of the above relation that

$$\lim_{n \to \infty} \frac{1}{n} (\mathrm{tr}\, A_n(G(f)) - \mathrm{tr}\, G(A_n(f))) = 0 .$$

As the trace of a matrix equals the sum of its eigenvalues, this implies

$$\lim_{n \to \infty} \frac{1}{n} \sum_{k=0}^{n-1} G(\Lambda_{kn}(f)) = \lim_{n \to \infty} \mathrm{tr}\, G(A_n(f))$$

$$= \lim_{n \to \infty} \frac{1}{n} \mathrm{tr}\, A_n(G(f)) = \frac{1}{\pi} \int_{-1}^{1} G(f)\, V ,$$

where the last equality follows from the second centered formula of Step 1. Writing $B = \|f\|_{L^\infty(d\alpha)}$, it is well known that all eigenvalues $\Lambda_{kn}(f)$ lie in the interval $[-B, B]$; the reason is that B is an obvious upper bound of the L^2-norm of the matrix $A_n(f)$ (cf. e.g. Nevai [8, Lemma 9 on p. 52]). Thus it follows from the Weierstrass Approximation Theorem that the extreme sides in the above formula are equal even if G is an arbitrary function continuous on $[-B, B]$. Thus Theorem 7 is established assuming $[-\|f\|_{L^\infty(d\alpha)}, \|f\|_{L^\infty(d\alpha)}]$ is the smallest closed interval containing the essential range of f. This assumption is harmless, since it can be made to hold true by adding an appropriate constant to f

(this necessitates also a corresponding transformation of G). Hence the proof of Theorem 7 is complete.

References

[1] Erdös, P. and Turán, P., On interpolation III, Ann. of Math. (2) 41 (1940), 510-553.

[2] Freud, G., Orthogonal Polynomials, Pergamon Press, New York, 1971.

[3] Grenander, U. and Szegö, G., Toeplitz Forms and Their Applications, University of California Press, Berkeley, 1958.

[4] Kolmogorov, A. N., Stationary sequences in Hilbert spaces, Bull. Moscow State University. 2 : 6 (1941), 1-40.

[5] Krein, M. G., Generalization of investigations of G. Szegö, V. I. Smirnov and A. N. Kolmogorov, Doklady Akad. Nauk SSSR 46 (1945), 91-94.

[6] Máté, A. and Nevai, P., Remarks on E. A. Rahmanov's paper "On the asymptotics of the ratio of orthogonal polynomials", J. Approx. Theory 36 (1982), 64-72.

[7] Máté, A., Nevai, P. and Totik, V., Asymptotics for the leading coefficients of polynomials orthonormal with respect to an almost everywhere nonvanishing weight on the unit circle, manuscript.

[8] Nevai, P., Orthogonal Polynomials, Memoirs of the Amer. Math. Soc. 213 (1979).

[9] Nevai, P., Eigenvalue distribution of Toeplitz matrices, Proc. Amer. Math. Soc. 80 (1980), 247-253.

[10] Nevai, P., Distribution of zeros of orthogonal polynomials, Trans. Amer. Math. Soc. 249 (1979), 341-351.

[11] Rahmanov, E. A., On the asymptotics of the ratio of orthogonal polynomials, Math. USSR Sbornik 32 (1977), 199-213.

[12] Rahmanov, E. A., On the asymptotics of the ratio of orthogonal polynomials, II, Math. USSR Sbornik 46 (1983), 105-117.

[13] Szegö, G., Orthogonal Polynomials, Amer. Math. Soc., Providence, 1967.

[14] Turán, P., On orthogonal polynomials, Analysis Math. 1 (1975), 297-311.

Polynomials with Laguerre Weights in L^p

H. N. Mhaskar and E. B. Saff
Department of Mathematics
California State University
Los Angeles, CA 90032

Department of Mathematics
University of South Florida
Tampa, FL 33620

Abstract. For each p $(0 < p \leq \infty)$, $s \geq 0$, and integer $m \geq 1$ we consider the extremal problem

$$E_{s,m,p} := \inf\{\| t^s e^{-t}(t^m - q_{m-1}(t))\|_{L^p} : q_{m-1} \in P_{m-1}\} ,$$

where the L^p-norm is taken over $[0,\infty)$ and P_{m-1} is the collection of polynomials of degree at most $m-1$. The asymptotic behavior of $E_{s,m,p}$ as $n := s+m \to \infty$ and $s/n \to \theta$ is determined along with the zero distribution for the associated Chebyshev polynomials. The paper includes the proofs of results announced in [7].

1. Introduction

Motivated by the theory of incomplete polynomials introduced by Lorentz [5], we proved in [6] that for every $\alpha > 0$ and integer $n \geq 0$, there is a constant $a(n,\alpha)$ such that for every polynomial P of degree at most n ,

$$\max_{x \in \mathbb{R}} |\exp(-|x|^\alpha)P(x)| = \max_{|y| \leq a(n,\alpha)} |\exp(-|y|^\alpha)P(y)| . \tag{1.1}$$

We found an explicit expression for $a(n,\alpha)$ in (1.1) which is asymptotically best possible (as $n \to \infty$) and also applied our results to the theory of certain orthogonal polynomials, now known as Freud polynomials. In order to develop a general theory unifying the previous results [3], [4], [5], [10], [11], [12] of Saff, Varga, Lachance, Lorentz, Kemperman, Ullman and others and our results in [6], we considered the case of the Laguerre weight function $x^s e^{-x}$ on $[0,\infty)$. The results for this weight were announced in [7]. In this paper, we shall provide the proofs.

During the last year, we did, in fact, make significant progress towards the development of a general theory ([8], [9]). Thus, instead of presenting the proofs which we had at the time of the publication of [7], we plan to use, as much as possible, the more general and more recent results in [8]. The novelty of the present paper is threefold. First, for the concrete case of the Laguerre weights, we can find explicitly the various quantities whose existence is asserted in [8]. In turn, we shall use these explicit expressions to prove the sharpness of our results. Second, we shall discuss some applications to the theory of Laguerre polynomials and other extremal polynomials. Third, the case of the Laguerre weights is, in a sense, "midway" between the case of the Jacobi weights studied in [10], [4] and the exponential weights investigated in [6]. Thus, it serves as an illustration of the connection between the two.

For the above reasons, we believe that Laguerre weights deserve a separate discussion, even in the presence of the general theory. In Section 2, we shall state the main theorems; the proofs are given in Section 3.

2. Main Results

For a Lebesgue measurable function g on an interval $I \subset [0,\infty)$, set

$$\|g\|_{p,I} := \begin{cases} \left(\int_I |g(x)|^p dx\right)^{1/p} & \text{if } 0 < p < \infty, \\ \operatorname*{ess\,sup}_{x\in I} |g(x)| & \text{if } p = \infty. \end{cases} \tag{2.1}$$

The suffix I will be omitted if $I = [0,\infty)$ and the suffix p will be omitted if $p = \infty$.

Our first theorem is an analogue of Theorem 2.7 of [6] and, in part, follows as a consequence of the more general Theorem 2.2 of [8].

Theorem 2.1: Let $m \geq 0$ be an integer, $s \geq 0$, $\mu > 0$, $n := s + m > 0$ and $0 \not\equiv P_m \in \mathcal{P}_m$ (the class of all polynomials of degree at most m). If $\xi \in [0,\infty)$ satisfies

$$|\xi^s e^{-\mu\xi} P_m(\xi)| = \|x^s e^{-\mu x} P_m(x)\|, \tag{2.2}$$

then

$$(2.3) \qquad a \le \xi \le b \quad ,$$

where

$$(2.3') \qquad \begin{aligned} a = a(s,n,\mu) &:= \frac{n}{\mu}\left(1 - \sqrt{1-(s/n)^2}\right) \quad , \\ b = b(s,n,\mu) &:= \frac{n}{\mu}\left(1 + \sqrt{1-(s/n)^2}\right) \quad . \end{aligned}$$

In particular,

$$(2.4) \qquad \|t^s e^{-\mu t} p_m(t)\|_{[a,b]} = \|x^s e^{-\mu x} p_m(x)\| \quad , \quad \forall\ p_m \in \mathcal{P}_m \quad .$$

In Section 3, we shall prove (2.4) using the more general results in [8]. However, a complete proof of Theorem 2.1 requires some explicit computations similar to those in [6].

Our next theorem concerns the asymptotic behavior of certain extremal polynomials. Before stating this result, we introduce some needed notation.

For each $p (0 < p \le \infty)$, $s \ge 0$ and integer $m \ge 1$, set

$$(2.5) \qquad E_{s,m,p} := \min\{\|t^s e^{-t}(t^m - g_{m-1}(t))\|_p \ : g_{m-1} \in \mathcal{P}_{m-1}\}$$

and let $T_{s,m,p}(t) = t^m + \ldots \in \mathcal{P}_m$ satisfy

$$(2.6) \qquad E_{s,m,p} = \|t^s e^{-t} T_{s,m,p}(t)\|_p \quad .$$

Theorem 2.2. Let θ $(0 \le \theta < 1)$ be fixed and suppose $\{s_i\}$ is a sequence of nonnegative real numbers and $\{m_i\}$ is a sequence of nonnegative integers such that $n_i := s_i + m_i > 0$ for each i , $n_i \to \infty$ and

$$(2.7) \qquad s_i/n_i \to \theta \quad \text{as} \quad i \to \infty \ .$$

Then,

(a) For each p $(0 < p \le \infty)$, the minimal error defined in (2.5) satisfies

$$(2.8) \qquad \lim_{i\to\infty} n_i^{-1} E_{s_i,m_i,p}^{1/n_i} = \left\{\frac{(1+\theta)^{1+\theta}(1-\theta)^{1-\theta}}{4e^2}\right\}^{1/2} \quad .$$

(b) *There are* $m_i + 1$ *nonnegative numbers* $\xi_{1i} < \dots < \xi_{m_i+1,i}$ *such that*

$$(2.9) \qquad |\xi_{ji}^{s_i} e^{-\xi_{ji}} T_{s_i,m_i,\infty}(\xi_{ji})| = E_{s_i,m_i,\infty}, \quad j = 1,\dots,m_i+1 .$$

Furthermore,

$$(2.10) \qquad \xi_{1i}/n_i \to 1 - \sqrt{1-\theta^2} \quad \text{and} \quad \xi_{m_i+1,i}/n_i \to 1 + \sqrt{1-\theta^2} \quad \text{as} \quad i \to \infty .$$

(*In this sense, Theorem* 2.1 *is sharp*.)

(c) *All the zeros of the extremal polynomials* $T_{s_i,m_i,p}$ *are real and, if* $p \geq 1$, *simple. For each* p $(0 < p \leq \infty)$ *and interval* $[c,d] \subset [0,\infty)$, *let* $N_{i,p}(c,d)$ *denote the number of zeros of the normalized polynomial* $T_{s_i,m_i,p}(n_i t)$ *which lie in* $[c,d]$. *Then*

$$(2.11) \qquad \lim_{i\to\infty} \frac{N_{i,p}(c,d)}{m_i} = \int_c^d h(\theta,x)\,dx ,$$

where

$$(2.12) \qquad h(\theta,x) := \frac{1}{(1-\theta)\pi} \frac{\sqrt{(b^* - x)(x - a^*)}}{x} \quad \text{if} \quad x \in [a^*,b^*] ,$$

$$a^* := 1 - \sqrt{1-\theta^2} , \quad b^* := 1 + \sqrt{1-\theta^2} ,$$

and $h(\theta,x) = 0$ *if* $x \notin [a^*,b^*]$.

While the case $\theta > 0$ and $p = \infty$ of Theorem 2.2 follows immediately from the results in [8], the assertion for arbitrary L^p-norms requires a new argument.

3. Proofs

It is convenient to prove (2.4) of Theorem 2.1 first.

Proof of (2.4). Since the case $m = 0$ of (2.4) is trivial to verify, we assume hereafter that $m > 0$. Our goal is to find a finite interval $[a,b]$ such that for every $P_m \in \mathcal{P}_m$

$$(3.1) \qquad \| [w(t)]^m P_m(t) \|_{[a,b]} = \| [w(x)]^m P_m(x) \| ,$$

where

(3.2) $$w(t) := t^{s/m} e^{-\mu t/m} \quad .$$

Since $Q(t) := \log(1/w(t)) = (\mu t - s \log t)/m$ is convex on $(0,\infty)$, we can now apply Theorem 2.2 of [8]. According to that theorem, the interval $[a,b]$ can be found by maximizing an "F-functional" defined by

(3.3) $$F(c,d) := \log\left(\frac{d-c}{4}\right) - \frac{1}{\pi} \int_c^d \frac{Q(t)\,dt}{\sqrt{(t-c)(d-t)}} \quad , \quad 0 \le c < d \quad .$$

To evaluate $F(c,d)$ explicitly, write

(3.4) $$t = \frac{c+d}{2} + \frac{d-c}{2} \cos\theta = \left(\frac{\sqrt{c}+\sqrt{d}}{2}\right)^2 \left(1 + \frac{e^{i\theta}}{\Phi}\right)\left(1 + \frac{e^{-i\theta}}{\Phi}\right) ,$$

where $\Phi := (\sqrt{d}+\sqrt{c})/(\sqrt{d}-\sqrt{c})$. Then, with $\alpha := s/m$, $\beta := \mu/m$, we have

(3.5) $$Q(t) = \beta t - \alpha \log t = \beta\left(\frac{c+d}{2}\right) - 2\alpha \log\left(\frac{\sqrt{c}+\sqrt{d}}{2}\right) + \beta\left(\frac{d-c}{2}\right) \cos\theta + 2\alpha \sum_{k=1}^{\infty} \frac{(-1)^k}{k\,\Phi^k} \cos k\theta \quad .$$

Substituting (3.5) and (3.4) into (3.3) we get

(3.6) $$F(c,d) = 2\alpha \log\left(\frac{\sqrt{c}+\sqrt{d}}{2}\right) - \beta\left(\frac{c+d}{2}\right) + \log\left(\frac{d-c}{4}\right) .$$

It is now elementary to check that the choice of c,d $(d > c)$ which maximizes $F(c,d)$ is given by

(3.7) $$c = a = \frac{1}{\mu}\left(n - \sqrt{n^2 - s^2}\right) \quad , \quad d = b = \frac{1}{\mu}\left(n + \sqrt{n^2 - s^2}\right) \quad ,$$

where $n := s + m$. □

A simple computation shows that the maximum value of F is

(3.8) $$\bar{F} := F(a,b) = \frac{1}{2(1-\gamma)} \log\left[\frac{n^2 (1+\gamma)^{1+\gamma} (1-\gamma)^{1-\gamma}}{4e^2\mu^2}\right] \quad , \quad \gamma := \frac{s}{n} \quad .$$

Theorem 2.3 of [8] and the remark following Lemma 4.3 of [8] assert that there exists a (necessarily unique) unit measure ν with support $[a,b]$ such that

$$(3.9)\qquad \int_a^b \log|t-x|\,d\nu(x) = \frac{\mu}{m}t - \frac{s}{m}\log t + \bar{F}\ , \quad \forall\ t\in[a,b]\ .$$

Further, if $P_m \in \mathcal{P}_m$ and

$$(3.10)\qquad |t^s e^{-\mu t} P_m(t)| \le M\ , \quad \forall\ t\in[a,b]\ ,$$

then

$$(3.11)\qquad |P_m(z)| \le M\exp\left(m\left[\int_a^b \log|z-t|\,d\nu(t) - \bar{F}\right]\right)\ , \quad \forall\ z\in\mathbb{C}\ .$$

In order to complete the proof of Theorem 2.1, we shall explictly compute $d\nu$ and then estimate the right-hand side of (3.11). These technical results are summarized in the following lemma.

<u>Lemma</u> 3.1. <u>Set</u>

$$(3.12)\qquad g(x) := g(s,m,\mu;x) := \frac{\mu(b-x)(x-a)}{mx}\ , \quad x\in[a,b]\ ,$$

<u>where</u> a,b <u>are given in</u> (3.7) <u>with</u> $n := s+m$. <u>Then</u>

$$(3.13)\qquad \frac{1}{\pi}\int_a^b \frac{g(x)\,dx}{\sqrt{(b-x)(x-a)}} = 1\ ,$$

$$(3.14)\qquad \frac{1}{\pi}\int_a^b \log|x-t|\ \frac{g(x)\,dx}{\sqrt{(b-x)(x-a)}} = \frac{\mu}{m}t - \frac{s}{m}\log t + \bar{F}\ , \quad \forall\ t\in[a,b]\ ,$$

$$(3.15)\qquad \frac{1}{\pi}\int_a^b \log|x-t|\ \frac{g(x)\,dx}{\sqrt{(b-x)(x-a)}} < \frac{\mu}{m}t - \frac{s}{m}\log t + \bar{F}\ , \forall\ t\in[0,\infty)\setminus[a,b]\ .$$

<u>Proof</u>. Let

$$(3.16)\qquad x =: \frac{a+b}{2} + \frac{b-a}{2}\cos\phi$$

and observe from (3.7) that with $\alpha := s/m$, $\beta := \mu/m$ we have

$$(3.17)\qquad a = \frac{1}{\beta}\,[1+\alpha-\sqrt{1+2\alpha}]\ , \quad b = \frac{1}{\beta}\,[1+\alpha+\sqrt{1+2\alpha}]\ ,$$

$$(3.18)\qquad \left(\frac{\sqrt{b}+\sqrt{a}}{2}\right)^2 = \frac{1+2\alpha}{2\beta} \quad , \quad \sqrt{ab} = \frac{\alpha}{\beta} \quad ,$$

$$(3.19)\qquad \Phi := \frac{\sqrt{b}+\sqrt{a}}{\sqrt{b}-\sqrt{a}} = \sqrt{1+2\alpha} = \beta\left(\frac{b-a}{2}\right) \quad ,$$

$$(3.20)\qquad x = \frac{1}{\beta}(1+\alpha+\Phi\cos\phi) = \frac{1+2\alpha}{2\beta}\left(1+\frac{e^{i\phi}}{\Phi}\right)\left(1+\frac{e^{-i\phi}}{\Phi}\right) \quad .$$

Substituting (3.17) into (3.12) we get

$$g(x) = 2+2\alpha-\beta x-\frac{\alpha^2}{\beta x} = 2+\alpha-\beta x-\alpha\left[\frac{(\alpha/\beta)-x}{x}\right]$$

which, with the aid of (3.19) and (3.20), can be written as

$$(3.21)\qquad g(x) = 1-\Phi\cos\phi+2\alpha\left(\frac{1+\Phi\cos\phi}{\Phi^2\left|1+\frac{e^{i\phi}}{\Phi}\right|^2}\right)$$

$$= 1-\Phi\cos\phi+2\alpha\ \mathrm{Re}\left\{\frac{e^{i\phi}/\Phi}{1+e^{i\phi}/\Phi}\right\}$$

$$= 1-\Phi\cos\phi-2\alpha\sum_{k=1}^{\infty}\frac{(-1)^k}{\Phi^k}\cos k\phi \quad .$$

Equation (3.13) is now clear. Also, (3.14) follows from the identity

$$(3.22)\qquad \frac{1}{\pi}\int_a^b \log|x-t|\ \frac{g(x)\,dx}{\sqrt{(b-x)(x-a)}} = \log\frac{b-a}{4}+\frac{1}{\pi}\int_{-\pi}^{\pi}\log|1-e^{i(\theta-\phi)}|\,g(x)\,d\phi \ ,$$

where θ is defined in (3.4), after comparing the Fourier series of the right-hand side of (3.22) with (3.5) for $c=a$, $d=b$.

To prove (3.15), let

$$(3.23)\qquad \psi(t) := \frac{d}{dt}\left\{\frac{1}{\pi}\int_a^b \log|x-t|\frac{g(x)\,dx}{\sqrt{(b-x)(x-a)}}-\beta t+\alpha\log t\right\}$$

$$= \frac{\beta}{\pi}\int_a^b \frac{\sqrt{(b-x)(x-a)}}{x(t-x)}\,dx-\beta+\frac{\alpha}{t} \quad .$$

We shall show that $\psi(t) > 0$ if $t < a$. Write

(3.24) $$x = \frac{a+b}{2} + \frac{b-a}{2}\cos\phi \quad , \quad t = \frac{a+b}{2} - \frac{b-a}{2}\lambda \quad , \quad \lambda > 1 \quad ,$$

and set $R := (b+a)/(b-a)$. It requires only elementary computations to see that

$$\psi(t) = \frac{\beta}{R-\lambda}\left[\sqrt{\lambda^2-1} - \sqrt{R^2-1}\right] + \frac{2\alpha}{(b-a)(R-\lambda)}$$

$$= \frac{\beta\sqrt{\lambda^2-1}}{R-\lambda} + \frac{1}{R-\lambda}\left[\frac{2\alpha}{b-a} - \beta\sqrt{R^2-1}\right].$$

But, in view of (3.18), the last expression can be simplified to get

$$\psi(t) = \frac{\beta\sqrt{\lambda^2-1}}{R-\lambda} > 0 \quad .$$

A similar computation shows that $\psi(t) < 0$ if $t > b$. Hence from (3.14) it follows that (3.15) holds. □

<u>Proof of Theorem</u> 2.1. In view of Lemma 3.1, the measure ν of (3.9) and (3.11) is given by $g(x)dx/\pi\sqrt{(b-x)(x-a)}$. Hence, the proof is complete, using (2.4), (3.11) and (3.15). □

To facilitate the proof of Theorem 2.2 for the case $p = \infty$, $0 < \theta < 1$, we introduce some abbreviations. Let

$$w_i(t) := t^{s_i/m_i}e^{-t} \quad , \quad w_\theta(t) := t^{\theta/(1-\theta)}e^{-t} \quad ,$$

(3.25)
$$\hat{T}_i(t) = t^{m_i} + \ldots \quad , \quad \hat{T}_{\theta,i}(t) = t^{m_i} + \ldots \in P_{m_i} \quad ,$$

$$E_{i,m_i} := \|[w_i(t)]^{m_i}\hat{T}_i(t)\|_\infty = \min_{P\in P_{m_i-1}} \|[w_i(t)]^{m_i}[t^{m_i}-P(t)]\|_\infty \, ,$$

$$E_{m_i} := \|[w_\theta(t)]^{m_i}\hat{T}_{\theta,i}(t)\|_\infty = \min_{P\in P_{m_i-1}} \|[w_\theta(t)]^{m_i}[t^{m_i}-P(t)]\|_\infty \quad .$$

Note that

(3.26) $$\hat{T}_i(t) = m_i^{-m_i}T_{s_i,m_i,\infty}(m_it) \quad , \quad E_{i,m_i} = m_i^{-n_i}E_{s_i,m_i,\infty} \quad .$$

Also observe that, by Theorem 2.1, there exists a finite interval

$[\bar{a},\bar{b}] \subset (0,\infty)$ such that (with $\mu = m_i$, $s_i/n_i \to \theta$) all the sup norms in (3.25) are attained on $[\bar{a},\bar{b}]$ for i sufficiently large.

We divide the proof of Theorem 2.2 into several special cases.

Proof of Theorem 2.2 ($p=\infty$, $0<\theta<1$): (a) In view of Theorem 2.2 of [8] and the formula for $\bar{F}$ in (3.8) we have

$$\lim_{i\to\infty} E_{m_i}^{1/m_i} = (1-\theta)^{\frac{-1}{1-\theta}} \left\{ \frac{(1+\theta)^{1+\theta}(1-\theta)^{1-\theta}}{4e^2} \right\}^{\frac{1}{2-2\theta}} =: \Delta \quad . \tag{3.27}$$

However, since $w_i(t)/w_\theta(t) \to 1$ uniformly on $[\bar{a},\bar{b}]$, where all the sup norms in (3.25) are actually attained, it is easy to see that (3.27) holds also for E_{i,m_i} replacing E_{m_i}. In view of (3.26), then, (2.8) follows by elementary computations.

(b) The existence of the extreme points is a consequence of the general theory for Haar systems. The limiting relations (2.10) will follow from part (c) together with Theorem 2.1.

(c) We first obtain the zero distribution for the polynomials $\hat{T}_i(t)$. For this purpose we show that

$$\limsup_{i\to\infty} \| [w_\theta(t)]^{m_i}\hat{T}_i(t)\|_\infty^{1/m_i} \le \Delta \, , \tag{3.28}$$

that is,

$$\limsup_{i\to\infty} \left\| [w_\theta(t)]^{m_i}\hat{T}_i(t) \right\|_{\infty,[\bar{a},\bar{b}]}^{1/m_i} \le \Delta \, . \tag{3.29}$$

Let $\varepsilon > 0$ and choose an integer I so large that $i \ge I$ implies

$$w_\theta(t)/w_i(t) \le 1+\varepsilon \, , \quad \forall\, t \in [\bar{a},\bar{b}] \, ,$$

and

$$E_{i,m_i}^{1/m_i} \le \Delta(1+\varepsilon) \quad .$$

Then, for $i \ge I$,

$$\| [w_\theta(t)]^{m_i}\hat{T}_i(t)\|_{\infty,[\bar{a},\bar{b}]} \le (1+\varepsilon)^{m_i} \| [w_i(t)]^{m_i}\hat{T}_i(t)\| \le (1+\varepsilon)^{2m_i}\Delta^{m_i}$$

from which (3.29) follows. Now, applying Theorem 2.4 of [8], we see from Lemma 3.1 that the limiting zero distribution of the $\hat{T}_i(t)$'s is given by $g(x)\,dx/\pi\sqrt{(b-x)(x-a)}$, where in the definition (3.12) of $g(x)$ we take $\mu = m$, $a = (1-\sqrt{1-\theta^2})/(1-\theta)$, $b = (1+\sqrt{1-\theta^2})/(1-\theta)$. Hence, from (3.26), this must also be the limiting zero distribution for the polynomials $T_{s_i,m_i,\infty}(m_i t)$. With a change of variable, we then obtain (2.11). □

<u>Proof of Theorem</u> 2.2: $(p=\infty,\ \theta = 0)$. In this case $\bar{a} = 0$ and $w_i(t)/w_\theta(t)$ need not converge to 1 <u>uniformly</u> on $[0,\bar{b}]$. However,

$$w_i(t) = t^{s_i/m_i}e^{-t} \le \bar{b}^{s_i/m_i}e^{-t} = \bar{b}^{s_i/m_i}w_\theta(t)\ , \quad t \in [0,\bar{b}]\ ,$$

and hence

$$\begin{aligned} E_{i,m_i} &= \|[w_i(t)]^{m_i}\hat{T}_i(t)\|_\infty \le \|[w_i(t)]^{m_i}\hat{T}_{\theta,i}(t)\|_\infty \\ &\le \bar{b}^{s_i}\|[w_\theta(t)]^{m_i}\hat{T}_{\theta,i}(t)\|_\infty = \bar{b}^{s_i}E_{m_i}\ . \end{aligned}$$

Since $s_i/m_i \to 0$ as $i \to \infty$, we get

$$(3.30) \qquad \limsup_{i\to\infty} E_{i,m_i}^{1/m_i} \le \lim_{i\to\infty} E_{m_i}^{1/m_i} = \frac{1}{2e}\ ,$$

where the equality follows from (3.27) with $\theta = 0$. Furthermore, Theorem 2.1 of [8] gives

$$\liminf_{i\to\infty} E_{i,m_i}^{1/m_i} \ge \frac{1}{2e}\ .$$

Hence $E_{i,m_i}^{1/m_i} \to 1/2e$ as $i \to \infty$ which is equivalent to assertion (2.8) when $\theta = 0$.

Once more, part (b) will follow from part (c). To prove part (c), observe that since $s_i/m_i \to 0$ as $i \to \infty$, the distribution of zeros for $\hat{q}_i(t) := t^{s_i}\hat{T}_i(t)$ is the same as that for $\hat{T}_i(t)$. From (3.30) we get (with $w_\theta(t) = e^{-t}$)

$$\begin{aligned} &\limsup_{i\to\infty} \left\| [w_\theta(t)]^{m_i+s_i} t^{s_i}\hat{T}_i(t) \right\|_{\infty,[0,\bar{b}]}^{1/(m_i+s_i)} \\ &\le \limsup_{i\to\infty} \left\{ \|[w_i(t)]^{m_i}\hat{T}_i(t)\|_{\infty,[0,\bar{b}]}^{1/m_i} \right\}^{m_i/(m_i+s_i)} \\ &\le \frac{1}{2e}\ . \end{aligned}$$

Hence Theorem 2.4 of [8] gives the zero distribution for the sequence $t^{s_i}\hat{T}_i(t)$ and therefore for the $\hat{T}_i(t)$. Just as in the case $\theta > 0$, this leads to the desired zero distribution for the polynomials $T_{s_i,m_i,\infty}(n_i t)$. □

We now turn to the case when $p < \infty$. As in [6], our major tool will be Nikolskii-type inequalities. In order to use these, we need an estimate on the Christoffel function for the Laguerre weights $x^{\alpha}e^{-x}$ which is independent of α if $0 \le \alpha \le 1$. To this end, we prove

Lemma 3.2. Suppose k is a positive integer and $0 \le \alpha \le 1$. Let $\{L_n\}$ be the sequence of orthonormal Laguerre polynomials with respect to the weight $x^{\alpha k}e^{-x}$. Then for $x \in [0,\infty)$,

$$(3.31)\qquad |x^{\alpha k}e^{-x}\sum_{j=0}^{n-1} L_j^2(x)| \le cn^{2k+1} ,$$

where c is a constant independent of α .

Proof. Clearly, it suffices to estimate L_j when $j \ge k$. So, let $j \ge k$. Then, by Theorem 2.1 (with $b = b(\alpha k, 2j+\alpha k, 1)$), we have for all $x \in [0,\infty)$,

$$(3.32)\qquad |x^{\alpha k}e^{-x}L_j^2(x)| \le b^{\alpha k}\,\|e^{-x}L_j^2(x)\|_{\infty} \le (4j+2\alpha k)^{\alpha k}\,\|e^{-x}L_j^2(x)\|_{\infty}$$
$$\le (6j)^k\,\|e^{-x}L_j^2(x)\|_{\infty} .$$

Also from [2, §10.12, §10.18] we have for $x \in [0,\infty)$,

$$e^{-x}L_j^2(x) \le \frac{\Gamma(j+\alpha k+1)}{j!}\cdot\frac{1}{[\Gamma(\alpha k+1)]^2} \le (2j)^k .$$

Together with (3.32) this leads to (3.31). □

Using Lemma 3.2, Theorem 2.1 and also Theorems 6.1 and 6.4 of [6], we get the following:

Proposition 3.3: Let $0 \le \alpha \le 1$, $n \ge 1$ be an integer, $0 < p,r \le \infty$ and $P_n \in \mathcal{P}_n$. Then there are constants c and d depending upon p and r but not on α , n or P_n such that

$$(3.33) \qquad \|x^\alpha e^{-x} P_n(x)\|_p \le cn^d \|x^\alpha e^{-x} P_n(x)\|_r \quad .$$

We are now in a position to complete the proof of Theorem 2.2.

Proof of Theorem 2.2: $(p < \infty)$. In the proof of part (a) for the case when $p < \infty$, we use Proposition 3.3 with $x^{[s_i]} T_{s_i,m_i,p}(x)$ in place of P_n , where $[s_i]$ is the greatest integer less than s_i , to see that

$$\lim_{i\to\infty} n_i^{-1} E_{s_i,m_i,p}^{1/n_i} = \lim_{i\to\infty} n_i^{-1} E_{s_i,m_i,r}^{1/n_i} \quad , \quad 0 < p,r \le \infty \quad .$$

Part (a) then follows from the previously proved case $p = \infty$.

The proofs of parts (b) and (c) are now exactly the same as the proofs in [6] of the analogues of the parts (b) and (c) of Theorem 2.2. Hence, we omit the details. □

References

1. J. B. Conway, Functions of one complex variable, Springer-Verlag, Berlin (1973).

2. A. Erdelyi, W. Magnus, F. Oberhettinger and F. G. Tricomi, Higher Transcendental Functions, Vol. II, McGraw-Hill Book Co., New York (1953).

3. J. H. B. Kemperman and G. G. Lorentz, Bounds for polynomials with applications, Nederl. Akad. Wetensch. Proc. Ser. A. 82 (1979), 13-26.

4. M. A. Lachance, E. B. Saff, and R. S. Varga, Bounds for incomplete polynomials vanishing at both endpoints of an interval, Constructive Approaches to Mathematical Models (C.V. Coffman and G.J. Fix, eds.), Academic Press, New York (1979), 421-437.

5. G. G. Lorentz, Approximation by incomplete polynomials (problems and results), Pade and Rational Approximation: Theory and Applications (E.B. Saff and R.S. Varga, eds.), Academic Press, New York (1977), 289-302.

6. H. N. Mhaskar and E. B. Saff, Extremal problems for polynomials with exponential weights, Trans. Amer. Math. Soc. 285 (1984), 203-234.

7. H. N. Mhaskar and E. B. Saff, Extremal problems for polynomials with Laguerre weights, Approximation Theory IV (C. K. Chui, L. L. Schumaker and J. D. Ward, eds.), Academic Press, New York (1983), 619-624.

8. H. N. Mhaskar and E. B. Saff, Where does the sup norm of a weighted polynomial live? (A generalization of incomplete polynomials), to appear in Constructive Approximation.

9. H. N. Mhaskar and E. B. Saff, Weighted polynomials on finite and infinite intervals: a unified approach, Bull. Amer. Math. Soc. 11 (1984).

10. E. B. Saff, J. L. Ullman, and R. S. Varga, Incomplete polynomials: an electrostatics approach, Approximation Theory III (E. W. Cheney, ed.), Academic Press, New York (1980), 769-782.

11. E. B. Saff and R. S. Varga, The sharpness of Lorentz's theorem on incomplete polynomials, Trans. Amer. Math. Soc. 249 (1979), 136-186.

12. E. B. Saff and R. S. Varga, On incomplete polynomials, Numerische Methoden der Approximationstheorie (L. Collatz, G. Meinardus, H. Werner, eds.) ISNM 42 Birkhauser Verlag, Basel (1978), 281-298.

ORTHOGONAL POLYNOMIALS FOR GENERAL MEASURES-I

Joseph L. Ullman
Department of Mathematics
University of Michigan
Ann Arbor, Michigan 48109-1003

Abstract. Associated with a unit Borel measure in the complex plane, α, whose support $K(\alpha)$ is compact and contains infinitely many points is a family of orthonormal polynomials $\{\phi_n(z)\}$, $n = 0,1,\ldots$. The family of potentials $\psi_n(z) = \frac{1}{n}\log|\phi_n(z)|$ will be studied. Conditions have previously been found which insure that $\psi_n(z)$ behaves like the Green's function of $O(K(\alpha))$, the unbounded component of the complement of $K(\alpha)$. We study the behavior of $\psi_n(z)$ when these conditions are not satisfied.

1. Introduction

After giving some definitions concerning weight measures and facts from potential theory, we state Theorem 3 concerning the behavior of orthonormal polynomials. Theorems concerning potential theory are referenced as they are introduced and Theorem 3 is proved in Section 2. The background of Theorem 3 and other related matters are discussed in Section 3.

Definition 1. Let $K(\alpha)$ designate the support of α, a Borel measure in the plane. A unit Borel measure in the plane, α, for which $K(\alpha)$ is compact and contains infinitely many points is called a weight measure. We let $O(K(\alpha))$ refer to the unbounded component of the complement of $K(\alpha)$.

Definition 2. If α is a weight measure, for each non-negative integer n there is a unique monic polynomial of degree n, say $P_n(z)$, such that $\int|P_n(z)|^2 d\alpha < \int|Q_n(z)|^2 d\alpha$, where $Q_n(z)$ is any other monic polynomial of degree n. These are the monic orthogonal polynomials associated with α. We let

$$N_n(\alpha) = \left(\int|P_n(z)|^2 d\alpha\right)^{1/2}, \quad \lambda_n(\alpha) = (N_n(\alpha))^{1/n},$$

and refer to these as the norm and linearized norm, respectively, of

$P_n(z)$. Finally, $\phi_n(z) = P_n(z)/N_n(\alpha)$ is called the orthonormal polynomial of degree n associated with α .

Definition 3. Let K be a compact set. By $C(K)$ we mean the logarithmic capacity of K . For a bounded Borel set E , $C(E)$ is the inner logarithmic capacity. For a Borel measure μ , let $U(z,\mu) = \int \log(|z-t|^{-1})d\mu$, and refer to this as the potential of μ .

Theorem 1. [1, p. 171]. Let E be a bounded Borel set with $C(E) > 0$. There is a unique Borel measure μ_E , known as the equilibrium measure of E , such that a) $K(\mu_E)$ is bounded, b) $U(z,\mu_E) \leqq \log \frac{1}{C(E)}$ for all z , c) $U(z,\mu_E) = \log \frac{1}{C(E)}$ on a set $E' \subset E$, where $C(E') = C(E)$. When (a), (b) and (c) are satisfied, then d) $U(z,\mu_E) = \log \frac{1}{C(E)}$ on E , except possibly for a set of capacity zero.

Definition 4. Let α be a weight measure. A Borel subset of $K(\alpha)$ with $\alpha(E) = 1$ is called a carrier of α . Let $\overline{C} = C(K(\alpha))$, $\underline{C} = \inf C(E)$, where E ranges over the carriers of α . These are referred to as the upper and lower carrier capacities of α , respectively. If $\underline{C} < \overline{C}$ we say α is an undetermined weight measure, and otherwise it is a determined weight measure. When $\overline{C} > 0$, let $\overline{\mu}$ be the equilibrium measure of $K(\alpha)$.

Theorem 2. [3, p. 121]. Let α be a weight measure with $\underline{C} > 0$. All minimal carriers have the same equilibrium measure, say $\underline{\mu}$, referred to as the inner carrier equilibrium measure. We have $\underline{\mu} = \overline{\mu}$ if and only if $\underline{C} = \overline{C}$. Let $G(z) = -U(z,\overline{\mu}) - \log \overline{C}$, $G_1(z) = -U(z,\underline{\mu}) - \log \underline{C}$. Then $G(z) < G_1(z)$ in $O(K(\underline{\alpha}))$ if $\underline{C} < \overline{C}$.

Theorem 3. Let α be a weight measure with $\underline{C} > 0$, and let $\{\phi_n(z)\}$, $\phi_n(z) = k_n z^n + \ldots$, $n = 0,1,\ldots$, be the associated orthonormal polynomials. Then a) $\overline{\lim} |\phi_n(z)|^{1/n} \leqq \exp G_1(z)$ for all z , b) equality in (a) holds for no z in $O(K(\alpha))$ if $\overline{\lim}\, k_n^{1/n} < 1/\underline{C}$, c) if $\overline{\lim}\, k_n^{1/n} = 1/\underline{C}$, equality in (a) holds for all z in $O(K(\alpha))$, except possibly for a set of capacity zero.

2. Proof of Theorem 3

We first present some needed preliminaries.

Theorem 4. [5]. Let ν be a unit Borel measure with bounded support and satisfy $U(t,\nu) \geqq \lambda$ on a Borel set E, $C(E) > 0$. Let μ_E be the equilibrium measure of E. Then

$$(1) \qquad U(t,\nu) - \lambda \geqq U(t,\mu_E) - \log \frac{1}{C(E)}$$

for all values of z. When (1) holds, $\log \frac{1}{C(E)} \geqq \lambda$. If $\log \frac{1}{C(E)} > \lambda$, inequality holds for all z in $O(E)$, the unbounded component of the complement of $\bar{E}$. If $\log \frac{1}{C(E)} = \lambda$, then equality holds for all z in $O(\bar{E})$.

Definition 5. Let α be a weight measure with associated orthonormal polynomials $\{\phi_n(z)\}$. For each positive integer n let $\nu_n(\alpha)$ be a unit atomic measure with mass p_i/n, $i = 1,\dots,k$, at the distinct zeros of $\phi_n(z)$, say $z_1,\dots,z_k$, where p_i is the multiplicity of the zero z_i. We say Borel measures ν_n, $n = 1,\dots$ converge to a Borel measure ν, if $\lim_{n\to\infty} \int f d\nu_n = \int f d\nu$ for all continuous functions that are zero outside of a compact set.

Theorem 5. [3, p. 133]. (Lower Envelope Theorem). If ν_n is a convergent sequence of unit Borel measures with uniformly bounded supports and limit measure ν, then $U(z,\nu) \leqq \underline{\lim}\, U(z,\nu_n)$, where inequality holds if at all, on a set of capacity zero.

Theorem 6. Let α be a weight measure and $\{\phi_n(z)\}$ the associated orthonormal polynomials. Let $\{n_k\}$ be an increasing sequence of integers for which $\nu_{n_k}(\alpha)$ converges, say to ν, and $\lambda_{n_k}(\alpha)$ converges, say to λ_0. Then $U(t,\nu) \geqq \log(1/\lambda_0)$ on $E\setminus Z$, where E is a carrier of α and Z is a set of capacity zero. In particular, if $C(E) > 0$ then $\lambda_0 > 0$.

Proof of Theorem 6.

a) Lemma 1. $\overline{\lim}\, |\phi_n(z)|^{1/n} \leqq 1$ for $z \in E$, where E is a carrier of α.

Proof. Since $\int \sum |\phi_n(z)|^2/n^2 = \sum 1/n^2 < \infty$, by the monotonic convergence theorem, $\sum_1^\infty |\phi_n(z)|^2/n^2$ converges on E, a carrier of α. Hence for $z \in E$, $n \geqq n_z$, $n_z < \infty$, $|\phi_n(z)|^2/n^2 \leqq 1$ from which the needed result follows.

b) With the notation of the theorem, $\overline{\lim}\, |P_{n_k}(z)|^{1/n_k} \leqq \lambda_0$ for $z \in E$ from which we get $\underline{\lim}\, U(z,\nu_{k_n}(\alpha)) \geqq \log(1/\lambda_0)$ for

$z \in E$. By Theorem 5, $U(z,\nu) \geqq \log(1/\lambda_0)$ on $E \setminus Z$, where Z is a set of capacity zero.

c) If $C(E) > 0$, then $C(E \setminus Z) = C(E) > 0$, [3, p. 125] , and since ν is a unit Borel measure, the infinities of $U(z,\nu)$ form a set of capacity zero, so $\lambda_0 > 0$ must follow. □

Proof of Theorem 3.

(A) Let $\{n_k\}$ be a sequence for which $\nu_{n_k}(\alpha)$ converges, say to λ_0 . By Theorem 6

$$(2) \qquad U(z,\nu) \geqq \log(1/\lambda) \quad \text{on} \quad E \setminus Z \ ,$$

where $C(Z) = 0$, and since $C(E) \geqq \underline{C} > 0$, $\lambda_0 > 0$. Let E_1 be a minimal carrier of α . Then $E_2 = E \cap E_1$ is also a minimal carrier and (2) holds on $E_2 \setminus Z$. Now the equilibrium measure of $E_2 \setminus Z$ and E_2 are the same by Theorem 1, so by Theorem 4, $U(z,\nu) - \log(1/\lambda_0) \geqq U(z,\mu) - \log(1/\underline{C})$ for all values of z , and $\lambda_0 \geqq \underline{C}$.

(B) For a fixed z , let $\beta = \overline{\lim}\, |\phi_n(z)|^{1/n}$. If $\beta = 0$ there is nothing to prove, so assume $\beta > 0$. Choose n_k so that $\lim |\phi_{n_k}(z)|^{1/n_k} = \beta$, $\nu_{n_k}(\alpha)$ converges, say to ν and $\lambda_{n_k}(\alpha)$ converges, say to λ_0 . Thus

$$(3) \qquad \log \frac{1}{\beta} = \lim\left[U(z,\nu_{n_k}) - \log \frac{1}{\lambda_{n_k}}\right] \geqq U(z,\nu) - \log \frac{1}{\lambda_0}$$

$$\geqq U(z,\underline{\mu}) - \log \frac{1}{\underline{C}}$$

by Theorem 5 and the results of paragraph (A). This yields part (a).

(C) If $\overline{\lim}\, k_n^{1/n} < 1/\underline{C}$, then $\underline{\lim}\, \lambda_n(\alpha) > \underline{C}$ and the last inequality in (3) is a strict inequality for z in $O(K(\alpha))$ by Theorem 4.

(D) Since we now know that $\overline{\lim} |\phi_n(z)|^{1/n} \leqq \exp G_1(z)$ for all z , to prove (c) it is sufficient to find a sequence $\{n_k\}$ for which $\lim |\phi_{n_k}(z)|^{1/n_k} = \exp G_1(z)$, except possibly for a set of capacity zero. Since $\overline{\lim}\, k_n^{1/n} = 1/\underline{C}$, choose $\{n_k\}$ so that $\lim \lambda_{n_k}(\alpha) = \underline{C}$ and $\nu_{n_k}(\alpha)$ converges. Then

$$\overline{\lim} \log |\phi_{n_k}(z)|^{1/n_k} = \overline{\lim}\left[-U(z,\nu_{n_k}(\alpha)) - \log \lambda_{n_k}(\alpha)\right]$$

$$= -\underline{\lim}\, U(z,\nu_{n_k}(\alpha)) - \log \underline{C} \leqq -U(z,\nu) - \log \underline{C} \leq -U(z,\underline{\mu}) + \log \underline{C} \ .$$

The second to last inequality is an equality, except for a set of capacity zero, and the last inequality is an equality for all $z \in O(E_2 \setminus Z)$, by paragraph (1) and Theorem 4. This completes part (c), since $O(\overline{E_2 \setminus Z}) = O(K(\alpha))$. □

3. Discussion

In [6] Widom defines admissible measures that have the following property:

(P) $\lim_{n\to\infty} |\phi_n(z)|^{1/n} = G(z)$, for z in $O(K)$ except possibly for a set of two-dimensional Lebesgue measure zero.

In a future paper we will show that admissible measures are determined measures (Definition 4) and that (P) holds for determined measures. Lower bounds for $\underline{\lim}|\phi_n(z)|^{1/n}$ are known when $K(\alpha)$ is a subset of R^1 [4], and [2] uses a method that yields partial results on this problem when $K(\alpha)$ is not a subset of R^1. The fact that the bound in Theorem 3 is the best that can be obtained from measurements on the carriers of α follows from the fact that there is a weight measure with the same carriers as α, say β, such that $\lim_{n\to\infty} \lambda_n(\beta) = \underline{C}$; see [3, p. 122].

References

1. Landkof, N. S., Foundations of Modern Potential Theory, Springer-Verlag, New York, Heidelberg, Berlin, 1972.

2. Nguyen, Thanh Van, Familles de polynômes ponctuellement bonnees, Annals Polonici Mathematici, XXXI (1975) 83-90.

3. Ullman, J. L., On the regular behavior of orthogonal polynomials, Proc. London Math. Soc. (3) 24 (1972) 119-148.

4. Ullman, J. L., A survey of exterior asymptotics for orthogonal polynomials, Proceedings of the N.A.T.O. Advanced Study Institute, St. Johns, Nfld, 1983.

5. Ullman, J. L., Orthogonal polynomials for general measures, II. In preparation.

6. Widom, Harold, Polynomials associated with measures in the complex plane, Journal of Math. and Mech., Vol. 16, No. 9 (1967) 997-1013.